Land Snails
of
Kentucky

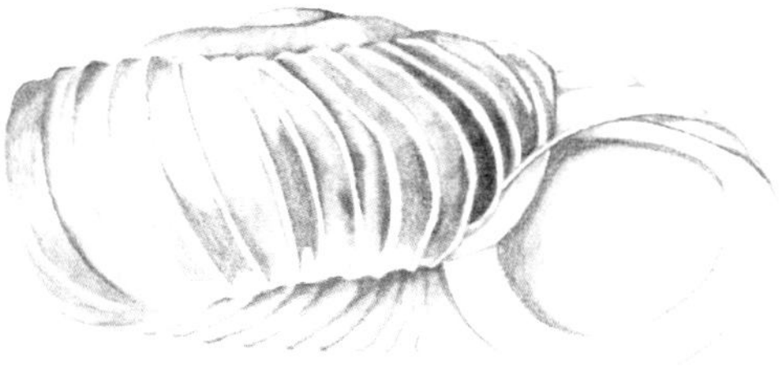

Daniel C. Dourson
Judy A. Dourson

Important Contributions by
Barbara Graham

SECOND EDITION

Developed in cooperation with

Office of Kentucky Natures Preserves (OKNP) &
Kentucky Heritage Land Conservation Fund (KHLC)

Illustrations & Photographs by the author unless otherwise stated

~2025~

Published by Goatslug Publications
1128 Gritter Ridge Road, Stanton, Kentucky, USA

Goatslug Publications

Current nomenclature based on the website: MolluscaBase eds. (2022). MolluscaBase. Accessed at https://www.molluscabase.org on 2022-08-21. doi:10.14284/448

CONTENTS

Acknowledgments

This manuscript would not have been possible without the support and assistance from numerous individuals and organizations. I offer my sincere gratitude for the contributions that allowed me to bring forward information about land snails, both historical and current.

Specimen Loans: Janet Voight, Rudiger Bieler, and Martin Pryzdia (*Field Museum of Natural History*) for specimen loans .

Collectors:
Office of Kentucky Nature Preserves: Zeb Weese, *Former Executive Director* ~ Zach Couch ~ Martina Hines, *Biological Assessment Branch Manager* ~ Shelby Fulton, *Terrestrial Zoologist* ~ Judy Cunningham, *Budget Specialist* ~ Courtney Hayes, *NCRS Environmental Scientist* ~ Josh Lillpop, *Natural Areas Branch Manager* ~ Kyle Napier, *Pine Mountain Manager* ~ Ryan Fortenberry, *Wild Rivers Manager* ~ Maddy Heredia, *KHLCF/Heritage Lands Outreach Specialist* ~ Doug Wilder, *Environmental Scientist & Trails Maintenance Supervisor* ~ Jessica Slade, *Bluegrass Manager* ~ Tara Littlefield, *Rare Plant Program Manager & Botanist* ~ Devin Rodgers, *Botanist* ~ Heidi Braunreiter, *Botanist* ~ Kendall McDonald, *Botanist* ~ Tony Romano, *Botanist* ~ Brian Yahn, *Heritage Biologist and Ecology Program Manager* ~ Cliff Hull, *Cumberland Plateau Manager* ~ Heather Drake, *Green River Manager* ~ Brainard Palmer-Ball, *Retired Terrestrial Zoologist* ~ Ellis Laudermilk, *Retired Invertebrate Biologist* ~ Joyce Bender, *Retired Branch Manager* Kentucky Natural Lands Trust: Derrick Lindsay Kentucky Energy & Environment Cabinet Division of Compliance Assistance: Robin Whitted, *Environmental Assistance Branch* Kentucky Department of Fish and Wildlife Resources: John MacGregor, *State Herpetologist* ~ Eric Williams, *Public Lands Wildlife Biologist* ~ Michaela Rogers, *Environmental Scientist* ~ Floracliff Nature Sanctuary ~ Beverly James, *Preserve Director* ~ Josie Miller, *Stewardship Director* ~ Bernheim Arboretum and Research Forest: Andrew Berry, *Director of Conservation* ~ Boone County Conservation District: Mark Jacobs, *Watershed Program Manager* ~ Susan Brown, *Conservation Education Specialist* Shaker Village of Pleasant Hill: Laura Baird, *Assistant Preserve Manager* Pennyrile State Resort Park: Haley Joseph, *Park Naturalist* Museums and Universities: John Slapcinsky, *Invertebrate Zoology Collection Manager, Florida Museum of Natural History* ~ Reham Fathey, *Associate Professor, Cairo University, Cairo, Egypt* ~ Alex Cooper, *Masters Student, University of Oklahoma* ~ Jedidiah Nixon, *Graduate Teaching Assistant, Western Kentucky University* ~ Jared Handley, *Adjunct Biology Instructor & Wildlife Biologist, US Army Fort Knox* ~ R. Wayne Van Devender, *Appalachian State University, Boone NC* ~ Jim Krupa, *Professor of Evolutionary Biology, University of Kentucky* ~ Allen Risk, *Professor of Biology, Morehead State University* ~ Byron Meade, *Staff Biology Lab Instructor, University of Pikeville* ~ Mary Whitson, *Department of Biological Sciences, Northern Kentucky University* ~ Drew White, *Graduate Student, University of Kentucky* ~ Micah Perkins, *Professor of Biology, Owensboro Community and Technical College* ~ John Abrams, *Program Coordinator, Berea College Forestry Outreach Center* ~ Jason Bried, *Professor of Biology, Murray State University* East Kentucky Power Cooperative, Inc: Chris Carpenter, *Natural Resources and Environmental Communications* US Forest Service: David Taylor, *Botanist* Copperhead Environmental Consulting: Mark Gumbert, *Principal Biologist, CEO* ~ Price Sewell *Biologist, Project Manager* ~ Theresa Wetzel, *Biologist* ~ Gregg Janos, *Biologist* ~ Ray Eaton, *Biologist* ~ Zachary Baer, *Biologist* ~Taylor Fagin, *Aquatic Biologist* ~ Stantec Environmental Services: James Kiser, *Senior Biologist* ~ Brittany White, *Environmental Scientist*

Illustrations & Photo Credits:
All illustrations and photos are those of Dan Dourson unless otherwise stated below:

Pg. 1-4: Dreamstime images **Pg. 21:** © Jaap Vermeulen **Pg. 22:** Great tit ©George Boorujy **Pg. 33:** © Chase & Blanchard **Pg. 34:** © Chase & Blanchard (2006) (top images) © Koene and Schulenburg (2005) **Pg. 39:** Breaks Interstate Park © Andrea Badgley **Pg. 43-45:** © Rachel G. Osborne **Pg. 47: Top Image:** © Rachel G. Osborne **Bottom Image:** John B. Burch (1962) **Pg. 48:** © Gordon Riley **Pg. 57-58:** John B. Burch (1962) **Pg. 71: Illustrations** © John B. Burch **Pg. 85: Illustrations ©** Henry Pilsbry (1948) **Pg. 89: Bottom image** © Piotr Naskrecki **Pg. 90: Illustrations** © © Nekola & Coles (2010) **Pg. 92: Illustrations ©** Henry Pilsbry (1948) **Pg. 98:** © Nekola & Coles (2010) **Pg. 100-107: Images** © Nekola & Coles (2010) **Pg. 101: Illustration ©** Henry Pilsbry (1948) **Pg. 108:** Figure (a) © Nekola & Coles (2010); Figure (b) © ANSP 78684 **Pg. 109-111:** © Nekola & Coles (2010) **Pg. 108: SEM image** © Jeff & Lori Schroeder **Pg. 115-117: Illustration** © Rachel G. Osborne **Pg. 119: Illustration** © Rachel G. Osborne **Pg. 127: Illustrations ©** Henry Pilsbry (1948) **Pg. 129: Illustration** © Henry Pilsbry (1948) **Pg. 132: Illustration ©** Henry Pilsbry (1948) **Pg. 133-134**: **©** Henry Pilsbry (1948) **Pg. 137 Illustrations ©** Aleta Kärstad **Pg. 138 Illustrations** © Rachel G. Osborne **Pg. 139-145, 147 Illustrations ©** Henry Pilsbry (1948) **Pg. 151: Illustrations ©** Henry Pilsbry (1948) **Pg. 156:** ©Aleta Kärstad & © Rachel G. Osborne **Pg. 170: SEM Image** © John Slapcinsky **Pg. 171-172**: © Rachel G. Osborne **Pg. 174, 185, 190:** © Henry Pilsbry (1946) **Pg. 191:** © John B. Burch (1962) **Pg. 192:** © John B. Burch (1962) **Pg. 193:** © Henry Pilsbry (1946) **Pg. 197: Illustration** © Henry Pilsbry (1946) **Pg. 201: Illustrations** © Henry Pilsbry (1946) **Pg. 229-230:** © Piotr Naskrecki **Pg. 236 Cartoons** Dreamstime **Pg. 245-246 Illustrations** © Henry Pilsbry (1946) **Pg. 251-253: ©** Larry Watrous **Pg. 254:** © Henry Pilsbry (1946) **Pg. 258: Illustrations** (top) © Rachel G. Osborne, (bottom) © Larry Watrous Kenneth Emberton **266: Cartoon** Dreamstime **Pg. 272: Images** © Larry Watrous **Illustrations** © Henry Pilsbry (1946) **Pg. 273: Images** © Larry Watrous **Pg. 275:** © John B. Burch **Pg. 278-280:** © John B. Burch **Pg. 281: Illustrations** © John B. Burch & © Henry Pilsbry (1946) **Pg. 283-284: Illustration** © John B. Burch **Pg. 298: Illustration ©** Kenneth Emberton **Pg. 328: Top Image** © Chrissy Mann **Pg. 381-382: Illustrations** © John B. Burch **Pg. 386-388**: © Henry Pilsbry (1948) **Pg. 390: Cartoon** Dreamstime **Pg. 399: ©** Gordon Riley **Pg. 403-408:** © Aleta Karstad **Pg. 409: Illustration:** Dreamstime **Pg. 411:** © Henry Pilsbry (1948) **Pg. 412:** © Nekola & Coles (2010) **Pg. 423:** © Jochen Gerber **Pg. 451-452:** Dreamstime

Maps:
Pg.16-18: Roger Banks, Kentucky Geological Survey, for his assistance in locating the detailed geological maps excerpted from the first edition. **Pg. 21:** snail images © Jaap Vermillion. **Pg. 36:** Piper Roby, Copperhead Consulting for recreating producing level IV ecoregion maps, adapted from the EPA Ecoregions of Kentucky maps (Woods *et al.* 2002).

Reviewers: A special note of gratitude to **Ronald S. Caldwell,** *retired director of Cumberland Mountain Research Center, Lincoln Memorial University*, for his extremely thorough and helpful review of the entire text. His input was invaluable.

Thanks also to **Jeff Nekola,** whose input on the Vertiginidae and Pupillids significantly improved the manuscript.

Finally, thanks to **John Slapcinsky**, Invertebrate Zoology Collections Manager, Florida Museum of Natural History, for his review of the Anguispira section. He is currently working on a revision of this genus.

Kentucky

Pine Mountain

From the eastern highlands of Pine and Cumberland Mountains to the watery lowlands along the Mississippi River, Kentucky is a land of varied expressions. While vast areas of forests remain in the eastern third of the state, much of the Commonwealth has been converted to pastures and woodlots created by centuries of human occupation. Extensive forests, small openings, wetlands, streams, cliffs, and caves are all natural communities. These biological ecosystems are clusters of organisms including plants, animals, fungi, and microbes which live together in a particular environment. Extraction of coal deposits found in both eastern and western Kentucky Coalfields has had a dramatic impact on the landscape and the biota that inhabit these areas.

Natural Wetland

Water is abundant with an average rainfall of around 60 inches a year. Kentucky has the greatest length of navigable waterways and streams in the contiguous United States and the two largest man-made lakes east of the Mississippi River (Lake Cumberland in volume and Kentucky Lake in surface area). A non-contiguous part known as Kentucky Bend, at the far west corner of the state, is an exclave surrounded completely by Missouri and Tennessee, and included in the boundaries of Fulton County. Road access to this small part of Kentucky on the Mississippi River requires a trip through Tennessee. The epicenter of the 1811–12 New Madrid earthquakes was near this area, causing the Mississippi River to flow backwards in some places. Though the series of quakes changed the area geologically and affected the small number of inhabitants of the area at the time, the Kentucky Bend is the result of a surveying error, not the New Madrid earthquakes.

Mammoth Cave

Kentucky is the only U.S. state to have a continuous border of rivers running along three of its sides—the Mississippi River to the west, the Ohio River to the north, and the Big Sandy River and Tug Fork to the east. These large waterways played a significant role in the presence or absence of species in the state acting as both barriers and vectors for flora and fauna. During the first Ice Age, the movement of glaciers deposited numerous Pleistocene species (including now extinct land snails) in loess deposits along the major river boundaries. The topography of the eastern half of the state is dotted with stunning waterfalls with the most notable being Bad Branch Falls and Cumberland Falls.

Rock strata in the state is also noteworthy. Mammoth Cave, located in southwestern part of the state is the longest cave system in the world with over 420 miles of mapped passageways. In eastern Kentucky, the state's famed bluegrass gently folds into the ruggedly scenic Appalachian Mountains. All around, natural beauty soars across narrow ridge tops, deep canyons, rim rock cliffs, and spec-

Eagle Falls

tacular spring and fall foliage. The Red River Gorge area boasts the highest concentration of arches on the planet with over 1200 documented natural arches and counting. Breaks Interstate Park located in extreme eastern Kentucky with its massive clifflines and deep canyons is known as the "Grand Canyon of the East". Counties eastward of where the Cumberland Plateau meets the Knobs and the Outer Bluegrass and to the eastward as part of Appalachia but the only counties that actually include the mountains of the Appalachian chain are Pike, Letcher, Harlan and Bell Counties.

Sky Bridge, Red River Gorge

The highest point in Kentucky at 4,145 ft in elevation is Black Mountain, part of the Cumberland Mountains and considered part of the Appalachian Mountains. It is also home to a wide diversity of land snail species, some that are endemic to that section of Kentucky. The lowest point in Kentucky is at the Mississippi River in Fulton County at an elevation of 257 feet above sea level.

Today, the Bluegrass region is best known for its rolling pastures and iconic horse farms but historically the region was mostly a wide savannah interspersed with massive burr oak trees. The limestone rock strata is loaded with available calcium that has produced thoroughbred horses raised here with stronger bones and created the economically profitable horse industry for the region. Coincidentally, the limestone also provides an excellent source of available calcium for the land snails of the area. The Bluegrass region is home to the Palisades, an area of steep limestone cliffs above the Kentucky River, known for its diversity of plants.

While 48% of Kentucky's terrain is forested with over 12.4 millions acres, 88% is under private landownership. Seventy-eight percent of the total forested acres are considered oak-hickory forest type with red maple being the most common tree species at 12% of the total tree species found. Kentucky's diversity of habitats, soil types, rock strata, and moderate climate are conducive to increased biodiversity, evident in the number of species of land snails documented in the state.

Breaks Interstate Park

Land Snail Studies in Kentucky (Historical & Recent)

Early land snail investigators in Kentucky included Henry R. Pilsbry, who compiled previous literature and described new taxa with his comprehensive two volume treatise, *Land Mollusca of North America (north of Mexico)*, published from 1940-1948. Branley Branson of Eastern Kentucky University along with his colleague D. L. Batch, conducted extensive snail surveys across the state and published numerous papers including Kentucky Land Mollusca: Checklist, Distribution and Keys (1973, 1988) followed by a later document in 1988. Leslie Hubricht, who conducted extensive land snail surveys for a number of years across the eastern US, published, *The Distribution Records of the Native Land Mollusks of the Eastern United States* (1985) which included a number of newly described land snails from Kentucky. From the late 1980s onward, significant contributions from John MacGregor of Kentucky Department of Fish and Wildlife Resources and James Kiser of Stantec, added many new state and county records. A study of Polygyrids of the eastern USA by Kenneth Emberton (1991, 1994) included land snails collected from Kentucky. More recent surveys at Bad Branch State Nature Preserve (Dourson & Beverly 2009), Breaks Interstate Park, Select Wildlife Management Areas (WMA) across Kentucky, Land Between the Lakes, Raven Run Nature Sanctuary, Blanton Forest, Lily Cornett Woods, Bernheim Arboretum and Research Forest, and Floracliff State Nature Preserve have added many new county and state records. In 2009, the author conducted a thorough review and compilation of Kentucky's land snail distributions from past literature, collections by others and museum records, and published *Land Snails of Kentucky and their Ecological Communities* (2010).

An 18-month statewide land snail survey in cooperation with the Office of Kentucky Nature Preserves (OKNP) & Kentucky Heritage Lands Conservation (KHLC) conducted by the authors from January of 2019 through June of 2020 was a collaborative effort that included biologists from state and federal agencies, state park employees, private biological consulting firms, and volunteers across Kentucky. Each of the 120 counties in Kentucky was surveyed with at least one collection site, resulting in over 1,787 new county records, 13 new state records and 5 species previously unknown to science and awaiting further study to describe. A staggering 58,000 specimens were examined by the author as a result of this statewide endeavor. There are currently 202 native land snail species confirmed from Kentucky, five of which are endemic. While the land snails in Kentucky are fairly well documented, there remains a dearth of basic biological and life history information for nearly all 202 species. A brief summary of results from the statewide survey and several predictions on additional discoveries follows. More information about this unprecedented effort can be found on page 7.

Species Undetermined/New to Science (?)

1. *Anguispira aff strongylodes*, Kentucky Tigersnail (page 225)
2. *Anguispira aff. mordax,* Craggy Tigersnail (page 227)
3. *Patera aff. appressa* Pine Mountain Bladetooth (page 300)
4. *Triodopsis aff. anteridon* (page 310)
5. *Triodopsis aff. tennesseensis* (page 314)

New State Records

1. *Gastrocopta riparia* (likely a non-native/disjunct population in Kentucky)
2. *Vertigo pygmaea*
3. *Vertigo meramecensis*
4. *Pupilla muscorum* (a European exotic, Nekola and Coles 2010)
5. *Pseudohyalina exigua*
6. *Striatura milium*
7. *Euconulus polygyratus*
8. *Strobilops cf. texasianus*
9. *Pallifera ohioensis*
10. *Ventridens virginicus*
11. *Hawaiia alachuana* (in KY previously known only from fossil records)
12. *Helicodiscus aldrichianus*
13. *Ventridens cf. acerra*

Land Snail Statistics in Kentucky

Results from Statewide Survey 2019-2020 &
Updated Data from Other Collectors from 2010 to present

Total Native Land Snail Species in Kentucky	**202**
Species Endemic to Kentucky	**5**
Globally Rare Species (G1/G2)	**19**
State Rare Species (S1/S2)	**55**
New County Records	**1,787**
New State Records	**13**
New Species To Science Documented But Not Yet Described	**5**
New Species Needing Additional Investigation (DNA)	**10**
# of Species Known Only from Fossil Records (Extinct)	**15**
Additional Species Close To Kentucky Borders But Not Yet Confirmed (see Part 2 of book)	**26**
Exotic Species (including slugs) in Kentucky	**21**
Total Number of Sites Surveyed During Statewide Study	**1,015**
Total Number of Specimens Examined & Identified During Statewide Survey 2019-2020	**58,725**

Statewide Land Snail Inventory (2019-2020)

The *Terrestrial Mollusk Biodiversity Natural Areas Inventory and Info Project*, conducted by Dan and Judy Dourson (the primary investigators), from January 2019 through June 2020, was implemented to provide a comprehensive survey of Kentucky's terrestrial mollusk fauna within the boundaries of the state with an emphasis on natural areas/lands managed by the Office of Kentucky Nature Preserves (OKNP). The project, modeled after a similar statewide effort conducted in West Virginia, sought to maximize research while minimizing costs. Professional biologists specializing in a wide variety of disciplines from state, federal and private consulting agencies along with a strong volunteer citizen science component participated in a statewide effort to survey for land snails in each county across the state. Special emphasis was given to public lands protected and managed by the Office of Kentucky Nature Preserves, Wildlife Management Areas and State Parks as well as biologically diverse hotspots like the Red River Gorge and the Pine Mountain Massif.

The survey effort largely targeted empty shells not live animals with minimal impact to the group as a whole. Shell collecting, a pastime often enjoyed by many (albeit, mostly of the marine variety), required little advanced knowledge and minimal time compared to surveys for more complex organisms like mammals as well as more specific and expensive survey equipment. Survey gear used to survey land snails utilized simple household items like Ziploc bags, cotton balls, and a phone that can give GPS coordinates.

Project Objectives:

* Conduct a statewide survey of land snail by utilizing a combination of state & federal agencies already conducting field work in other taxa as well as citizen science organizations like the Master Naturalist Program, Sierra Clubs, Lions Clubs, Scouts of America, and other civic organizations and clubs to assist in surveys for terrestrial mollusks.
* Offer two-day workshops to train biologists, citizen scientists, and students to survey for land snails.
* Identify and map land snail species collected to update the existing Land Snails of Kentucky book.
* Describe any species new to science in peer-reviewed journal.
* Review and Revise the Office of Kentucky Nature Preserves (OKNP) threatened and endangered list for land snails based on results of survey as well as revisions to the NatureServe data base rankings.

Review of Literature, Museum Collections, and Other Data

Past surveys, museum collections, and *Kentucky's Land Snails and their Ecological Communities* (Dourson, 2010) were reviewed prior to the commencement of field survey work to ensure a comprehensive overview of the status of terrestrial land snails. A project database was developed in cooperation with OKNP staff in January 2019 and updated regularly as specimens were received and identified.

Museums: Databases provided by the following museums were reviewed: Carnegie Museum of Natural History (Pittsburgh) , Field Museum of Natural History (Chicago), Florida Museum of Natural History (Gainesville), Smithsonian Museum of Natural History (Washington, DC), Academy of Natural Sciences Philadelphia, University of Michigan Museum of Zoology (Lansing).

Publications: The following publications were utilized in identification of species documented in the state of Kentucky: *How to Know the Eastern Land Snails*, (Burch 1988), *Land Mollusca of Northern and Central America* (Pilsbry 1946, 1948), *Distribution of Land Snails of Eastern North America* (Hubricht 1985), and *Kentucky's Land Snails and their Ecological Communities* (Dourson 2010).

Land Snail Workshops

Four land snail workshops were conducted throughout the duration of the project. Biologists, park naturalists, and citizen scientists were introduced to the importance of land snails, methods for collecting land snail specimens, and how to process specimens to be sent to Dourson Biological Consulting LLC for identification.

The following workshops/presentations were conducted:

* Carter Caves State Resort Park on March 3-4, 2019. (22 participants)
* Pennyrile State Resort Park on April 3-4, 2019. (10 participants)
* Bluegrass Homeschool Learning Cooperative May 17, 2019 (16 participants)
* Floracliff State Nature Preserve on June 4, 2019. (23 participants)
* Louisville Youth Science Summit hosted by Kentucky Science Center for the Governor's Scholar Program @ Bellermine University in Louisville, KY on July 9, 2019. (25 participants)
* Murray State University Ecology Club at Hancock Biological Field Station in Murray, KY was held on March 6, 2019. (9 participants)
* Additional Identification Session was held at Dourson Biological Lab for volunteers from Floracliff. (6 participants)

A **handbook** was developed for participants of each workshop that included information about how to collect land snails, preserve the specimens for future identification and other pertinent information about the project.

An **informational laminated doubled-sided card** that included samples of all land snail genera found in Kentucky was developed by the Doursons (see cards on next page). Four hundred copies were printed and disseminated to volunteers and other agencies across the state.

Instructions on selecting habitats to search as well as how to search for land snails was provided to all persons who attended the land snail workshops or signed up to volunteer for the project.

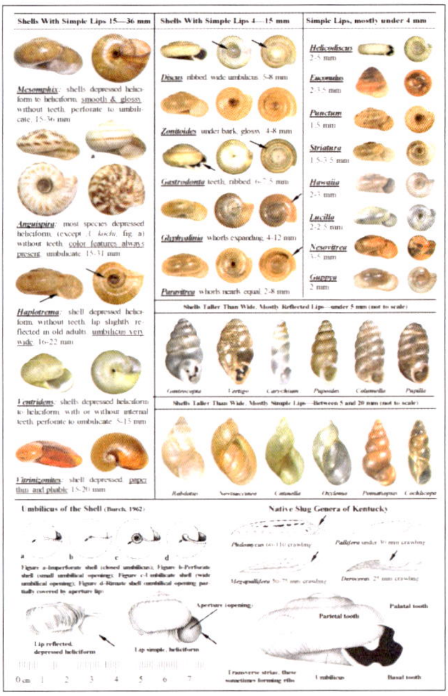

Targeted Research Trips

Research trips funded by the project grant were conducted in target areas where there was uncertainty of the status of a particular species or in areas where potential new species may occur. In August of 2019, a cadre of professional malacologists convened at Pine Mountain in southeastern Kentucky to conduct surveys for live specimens of *Anguispira rugoderma*, a land snail that had not been seen live in over 20 years. Live specimens of the target species were observed and collected for the purpose of conducting DNA sequencing of the entire genus of *Anguispira*, currently under review by Reham Fathey, Cairo University, Cairo, Egypt and John Slapcinsky, Florida Museum of Natural History, Florida, USA. The research team included Dan and Judy Dourson, Rahem Fathey, University of Cairo, Egypt, John Slapcinsky, Florida Museum of Natural History, Amy Van Devender and Wayne Van Devender, professional malacologists from North Carolina, and volunteer, Zachary Christensen. Specimens collected during this trip are currently undergoing DNA analysis. According to John Slapcinsky, the entire genus of *Anguispira* will be revised with several species previously identified as more common species in Kentucky likely to be new species.

In November 2019, a targeted collecting trip to southeastern Kentucky focused on the high elevations of Black Mountain searching for a diminutive species of land snail, (less than 5 mm) in the genus *Pilsbryna.* This species had previously been documented by John Petranka in the early 1980's but remained as an undescribed species. John Slapcinsky (pictured right) of the Florida Museum of Natural History joined Dan Dourson, Judy Dourson, and Barbara Graham in the search for live material for DNA analysis. Colder temperatures and 4 inches of fingertip-numbing snow on Black Mountain hindered but did not deter the research team. Both live and dead specimens were collected. Other sites in the area were also examined and collected.

Dan and Judy Dourson conducted collecting trips throughout the state focused on unsampled areas or counties and revisits to sites collected by volunteers where interesting or undetermined species were documented. Barbara Graham, independent biologist, conducted extensive surveys throughout Kentucky during the project. The Office of Kentucky Nature Preserves personnel contributed significantly to the overall survey effort, conducting targeted visits and revisits to sites where species of interest were documented based on recommendations by Dan Dourson, project manager.

Alex Cooper, University of Oklahoma graduate student, spent several days in Kentucky and at the Dourson Lab, surveying for species in the genus *Stenotrema* as part of his research and review of the entire genus. In July of 2019, he joined Dan and Judy Dourson on a collecting trip in the Red River Gorge to search for live specimens in this genus. Specimens collected in the in the area included the newly described *Stenotrema macgregori* Dourson, 2011. Preliminary results of DNA sequencing support the status of *Stenotrema macgregori* as a distinct species from *Stenotrema stenotrema*.

Fraudulent Slitmouth, *Stenotrema macgregori*

Nicholas Gladstone, University of Tennessee, also joined the Doursons for consultation and the sharing of results of cave invertebrate surveys in Kentucky with an emphasis on the cave species.

Citizen Science

The citizen science component of the project generated a great deal of interest in an otherwise often over-looked group of organisms. Informational packets that included collecting materials like sampling vials, cotton balls, and soil/leaf litter sample bags were mailed to 75 individuals. Forty individuals collected and returned land snails in their respective counties or areas and returned them to the Dourson Lab for identification. An unprecedented total 110 individuals participated in land snail workshops across the state. The comprehensive collection effort far exceeded a similar collection statewide effort in West Virginia (which served as a model for this project).

Dan sorting land snails at lab

Judy and grandson Jude sorting snails

Judy searches for land snails around an old car on top of Loess Buffs near the Mississippi River in Fulton County, Kentucky.

Soils sample bags containing tiny land snails drying

Barbara Graham searching for cave snails in Stash Cave, Wayne County Kentucky

Christina Fizer searches for cave snails

Molly Gooden & Dan in a very tiny cave

Land Snails of Kentucky

The book is presented as follows: **Part I:** Species Accounts of Native Land Snails in Kentucky **Part II:** Pleistocene (Fossil) Land Snails in Kentucky **Part III:** Exotic Land Snails of Kentucky **Part IV:** Land Snails of Bordering States (illustrated below inside the dashed lines) found in border counties from Ohio, West Virginia, Virginia, Tennessee, Missouri, Illinois and Indiana. The text is fully illustrated with hundreds of color photographs and drawings and a pictorial key designed for the beginner and advanced malacologist alike. The table below indicates the number of species known to occur in each county as of 2022.

Native Species of Land Snails Occurring in Each County from Largest to Smallest

Harlan: 112	Jackson: 63	Russell: 49	Marion: 36	Lincoln: 29	Spencer: 22
Letcher: 95	Bullitt: 61	Knox: 48	Clinton: 35	Rowan: 29	Ballard: 21
Edmonson: 93	Cumberland: 61	Barren: 47	Fulton: 35	Bath: 28	Caldwell: 21
Fayette: 92	Jessamine: 61	Trimble: 47	Green: 35	Campbell: 28	Crittenden: 21
Powell: 92	Warren: 61	Allen: 46	LaRue: 35	Grayson: 28	Magoffin: 21
Clark: 89	Hardin: 60	Lee: 45	Morgan: 35	Fleming: 27	Marshall: 21
Pike: 89	Meade: 60	Scott: 44	Boyle: 33	Ohio: 27	Hancock: 20
Bell: 88	Carter: 57	Owen: 42	Carroll: 33	Calloway: 26	Hopkins: 20
Pulaski: 87	Garrard: 57	Woodford: 42	Christian: 33	Mason: 26	Muhlenburg: 20
Whitley: 83	Hart: 57	Anderson: 40	Casey: 32	Nicholas: 26	Daviess: 19
Madison: 81	Leslie: 55	Bourbon: 40	Simpson: 32	Pendleton: 26	Bracken: 18
McCreary: 78	Adair: 53	Knott: 40	Martin: 31	Butler: 25	Union: 18
Estill: 76	Livingston: 53	Perry: 40	Trigg: 31	Oldham: 25	McCracken: 17
Franklin: 76	Mercer: 53	Todd: 39	Harrison: 30	Shelby: 25	Boyd: 16
Wolfe: 73	Breathitt: 52	Boone: 38	Hickman: 30	Greenup: 24	Robertson: 16
Nelson: 70	Breckinridge: 52	Taylor: 38	Lewis: 30	Lawrence: 24	Lyon: 15
Jefferson: 66	Clay: 51	Logan: 37	Montgomery: 30	Elliott: 23	Webster: 15
Menifee: 65	Laurel: 51	Monroe: 37	Carlisle: 29	Washington: 22	Graves: 14
Rockcastle: 64	Metcalfe: 50	Floyd: 36	Gallatin: 29	Grant: 22	Johnson: 13
Wayne: 64	Owsley: 49	Henry: 36	Henderson: 29	Kenton: 22	McClean: 5

Species Diversity Across Kentucky

Counties Shaded Gray = # of species documented > 70

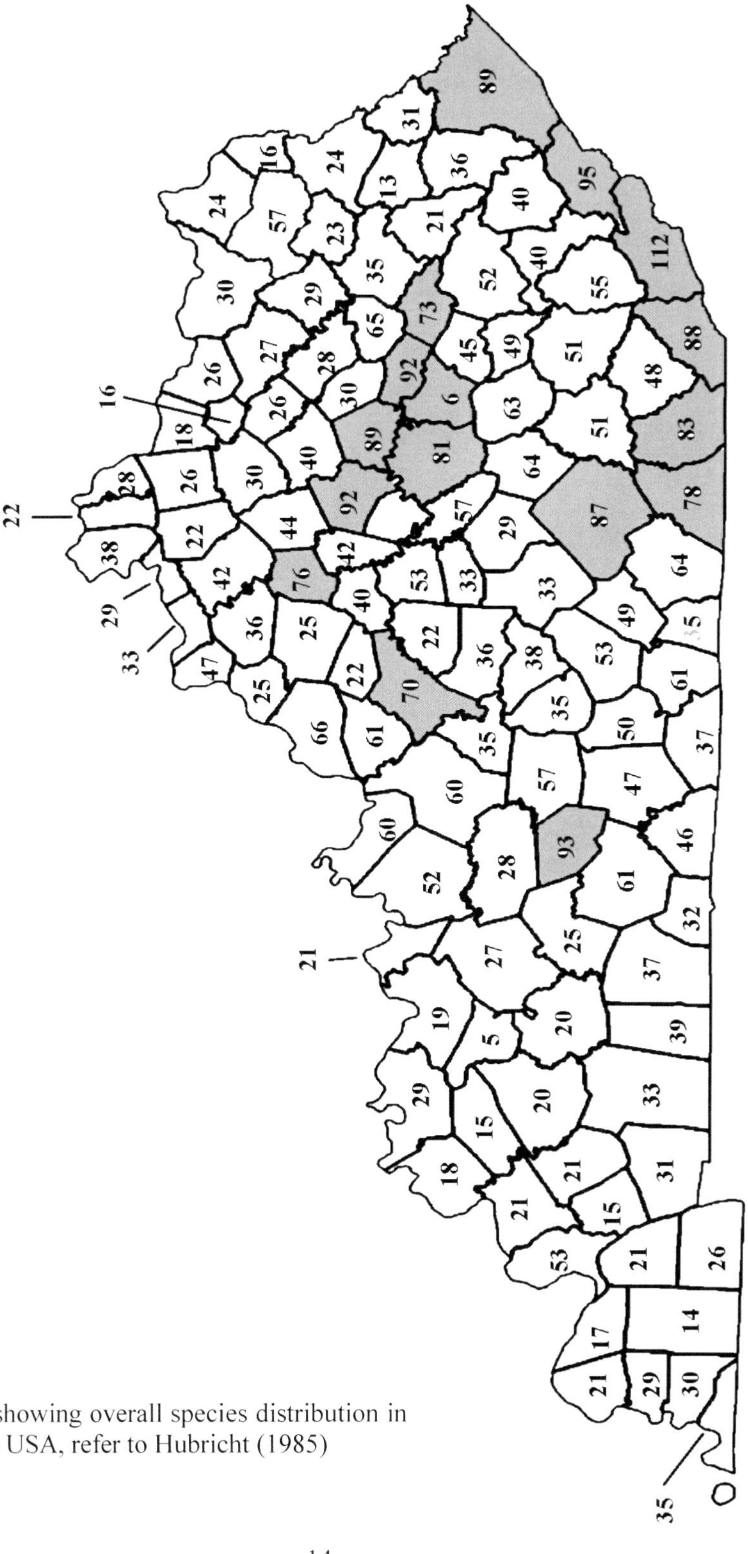

For maps showing overall species distribution in the eastern USA, refer to Hubricht (1985)

Kentucky Counties

General Eco-Regions of Kentucky

KENTUCKY GEOLOGICAL SURVEY
Donald C. Haney, State Geologist and Director
UNIVERSITY OF KENTUCKY, LEXINGTON

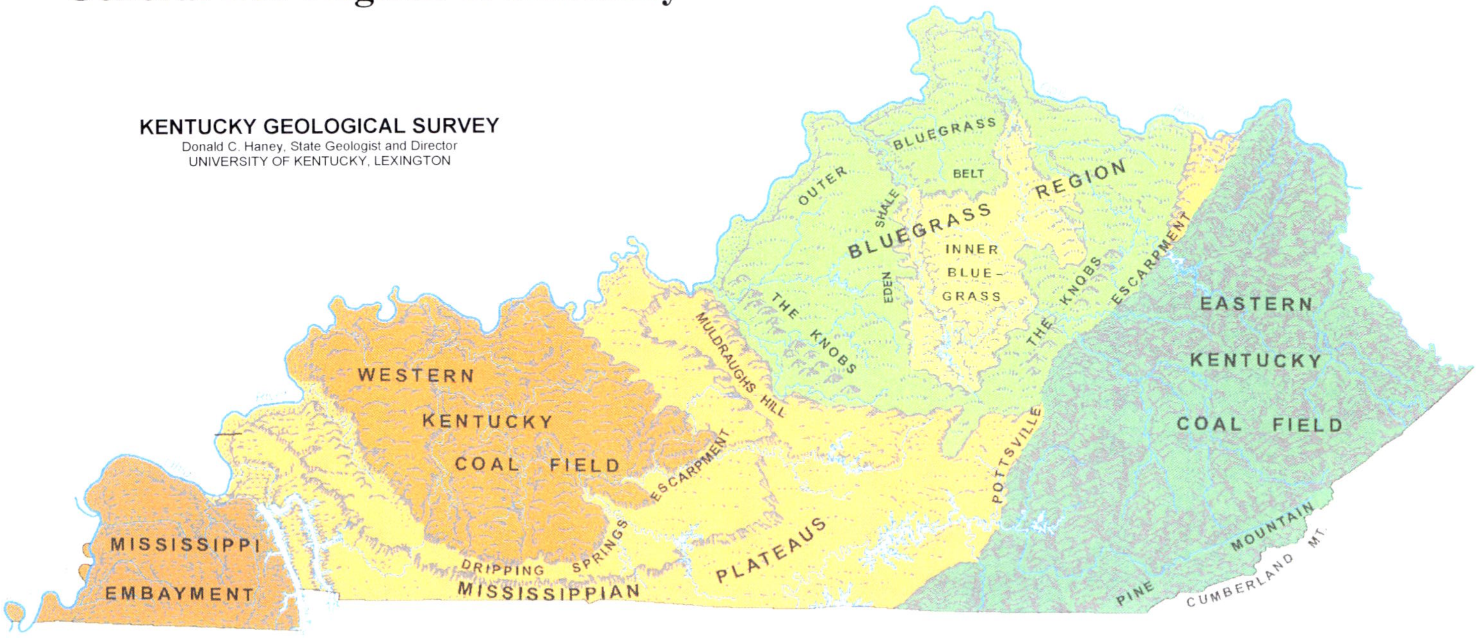

KENTUCKY GEOLOGICAL SURVEY
Donald C. Haney, State Geologist and Director
UNIVERSITY OF KENTUCKY, LEXINGTON

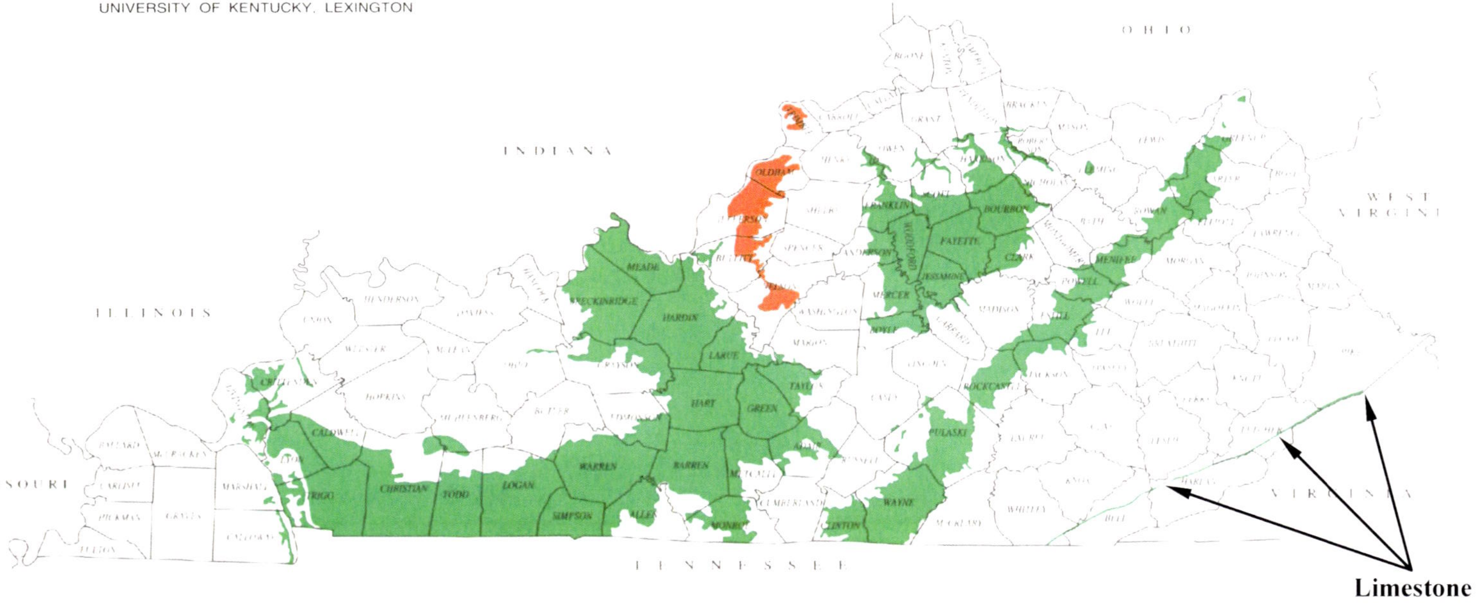

PRINCIPAL OUTCROP OF LIMESTONE AND DOLOMITE RESOURCES IN KENTUCKY

Compiled by Garland R. Dever, Jr.

EXPLANATION

Limestone

Dolomite and dolomitic limestone

0 40 Miles
0 50 Kilometers
SCALE

N

KENTUCKY GEOLOGICAL SURVEY
Donald C. Haney, State Geologist and Director
UNIVERSITY OF KENTUCKY, LEXINGTON

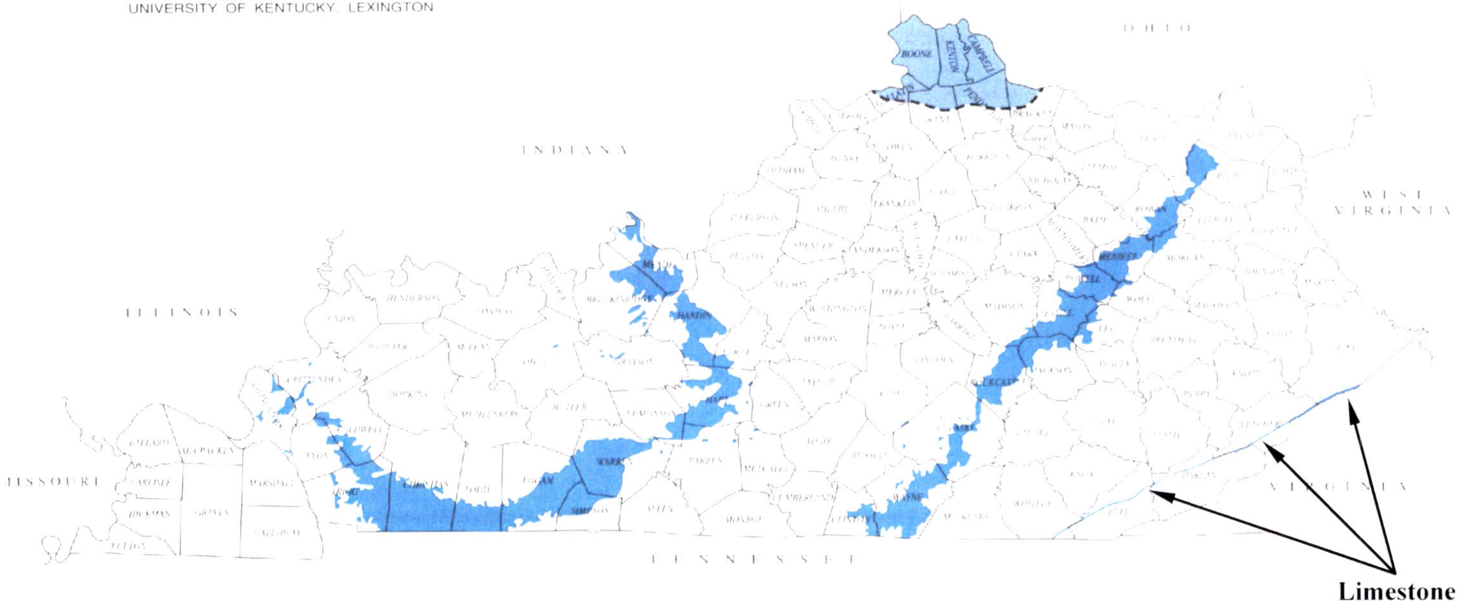

HIGH-CALCIUM LIMESTONE RESOURCES IN KENTUCKY
Compiled by Garland R. Dever, Jr.

HOW TO USE THE BOOK

Pictorial Key to Genera

The book is designed to bridge the gap between biologists and armchair naturalists. The key will be useful in determining the genus and from there, the user will be guided to species accounts of the same genera. Further details are provided in species accounts.

Species Accounts

For each species illustrated, the **Common Name**, **Scientific Name**, **Description**, **Similar Species**, **Habitat**, **Status** (both state and global rankings) and **Specimen** (information as to the location of the particular specimen photographed) are included, making the text useful and comprehensive for Natural Resources agencies and other land managers whose lands harbor terrestrial gastropods. Current nomenclature based on the website: MolluscaBase eds. (2022). MolluscaBase. Accessed at https://www.molluscabase.org on 2022-08-21. doi:10.14284/448

Maps

Blue= Historical Records Prior to 2019
Yellow= Records from Statewide Land Snail Survey 2019-2020 (funded by Office of Kentucky Nature Preserves and Kentucky)

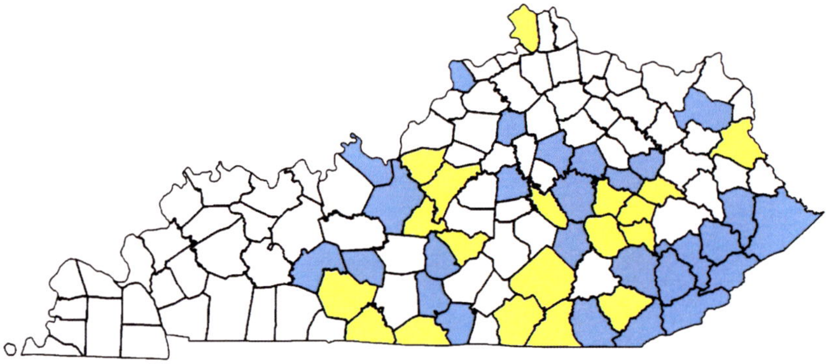

Abundance Categories of Kentucky's Land Snails

The abundance of land snails within Kentucky was determined by investigating records for each species and assigning a category based on that examination. The categories used in the book are as follows:

1. **Common:** Species is found in quantities in its habitat.
2. **Relatively Common:** Species is usually encountered in its habitat in smaller numbers, but with regularity.
3. **Uncommon:** Species not encountered frequently within its habitat, but widespread in small numbers.
4. **Rare:** Rarely encountered even in suitable habitat.

Global & State Rankings and Status

Each land snail species has two ranks: **Global Rank** and **State Rank** (i.e. G1/S3). **State Status** is given as an additional ranking for species that are ranked S1 or S2. Each species is evaluated based on multiple factors with stability of habitat needed to sustain the species. The state status may or may not correlate to the State Rank assigned.

Global Ranks (NatureServe 2024): Global ranks are assigned by NatureServe and updated periodically based on current research.

G1-Critically Imperiled At very high risk of extinction or collapse due to very restricted range, very few populations or occurrences, very steep declines, very severe threats, or other factors.
G2-Imperiled: At high risk of extinction or collapse due to restricted range, few populations or occurrences, steep declines, severe threats, or other factors.
G3-Vulnerable At moderate risk of extinction or elimination due to a fairly restricted range, relatively few populations or occurrences, recent and widespread declines, threats, or other factors.
G4-Apparently Secure: At fairly low risk of extinction or elimination due to an extensive range and/or many populations or occurrences, but with possible cause for some concern as a result of local recent declines, threats, or other factors such as climate change.
G5: Secure: At very low risk or extinction or elimination due to a very extensive range, abundant populations or occurrences, and little to no concern from declines or threats.

State Ranks: State ranks are listed below and in most cases, parallel the Global rankings in terms of their connotation. Ranks listed are considered "provisional" until each species can be thoroughly reviewed and the state ranks updated.

S1 - Extremely Rare
S2 - Rare
S3 - Uncommon
S4 - Many Occurrences
S5 - Very Common
SX - Extirpated from Kentucky
EX- Exotic

State Status:
E-Endangered A taxon in danger of extirpation and/or extinction throughout all or a significant part of its range in Kentucky

T-Threatened A taxon likely to become endangered within the foreseeable future throughout all or a significant part of its range in Kentucky

S-Special Concern A taxon that should be monitored because (1) it exists in a limited geographic area in KY (2) it may become threatened or endangered due to modification or destruction of habitat (3) certain characteristics or requirements make it especially vulnerable to specific pressures (4) experienced researchers have identified other factors that may jeopardize it or (5) it is thought to be rare or declining in KY but insufficient information exists for assignment to the threatened or endangered status categories

H-Historic A taxon documented from KY but not observed reliably since 1980 but not considered extinct or extirpated

X-Extinct/Extirpated A taxon for which habitat loss has been pervasive and/or concerted efforts by knowledgeable biologists to collect or observe specimens within appropriate habitat have failed.

Introduction to the
Land Snails

© Jaap Vermeulen

Land snails come in every color, form, and size imaginable, some almost beyond belief. Eye-catching gems in the genus *Opisthostoma* (illustrated above) have taken calcium carbonate sculpturing to extremes. No less complicated than a Roman Cathedral, their architecture represents spectacular achievements in convoluted evolution. This genus is known only from limestone outcrops and caves of Borneo .

As part of a large and diverse group of organisms known as Gastropods, land snails fall within the prodigious and immense Phylum Mollusca. Containing around 85,000 described and living species world-wide, mollusks include aquatic snails (both marine and freshwater), land snails, terrestrial slugs, sea-slugs, sea-hares, limpets, bivalves (clams, oysters, and mussels), squids, octopuses, and the famous nautiluses. Octopuses are considered the most neurologically advanced invertebrates on earth, having feet divided into a number of prehensile and skillful tentacles capable of twisting the lid off a glass jar to access the food within.

Although snails are slow moving, squids are among the fastest animals known, exceeding underwater speeds of more than 70 mph in short bursts, a result of their water-propelled jet propulsion. At 18 meters long and weighing around two tons, the colossal squid is the largest living invertebrate. Its 15-inch eyes are the biggest of any animal.

Mollusks are also the longest lived, multi-celled animals. One species of bivalve, the ocean quahog, *Arctica islandica* can live more than 500 years.

There is scarcely a place on the dry surface of the world, outside the polar regions, where one cannot find at least a few examples of gastropods (Abbott 1989). From hot, nearly waterless deserts to cold mountain tops, land snails are thriving. Despite their biological and physiological limitations, land snails evolved efficient mechanisms for coping with freezing, starvation, and desiccation. When conditions become increasingly dry, a snail may cover the aperture of its shell with an epiphragm (a hardened mucous sheet) sealing in critical moisture and slowing desiccation. Others have "trap doors" called operculums that function in much the same way. Some snails can remain dormant for years, when favorable conditions return. Land snails in North America are typically nocturnal and are most active in wet weather when temperatures are between 50-75 degrees Fahrenheit.

Even though mollusks rank as one of the most numerous and speciose groups of organisms on Earth, they remain largely unstudied. As a result, little is known of their importance in ecosystems. Land snails, like most invertebrates, suffer from being in a conservation "blind-spot". As snail research moves forward, however, our understanding of the value of these organisms is increasing. Research has shown, for example, that land snails play an important role in micronutrient cycling in terrestrial ecosystems (Dallinger *et al.* 2000) and in dispersing plant seeds and fungal spores (Richter 1980). Land snail diversity has been shown to predict vertebrate conservation priorities (Moritz *et al.* 2001). Further, live snails and their vacant shells provide a food and calcium carbonate source to animals in many systematic groups. These include ants, firefly larva, snail-killing flies (Foote 1959), *Cychrine* beetles (which feed chiefly on land snails) (Symondson 2004), Arachnids including harvestmen, carnivorous snails, numerous species of salamanders (Petranka 1998), turtles and frogs (Burch 1962), a variety of small mammals including shrews, mice and moles (Reid 2006), snakes (Lee 1994, Dourson 2012), a variety of passerine birds (Graveland *et al.* 1994; Graveland 1996), thrushes, ruffed grouse and wild turkey (Martin *et al.* 1951), bats, (Bonato *et al.* 2004; Thabah *et al.* 2007; Dourson 2012), and primates, including humans.

While a building body of evidence suggests the importance of mollusks in present-day ecosystems, their historical value is less well known; namely their contributions made to existing plant communities, animals and, in particular, caves. In the past, the colossal accumulation of deceased mollusks, corals, and tiny creatures known as Foraminifera (which have calcareous skeletons), provided the necessary building material to create the limestone where caves are formed. Many species uniquely adapted to caves or otherwise, live in these vast underworlds, a number of species occurring nowhere else. These ancient shells have also provided the necessary limestone (in cement) to form the foundations of our cities and homes.

Declining land snail populations can have ripple effects to surrounding ecosystems. The great tit, *Parus major*, in the Netherlands, for example, has declined precipitously with declining land snails as a result of acid rain (Graveland *et al.* 1994; Graveland 1996). A lack of snail shells in the bird's diet causes the bird to lay eggs with thin shells which break, and reduces the reproductive success rates. In North America, Hames *et al.* (2002) documented a correlation between a reduced

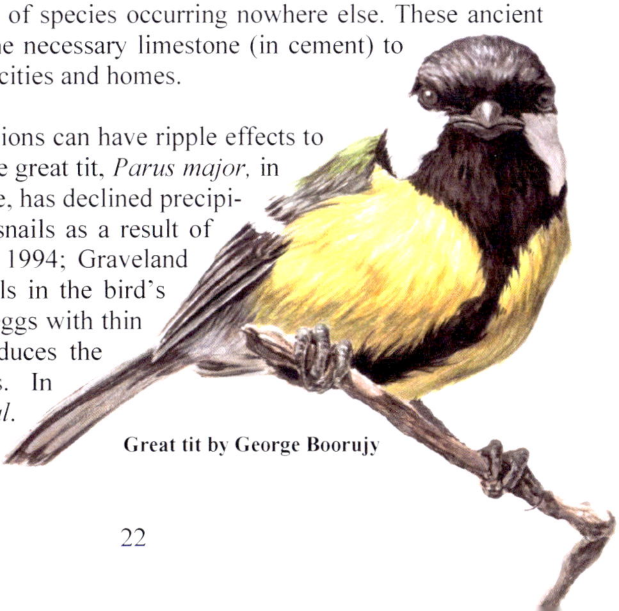

Great tit by George Boorujy

number of wood thrushes and acid rain, hypothesizing a connection to reduced land snail populations.

Sensitive to changes in the environment, native land snails could provide an early warning to impending habitat deterioration, similar to the way that freshwater mussels found in streams and rivers are used to determine the quality of waterways. For example, research has shown that when snails feed on various foods such as mushrooms, green vegetation, or forest litter (detritus), environmental contaminants present are ingested and sequestered in their tissues (Dallinger and Wieser 1984a), the midgut gland being the main accumulation site of these trace elements (Dallinger 1993). Further laboratory experiments by Dallinger and Wieser (1984) have shown that land snails that ate lettuce laced with zinc, cadmium, lead, and copper readily sequestered these elements. More concerning though, snails quickly become poisoned when simply raised on soils contaminated with cadmium, raising fear that toxins in polluted soils may be more bio-active than previously believed (Scheifler *et al.* 2003).

The environmental concerns to ecosystems, and the consequence to snails and higher organisms which feed on contaminated gastropods are valid. Rimmer *et al.* (2005) found elevated levels of mercury in the blood of Bicknell's thrush on mountains in New England, and thrushes are reported to eat snails (Martin *et al.* 1951), presumably to obtain calcium for egg laying. A recent study, conducted in Iraq in 2021, conducted experiments with the land snail, *Eobania vermiculata*, to determine if it would be an effective bioindicator heavy metal contamination in soil. The study determined that *E. vermiculata*, is an appropriate sentinel organism for some metals, mainly for Cr, and the bioindicator monitoring with this snail should be extended to mixtures of heavy metals, since such relationships frequently occur in soil (Salih *et al.* 2021). Native land snails could, therefore, be used to forecast impending problems created by anthropogenic pollutants reported to be accumulating in forests of Kentucky.

Human Health and Snails

Mollusks are providing a number of life-supporting contributions to humans. Marine mollusks may help fight cancer; the drug Kahalalide F, a protein extracted from a species of mollusk that eats sea slugs in the Pacific Ocean, has shown great promise as chemotherapy for the treatment of liver cancer (Satheeshkumar *et al.* 2010). Lethal toxins produced by cone snails were used to develop a drug called Ziconitide for patients with cancer and AIDS who suffer from chronic pain that cannot be relieved by opiates, and Ziconitide is not addictive (Wallace *et al.* 2008). Slime from the land snail, *Cornu aspersum* (one of the commonly eaten snails referred to as escargot) is now used to treat many different types of skin disorders. This snail slime repairs skin damage from overexposure to the sun and reduces scarring caused by severe acne. Land snail mucus is also known to contain natural antibacterial properties and, some scientists speculate, may be the next generation of human antibiotics. Recent research conducted by Dr. Sarah Pitt has demonstrated efficacy of the use of antibiotics made from snail slime can destroy an antibiotic resistant bacteria called *Pseudomonas aeruginos.* The bacteria, which occurs in soil and standing water, wreaks havoc on the lungs of patients with cystic fibrosis and other pulmonary diseases (Pitt *et al.* 2019). In Central America, the Maya use gastropods to treat a number of ailments including skin disorders, glaucoma, and whooping cough (Dourson 2009).

The Diet of Land Snails

Most snails are dietary generalists (Burch and Pearce 1990) consuming a wide variety of herbaceous plant leaves or stems, decaying vegetation and leaf litter (detritus), wood or bark, and fungal fruiting bodies (such as mushrooms), animal scat, carrion, and shelf or bracket fungi (Burch and Pearce 1990; Dourson 2008). Land snails in the genus *Haplotrema*, *Ventridens*, *Vitrinizonites*, and *Mesomphix* are documented carnivores, feeding on a variety of other gastropods (Atkinson 1998). Gastropods sample and judge potential food by using the chemoreceptors located on the lower two tentacles (Shearer and Atkinson 2001). Once a food source is discovered, the snail begins the feeding process by first touching the food with its foot and mouth followed by rasping (rasping signs can be seen on the fungi illustrated to the right) and dislodging bits of food with the radula structure located in the mouth (Machensted and Markel 2001). The meal is then swallowed, and muscular contractions move the food along the esophageal tract mixing it with saliva. Feeding episodes can last anywhere from a few minutes to nearly an hour, depending on the durability of the food being consumed (Dourson 2008).

Land snail feeding signs on a bracket fungi

A Brown-spotted mantleslug, *Philomycus venustus*, feeding on Golden trumpets, Roan Mountain, NC

Fruiting bodies of fungi are a favorite food of land snails, especially native slugs. As slugs consume the tissues of mushrooms, toxins sequestered from the fungi are suspected to bolster defensive properties of their mucus. Slugs in the genus *Philomycus* have been observed feeding on the destroying angel, *Amanita virosa* (pers. obs.). This is one of the most poisonous mushrooms known in North America, and one that has caused the majority of mushroom fatalities in humans (Roody 2003), yet slugs seem impervious to the mushroom's lethal effects. Little is known of symbiotic relationships between mushrooms and slugs. What is known about these potential mutualistic behaviors is that slugs feed on the mushroom tissues, ingesting numerous spores, which are carried for a time and excreted elsewhere, spreading the fungi spores to new locations.

Three Glassy grapeskins, *Vitrinizonites latissimus,* and a Black mantleslug, *Pallifera hemphilli,* feeding on a fresh dead Balsam globe, *Mesodon andrewsae*, Roan Mountain, North Carolina.

Empty Snail Shells

When snails die, the shells do not immediately decompose. Some research suggests that shells can remain intact for years (Pearce 2008), and empty shells in some limestone locations can reach exceptional numbers. But these discarded shells are anything but vacant and actually provide a secure and protected refuge for a whole host of living micro-invertebrates including pseudoscorpions, ants, millipedes, tardigrades, and smaller land snails. Some invertebrates even deposit their eggs to be incubated in the security of shells. The translucent 3 mm wide shell of *Glyphyalinia indentata* (figure a) was the depository site for eggs of an unknown organism, not an uncommon event (pers. obs.).

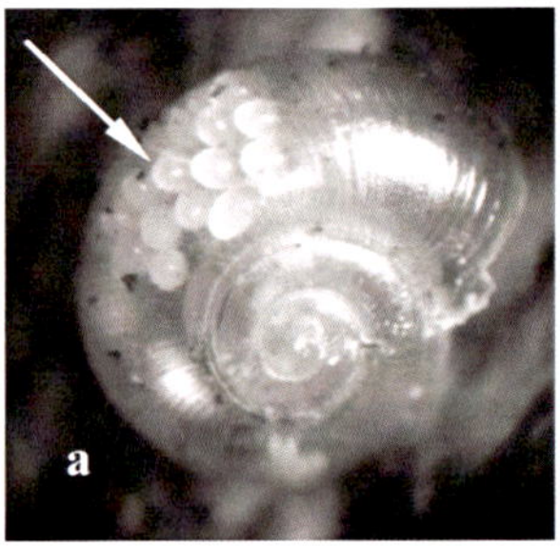

Parasites in Land Snails

Nearly species of mollusk is inhabited by some form of parasite, only a few are of medical or veterinary importance (Burch 1962). Of these, almost all live in fresh water environments. Snails are required hosts in the life cycles of parasitic trematode worms. A few land snails, such as *Cochlicopa lubrica,* are vectors of lancet liver flukes in sheep, cattle, deer, and groundhogs (Burch 1962). *Zonitoides arboreus* and *Anguispira alternata*, both native land snails in Kentucky, are implicated in the spread of lungworm in domestic sheep (Burch 1962). Multiple species of land snails and white-tailed deer play a significant role in the transmission of a parasitic nematode known as brainworm, *Parelaphostrongylus tenuis*. The initial host of the nematode is the white-tailed deer with land snails and slugs serving as intermediate hosts. Interestingly, brainworm nematodes do not appear to significantly affect white-tailed deer populations, but may cause debilitation or even fatality in elk, moose, goats, and sheep. The life cycle of the nematode begins with adult worms which are normally located between the membranes (meninges) that cover the brain and in the spinal cord of the white-tailed deer. Eggs are deposited either on these membranes or directly into blood vessels. Those deposited on the membranes hatch, and the larvae enter small blood vessels to be carried to the lungs where they enter the alveoli. Eggs deposited into blood vessels are carried to the lungs and eventually hatch with larval penetration of the alveoli. Activity in the lung tissue produces an interstitial pneumonia. The larvae pass up the respiratory tract from the alveoli, are swallowed, and then are eliminated in the feces of the deer. Larvae appear in the feces about three months after the host becomes infected. The larvae then enters into a snail or slug (the intermediate host) through the gastropods foot while crawling over infected deer feces or direct ingestion of deer feces by the gastropod. Development of the larvae in the gastropod to a stage when they are infective to the vertebrate host takes about three weeks.

Cervids become parasitized by ingesting infected gastropods that are clinging to grass or other browse foods. In the final host—cervids—development of the larva to the adult worm takes place in tissues of the central nervous system, particularly the spinal cord. Parasites leave the tissues of the spinal cord after about 20-40 days and locate between the spinal membranes where they mature. Subsequently, they tend to accumulate in the cranial region. The adult worms are about 50 mm in length and may be seen fairly readily when free in the cranial cavity. From one to 20 worms have been found in the crania of infected deer, yet as previously stated, *P. tenuis* seldom causes damage in white-tailed deer. In other cervids, there is often extensive damage to brain and spinal cord tissues. The resulting neurologic disease is characterized by weakness, fearlessness, lack of coordination of movement, circling, deafness, impaired vision, paralysis, and death. When found in moose, this disease is often called "moose sickness" or "moose disease." A correlation to prevalence of the disease in elk, moose, and other cervids has been linked to an increase in population of the white-tailed deer. In Canada, studies by Anderson and Prestwood (1981) have substantiated the importance of this problem in management of big game and the dynamics of the host/parasite relationships among wild populations. Land snail species implicated in the spread of brainworm include: *Anguispira alternata, Deroceras leave* (a slug), *Deroceras reticulatum* (a slug), *Discus catskillensis, Euchemotrema fraternum, Philomycus carolinianus* (a slug), *Neohelix albolabris, Pallifera dorsalis* (a slug), *Pseudohyalina exigua, Triodopsis tridentata, Ventridens intertextus,* and *Zonitoides arboreus.* Any species of land snail that occurs where there is a high prevalence of white-tailed deer could potentially be a host and contribute to the spread of this disease (Anderson and Prestwood 1981).

Land Snails as Pests

Land snails, including slugs, can be agricultural pests. By and large, snails and slugs that are problematic in gardens are species that are non-native introductions. Most have been accidently released into North America by way of plants, potting soils, or shipping crates, but a few species, like *Cepaea nemoralis,* were introduced for their colorful shells. Exotic gastropods naturalize quickly and multiply. Degraded native habitats only make things worse by providing the conduit for dispersal and movement into new areas. These exotics can carry molluscan diseases and problematic parasites which can effect native land snails and other wildlife. Exotic slugs can be especially damaging pests in greenhouses and agricultural lands, costing millions of dollars worth of crop damage every year. Gardeners are most familiar with those primarily exotic species, the slugs, due to their insatiable appetites for tender vegetables. While foreign gastropods quickly become overpopulated without natural controls in place, native snails and slugs are rarely problematic, and most species actually become scarce or disappear entirely in areas where the natural vegetation has been eliminated. The best defense against exotic gastropod infestations is to keep natural forests intact.

Fluorescence in Land Snails

Not to be confused with bioluminescence (the natural light observed in fireflies), fluorescence is the term used to describe the absorption of light at one wavelength and its emission at another. Only one species of land snail is known to produce bioluminescence, *Quantula striata,* from Malaysia. Research has shown, however, that the mucus of several North American land snail families including Discidae and Helicodiscidae have fluorescent slime under ultraviolet light. *Anguispira alternata* and *A. kochi* slime are a brilliant, bluish fluorescence (Rawls and Yates 1971). Under laboratory conditions, the mucus on the foot and body and mucus trails of *Anguispira* and *Discus* species glowed brightly when exposed to UV light. *Discus patulus* and *Helicodiscus parallelus* mucus appears to have fluorescence that is unique to each species. Other species of polygyrids were tested and failed to exhibit any sign of fluorescence. The fluorescence observed in the specimens of the three genera noted above is extremely long-lived, being as bright and as distinctive in specimens preserved for twenty years or more as it is in living snails (Rawls and Yates 1971). The crawling slime of *Anguispira jessica* (typically clear) did not fluoresce under UV exposure, but the defense slime, produced by the species under attack (which is orange-yellow), did fluoresce (Dourson pers. comm. 2010).

The function of fluorescence in land snails remains a mystery. There is some speculation that fluorescent slime has no real function, but simply a random act of evolution. The fluorescence in land snails may be explained by the hypothesis that moonlight has a component of UV light and *Anguispira* species are most active at night. Could the fluorescence be an evolved strategy of "avoidance conditioning" of nocturnal land snail predators (i.e. snail-hunting beetles) the same way that predators are conditioned to steer clear of the bright colors of coral snakes? Snail slime, especially that of *Anguispira* species, is pungent and distasteful to animals, including humans. *Anguispira* slime has a disagreeable numbing affect in one's mouth (author's personal experience). Like the coral snake which uses color to save its venom, snails may be using florescence to save precious mucus reserves. The more snails are harassed, the more mucus is produced, and losing copious slime (during an attack) could put the snail at risk of dehydration.

A Flamed tigersnail, *Anguispira alternata,* (above) in defensive posture, its aperture blocked by a noxious orange mucus and (below) with UV light only. The defense-slime appears as a bluish glow through camera lens but to the human eye, a fluorescent green. Images taken in the Red River Gorge, Powell County, Kentucky.

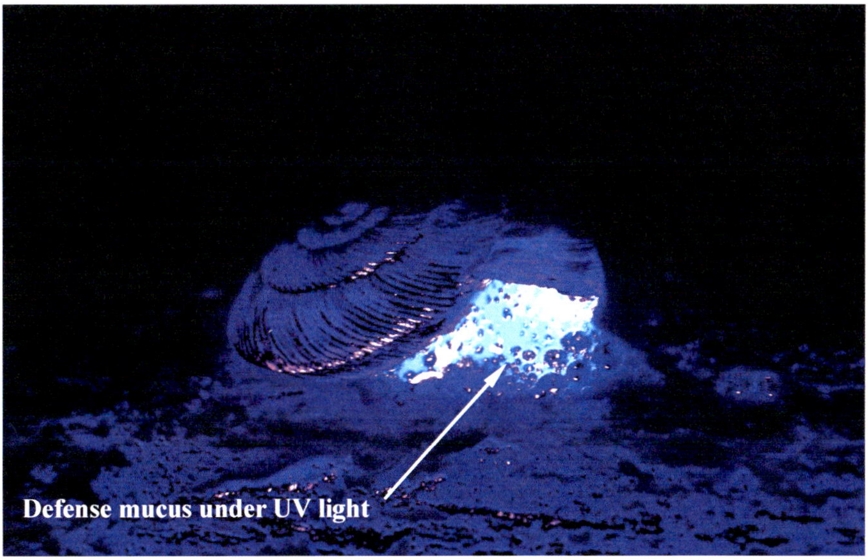

Shells, Ribs, Teeth, Hairs, Fringes, and Viscous Slime

Teeth

Hairs

Fringes

Land snails use a variety of strategies to protect themselves from harm. The shell is the first line of defense to ward off predators. Opercula (snail doors) can prevent invertebrate attacks or help conserve critical body moisture but few snails in Kentucky possess these protective structures. Teeth and lamellae act as barricades, preventing the entry of predators such as predacious beetle larvae. Some research suggests that aperture teeth may also provide a calcium storehouse which can be used to repair damaged shells, act as pivotal points for balancing the shell during movement, or trap air if the animal becomes immersed in water, keeping it afloat (Emberton 1994). Snails preferring talus or rocky areas many times have periostracal hairs and fringes like those found on *Inflectarius, Stenotrema,* and *Helicodiscus.* These features may cut down on abrasions of shells or may act as natural shock absorbers when shells fall through talus ornamentation or increase shell-crypsis to avoid predation. Hairs, fringes, and ribs do in fact function as water retention features, which hold and uniformly transfer (wick) water across the entire shell surface (pers. obs.). Although speculation, shells with these features may help facilitate snail movement across drier surfaces and/or help keep the snail cooler. Several species (like those in the genus *Anguispira*) contain colorful defense mucus which has an anesthetizing affect in the mouth (personal experience), no doubt an effective predator repellant. Although slugs are without protective shells, they are anything but defenseless. Slugs have copious, viscous, water-insoluble mucus (a), which can gum-up the mandibles of beetles or cause antagonistic garter snakes to vomit (pers. obs.).

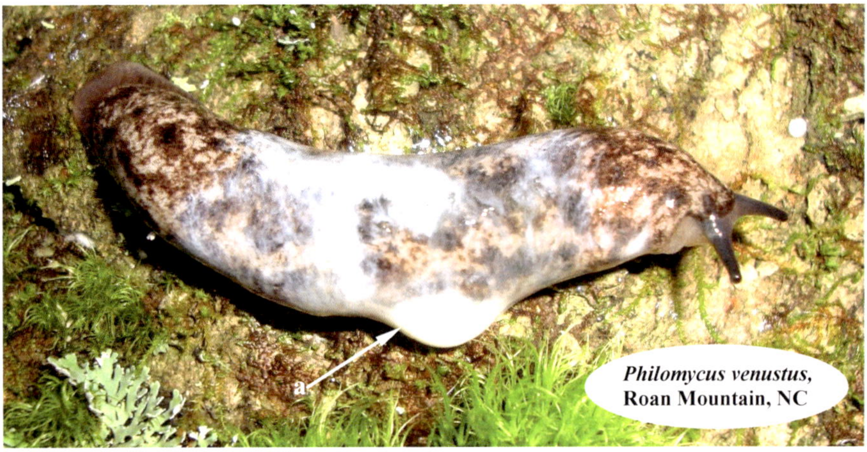

Philomycus venustus,
Roan Mountain, NC

Land Snails and Spider Webs

While crawling across a spider web, a Flat bladetooth, *Patera appressa,* appears to glean precious condensed water during a dry period in late July, (Red River Gorge, Powell County, Kentucky). Land snails are often seen on spider webs, crawling effortlessly across the web's sticky surface without becoming entangled, an act that would imprison most other invertebrates of similar size. There is some evidence that snails are attracted to the silk nets that trap condensed moisture as a source of drinking water.

Land Snail Predators

With regard to land snail predators, small mammals such as shrews and moles are at the top of the list. Above image of a Hairy-tail mole preying on a Magnolia threetooth, *Triodopsis tridentata*, Roan Mountain, NC. Smaller than moles, shrews are tiny, venomous mammals that chew through the apex and sides of the shell, eating the snail flesh within. Piles of empty shells under rocks and logs are a sure sign of small mammal feeding. Salamanders are major predators on a variety of small to medium-sized land snails (Petranka 1998; Van Devender and Van Devender 2003), devouring the entire animal, shell and

A spotted salamander stalking a slitmouth, Red River Gorge, Kentucky

all. By and large, the shell passes through the salamander's digestive tract with little harm, the contents partially or entirely digested. Van Devender and Van Devender (2003) collected a number of land snail species from fecal matter of the Red eft, *Notophthalmus viridescens* and found the following genera: *Cochlicopa, Euconulus, Gastrodonta, Hawaiia, Glyphyalinia, Novisuccinea, Triodopsis, Ventridens, Vertigo, Vallonia,* and *Zonitoides. Vallonia excentrica* was a new record for North Carolina. Few studies have focused on the importance of gastropods in the diet of salamanders, especially juvenile salamanders which have easy and safe access to small land snail species (under 5 mm) living under the leaf litter. Spotted salamanders (page 31) are also reported to include land snails as part of their diet (Van Devender pers. comm.).

In the above image, a snail-hunting *Cychrine* beetle dines on snail flesh of a golden dome, *Ventridens arcellus* (Spruce Knob, West Virginia.) The head and thorax regions of beetles in this group are significantly smaller than other beetles of similar size, allowing the insects to enter through the apertures of live snails to extract the slimy flesh with their large serrated jaws.

Reproduction

The majority of land snails found in the Kentucky are hermaphrodites, each individual having ovotestis in which both sperm and eggs are produced. When two individuals of the same species, in search of propagation, find one another (by following scent trails), they typically exchange sperm. Sperm can be stored for months to years by each individual snail until conditions are favorable for fertilization and egg laying, at which time, eggs are deposited under logs or in moist leaf litter. Interestingly, many land snails, including several native slugs in the genus *Philomycus,* are characterized by the presence of "love darts".

Dr. Ron Chase of McGill University, Canada, studied the complicated sex life of the European land snail commonly found on people's plates, *Cornu aspersum.* This species is one of many that employ calcareous love darts as part of courtship rituals. The love darts are not shot as one might imagine, but are forcefully expelled through the body wall of the partner. The dart itself appears in a variety of sizes and shapes, and most species have a single dart which is used only once (Chase *et al.* 2006). In his research on this fascinating sexual behavior, Dr. Joris Koene learned that some land snails employ multiple darts, while others use the same dart to stab partners repeatedly, as many as 3000 times (Koene *et al.* 2013). But what exactly is the function of the love dart?

Recent studies suggest that after copulation the dart functions to increase the reproductive fitness of the shooter. Snails are promiscuous and store sperm from multiple donors for several years before they use it to fertilize eggs. Thus, sperm donors (males) must compete to fertilize eggs. Dart receipt promotes the safe storage of the shooter's sperm, so there will be more sperm from successful shooters available for fertilization than from unsuccessful shooters. Since the female function chooses the sperm by a lottery-like mechanism, successful dart shooters sire more offspring than unsuccessful dart shooters.

Chase and Blanchard (2006) tested whether the dart works by simply rupturing the skin or by injecting a bioactive agent. Just before the dart is thrust into a partner, it is covered with mucus (Fig. a, page 34) from a special gland located near the dart's launching site. Koene et al. (2013) conducted an interesting test in which needle stabbings were substituted for dart shooting. In one mating, saline was injected through the needle, in the other, mating mucus was injected. They found that the matings that were associated with mucus injections were responsible for more than twice the number of offspring as were the matings associated with saline injections. Thus, mucus is the agent of the dart's effect on reproductive success.

Chase and Blanchard (2006)

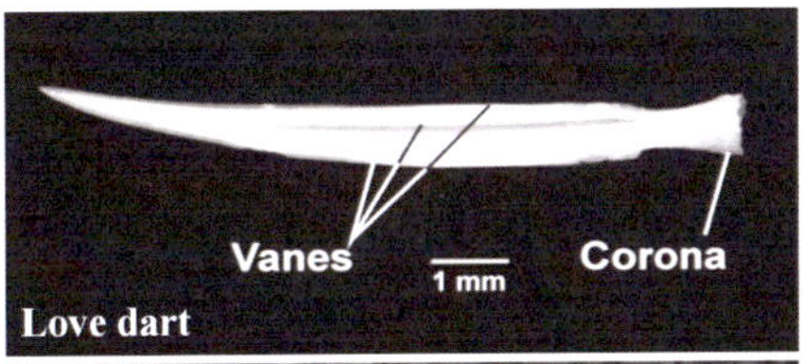

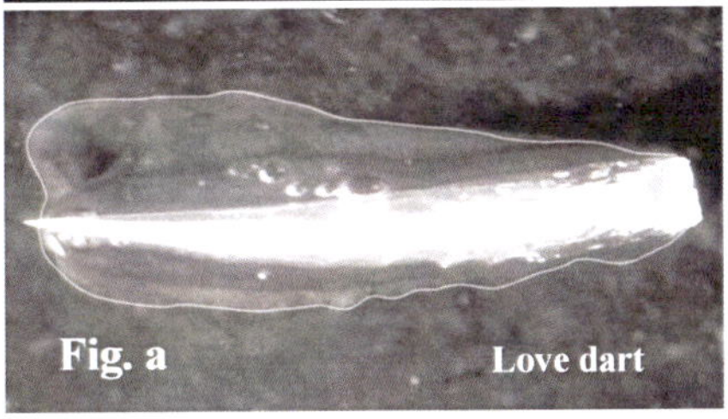

© Chase and Blanchard (2006)

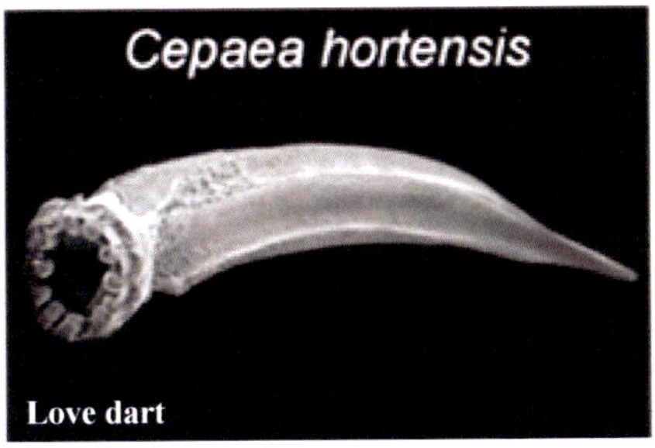

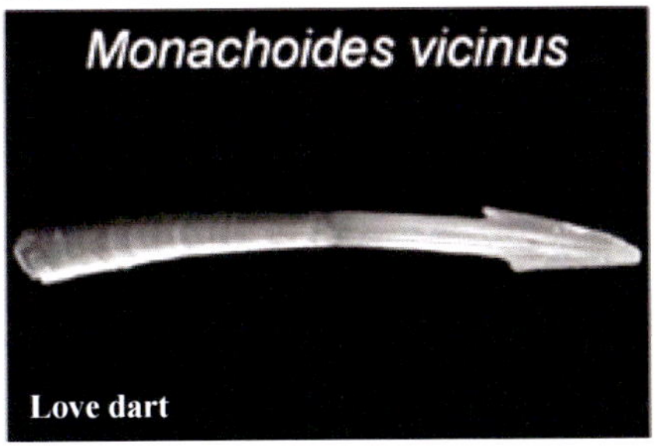

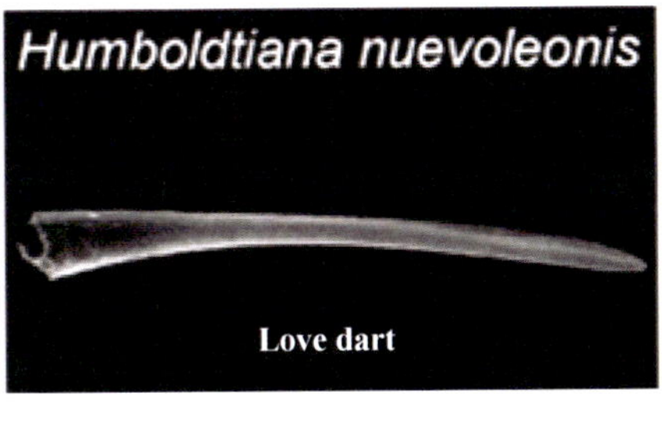

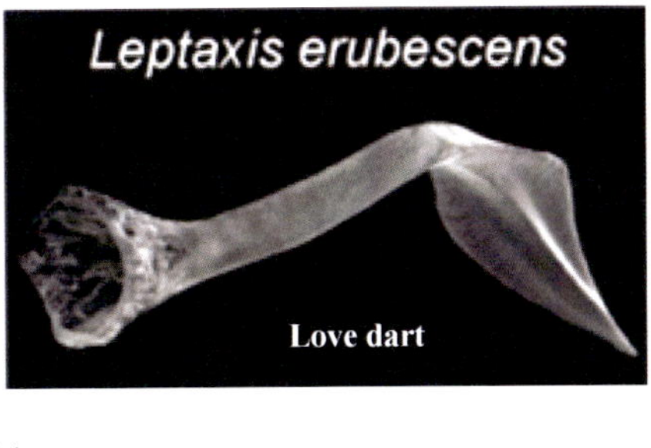

© Koene and Schulenburg (2005)

Environment and Land Snail Distribution

In 1962, Burch affirmed that the most distinctive molluscan province east of the Rocky Mountains was the **Cumberland** which has several unique genera (*Clappiella, Gastrodonta, Pilsbryna* and *Vitrinizonites*) and many endemic species. This diverse geographic area includes eastern and southeastern regions of West Virginia, western Virginia, southeastern Kentucky, eastern Tennessee, western North Carolina, and small portions of northern Alabama, Georgia, and South Carolina (page 36). Today, 293 land snail species are reported within this territory, representing roughly 56 percent of the total land snail fauna of eastern North America. A surprising 114 species are restricted or nearly restricted to the Cumberland Province, and, consequently, endemic to the Southern Appalachian Mountains including Pine Mountain and Black Mountain in Kentucky.

Two exceptionally high land snail endemism sub-regions exist within the Cumberland Province. The most northern of the two, labeled as **Region (a)**, occurs mostly along the West Virginia and Virginia border and a small sliver of extreme southeastern Kentucky in Pike County, containing no less than 165 species, 18 of which are endemic. **Region (b)** includes the higher mountain ranges of Tennessee and North Carolina, where 192 land snail species, 55 endemic, have been documented. Region (b) is essentially the southern portion of a greater area extending as far north as southeastern Pennsylvania where it merges with the Ridge and Valley sub-region. Positioned between the two provinces rests the **Great Appalachian Valley**, a lowland trough that begins in northern Alabama and ends in Canada; the whole conglomerate is sandwiched between the Appalachian Plateaus and the great costal plains of the Piedmont. While there are shared land snail species between the Ridge and Valley and Blue Ridge provinces, both contain numerous taxa exclusive to each. For its size, the Great Smokies area of the Blue Ridge or Region (b) harbors the richest land snail fauna in North America, and new species continue to be discovered and described (Slapcinsky and Coles 2004; Dourson 2012).

These continental and global hot-spots have also been shown to harbor high salamander, invertebrate (i.e. crayfish and mussels), plant, lichen, bryophyte, fungi and slime-mold endemism as well. One reason the area holds a myriad of organisms is thought to be a result of its steep topography. Precipitous relief (contained within deep gorges to soaring peaks) produce two or three times the surface area of flat land of comparable size. As a result, this relief has more than doubled or tripled the biomass-holding capability of the region. Steep topography in concert with geology and geophysical aspects of the land (caves, cliffs and talus), microclimates created by aspect (south and north facing slopes), all contribute to the area's remarkable biodiversity. Is it any wonder land snails have been so successful, finding refuge from the deepest caves to the highest mountain tops?

Without question, calcium carbonate is an essential mineral to land snails for regulation of bodily processes and reproduction, and, most importantly, shell-building (Burch 1962; Fournie et al. 1984; Hickman et al. 2003). Land snails obtain calcium in several ways including consuming soil particles from calcareous substrates, eating decaying leaf matter (Wareborn 1970; Burch and

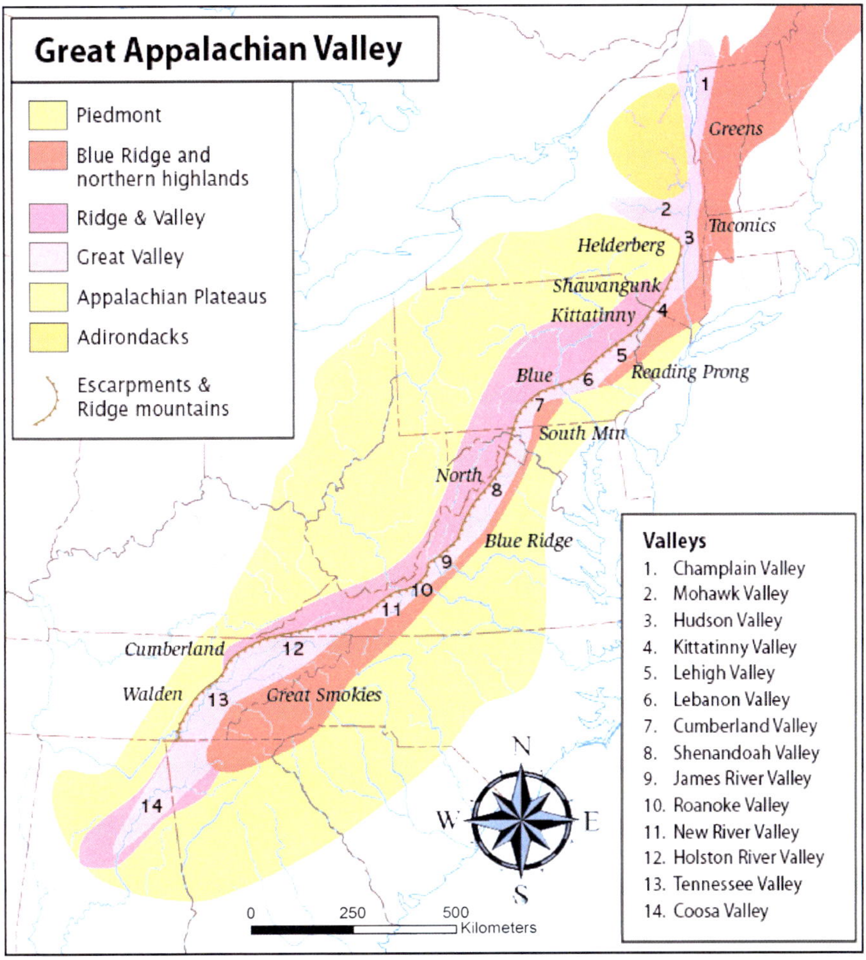

Map source from Wikipedia

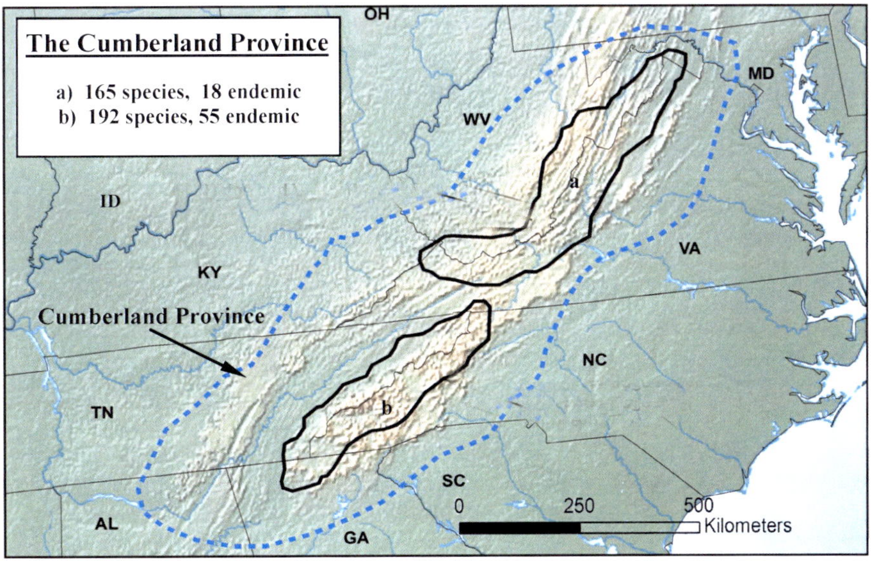

Pearce 1990; Nation 2005), almost certainly by ingesting Physarales slime molds (which precipitate amorphous calcium carbonate), and by gleaning calcium from the shells and bones of deceased animals. *Triodopsis platysayoides* (an endemic West Virginia land snail) has been documented feeding on the vacant shells of land snails, including its own kind (Dourson 2008).

Land snail abundance (number of shells) and land snail diversity (number of species) have long been associated with a variety of geological and ecological factors. Studies have shown, for example, that terrestrial gastropods living around carbonate cliffs can exhibit large and diverse populations (Nekola 1999), but show significant declines in abundance in as little as 50 m from a calcareous source (Kalisz and Powell 2003) or limestone cliffline (Dourson 2007). Other research has demonstrated that while limestone may impact abundance, it has little affect on diversity.

Burch (1955) has reported land snail scarcity associated with low soil pH (Burch 1955), declining soil cations, specifically calcium (Petranka 1982), increasing coniferous presence (Jacot 1935; Karlin 1961), and increasing elevation (Petranka 1982). The influence of pH on land snails is thought to be indirect, the main effect of low Ph being a lowering of the amounts of soil cations, principally calcium (Karlin 1961; Cameron 1970). However, low abundance in non calcareous (acidic) areas may only give the illusion of low diversity. Douglas (2011) found that land snail diversity on acid soils covered by old growth forests at Lilley Cornett Woods in Kentucky were analogous to limestone soils at Floracliff Nature Sanctuary along the Kentucky River Palisades.

In the Southern Appalachian Mountains, geological sources of calcium occur in limestone, dolomite, calcareous shale, sandstone, and the igneous-derived amphibolite, all which have calcium and/or Mg in varying amounts, supplying the necessary calcium for shell building. But what about land snails thriving on non-calcareous substrate like Kentucky's highlands? Where and how these land snails obtain sources of calcium are poorly understood. In regions lacking calcareous substrates, land snails may rely on abscissed leaves of deciduous trees or herbaceous plants (e.g. stinging nettle) as a primary source for calcium.

McHargue and Roy (1932), in a study of several species of deciduous forest trees, found that the amount of calcium in leaves expressed as a percentage of dry weight ranged from 1.64 to 7.8%, with the higher values occurring towards the end of the growing season. Other studies have shown that, unlike other macronutrients which are reabsorbed by trees prior to leaf abscission, foliar calcium concentrations in deciduous trees increase throughout the growing season and peak at senescence (Guha and Mitchell 1966; Potter et al. 1987). Calcium is a relatively immobile nutrient and the reabsorption of calcium may not be as high a priority to deciduous species as other nutrients. Further, Gosz et al. (1973) reported that dead birch leaves could form a significant pool of calcium on the forest floor because the concentrations of calcium remained high in dead leaves 12 months after abscission. Vesterdal and Raulund-Rasmussen (1998) found that the nutrient content of the forest floor under a single species of tree varied with the soil type, but they also found variation between different species on the same soil type. The idea that certain deciduous species contain higher levels of foliar calcium than others was supported by Arthur *et al.* (1993) who reported that yellow birch had relatively high levels of foliar calcium. Gosz et al. (1972) found foliar calcium concentrations were higher in yellow birch than in maple or beech. Ricklefs and Matthews (1982) looked at the leaf chemistry of 34 species of broad-leafed deciduous trees and found that yellow birch had higher than average calcium concentrations. Jenkins (2007et al.) found that calcium levels in dogwood leaves growing in the Smoky Mountains were the most significant source of calcium in acidic forests, although many have since died due to dogwood anthracnose.

Additional factors such as gradient (slope), litter moisture, elevation, and microhabitat (leaf litter, moss and logs) can significantly affect the presence or absence of land snails. Coney et al. (1982) found more species of land snails on steep slopes than on more moderate ones. Petranka (1982) found that 15 of the 56 land snail species found on Black Mountain, Kentucky showed some preference for slope, with 9 species showing an affinity for increasing slope. The importance of leaf litter moisture (thought to be a factor of slope) to land snails was emphasized by Boycott (1934), Getz (1974), Pollard (1975), and others. Although aspect was reported to markedly affect microclimate (Geiger 1965; Braun 1940), Petranka's (1982) study found no environmental variable to be significantly correlated with aspect. With respect to elevation, Petranka (1982) reported that pH, potassium, calcium, and magnesium levels decreased with increasing elevation, and that the number of snail species and the number of individuals found per site decreased with elevation. In a study by Coney et al. (1982) conducted in the Hiwassee River Basin of Tennessee, the most important environmental factors influencing the presence or absence of land snail species was microhabitat (leaf litter, moss, and logs, $P<0.05$ for 27 species), followed, in decreasing order of importance, by slope ($P<0.05$ for 15 species), rock type ($P<0.05$ for 13 species), stages of forest succession ($P<0.05$ for 12 species), soil pH ($P<0.05$ for 8 species), elevation ($P<0.05$ for 7 species), and soil moisture ($P<0.05$ for 6 species).

Less well-known is how the convergence of large physiographic and geophysical landscape edges serve to bridge distinctive regions and their allied terrestrial gastropod communities (Dourson and Beverly 2009; Douglas 2013 et al.). Acting like travel corridors for dispersal, neighboring land masses may concentrate the distribution and mixing of snail faunas resulting in remarkably high land snail diversity in comparatively small places (Dourson 2007). The two rather minor regions (a and b) on page 36 (bottom map) are both positioned between the much larger provinces of Appalachian Plateau, Great Valley and Piedmont. These two minor provinces, although relatively small, harbor around half (256 species) the reported land snail fauna east of the Rocky Mountains, and 73 endemic species.

Further evidence that converging physiographic and geophysical landscape edges are driving land snail diversity "hot spots" is supported by recent inventories at Bad Branch State Nature Preserve in Letcher County, Kentucky and

© Andrea Badgley

Breaks Interstate Park bordering Kentucky and Virginia. Bad Branch (approximately 2600 acres) harbors no less than 64 species, eleven representative of more southern mountain ranges, eight representative of West Virginia and Virginia, and nine having affinities with more northerly faunas (Dourson and Beverly 2009). Breaks Interstate Park, with its copious 84 species (above image of the 4600 acre park), represents the richest single site in North America and demonstrates still another great gastropod mixing pot located along a vast physiographic edge (Dourson and Beverly, unpublished data). Both Bad Branch State Nature Preserve and Breaks Interstate Park are positioned along the Pine Mountain massif of southeastern Kentucky, close to the Ridge and Valley Province to the south and to the north by the Cumberland Plateau. All three regions contain their own snail affiliations. Moreover, the amalgamation of these snail rich ecoregions provide a number of terrestrial gastropods an opportunity to coexist and represents a fascinating assemblage of species not occurring elsewhere in the states of Kentucky and Virginia.

The richest single "micro-site" reported in North America rests on a mesic hillside of Furnace Mountain in Powell County, Kentucky, located along the Central Knobstone Escarpment. Forming another large physiographic landscape edge, Furnace Mountain is positioned at the coalition of the Cumberland Plateau, the Knobs, and the Outer Bluegrass regions of Kentucky. The merging of these distinct regions brought together no fewer than 61 species, all packed into just 5 acres (Dourson 2007). A number of land snails documented at Furnace Mountain are established well beyond their eastern and western limits in the state (Branson 1973; Hubricht 1985; Branson and Batch 1988).

In Kentucky, land snails can be found in every type of habitat, even your own backyard. The most successful searches begin by turning over leaf litter; the

more you uncover, the greater number of gastropods you'll find, especially around outcrops of north-facing limestone (above image of the Furnace Mountain Site harboring 61 species). Using a three or four-prong garden hand rake works best and saves your finger tips from the sharp conveyers of pain (thorns and broken glass) and the occasional leaf-covered copperhead. Tree crotches that contain abundant litter-fill will often support large numbers and diverse populations of small species, where upwards of fourteen taxa have been found. The base of large diameter trees such as Black walnut, Butternut, Basswood, Buckeye, and Birch (the 5 "Bs") also appear particularly rich (pers. obs.). Prime real estate for native slugs is found in natural forests under exfoliating bark of rotting hardwood logs (below image) in advanced stages of decay, particularly those logs that form log bridges over small ravines and streams.

HOW TO COLLECT LAND SNAILS

Field Equipment

- Ziploc bags and plastic vials (for collecting macro snails >5mm)
- Quart-size cotton soil sample bags (best source: Hubco)
- Permanent marker
- GPS unit/Smartphone GPS capabilities
- Field notebook
- Hand lens of 10X
- Small hand rake (like used in gardening)
- Rite-in-the-Rain field notebook
- Pencil
- Field Sieves (optional)

Selecting Sites to Survey

Land snails are most abundant in areas where limestone, dolomite, or other calcareous strata occur but greater species diversity often occurs in more acidic environments where snails have adapted to less conducive conditions.

Specific habitats that should be targeted in order to comprehensively cover an area: under leaf litter, rocky outcrops, rock crevices, and logs; under exfoliating bark of standing and/or down dead trees; hollow trees like American beech and sycamore; damaged trees oozing sap which attracts snails; under and on top of caps of fungi; under moss mats and the flaps of rock tripe; bases of black walnut trees; crotches of trees where leaf litter has accumulated; human-made features such as roadsides, steep banks, retaining walls, cement structures, spring houses, discarded bottles (neck facing upward), or other discarded refuse; cliffline features, caves, and rock talus. Once a site is selected, record the following data in a field notebook.

Field Notes

In **field notebook**, record the following:

a. Site #
b. Date
c. Time
d. Weather Conditions
e. Habitat Description: Main overstory tree species (if known); dominant ground cover species, solar aspect; presence and abundance of coarse woody debris (CWD)
f. GPS (including datum) or UTM
g. Elevation
h. Record any live material
i. Site Image
j. Record the number of litter bags collected.

If you are conducting a survey that needs to be repeatable, timed searches are recommended. In order to provide optimum coverage of a site, a total of 30 person minutes is standard protocol. If there are two persons, 15 minutes with 2 persons collecting is adequate and so forth.

Collecting Methods

Collecting Macro Snails > 5 mm:

Using a hand rake (or stick), scrape the leaf litter to expose detritus. Hand pick snail shells observed. Place in a Ziploc bag or plastic vial with date, site number, GPS coordinates, and collector name (using a Sharpie). Do not put paper labels in bags with live snails; they will most likely eat the paper. Don't forget to check any rock talus or turn over logs at the site (making sure to return any turned structure to its original position to minimize disturbance).

Collecting Micro Snails < 5 mm:

Smaller (micro) specimens are best collected from leaf/soil collections. Micro habitats to search and collect detritus and/or leaf litter include bases of black walnut and butternut trees, bases of large mature hardwood trees, tree crotches (pockets of detritus), leaf litter along the edges of seeps, talus and along rock faces. Using hand rake, scrape leaf litter to expose detritus. Collect a handful of soil/leaf litter, then scan the litter with a hand lens for evidence of micro specimens. If any snails are observed, fill a quart-sized untreated cotton litter bag (Hubco) with the material from the site, labeled with the date, site number, collector's name and GPS coordinates.

Processing Specimens and Leaf Litter Processing

The leaf samples are taken back to the lab and dried for approximately two weeks. Dried samples should be sifted through a series of sieves ranging from 4.76 mm down to 500 micrometers. The subsequent debris that remains after this sifting process is then searched with the aid of an Optivisor or other magnification device. It will be necessary to use a zoom dissecting-scope to determine the species of these small snails. Many of them have microscopic ornamentation that can be seen only under high magnification.

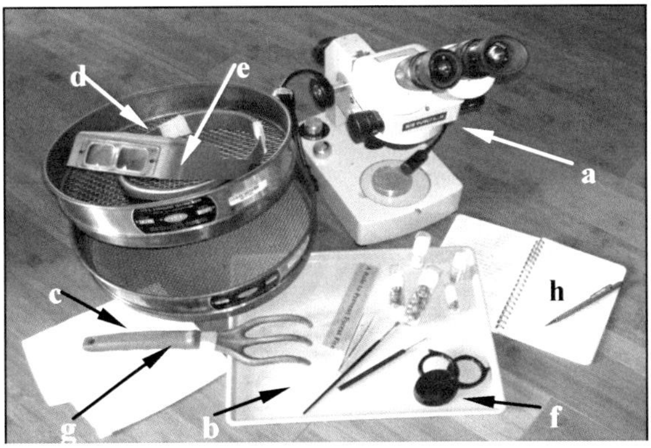

Important equipment for the study of land snails: a dissecting microscope (a), sorting tray (b), 1-liter litter bags (for the soil and leaf litter collections) (c), a series of soil sieves sizes #4 (4.75mm), #10 (2.00mm), #18 (1.00mm), #14 (1.40mm) (d), magnifying head loop (e), forceps, vials, labels, paintbrush to pick up and transfer small snails (all pictured on tray), hand lens (f), hand rake (g), and notebook (h).

Shell Morphology

Pictured (right) are three standard shell views (a, b, & c) used by malacologists to identify land snails. Always observe snails in these views. The frontal view (a) shows the shell's general form and aperture shape. The bottom view (b) shows one of the most important diagnostic features of any land shell, the umbilicus region. The top view (c) shows the apex or embryonic whorl, the number of whorls, and the width of the whorls. In general, the frontal and bottom views are the most important diagnostic views of land snail shells. While the top view has limited value for separating snails within the same genus, it is a reliable view for separating snails of different families. Short black arrows or lines indicate key features of shells. All measurements used in this book are for greater diameter or height of adult shells.

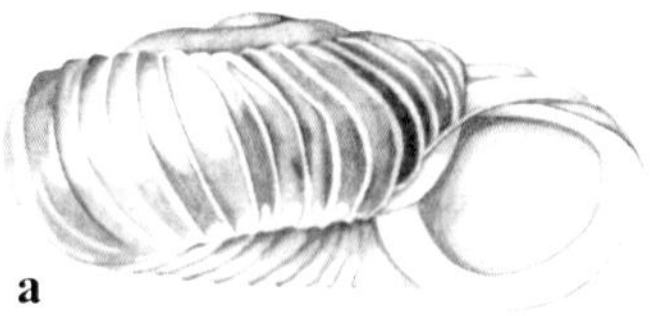

a

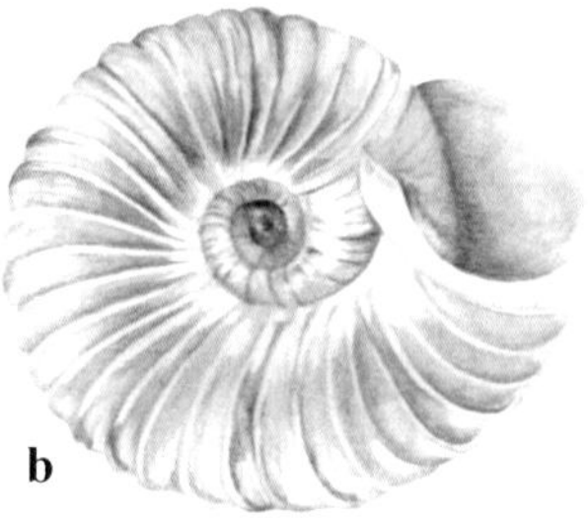

b

c

Terminology of the Shell

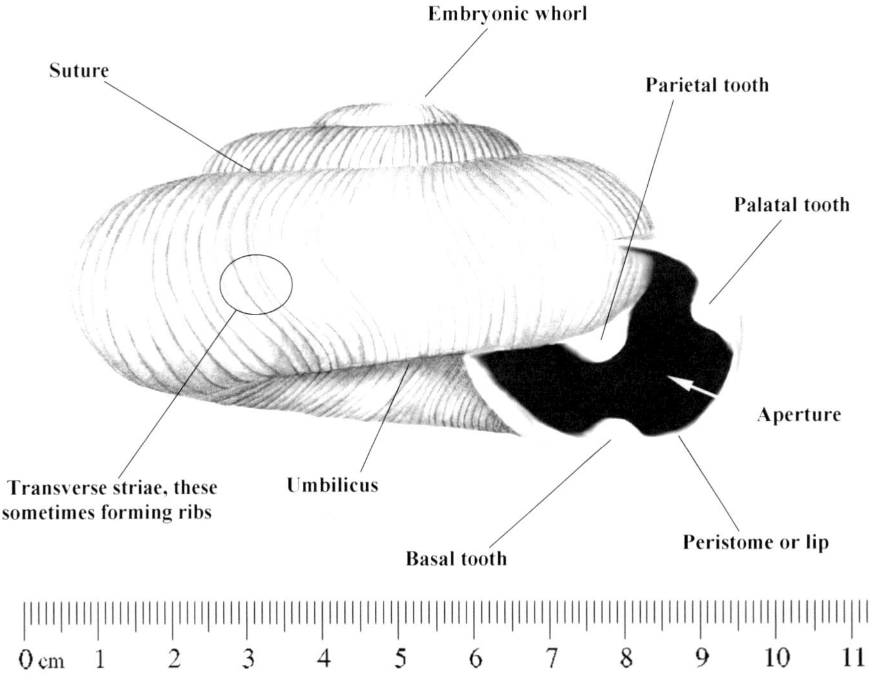

Umbilicus of the Shell (Burch, 1962)

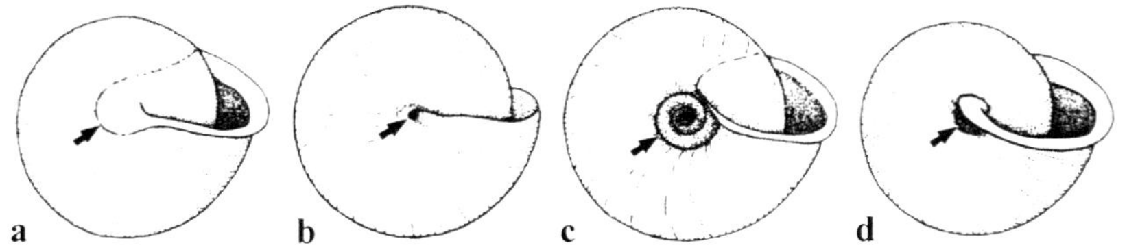

Figure a-Imperforate shell (closed umbilicus); Figure b-Perforate shell (small umbilical opening); Figure c-Umbilicate shell (wide umbilical opening); Figure d-Rimate shell (umbilical opening partially covered by aperture lip)

Shell Measurements

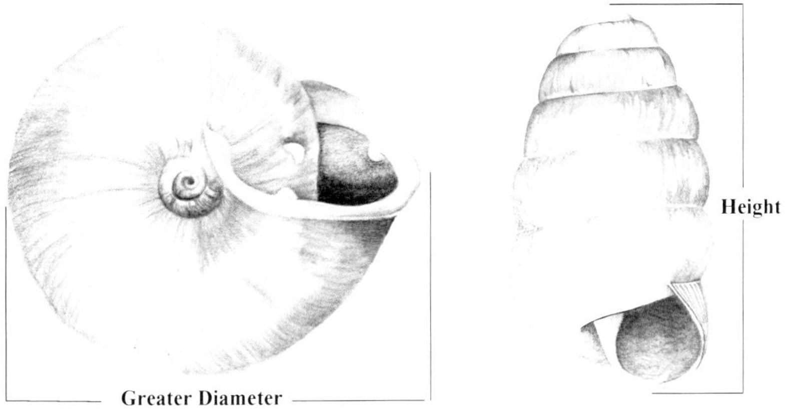

Shell Periphery

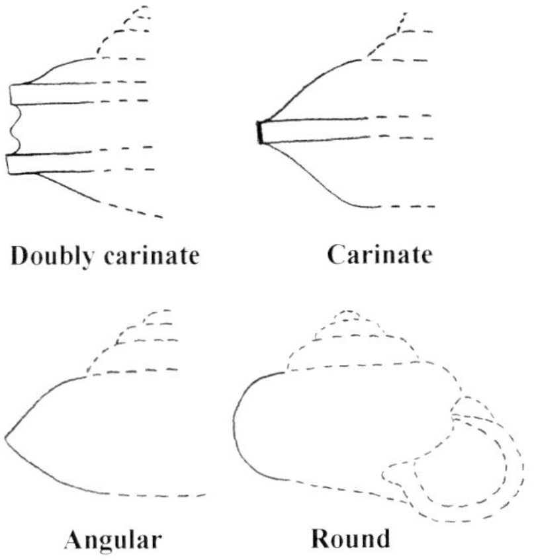

Peristome or Lip

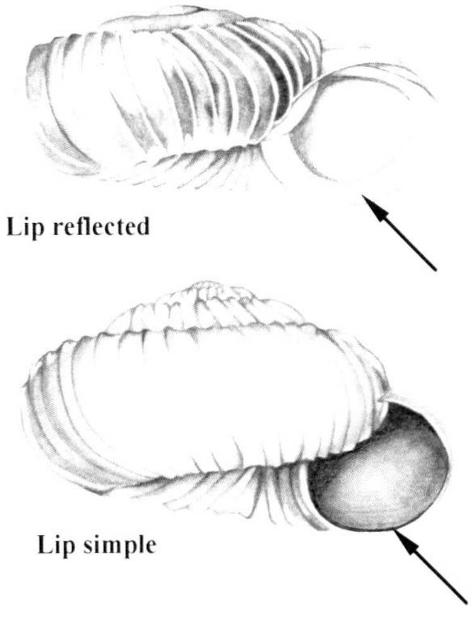

Counting Whorls

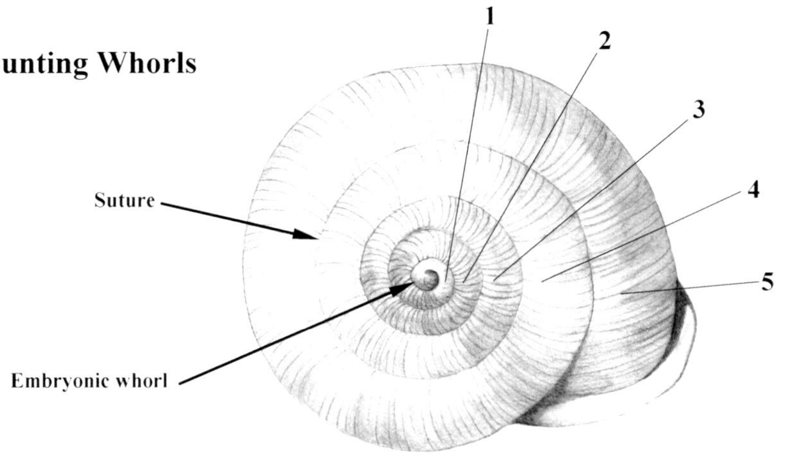

Micro-features of the Shell Surface

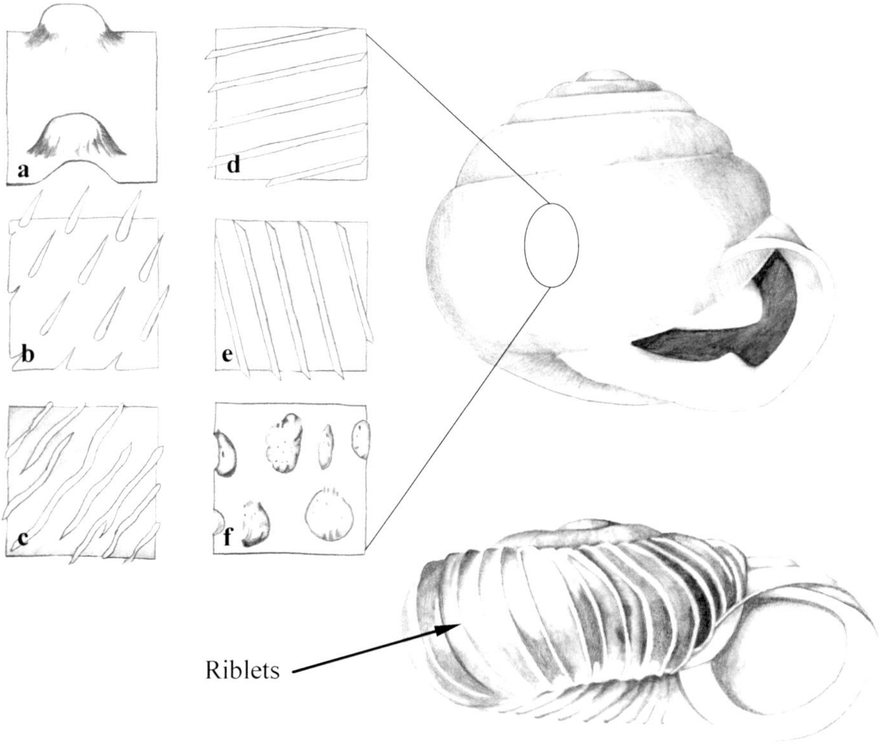

(a) Papillae (raised bumps); (b) hairs; (c) wrinkles; (d) spiral striae (lines that are indented or engraved into shell surface) or spiral lirae (lines raised above shell surface) both features running parallel with the shell spire; (e) transverse striae (can be transverse lines, ridges or riblets) which are perpendicular (vertical) to the shell spire and (f) pits in the shell.

Shell Growth

Illustrations by Dan Dourson

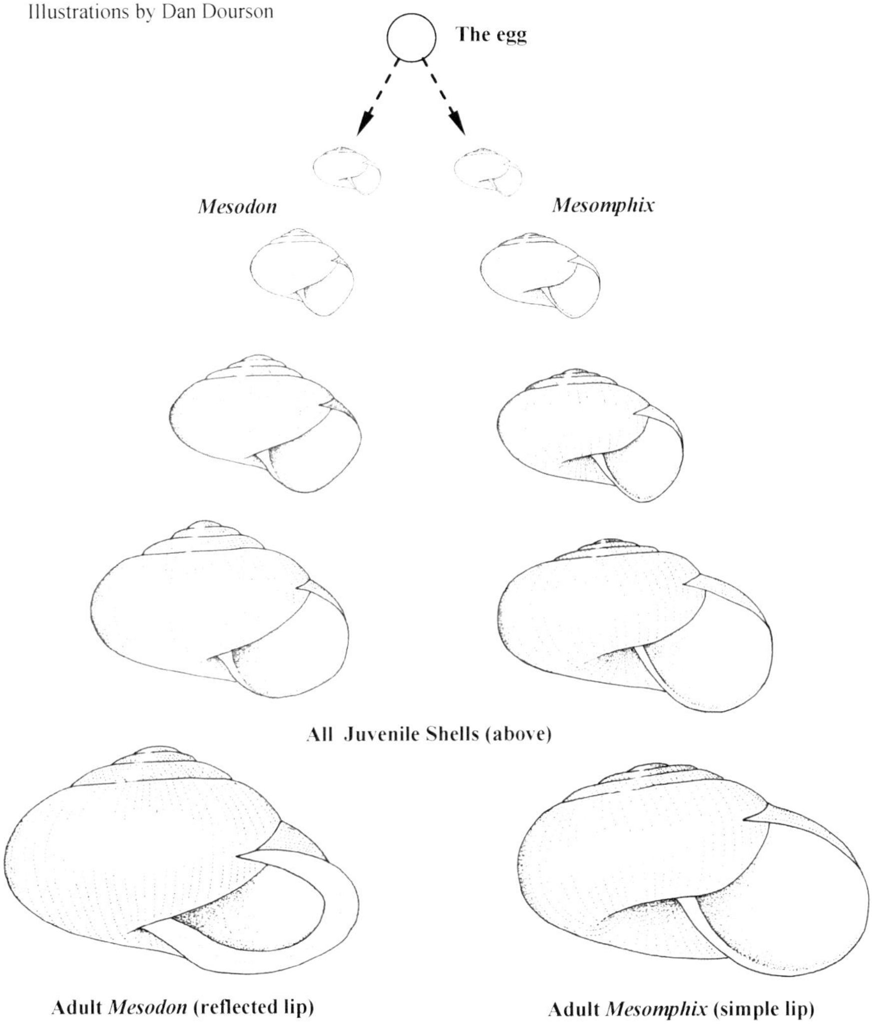

Adult *Mesodon* (reflected lip) **Adult *Mesomphix* (simple lip)**

Immature shells of any species are difficult to identify. Determining the maturity of a shell can often be accomplished by examining the aperture and number of whorls. As shells mature, the shape of the aperture changes. Note the apertures of the juvenile shells of *Mesodon* and *Mesomphix*. The bottom of the aperture of each species appears to droop as if an invisible weight is attached. As shells of both species mature, the shape of the aperture changes to a more horizontally oval shape. For adult *Mesodon* species, a reflected lip forms and for adult *Mesomphix* species, the lip remains simple. Immature *Mesodon* species are easily confused with *Mesomphix* species. *Mesodon* and *Triodopsis* species do not form reflected lips until they reach maturity. Other species such as *Mesomphix, Anguispira* and *Ventridens* do not possess reflected lips at maturity. Finally, most mature shells should contain no less than 4 to 5 whorls at maturity.

Basic Land Snail Anatomy (redrawn from Burch, 1962)

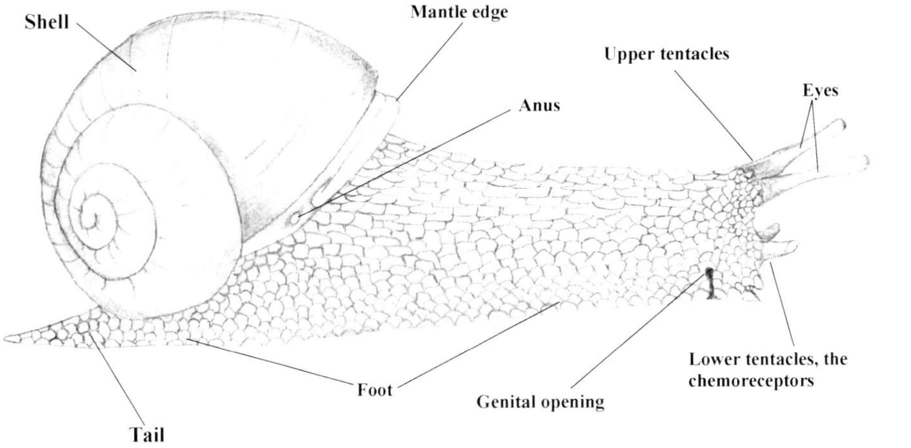

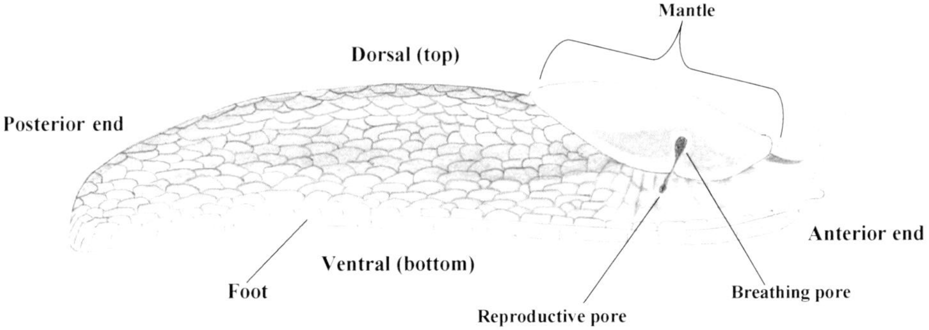

Internal Anatomy

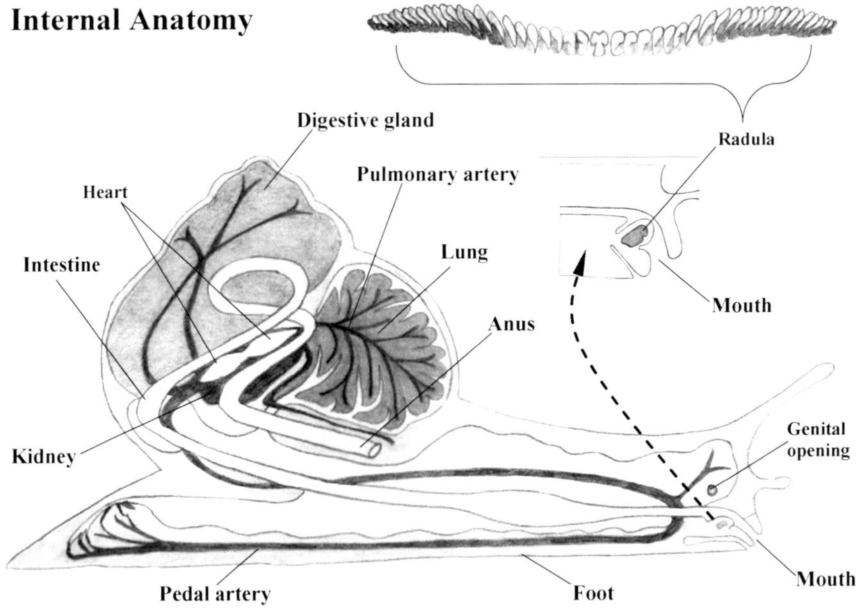

Detailed Anatomy of a Land Snail (*Cornu aspersum*) Illustrations © Gordon Riley (Kearney & Cameron 1979; reprinted with permission from copyright holders)

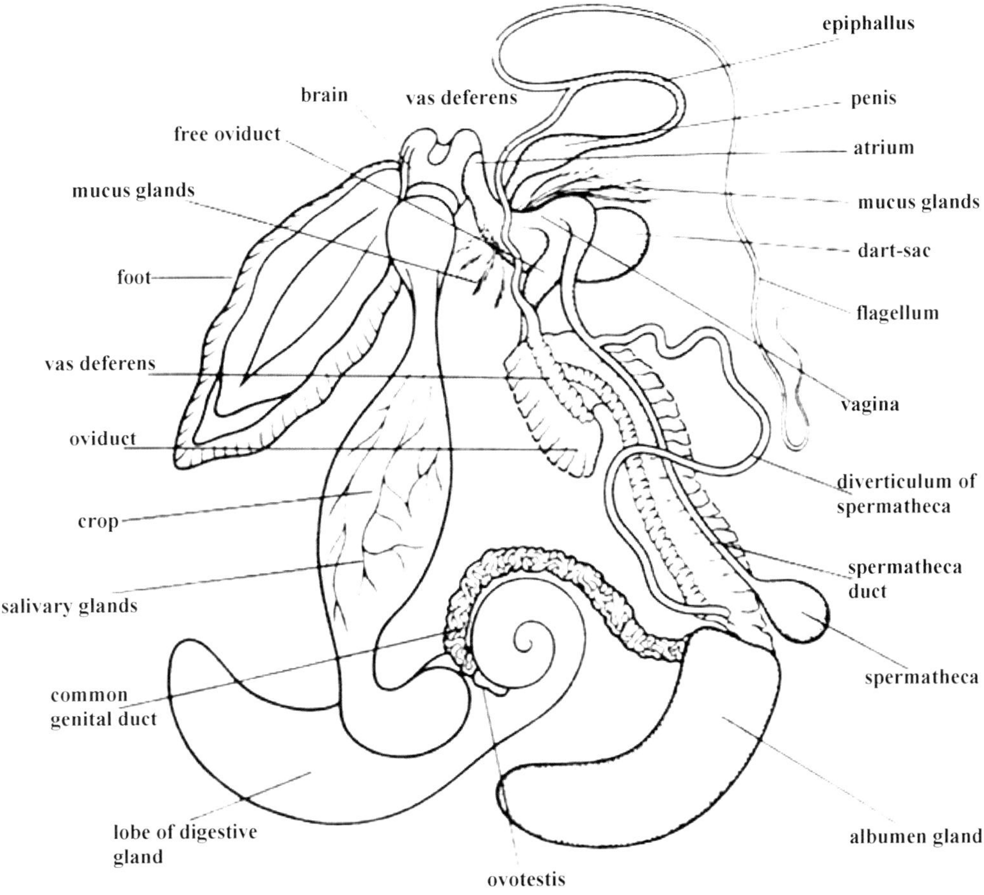

Basic Anatomy of a Slug

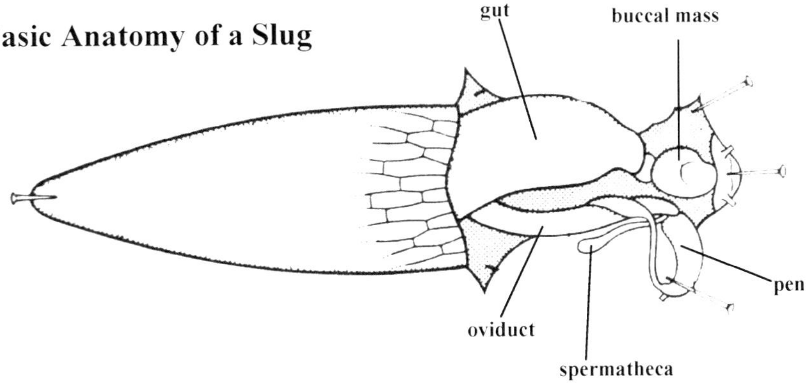

Identifying Land Snails

Land snails are a study in subtlety, and successful identification will depend on your aptitude for recognition of shell detail. The principal obstacle for beginners are the dichotomous keys. In an attempt to ascertain every species in one place, many keys overwhelm the reader with mind-numbing particulars. The reader can become frustrated quickly and lose interest. In this snail manual, the pictorial key takes you to genera only, keeping the finer details of each species in the text of the species accounts. Consequently, the keys are less cluttered with scientific muddle and, hopefully, easier to follow.

The keys are fundamentally based on two strategic characteristics: **shape** and **size** of the shell, with five additional shell distinctions or features. These include: number of whorls, umbilicus, teeth or the absence of teeth, simple or reflected lip, and spiral striae or lack of this micro-feature. If you move through the keys but are uncomfortable with your final deliberation, go back and reconsider the shape (form) of the shell; this feature alone will most often lead to a misidentification. While most shells fit easily into certain shapes or categories, some will prove more difficult to discern, so allow for these discrepancies. It is not uncommon for the shell shape of some species to vary slightly from population to population and, on occasion, within the same population. The text includes shell color but color can vary greatly from site to site and some of us (me included) are colorblind. Additionally, minute species were photographed through a microscope sometimes changing their natural color. Many shells used in the book have flaws such as cracks, blemishes or are bleached from weathering. To exemplify bottom and aperture views of some species, certain features were "digitally enhanced" in Photoshop. Remember, shell shape or form and micro-features are the most important diagnostic features. In aged shells that are bleached, slightly wetting the surface will often bring out the micro-features.

Record observations 1-8, then proceed with the shell identification.

1. Shape: Determine the general shape of the shell (page 50); same-species shells will sometimes vary slightly from site to site, and occasionally within the same population.

2. Greater Diameter or Height: Measure the greater diameter for Heliciform shells and the height for Succiniform, Conical, and Pupa-shaped shells. Sizes may sometimes vary slightly from measurements given in species accounts. (page 44)

3. Whorls: Count the number of whorls. (page 45)

4. Umbilicus: Imperforate, perforate, umbilicate, or rimate? (page 44)

5. Teeth: See the three types of teeth. Make note of any presence or absence of teeth in the aperture, size, and spatial arrangement. (page 43)

6. Lip: Is the lip simple or reflected (page 44)

7. Spiral Striae: To determine if spiral striae exist, *examine shell surface* under a dissecting microscope (page 45)

8. Other Micro-Surface Features: Hairs, riblets, papillae etc. (page 45)

This symbol indicates the use of a dissecting microscope is necessary or useful to view micro-features of the shell.

◆———◆ A scale bar is included near each shell to represent the actual size.

Basic Shell Shapes of Terrestrial Land Snails

Heliciform (globose)

Depressed heliciform

Shape of this shell lingers between depressed heliciform and heliciform; when this happens try both options.

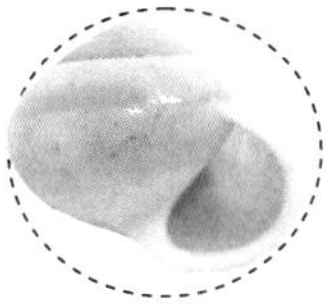

Roundish shape

Dome-shaped

Pill-shaped

Vitriniform

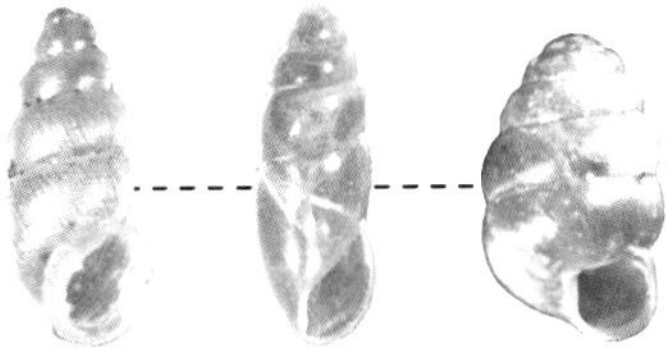

Pupa-shaped

Conical

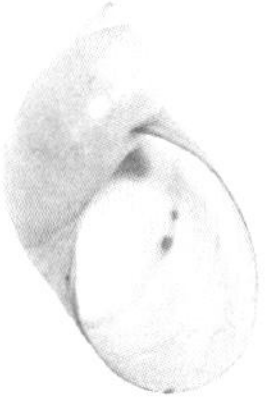

Succiniform

General Groups of Land Snails: primarily based on shell shape, lip (peristome), and/or presence or absence of teeth.

Pictorial Key to Kentucky's Native Land Snail Genera

Keys work for adult shells only

(Shells are generally arranged from smallest to the largest)

1) Shell taller than wide, simple lip or slightly reflected (teeth always absent) 2-26 mm

Columella: shell pupa-shaped, 2-3 mm tall, page 64

Cochlicopa: shell pupa-shaped and translucent 4-7.5 mm tall, pages 65-67

Pupoides: shell pupa-shaped, lip slightly reflected, 4-5 mm tall, page 68

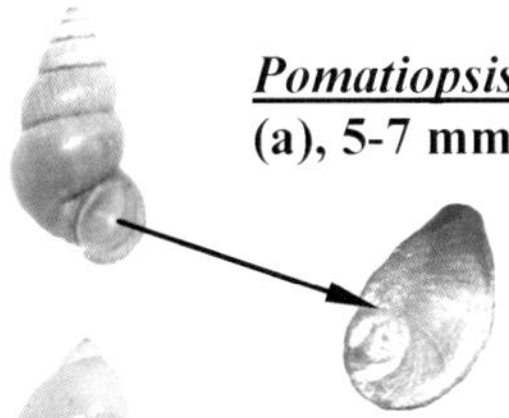

Pomatiopsis: shell conical, live snail with an operculum (a), 5-7 mm tall, pages 69-70

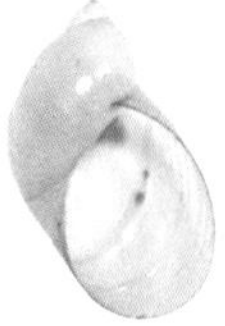

Novisuccinea, Oxyloma, Mediappendix, Succinea: shell succiniform and paper thin, without an operculum, 7-20 mm tall, pages 71-76

Rabdotus: shell pupa-shaped, lip slightly reflected, 15-26 mm tall, page 77

2) Shell taller than wide, reflected lip (teeth always present) under 4 mm

Carychium: shell pupa-shaped, reflected lip, tiny parietal tooth always present (a), 1.5-2 mm tall, pages 80-88

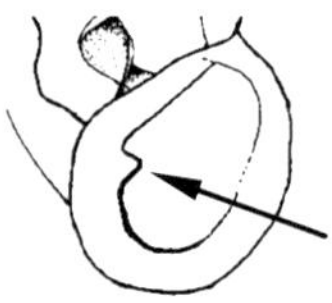

Gastrocopta: shell pupa-shaped, multiple teeth, lip notably reflected, 2-4.8 mm tall, pages 90-100

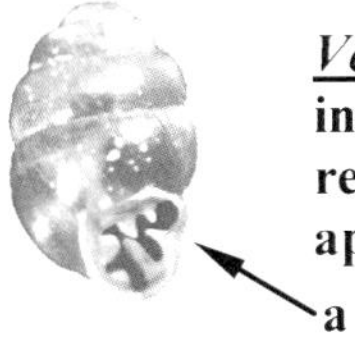

Vertigo & Pupilla: shell pupa-shaped, aperture containing multiple and sometimes crowded teeth, lip narrowly reflected with a slight indentation on the right side of aperture (a), 1.5-2.3 mm tall, pages 101-111

3) Shells tiny, depressed heliciform, umbilicate (a) except for *Guppya* which is perforate, simple lip (usually without teeth) under 4 mm

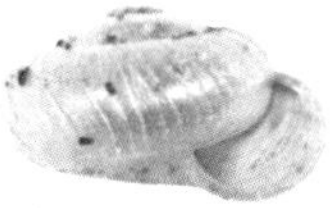

Punctum: shell minute with strong transverse and spiral crisscrossing sculpture (b) (strong scope required), umbilicate, 1.5 mm, pages 112-119

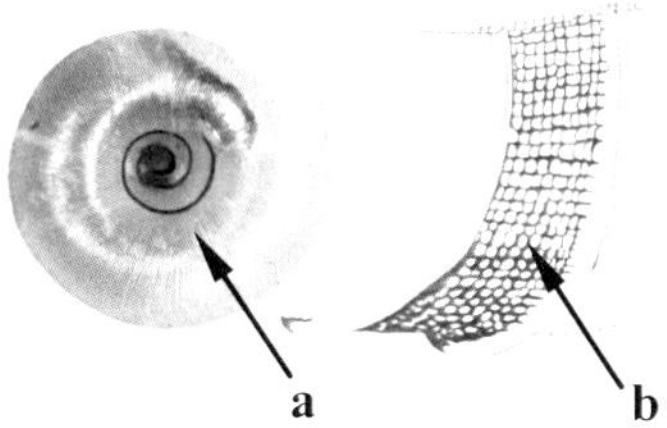

Pseudohyalina & Striatura: shell minute with strong transverse and spiral sculpture (strong scope required), umbilicate, 1.5-3.5 mm, pages 114, 120-123

Hawaiia: shell minute with weak spiral sculpture (scope required), umbilicate, 2-3 mm; pages 114, 124-126

Lucilla (formerly *Helicodiscus*): shell minute with a nearly smooth surface (scope required), umbilicate, 2-2.5 mm, pages 114, 127-128

Perpolita (formerly Nesovitrea): shell minute with a nearly smooth surface, umbilicate, 3-5 mm, 379-380

Guppya: shell minute with a nearly smooth surface, but showing traces of spiral ornamentation that are strongest around the apex of shell (scope required), perforate, (a) 2 mm, pages 114, 129

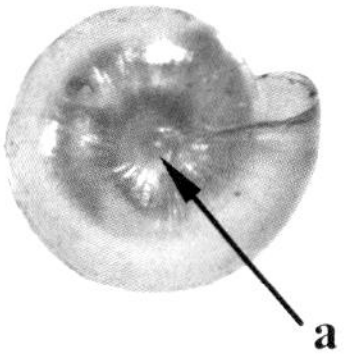

5) Shell depressed heliciform, simple lip (typically without teeth) 4 to 12 mm

Glyphyalinia: shell thin with distinctive indented transverse striae (a), imperforate to umbilicate, 4-12 mm, pages 137-154

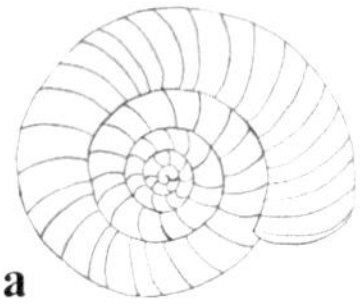

a

Zonitoides: shell surface smooth, transverse striae poorly developed, umbilicate, 4-8 mm, usually found under bark of dead trees, pages 130-134, 419

Discus: shell surface minutely ribbed (can be seen with hand lens of 10X), without teeth except for *Discus patu-lus*, widely umbilicate, 5-8 mm. pages 172-175, 384

6) Shell domed-shaped, simple lip (teeth usually absent) 2-3.5 mm

Euconulus: shell dome or bee-hive shaped and thin, without teeth except for *E. dentatus* which has one or two inner teeth, perforate, 2-3.5 mm, pages 236-246

7) Shell heliciform to depressed heliciform, simple lip (teeth usually present, except for *Hendersonia and Helicina*) 2-15 mm

Paravitrea: shell thin with either regular or irregular transverse striae on entire surface, strongest on the top, perforate to umbilicate, most species with internal teeth (a & b) at some stage of growth, 2-8 mm, pages 155-169, 414-418

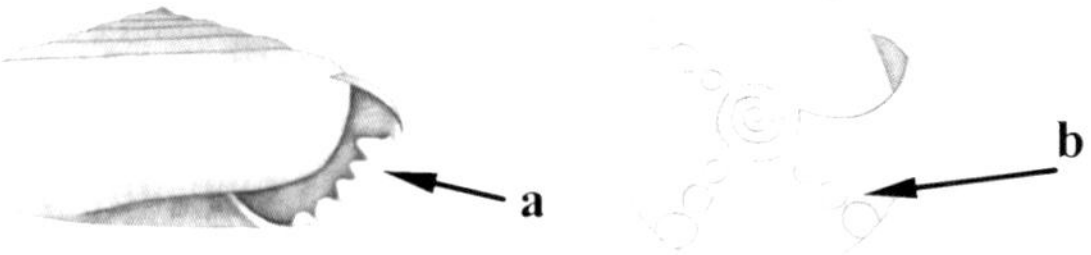

Gastrodonta: shell with two internal teeth, top surface boldly ribbed (a) and visible with a hand lens of 10X, perforate, 6-7.5 mm, pages 130, 135-136

a

Helicodiscus: shell flat, internal teeth, spiral striae (sometimes fringed or haired) a strong feature, looking like the tread of a car tire, widely umbilicate, 2-5 mm, pages 172, 177-184

Ventridens: shell can be flat or globose, glossy, with or without internal teeth, perforate to umbilicate, 5-15 mm, important– note color of live animal, pages 185-201, 422-423

Hendersonia: shell heliciform, thick for its small size with a callus on outer lip, imperforate, live snail with an operculum, 6-8 mm, pages 236, 248-249

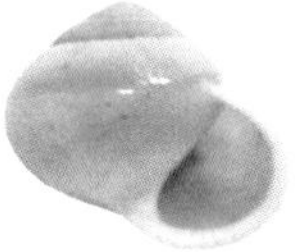

Helicina: shell roundish in form, thick for its small size, imperforate, live snail with an operculum, 6-8 mm, pages 237, 249-250

8) Shell depressed heliciform, reflected lip (without teeth) under 3 mm

Vallonia: shell small with a widely reflected lip, widely umbilicate, 1.6-3 mm, pages 270-276, 381-382

9) Shell domed-shaped, reflected lip (lamellae present) under 3 mm

Strobilops: shell small and beautifully sculptured with fine ribs, with two or more elongated lamellae, perforate, 2-2.8 mm, pages 238, 251-254

10) Shell depressed, lip reflected, teeth crowded & distorted, 5-10 mm

Lobosculum: rimate to umbilicate, parietal tooth with unique upper fold (a), umbilicate, hairy, pages 278, 282, 284

Daedalochila very widely umbilicate (b), parietal tooth with unique upper fold (c), rimate, hairy, pages 278-283

11) Shell pill-shaped, (aperture opening slit-like or open) 5-12 mm

Stenotrema: shell small and pill-shaped, lip opening restricted and narrow (a), im-perforate (b), 5-15 mm, pages 267, 289, 425-426

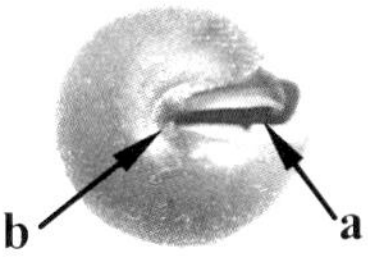

Euchemotrema: shell like *Stenotrema* but with a reflected lip, more open aperture and with-out a basal notch, rimate (c), 7-12 mm, pages 286-289, 385, 427

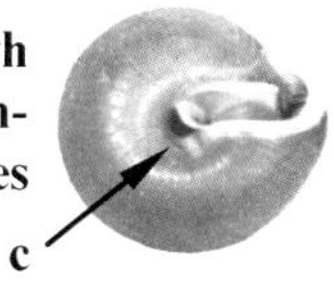

12) Shell depressed heliciform, reflected lip, imperforate, (large parietal tooth present) 7.5-14 mm

Inflectarius: shells like *Stenotrema* but more open aperture,, in KY, parietal, basal and palatal tooth always present, imperforate (a), umbilicus completely sealed; 7-14 mm, pages 292-295

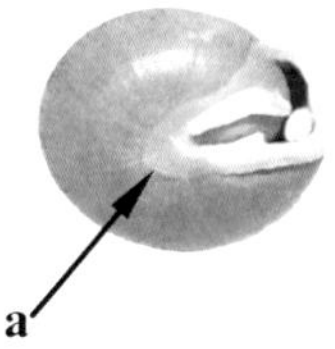

13) Shell heliciform to depressed heliciform, reflected lip, rimate (with or without teeth) 7.8-30 mm

Euchemotrema: shells small like *Inflectarius* but rimate (a), umbilicus never completely sealed, 7.8-11 mm, pages 286-289, 385, 427

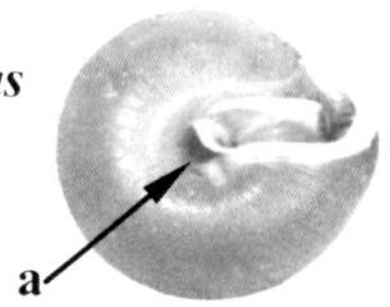

Mesodon (in part), shells globose, rimate, umbilicus never completely sealed, 15-30 mm, pages 290-291

14) Shell heliciform to depressed heliciform, simple lip, perforate to umbilicate, (without teeth) 15-36 mm

Mesomphix: shell generally glossy, without teeth in all stages of growth, perforate to umbilicate, 15-36 mm, pages 203-215

Anguispira: shell surface generally dull, without teeth in all stages of growth, color features almost always present, widely umbilicate, 15-31 mm, pages 217-228

Haplotrema: shell glossy, without teeth, lip slightly reflected in old adults, umbilicate, umbilicus one of the widest of any large snail found in Kentucky (a), 16-22 mm, pages 233-234

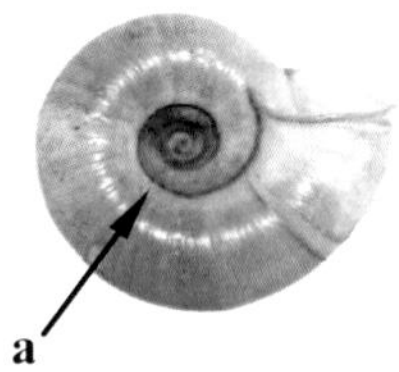

15) Shell mostly depressed heliciform, reflected lip, imperforate, (long, low basal tooth usually present (a) 13-23 mm

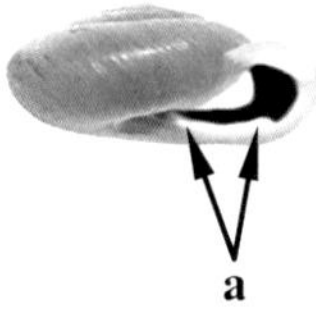

Patera: shell depressed (except for *P. pennsylvanica*) with a large parietal tooth, imperforate, 13-23 mm, pages 297-304, 432

16) Shell depressed heliciform, reflected lip (three distinct teeth) 9-25 mm

Triodopsis: mostly depressed, three teeth, parietal tooth largest of the three, umbilicate (a), 9-25 mm, pages 305-316, 427-429

a

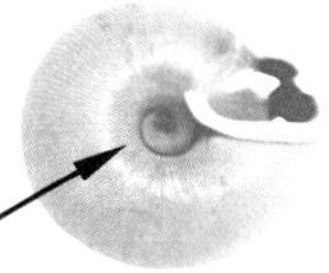

Xolotrema: mostly depressed, three teeth, parietal tooth largest of the three, imperfo-rate (b), 19-25 mm, pages 305, 319-

b

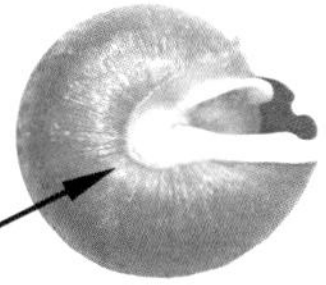

17) Shell large, depressed heliciform, reflected lip, umbilicate, 20-30 mm

Allogona: shell with or without multiple color bands, with a single basal tooth, around 30 mm, umbilicate, pages 321, 333

Appalachina: shell thin and somewhat fragile, with a small parietal and basal tooth, umbilicate, 20-30 mm, pages 321, 335-337

18) Shell mostly large and heliciform (globose), reflected lip, imperforate, 15-45 mm, but most shells here are greater than 20 mm

Fumonelix: shell globose, with a medium parietal tooth, prostrate hairs covering entire shell surface imperforate, 17-19 mm, pages 321-323

Mesodon: shell globose, with or without a small parietal tooth, imperforate, 15-38 mm, most greater than 20 mm, pages 321-322, 324-328

Neohelix shell more depressed, solid, with or without a large parietal tooth, 20-45 mm, imperforate, pages 321-322, 329-331, 430

Webbhelix shell with multiple color bands, thin, without a parietal tooth, around 20 mm, imperforate, pages 321-322, 332

Key to Kentucky's Native Slugs

(Drawings by John B. Burch, 1962)

Generally, slugs found in your garden are exotic species that can quickly become over-populated in degraded habitat. Conversely, native slugs like ***Pallifera***, ***Megapallifera,*** and ***Philomycus*** species are generally rare or disappear altogether when forests are cleared. Even in natural areas, native slugs are largely uncommon, rarely found in numbers. Separating native species from invasive ones is relatively simple and straightforward. Non-native slugs have a mantle that covers only the anterior part of the body and is saddle-like.

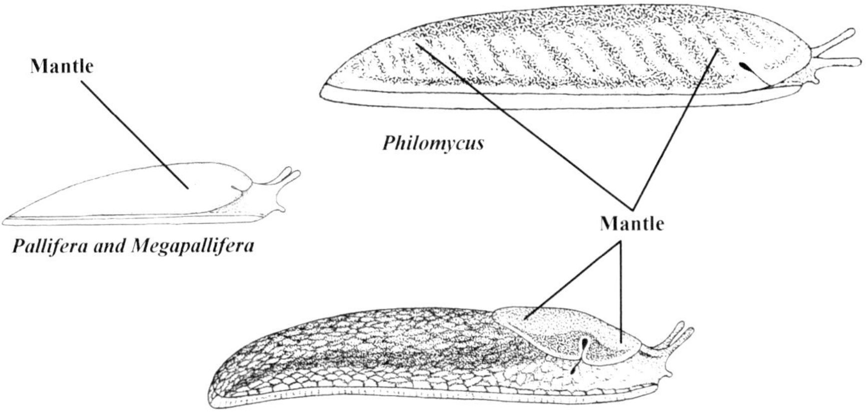

Arion **(exotic)**

Mantle does not cover the head region when crawling, under 30 mm

1) ***Pallifera:*** small thin slugs, mantle does not cover the head region when crawling (a) **pages 338-339, 363-370**

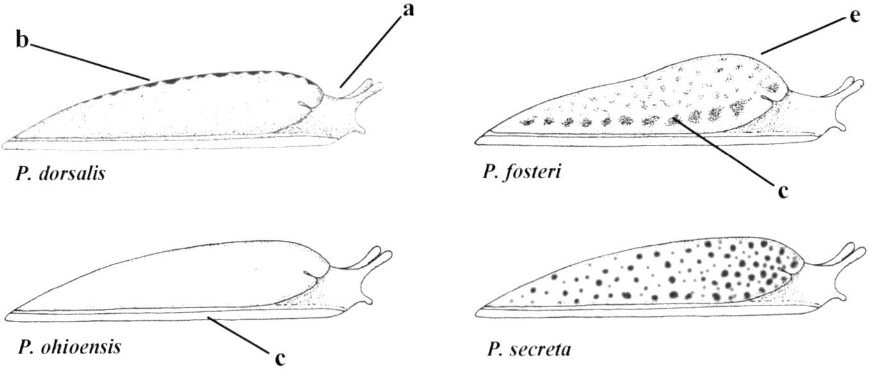

With or without a interrupted line down back (b), 6-15 mm (infrequent)....*Pallifera dorsalis* **pg. 363-364**

Without black line, foot with two red bands (c), 15-30 mm (rare).......... *Pallifera ohioensis* **pg. 365-366**

Spots profuse at mantle edge (d), mantle humped (e), 15-25 mm (common) *Pallifera fosteri* **pg. 367-368**

Well-scattered, small round black spots, 20-30 mm (infrequent)..............*Pallifera secreta* **pg. 369-370**

Mantle does not cover the head region when crawling, 50 to 75 mm

2) *Megapallifera:* medium size slugs, mantle does not cover the head region (a) **pages 338-339, 359-362**

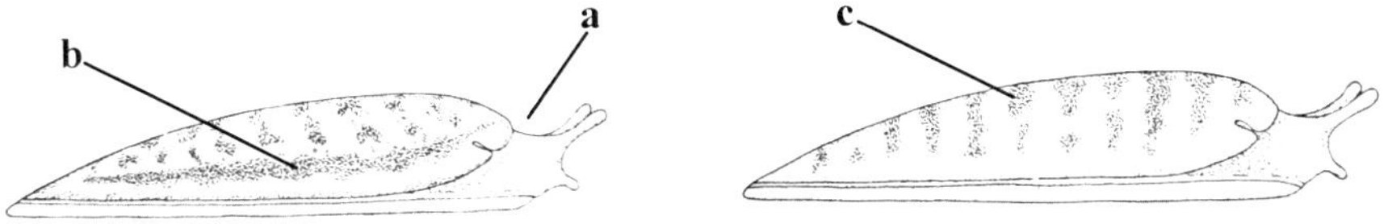

Continuous wavy lateral band (b), 60 mm (common)........................*Megapallifera mutabilis* **359-360**

Broken dorsal bands or spots (c), 75 mm (uncommon)....................*Megapallifera wetherbyi* **361-362**

Mantle covers entire head region when crawling, 60 to 110 mm

3) *Philomycus:* large slugs, mantle covers the entire head region (d) **pg. 338-358**

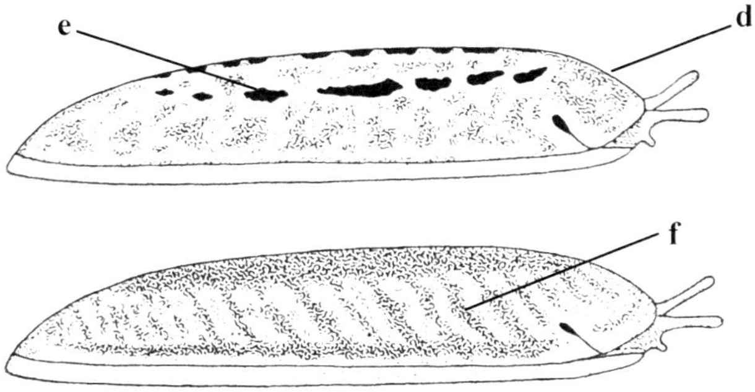

Two longitudinal rows of black spots (e), 50-100 mm (common) *Philomycus carolinianus* **341-344**

Three longitudinal mottled stripes, 50-75 mm (uncommon)..........*Philomycus flexuolaris* **345-348**

Dark and heavily mottled, slime orange, 50-110 mm (uncommon)....*Philomycus togatus* **349-350**

Dark oblique spots, forming chevrons, 50-100 mm (rare)...................*Philomycus venustus* **351-352**

Light oblique bands, forming chevrons (f), 50-100 mm (rare)............*Philomycus virginicus* **353-354**

Brown longitudinal band, starting mid-dorsally, 40-85 mm (rare)......*Philomycus bisdosus* **355-356**

Mantle light taupe to black, 28-60 mm (uncommon)..........................*Philomycus batchi* **357-358**

Mantle located in the middle of the body, 25 mm

4) *Deroceras:* the one odd-ball native slug in Kentucky that resembles exotic slugs, having the mantle more saddle-like and centered (g)

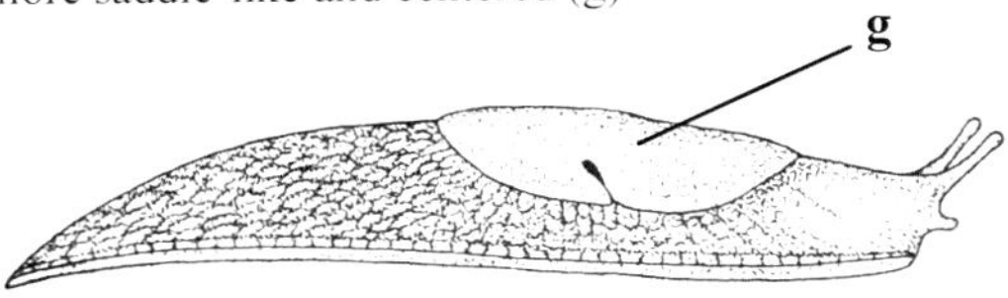

Mantle situated in middle of back (j), 25 mm.................................*Deroceras laeve*.................. **371-372**

The Junkyard Bug

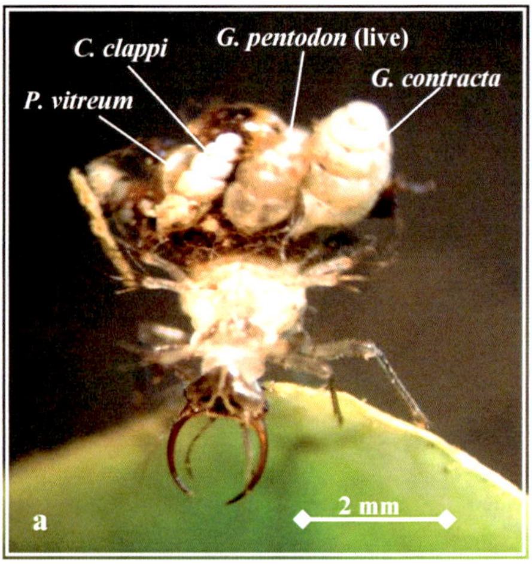

The notorious "Junkyard Bug" is rarely observed because of its small size and cryptic behavior. Actually a larva of a lacewing in the family Chrysopidae, this interesting insect carries an unusual cargo of forest rubbish on its backside. The one pictured here was found in the Great Smoky Mountains National Park in 2006 and was hauling six species of land snails including: *Punctum vitreum* (new record for North Carolina), *Punctum minutissimum, Punctum blandianum, Carychium clappi, Gastrocopta contracta,* and *Gastrocopta pentodon* (figure a). Remarkably, one species, *G. pentodon,* was still alive but unable to detach itself from the back of the insect larva. The bug is reported to use forest litter, insect parts and now land snails to camouflage itself from potential predators. Figure (a) shows four of the six land snails (two remain hidden from view). Figure (b) is the naked lacewing larva showing the hairs that hold the "junk" in place. Other researchers have also reported finding lacewing larva with attached land snails.

Part I
Land Snails of Kentucky

With more than 200 species of terrestrial gastropods, Kentucky's land snails are as varied and interesting as any group of animals in the state.

Land snails in this section are species that have shells taller than wide (Conical, Succiniform, or Pupa-shaped) and that have simple lips; although some may have a portion of the lip (usually on the left side) that is slightly reflected. They include small (under 5 mm), but mostly mid-sized (10-20 mm), snails and are found throughout Kentucky. At first glance, many of these snails look aquatic, and some are even referred to as amphibious, living close to, but not directly in water. Many have a glass-like luster, transparent and a near paper thin, but rigid shell. Consequently, after death these shells are quick to decompose and vanish from sight, therefore making their detection more challenging than other thicker-shelled species. All taxa in this section are without internal lamellae or teeth in the aperture.

Genera Included:
(in order of appearance in text)

Columella
Cochlicopa
Pupoides
Pomatiopsis
Novisuccinea
Mediappendix
Oxyloma
Succinea
Rabdotus

Oxyloma

The study of snails is an inquiry into subtlety and masterful sculpture.

Toothless Pupa

Pupillidae

Columella simplex (Gould, 1841)

Height: 1.75-2.5 mm, width 1.3 mm

Description: Pupa-shaped; shell with a slight taper; left side of lip slightly reflected, while the right side is simple; shell with 5-7 whorls that taper over the first 4-5 whorls; perforate; no teeth present in aperture; transverse striae weakly developed; two forms are illustrated, both found in Kentucky.

Similar Species: *Vertigo* and *Gastrocopta* species have teeth in their aperture and have notably reflected lips; *Carychium* are smaller, more narrow in form, and with small teeth.

Habitat: Found across a wide variety of open and forested habitats, ranging from subtropical to taiga, xeric to wet, and acidic to calcareous and on ferns and low shrubs, ubiquitous in most upland woods (Nekola and Coles 2010).

Status: G5/S5; Common; widespread throughout Kentucky; most likely occurring in every county.

Specimen: Figure (a) from Rock Creek, Cedar County, Iowa and figure (b) from Haywood Landing, Jones County, North Carolina (Nekola and Coles image 2010).

a

Two forms illustrated

b

Shells proportionate

P. muscorum

Columella simplex

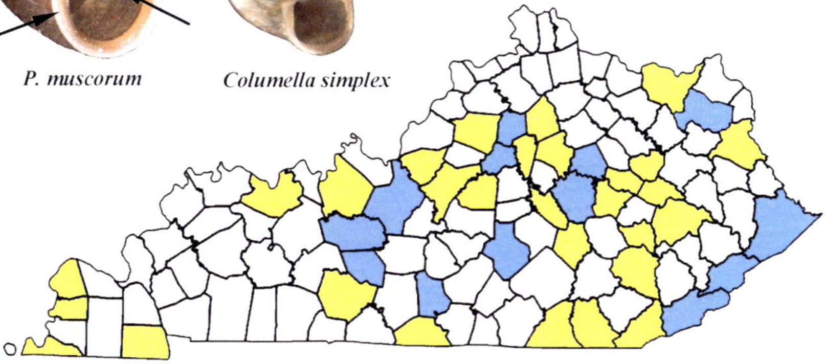

Glossy Pillar

Cochlicopidae

Cochlicopa lubrica (Müller, 1774)

Height: 5-7.5 mm, width 2.4-2.9 mm

Description: Pupa-shaped; shell with 5.5-6 whorls; sutures slightly impressed; no teeth present at any stage of growth; shell smooth, glossy and translucent with a glass-like surface.

Similar Species: *Cochlicopa morseana* is more narrow in form, its aperture more elongate with a thinner callus rim of the outer lip and is a lighter, more translucent honey color; *Gastrocopta* and *Vertigo* species are much smaller and contain teeth in apertures.

Habitat: Thought to be an exotic by some authors; an Holarctic species (Hubricht 1985); in North America, prefers disturbed sites (yards and road verges) and hard clay soils with thin litter (Nekola 2003).

Status: G5/S5; Common where it occurs, most likely transported to new locations via potting soils of yard plants and construction lumber.

Specimen: Image to the right from Larry Watrous photo collection.

C. lubrica **(two shell forms), both figures from West Virginia, Tucker County, Canaan Valley**

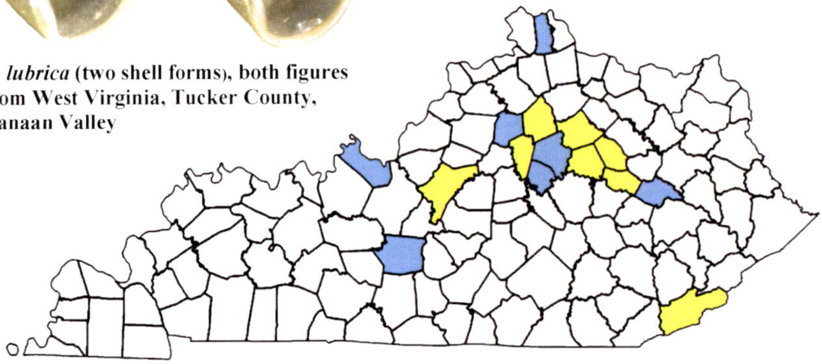

Thin Pillar

Cochlicopidae

Cochlicopa lubricella (Porro, 1838)

Height: 4.5-6.8 mm, width 2.1-2.5 mm

Description: Pupa-shaped; shell with 5.5-6 whorls; sutures slightly impressed; no teeth present at any stage of growth; shell smooth, glossy, translucent with a glass-like surface.

Similar Species: *Cochlicopa morseana* is larger, narrower in form, its aperture more elongate with a thinner callus rim of the outer lip and is a lighter, more translucent honey color; *C. lubrica* is larger and less compact in build.

Habitat: Thought to be an exotic by some authors; an Holarctic species (Hubricht 1985); in North America, favoring drier places than *C. lubrica*, such as limestone, grasslands, calcareous sand dunes and screes (Kerney and Cameron 1979); also a species of developed ground such as yards and road verges.

Status: G5/S3; Uncommon; reported from only seven Kentucky counties.

Specimen: Kentucky, Wolfe County, yard in town of Zachariah (author's collection).

Shells proportionate

C. morseana | *C. lubrica* | *C. lubricella*

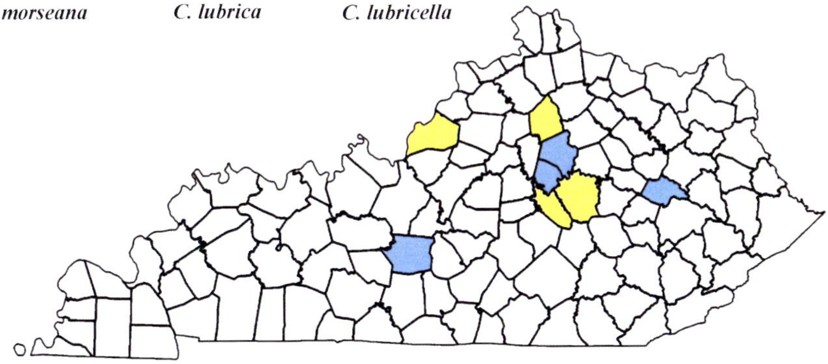

Appalachian Pillar

Cochlicopa morseana (Doherty, 1878)

Height: 6.6-7.2 mm width 2.3-2.5 mm

Description: Pupa-shaped; shell with 5.5 whorls; sutures slightly impressed; no teeth present at any stage of growth; shell smooth, glossy, translucent with a glass like surface; foot of live animal white with a grayish head.

Similar Species: *Cochlicopa lubrica* is similar in shape and luster, differing mainly in its broader stature, tighter less rounded whorls, and by its slightly wider and thicker callus rim of the outer lip; *Gastrocopta* and *Vertigo* species are much smaller and contain teeth in their aperture; *Columella simplex* is around 4mm smaller and does not have a translucent shell.

Habitat: A species that hides well in upland woods among deep moist leaf litter; it is rarely found on top of the litter even in wet weather, staying below or between the layers of leaves.

Status: G5/S5; Common; restricted to central and eastern Kentucky counties; an under-collected snail, likely due to its diminutive size.

Specimen: Kentucky, Powell County, Furnace Mountain (author's collection).

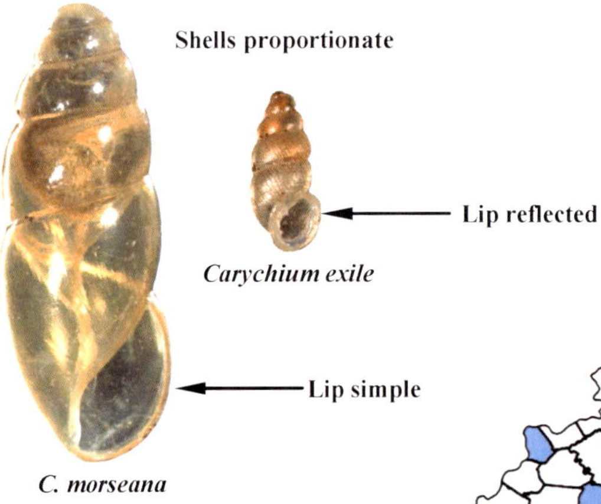

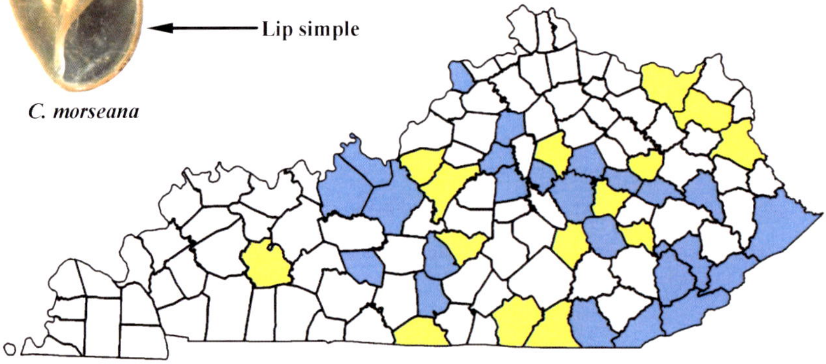

White-lip Dagger

Pupoides nitidulus (C. B. Adams, 1821)

Height: 4.2-5 mm, width 2 mm

Description: Pupa-shaped; lip reflected; shell with 6-6.5 whorls; with or without a small angular lamella in the form of a callus located in the upper right hand corner of the aperture; surface of shell mostly smooth or with weak transverse striae; without spiral striae.

Similar Species: In Kentucky, other pupa-shaped shells such as *Carychium, Vertigo, Gastrocopta* are notably smaller; the shell of *Cochlicopa* species are larger and have a glass-like, translucent surface; *Pupilla muscorum* is more drum-shaped (see below).

Habitat: A species of bare ground around natural glades but also found in waste places such as old quarries and parking lots; in wet weather, the snail can be observed crawling up the stems of plants; may be especially abundant in limestone areas (Hubricht 1985).

Status: G5/S5; Very Common; found across the entire state.

Specimen: Figure (a) from Ron Caldwell collection (CMRC 1594) and figure (b) from Missouri, Wright County, Gettle Farm (Nekola and Coles, 2010).

a

b

P. muscorum ***P. nitidulus***

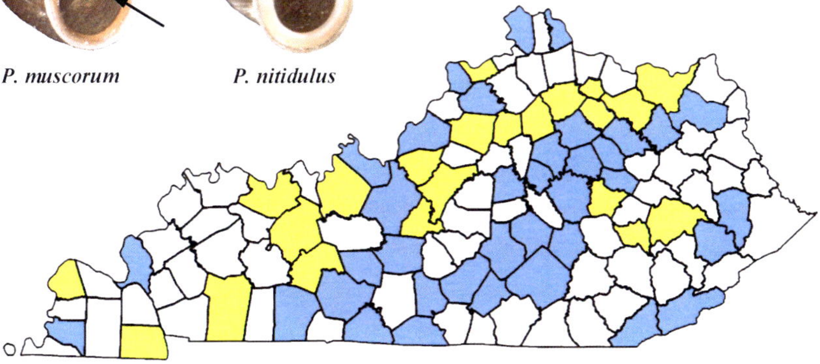

Brown Walker

Pomatiopsidae

Pomatiopsis cincinnatiensis (I. Lea, 1840)

Height: 4.5 mm, width 3.5 mm

Description: Conical shaped; lip simple; shell with 5 whorls; no teeth present at any stage of growth; fresh shells are a rich brown; live animals with an operculum.

Similar Species: *P. lapidaria* is larger, has narrower shell profile, a more elongated aperture, is more rounded in form, and has one more whorl.

Habitat: A calciphile snail that is often referred to as amphibious; the species appears to prefer wetter sites than *P. lapidaria* but they are sometimes found together; usually close to muddy stream banks, generally a foot above the normal low-water line (Hubricht 1985).

Status: G4/S3; Uncommon; although only reported from seven Kentucky counties, the species is likely more widespread.

Specimen: Kentucky, Edmonson County, Mammoth Cave National Park (FM 229105).

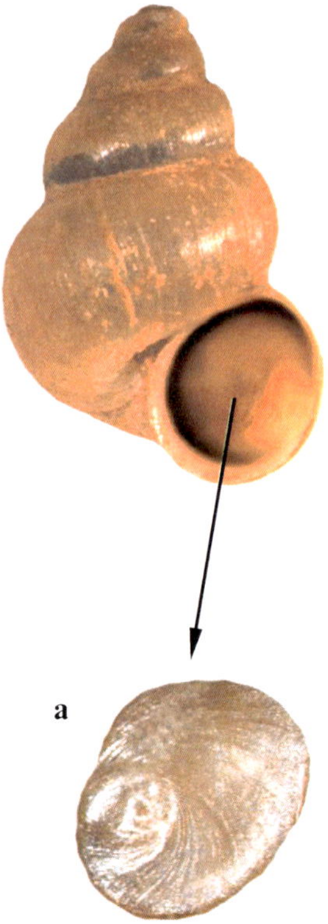

With an operculum in live individuals

Shells proportionate

P. lapidaria

P. cincinnatiensis

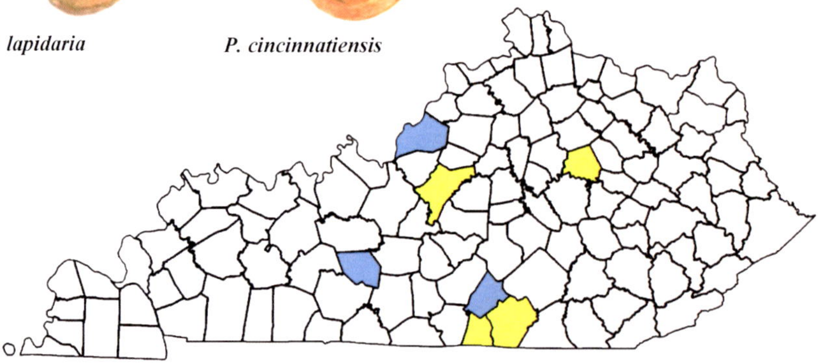

Slender Walker

Pomatiopsis lapidaria (Say, 1817)

Height: 5-7 mm, width 3-3.5 mm

Description: Conical; lip simple; shell with 6-7 whorls; no teeth present at any stage of growth; fresh shells are a rich brown; live animal with an operculum (a), a calcareous door.

Similar Species: *P. cincinnatiensis* is smaller, has a more inflated profile and has one less whorl.

Habitat: A calciphile often referred to as amphibious or even aquatic but this species does not live under water; common on wet limestone rock faces, in dripping seeps, and in mats of algae along small streams but also found on hillsides away from water in drier and more acidic situations.

Status: G5/S5; Very Common, found across the state in suitable habitat.

Specimen: Kentucky, Pike County, Breaks Interstate Park (author's collection).

With an operculum in live individuals

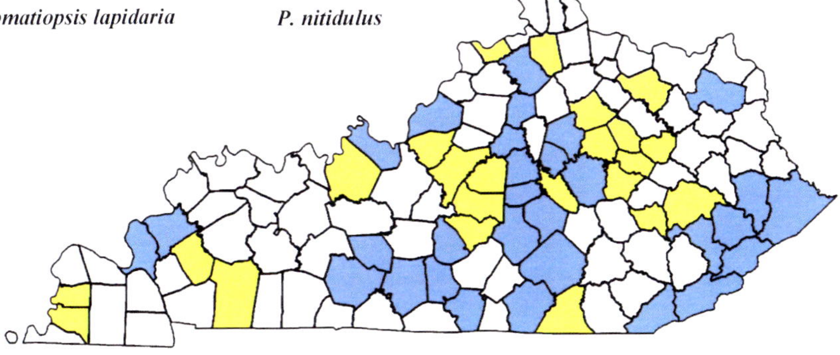

Oval Ambersnail

Novisuccinea ovalis (Say, 1817)

Height: 10-20 mm, width 10.8-11 mm

Description: Succiniform; lip simple; shell with 2.5-4.5 whorls; imperforate; no teeth; shell paper thin, greenish-yellow to amber, often with darker streaks; shell surface weakly striate; sole of foot bluish-gray shading to orange at the edges; margin of genital opening noticeably swollen (a); the genital opening of *Oxyloma* species is not swollen (b); Figure (c) shows live animal through translucent shell.

Similar Species: *Mediappendix oklahomarum* is smaller, more compact and frequently covered in dried soil; *Oxyloma retusum* is more narrow in build with a notably more elongated aperture.

Habitat: Southern mountains, found from low wet valleys to high mountain tops (up to 2000 meters); often seen climbing on jewelweed, stinging nettle, and false hellebore.

Status: G5/S5; Common; scattered records from across the state, under-reported due to its thin shell.

Specimen: North Carolina, Mitchell County, near Carvers Gap (author's collection).

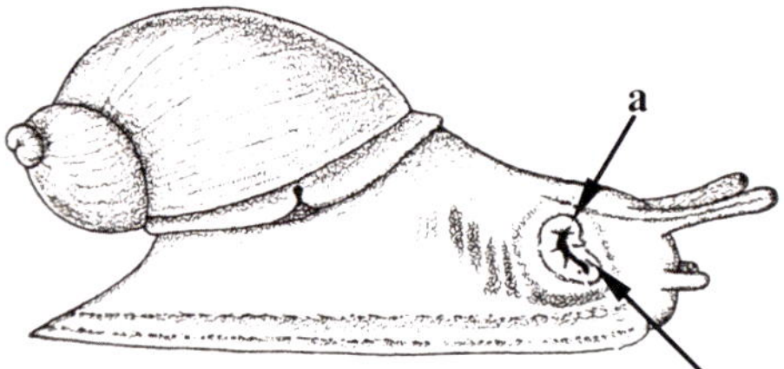

Novisuccinea ovalis

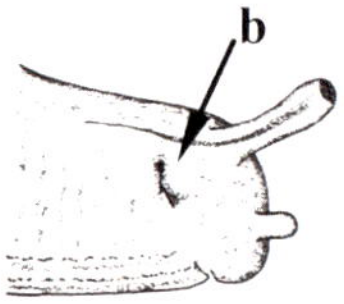

***Oxyloma* species**

© Burch 1962)

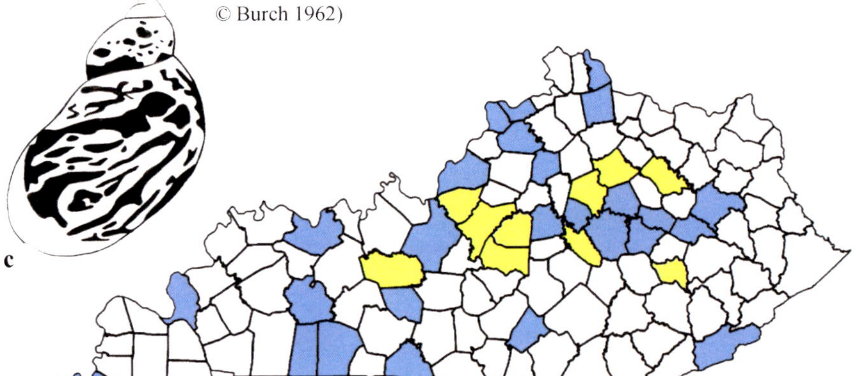

The live animal of Oval Ambersnail, *Novisuccinea ovalis,* can be seen through its thin, translucent shell. Canaan Valley, West Virginia.

Blunt Ambersnail

Oxyloma retusum (I. Lea, 1834)

Height: 7.5-20 mm, width 7.7-8.7 mm

Description: Succiniform; lip simple; shell with 2.5-4.5 whorls; perforate; no teeth or lamellae present in the aperture at any stage of growth; shell weakly striate, near paper thin and fragile; surface smooth; margin of genital aperture or opening not swollen as in *Novisuccinea ovalis*.

Similar Species: *Novisuccinea ovalis* has a wider shell and on the live animal, the margin of the genital opening is notably swollen.

Habitat: Found in low wet places around marshes and ponds crawling on mud or on the leaves of cattails and other aquatic plants.

Status: G5/S1; State Special Concern, this species in known from two river counties, Meade and Fulton where it was found along river drift, perhaps carried in during flood events. Extant populations in Kentucky remain questionable.

Specimen: West Virginia, Wood County, Muskingum Island (author's collection).

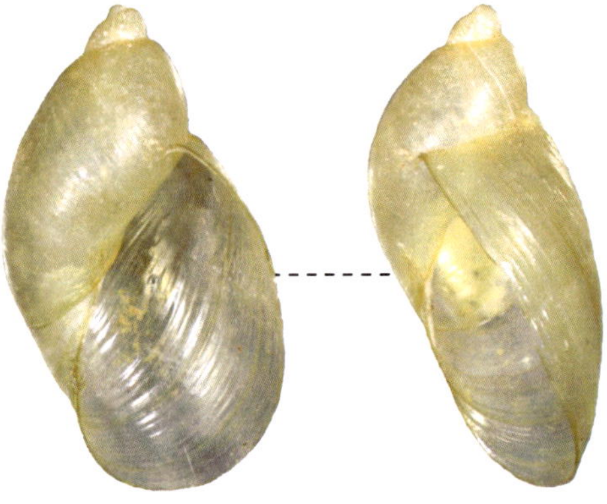

Bottom view

Top view

Oxyloma retusum, **West Virginia, Pendleton County at a spring on watercress (FMNH 235482)**

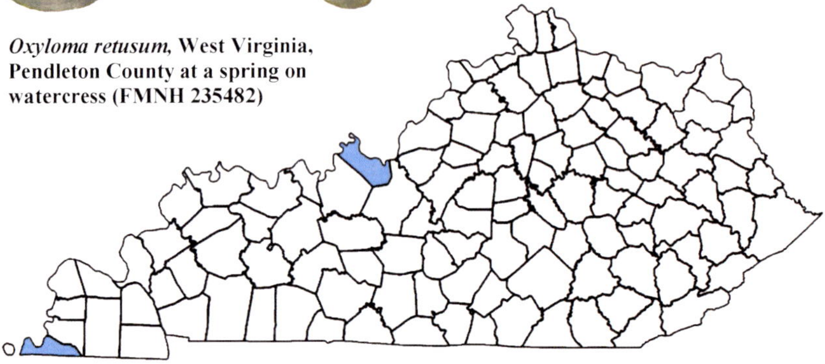

Detritus Ambersnail

Mediappendix oklahomarum (Webb, 1953)

Height: 7-8.8 mm tall, width 4mm

Description: Succiniform; lip simple; shell compact and obese, with 2-2.5 whorls; no teeth or lamellae present in the aperture in any stage of growth; shell surface weakly striate and glossy, often coated with soil particles (as seen in specimens illustrated to the right); shell color is a greenish-yellow, but this feature is variable; sides of the foot smoky.

Similar Species: *Novisuccinea ovalis* is taller with a notably larger aperture; *M. vermeta* is larger and less compact in form; *Succinea concordialis* from nearby states is larger with a capacious aperture.

Habitat: Usually found in the leaf litter of wooded hillsides or in pine woods, on acidic soils; the species is rarely very abundant (Hubricht 1985).

Status: G5/S4; Fairly Common; records are scattered across the state. Formerly known as *Catinella oklahomarum.*

Specimen: Kentucky, Wayne County, Monticello (FMNH 236387).

Shells proportionate

N. ovalis

M. oklahomarum

Shell small and generally covered in soil

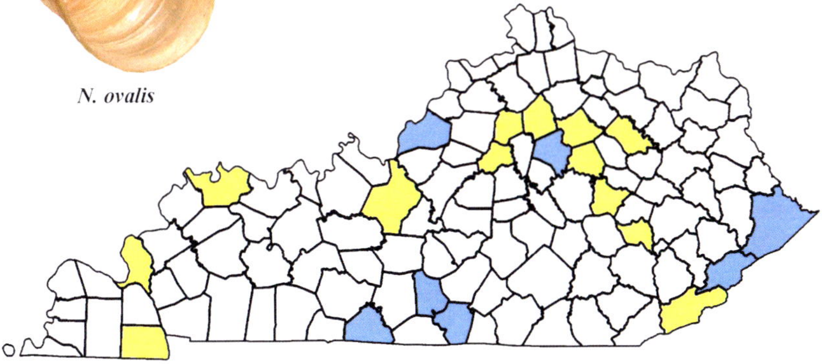

Suboval Ambersnail Succineidae

Mediappendix vermeta (Say, 1829)

Height: 7-13 mm tall, width 4-6.8 mm

Description: Succiniform; lip simple; shell with 2.5-3.5 whorls; perforate; shell surface weakly striate, pale yellow-olive, often coated with soil particles.

Similar Species: *Mediappendix oklahomarum* is more compact and generally found in drier situations; *Novisuccinea ovalis* is larger with a notably larger aperture.

Habitat: An amphibious land snail that appears restricted to wet ground found around ponds, marshes, muddy banks of open ditches, and swamps in both open and shaded conditions; in upland sites found under stones.

Status: G5/S4; Fairly Common; reported from many locations across the state, most likely occurring in every county. Formerly known as *Catinella vermeta,* and before that as *Catinella avara* (Turgeon et al. 1998).

Specimen: Pennsylvania, Allegheny County (CM 75744).

***Mediappendix vermeta,* Kentucky, Fayette County, Raven Run Nature Sanctuary**

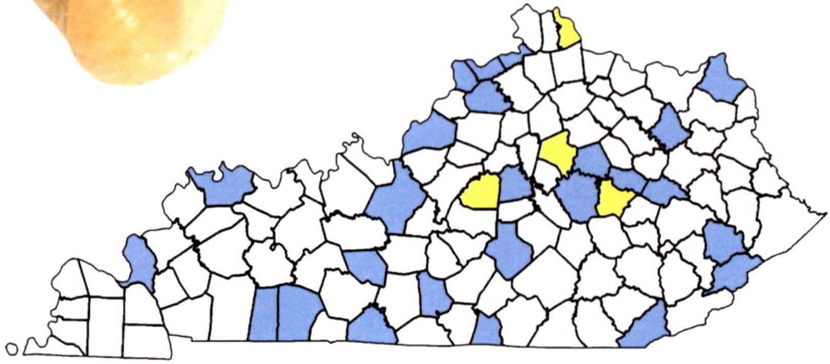

Spotted Ambersnail

Succineidae

Succinea concordialis I. Lea, 1864

Height: 12-16.8 mm, width 7.6 mm

Description: Succiniform; lip simple; shell with 3 whorls; perforate; no teeth or lamellae present in the aperture at any stage of growth; shell surface weakly striate and near paper thin but firm; color of shell a pale honey-yellow; the foot, including head and eye-stalks, is a grayish -white, speckled with irregular grayish-black spots; sole of foot pale yellow (Pilsbry 1948).

Similar Species: *Novisuccinea ovalis* is larger and usually more inflated in its form, but this feature varies among populations.

Habitat: An amphibious gastropod found on moist earth near water such as the banks of small creeks and gravel bars.

Status: G5/S2; Rare; in Kentucky, this species is reported from just two counties. Formerly known as *Succinea forsheyi.*

Specimen: Kentucky, Crittenden County, near Preacher Creek, 3 miles northwest of Crayne (FMNH, 235656).

Oxyloma retusum

Novisuccinea ovalis

Succinea concordialis

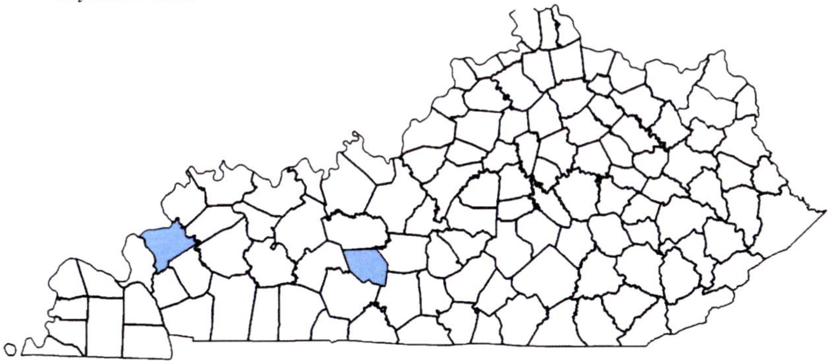

Whitewashed Rabdotus

Rabdotus dealbatus (Say, 1821)

Height: 15-26 mm tall, width 12.6 mm

Description: Bulimoid shaped; left side and bottom of lip notably reflected, the upper right side of lip is simple; shell with 6-7 whorls; shell solid, with or without streaked opaque color bands; no teeth or lamellae present in the aperture at any stage of growth. Most snails in this family are tropical and therefore restricted to South and Central America and islands within the Caribbean, but a few dozen species occur here in the United States including the extraordinarily colorful *Liguus* tree snails of the Florida Everglades. The map on page 78 shows in black the American distribution of some remarkable species of Orthalicidae (map from Pilsbry 1946, with permission).

Similar Species: No other native land snail in Kentucky has the Bulimoid-form and light color features.

Habitat: A calciphile found in open glades and meadows; often seen on the ground or on low vegetation during moist weather.

Status: G5/S1; State Endangered, in Kentucky, currently restricted to three south-central counties.

Specimen: Kentucky, Edmonson County, Mammoth Cave area (author's collection).

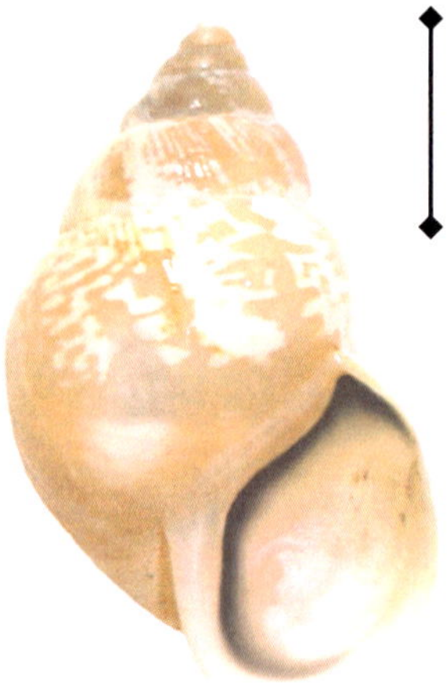

Specimens south of Kentucky are larger and have darker color stripes

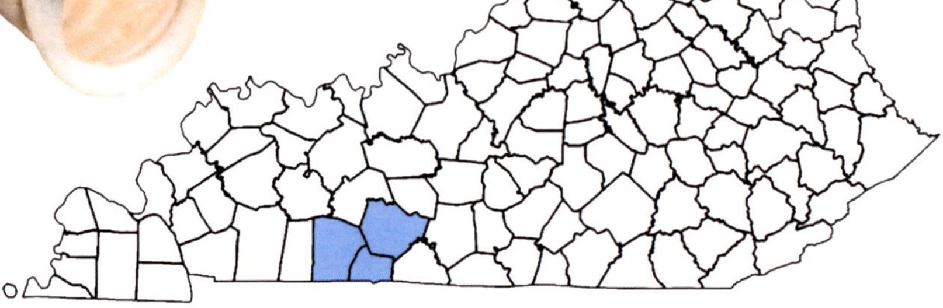

Distribution of ORTHALICIDAE in the Americas

Charmers of the Land Snail World

With nearly 1500 described species, snails in the family ORTHALICIDAE are some of the most diverse and colorful gastropods in the Americas, if not the world. These arboreal snails possess ornately decorated shells sporting every color imaginable in a staggering number of pattern configurations, painted on an overwhelming number of shell forms. Most species are cone-shape and found living on elevated structures such as cliff faces, trees and shrubs from sea level to steamy cloud forests. Several early cultures traded these vivid shells for currency or used them as decorative neck wear. Species in the genus *Drymaeus* are the principal diet of the co-evolved snail-eating snakes in the genus *Sibon* of Central and South American rainforests.

This section includes land snails that have shells that are taller than wide, pupa -shaped, and have reflected lips. These are the miniatures of terrestrial gastropods, most species under 3 mm as adults. In some habitats such as glades, dry outcrops of limestone or wet seeps, they may represent as much as 75% of the species found. Consequently, gathering leaf litter at all sites is essential for a accurate register of species. In general, fresh dead and live shells of this group are translucent and dead shells, with age, become bleached. *Vertigo* and *Gastrocopta* snails are cryptic, remaining well-hidden in the detritus and leaf litter. These two genera also have teeth in their apertures which are thought to prevent attacks from predacious beetle larva. Although tiny, shells of *Carychium* species tend to be more easily spotted, a result of their lighter color.

Genera Included:

(in order of appearance in text)

Carychium
Gastrocopta
Vertigo
Pupilla

Gastrocopta

Ice Thorn

Ellobiidae

Carychium exile H. C. Lea, 1842

Height: 1.5-1.8 mm, width 0.6 mm

Description: Pupa-shaped; lip reflected and somewhat thickened within; shell with 4.5-5 whorls; there are two parietal lamellae on the left side of the aperture, below enlarged aperture view with these features which at times may difficult to see; are transverse striae are variable in their spacing and their depth; no visible spiral striae; shell morphology highly variable in Kentucky.

Similar Species: *C. clappi* is taller, more narrow in build and the finer transverse striae are more closely spaced, although this distinction is variable and only seen under a strong lens; *C. nannodes* is smaller and lacks prominent transverse striae.

Habitat: The species is found living between moist, but not wet, matted leaves that accumulate in pockets of low wet depressions, next to logs and adjacent to headwater seeps.

Status: G5/S5; Very Common; the most common *Carychium* in Kentucky; most likely occurs in every county.

Specimen: West Virginia, Greenbrier County, Greenbrier State Forest (author's collection).

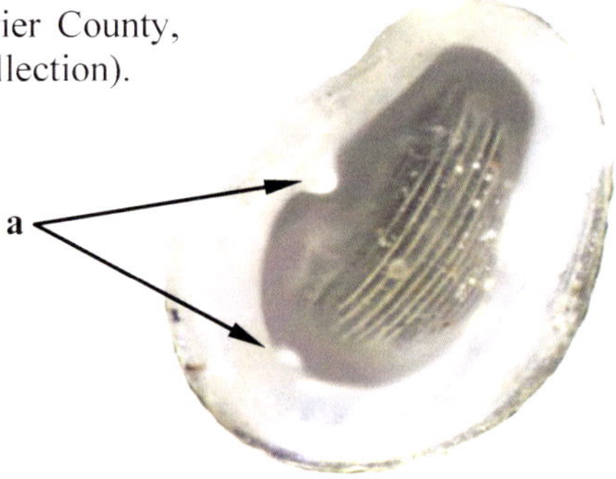

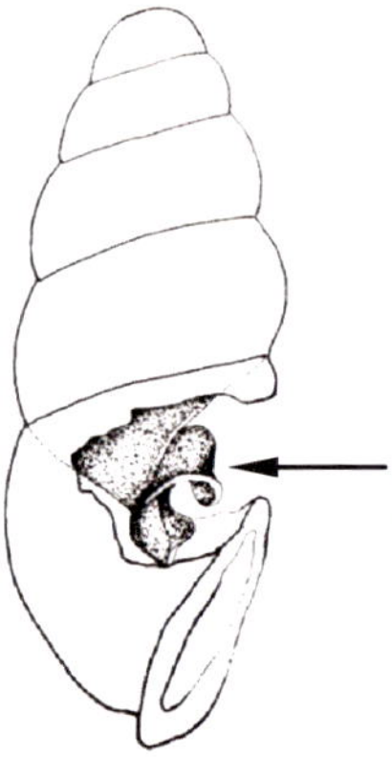

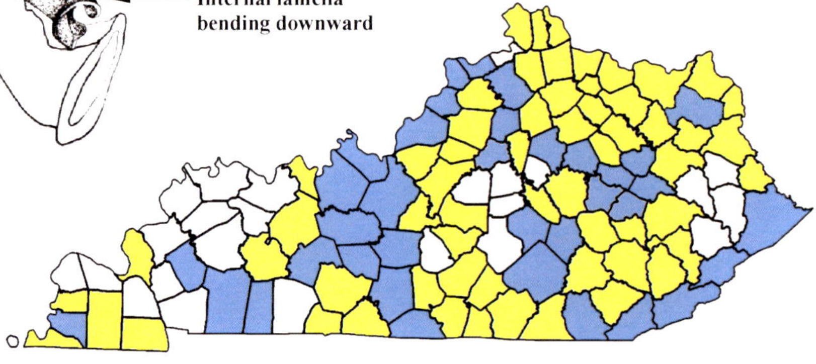

***Carychium clappi*, Pendleton County, WV**

***Carychium exile*, Greenbrier County, WV**

Appalachian Thorn

Carychium clappi (Hubricht, 1959)

Height: 1.7-1.8 mm, width 0.8 mm

Description: Pupa-shaped; lip reflected; shell tiny with 4.5 whorls; two parietal lamellae on the left side of aperture, a feature that is sometimes difficult to observe; transverse striae well developed and closely spaced; shells translucent in live individuals, becoming bleached white with age.

Similar Species: *Carychium exile* is shorter in height but is wider in stature and has less distinct transverse striae on the first three whorls; *C. nannodes* is smaller and has a smoother surface.

Habitat: Found living between layers of moist leaves located in moist depressions on hillsides; it is often found with *C. exile* and *C. nannodes.*

Status: G5/S5; Common where it occurs; found throughout most of the central and eastern Ken-tucky.

Specimen: West Virginia, Pendleton County, UTM: 625389, 4283122 (author's collection).

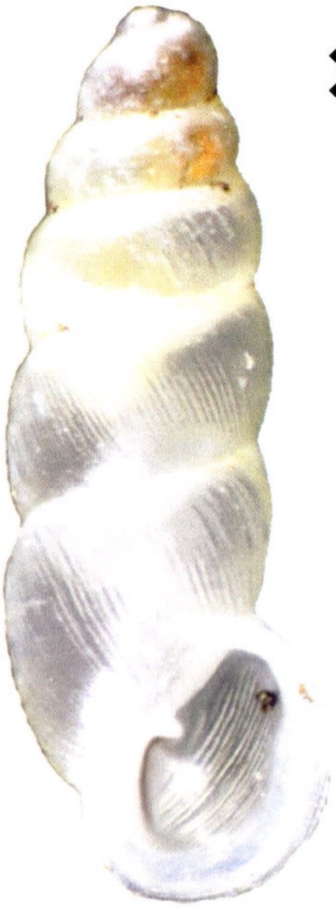

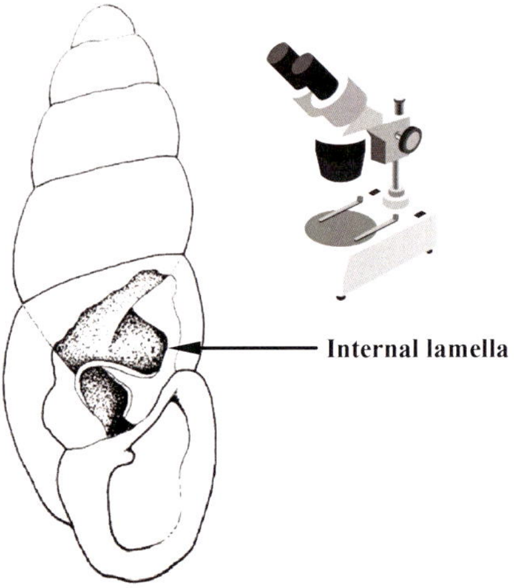

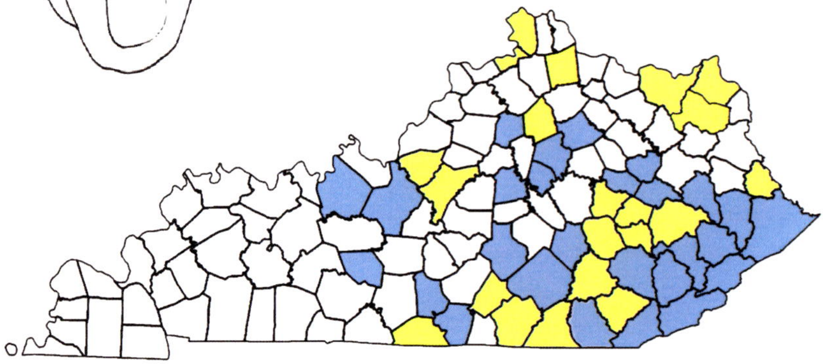

Obese Thorn

Ellobiidae

Carychium exiguum (Say, 1822)

Height: 1.5-1.6 mm, width 0.75 mm

Description: Pupa-shaped; lip widely reflected; shell with 4.5-5 whorls; there are two parietal lamellae seen on the left side of the aperture; transverse striae are poorly developed and at times, hardly perceptible; no spiral striae; shells are translucent in live individuals.

Similar Species: *Carychium clappi* is slightly larger and more narrow in form with transverse striae which are more distinct; *C. exile* is around the same height, but is narrower (the most important feature separating the two species) and has more distinct transverse striae.

Habitat: Found in pockets of moist decaying leaves around and in sinkholes or around the entrances of caves.

Status: G5/S3; Uncommon, found in scattered locations, with an outlier in western KY.

Specimen: Tennessee, Blount County, Cades Cove, GSMNP (GSMNP collection).

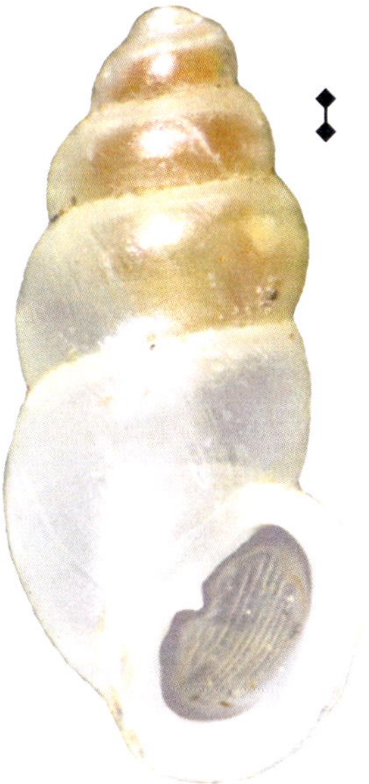

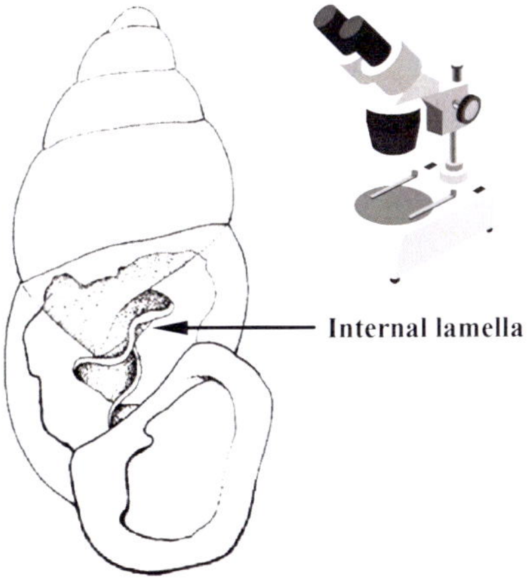

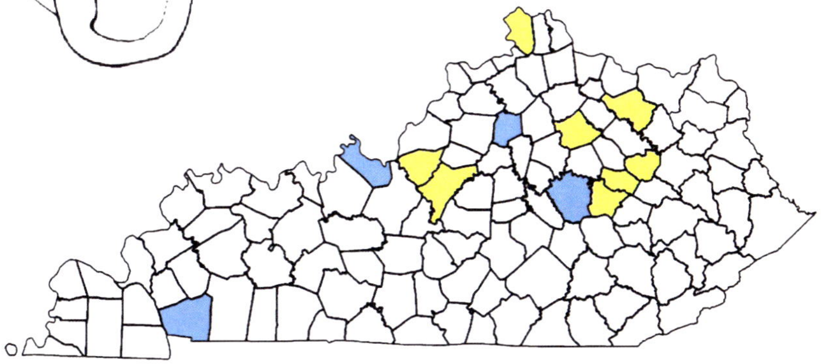

Cave Thorn

Carychium stygium (Call, 1897)

Height: 1.5-2.6 mm, width 1 mm

Description: Pupa-shaped; lip reflected; shell tiny with 4.5 whorls; there are two parietal lamella seen on the left side of the aperture; traverse striae are poorly developed, shell surface nearly smooth; no spiral striae; shells are translucent in live individuals and fresh shells but become bleached with age.

Similar Species: *Carychium exile* is shorter in height, more narrow in form with distinct traverse striae; *C. clappi* has distinct transverse striae; the shell of *C. riparium* is essentially identical to *C. stygium* and therefore can not be reliably separated by shell morphology.

Habitat: Reported from the dark zones of deep limestone cave systems where it feeds primarily on cave cricket (*Hadenoecus* sp.) guano; in most caves, there is a direct correlation between the number of cave crickets and the abundance of *Carychium stygium*.

Status: G3/S2; State Special Concern; recent DNA studies by Weigand et al. (2010) treats *C. stygium/ C. exile* cf. *mexicanum* clade as a single taxon and neither the species status of *C. stygium* nor the subspecies rank of *C. exile* cf. *mexicanum* is supported. Nevertheless, histological observations of the somewhat reduced eyes of *C. stygium* and shell morphology as treated by Hubricht (1960) very well suggest either a relatively young species formation, incipient speciation or a broad range of suitable habitats for a *C. stygium* + *C. exile* cf. *mexicanum* taxa (Weigand et al. 2010).

Specimen: Tennessee, Franklin County, Cove Cave, (author's collection).

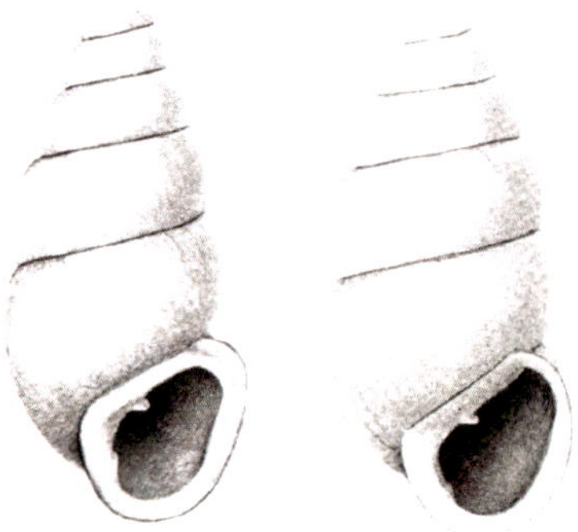

***Carychium stygium* (two forms), Mammoth Cave, Kentucky**
© Pilsbry 1948

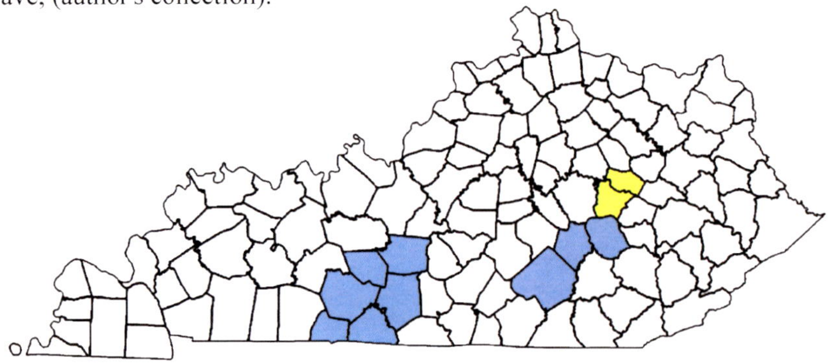

Floodplain Thorn Snail

Ellobiidae

Carychium riparium

(Hubricht, 1978)

Height: 1.5-2.0 mm, width 1 mm

Description: Pupa-shaped; lip reflected; shell with 5 -5.5 whorls; there are two parietal lamellae that can be seen on the left side of the aperture (strong lens required); traverse striae are poorly developed and at times hardly perceptible; no spiral striae; shells are translucent in live individuals and fresh shells but become bleached with age.

Similar Species: *Carychium riparium* differs from *C. exile* in its more opaque shell, more conical spire, weaker sculpture and expanded rather than reflected outer lip; *C nannodes* is notably smaller; *C. riparium* is essentially identical to *C. stygium* and therefore can not be reliably separated by shell morphology, only by habitat.

Habitat: Found in pockets of moist decaying leaves found next to forest structure such as logs in floodplains of streams and rivers.

Status: G2/S2; Rare, there is some doubt as to the validity of *C. riparium* and some hypothesized that it is a member of the *C. stygium/C. exile* cf. *mexicanum* clade (pers. comm. Nekola, 2011), but DNA sampling will be necessary to make a final determination.

Specimen: Kentucky, Meade County (FMNH 171659 Holotype).

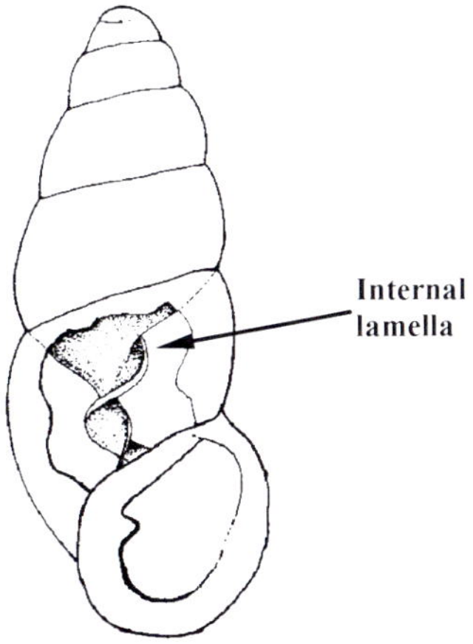

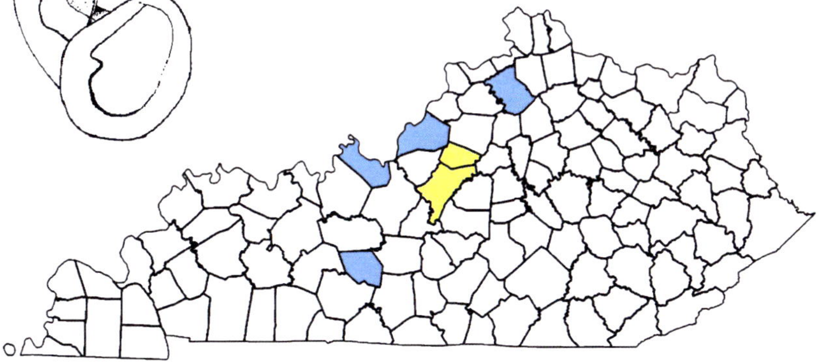

File Thorn

Ellobiidae

Carychium nannodes (Clapp, 1905)

Height: 1.3-1.5 mm, width 0.5 mm

Description: Pupa-shaped; lip reflected; shell with 4.5-5 whorls; sutures notably deep; there are two parietal lamellae that can be seen on the left side of the aperture (strong lens required); transverse striae are poorly developed and at times, hardly discernible; shells are translucent when fresh but become bleached with age. This is the smallest *Carychium* species in Kentucky and the southern Appalachian Mountains.

Similar Species: *Carychium exile* and *C. clappi* are both larger with conspicuous transverse striae.

Habitat: Found in pockets of moist decaying leaves in low depressions and talus slopes; also common around large walnut, butternut, basswood, and buckeye trees.

Status: G5/S5; Very Common; found throughout much of central and eastern Kentucky in appropriate habitat; but absent from western countries of the state.

Specimen: West Virginia, Greenbrier County, near Davis Spring (author's collection).

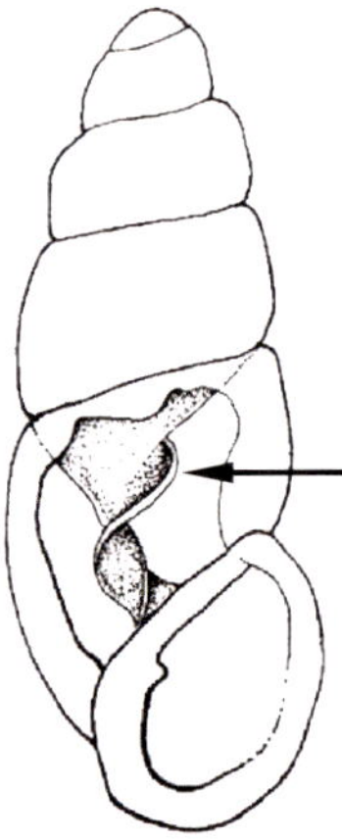

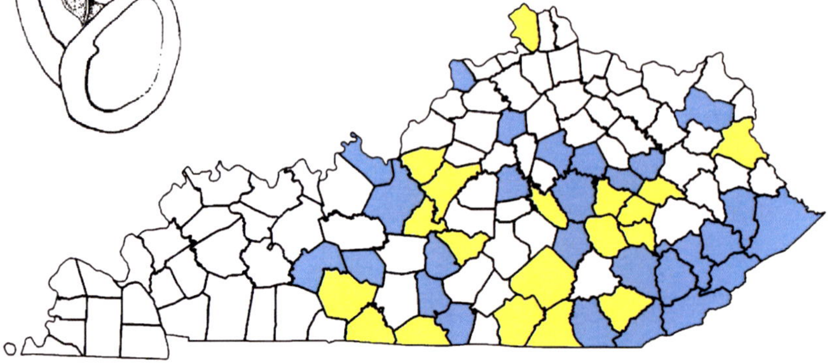

Kentucky *Carychium* Compared (proportionate)

Some Rather Remarkable Predators of Land Snails

Above image of a Speckled snail-sucker, *Sibon nebulata,* eating a Banded cone, *Antidrymaeus sulphureus,* Bladen Nature Reserve, Belize, Central America. In North America, brown snakes are known to hunt and consume slugs, especially the exotic *Arion* species (pers. comm. John MacGregor). Ring-necked snakes are also reported to eat slugs while red-bellied snakes will consume both slugs and shelled land snails as part of their overall diet. Other fascinating predators of land snails are the ants. As with our native snakes, ants may also be hunting minuscule snail species in the leaf litter, but this remains entirely unstudied. Below picture of African weaver ants carrying a land snail to the nest for consumption.

Key to *Gastrocopta* and *Vertigo* Species

Gastrocopta and ***Vertigo*** are tiny snails with pupa-shaped shells and well developed teeth in the aperture. In general, they differ from *Carychium* by having larger, more obese, and 'toothier' shells. *Gastrocopta* species are generally larger than *Vertigo*, usually have wider, more reflected lips (a) and are without a small indentation (b) in the right side of the aperture. The angular & parietal lamellae are fused in *Gastrocopta*, simple in *Vertigo*. Shell surface for either group can be smooth or with fine transverse striae, but are typically without spiral striae. The shell apertures in this section have been darkened behind the teeth to better illustrate the teeth and their spatial arrangement.

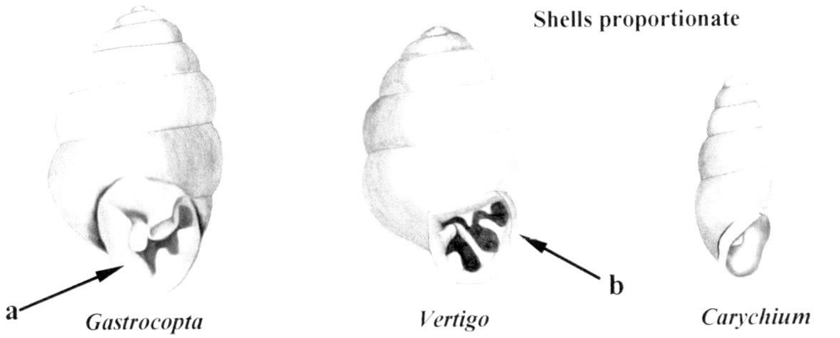

Terminology Used in Vertiginidae Identification (Nekola and Coles 2010).

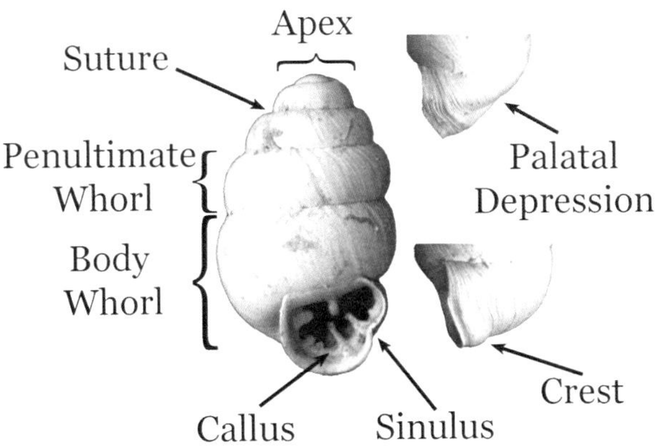

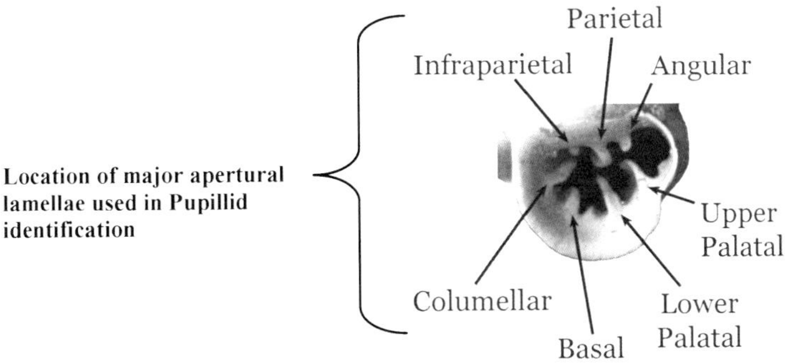

Armed Snaggletooth

Gastrocopta armifera (Say, 1821)

Height: 3-4.8 mm, width 2.2 mm

Description: Pupa-shaped; lip reflected; shell with 6.5-7.5 whorls; aperture crowded with large teeth, the parietal tooth large and forked; columellar lamella with both forward and basally pointing components (a), appearing more or less pyramidal in apertural view; the internal lower columellar lamella branching or forked (can only be seen by breaking open the shell).

Similar Species: This is a giant among *Gastrocopta,* usually double the size of most species in this genera; *Gastrocopta contracta* is smaller, having only 5.5 whorls and the aperture teeth are notably larger, choking the aperture; other *Gastrocopta* species are notably smaller.

Habitat: A calciphile found in more open, sunny areas such as cedar glades and limestone outcrops.

Status: G5/S5; Very Common; although not as common as *G. contracta,* it is a close contender in its abundance and distribution in Kentucky.

Specimen: Figure (b) from Larry Watrous photo collection; figure (c) from Vermont, Rutland County, Poultney River (Nekola and Coles image 2010) and figure (d) from Kentucky, Bourbon County illustrating an uncharacteristic shell completely missing lower palatal tooth

b

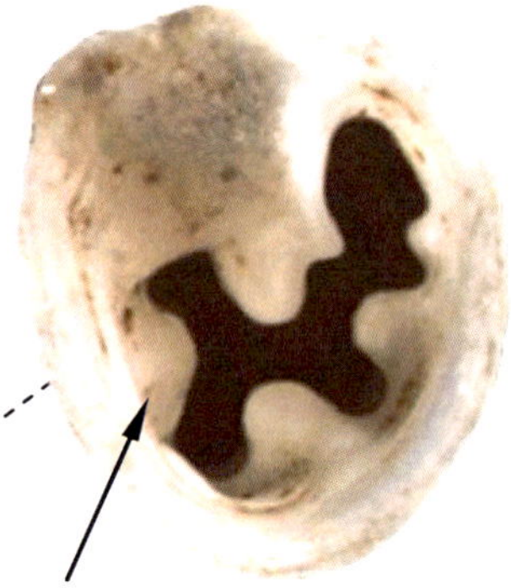

a

d

c

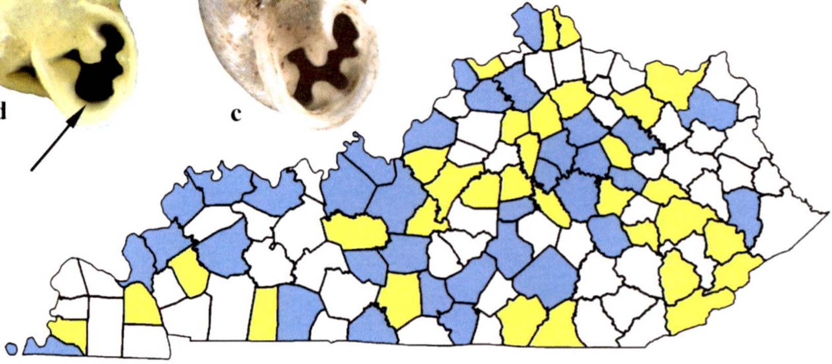

Bluegrass Snaggletooth

Gastrocopta clappi (Sterki, 1909)

Height: 3.5-4 mm, width 2 mm

Description: Pupa-shaped; lip reflected; shell with 6.5-7.5 whorls; angulo-parietal lamella forked; aperture crowded with large teeth; the internal lower columellar lamella not pyramidal and branching as in *G. armifera* (a) but straight plate (b); shells are somewhat translucent in live individuals and fresh shells; transverse striae finely developed.

Similar Species: *Gastrocopta armifera* will average a larger size, possess a coarser surface and its lower columellar lamella branches (a), a feature determined by carefully breaking open the shell using something like a syringe needle.

Habitat: A xenophile and calciphile found in more open sunny areas such as cedar glades, limestone outcrops, along railroads and roadways, sometimes found in company with *G. armifera* (Hubricht 1985).

Status:G5/S4; Common, but not as widespread as *Gastrocopta armifera*.

Specimen: Tennessee, Wilson County, Cedars of Lebanon State Park (Nekola and Coles image 2010).

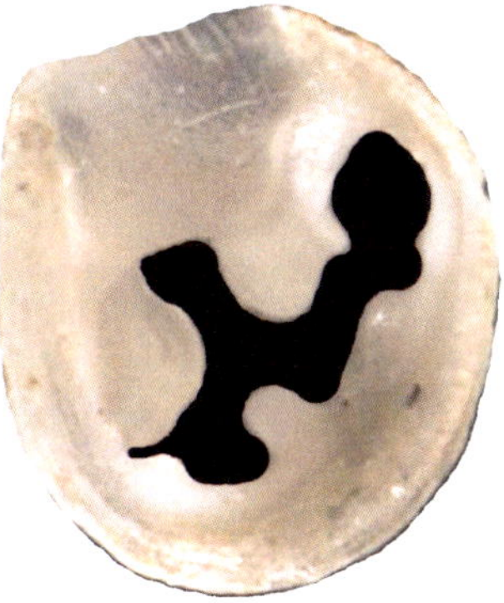

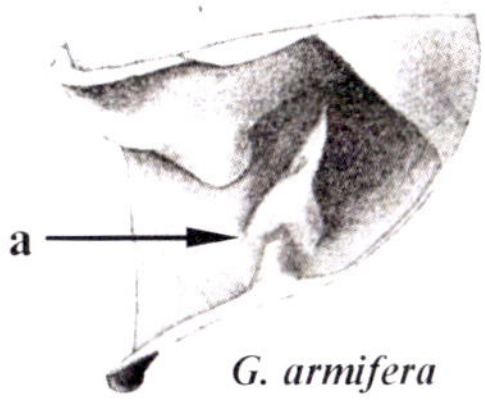

G. armifera

© Pilsbry, 1948)

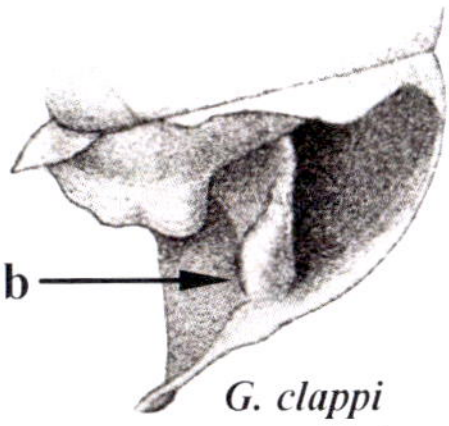

G. clappi

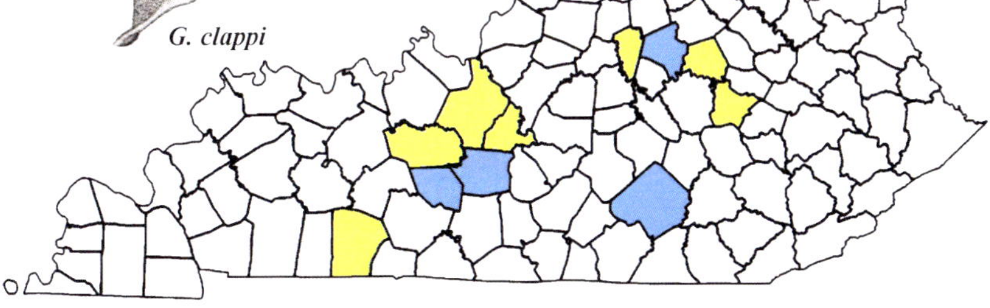

White Snaggletooth

Gastrocopta tappaniana (C.B. Adams, 1842)

Height: 1.6-2 mm, width 1.1-1.2 mm

Description: Pupa-shaped; lip reflected; shell with 4-5 whorls; outer lip with a distinct palatal callus or ridge (a) which support multiple teeth; shells are somewhat translucent in live individuals and fresh shells but become bleached with age.

Similar Species: *Gastrocopta pentodon* is only slightly smaller, less oval, thinner in form, and has a more tapered shell.

Habitat: A calciphile found around moist areas including marshes, floodplains of streams, and margins of ponds, but also in drift piles along small streams and in accumulations of decomposing leaf litter.

Status: G5/S3; Fairly Common; reported in scattered locations across the state; this species has been confused with *G. pentodon*, even though intermediates do not exist in places of co-occurrence.

Specimen: Minnesota, Norman County, Faith Fen (Nekola and Coles image 2010).

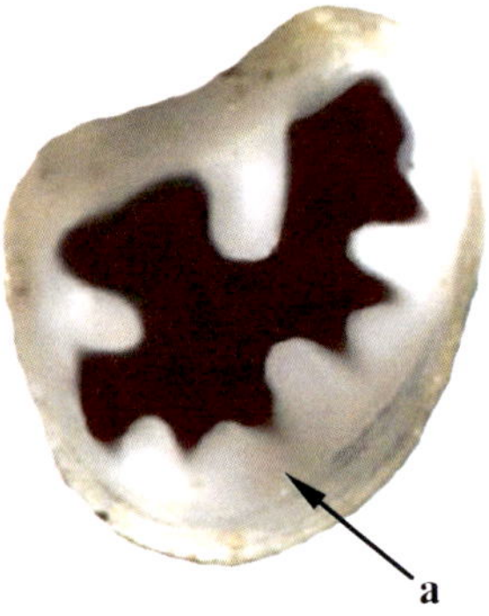

G. tappaniana

G. pentodon

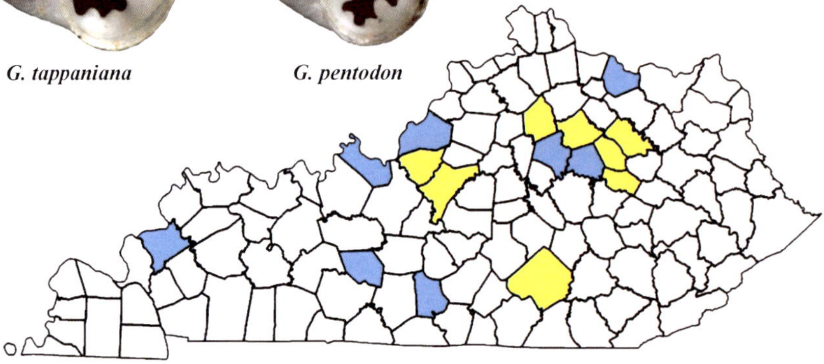

Comb Snaggletooth

Gastrocopta pentodon (Say, 1821)

Height: 1.5-1.8 mm, width 0.8-1.1 mm

Description: Pupa-shaped; lip narrowly reflected; shell with 5 whorls; parietal tooth is the largest (a); outer lip with a distinct palatal callus (b) or ridge containing 2 to 6 teeth but sometimes up to 9 teeth; also a rounded ridge behind the lip (c); shells are somewhat translucent in live individuals and fresh shells are typically covered in soil; two size classes (tall and short form, figures d) of *G. pentodon* are found in Kentucky, often in the same location. DNA sequence data suggest these represent distinct species, perhaps as many as 3 to 4 taxa across its entire range (Pers. comm. Nekola, 2019) .

Similar Species: *Gastrocopta contracta* and *G. armifera* are both larger, containing fewer but larger teeth in their aperture.

Habitat: A species that is found in a variety of habitats including dry upland hardwood forests, acidic forests such as pine savannas, around limestone outcrops, and occasionally, in low wet places.

Status: G5/S5; Very Common; one of the most common and widespread *Gastrocopta* in Kentucky and most likely occurs in every county.

Specimen: Figure (c) from New Jersey, Burlington County, Lebanon State Forest (Nekola and Coles image 2010) and figures (d) illustrating two size classes from Kentucky, Powell County.

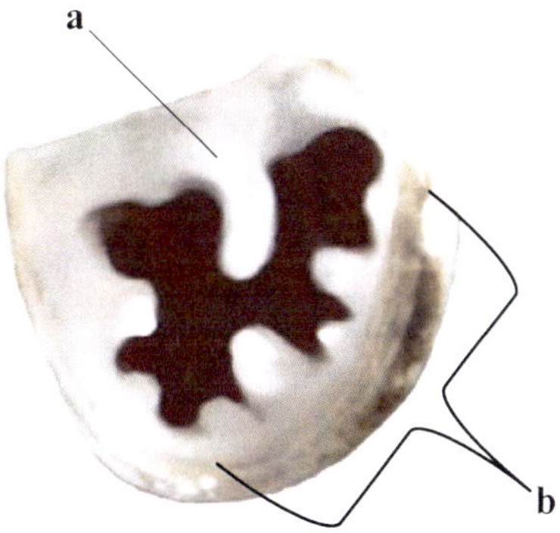

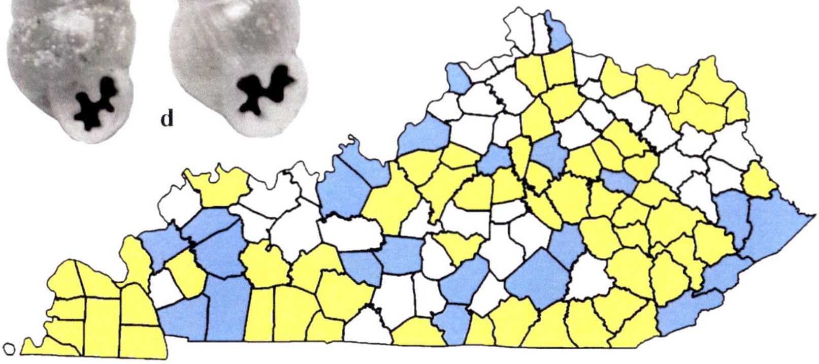

Bottleneck Snaggletooth

Gastrocopta contracta (Say, 1822)

Height: 2.2-2.5 mm, width 1.3-1.4 mm

Description: Pupa-shaped; lip widely reflected, even detached in last whorl (b); shell with 5.5 whorls; with large teeth, the parietal tooth the largest (a); spire of shell tapering; aperture more or less triangular in shape; shells are somewhat translucent in live individuals and fresh shells are frequently covered with dried soil; old shells quickly become bleached.

Similar Species: *Gastrocopta armifera* is larger, adding 2 additional whorls to its shell and the aperture teeth are less crowded; the lamellae configuration of *G. contracta* is also very different than *G. armifera* and has a tapered (not ovate) shell shape.

Habitat: This species is found in nearly all terrestrial habitats including, but not limited to, low wet places to dry mountainsides and acidic forests.

Status: G5/S5; Very Common; this widespread species in Kentucky is found in nearly every county; leaf litter collection will no doubt close in the remaining gaps.

Specimen: Iowa, Buchanan County, Rowley Fen (Nekola and Coles image 2010).

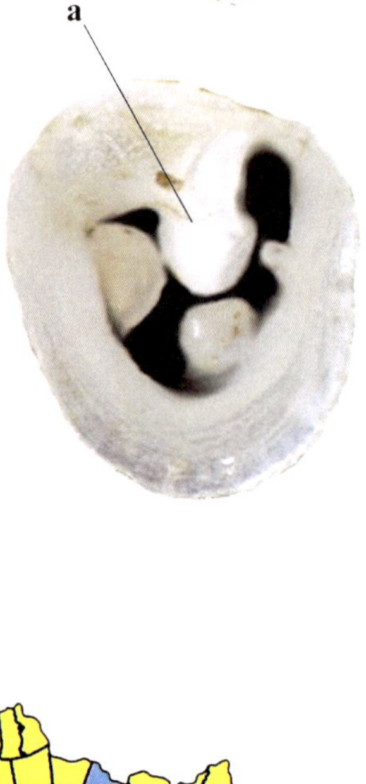

***Gastrocopta contracta*, Greenbrier State Forest, Greenbrier County, West Virginia**

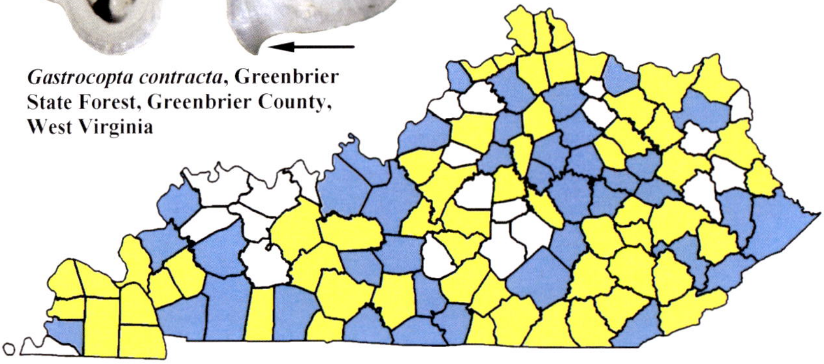

Bark Snaggletooth

Gastrocopta corticaria (Say, 1816)

Height: 2.5 mm, width 1 mm

Description: Pupa-shaped; lip reflected; shell with 5.5 whorls; one small forked tooth present on the parietal wall (a) and one small columellar tooth (b); outer palatal wall without teeth (c); shells are somewhat translucent in live individuals and fresh shells.

Similar Species: This is the only *Gastrocopta* species in Kentucky to possess as few as two teeth, lacking any lamellae or teeth on the outer (palatal) wall of the aperture; *Columella simplex* is without any teeth in the aperture, is more compact in form and has a simple non-reflected lip.

Habitat: A calciphile found in mixed hardwood forests with limestone outcrops but also found in acidic forests where it may be restricted to mossy tree trunks; in wet weather it can be found crawling on logs and the trunks of trees (Hubricht 1985).

Status: G5/S5; Very Common; reported from counties across Kentucky.

Specimen: West Virginia, Roane County, Barnes Run (author's collection).

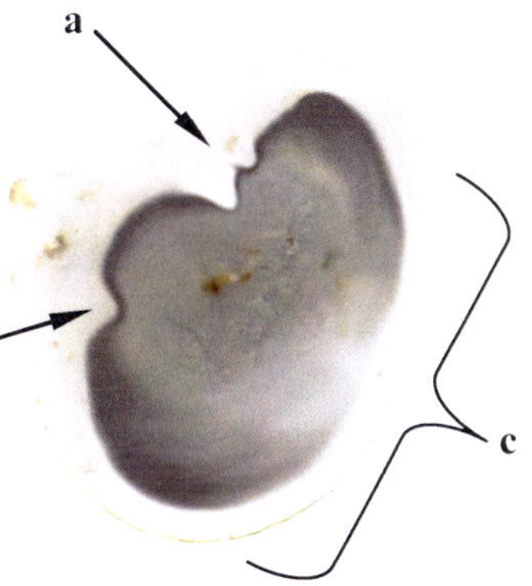

Aperture enlargement

Gastrocopta corticaria *Columella simplex*

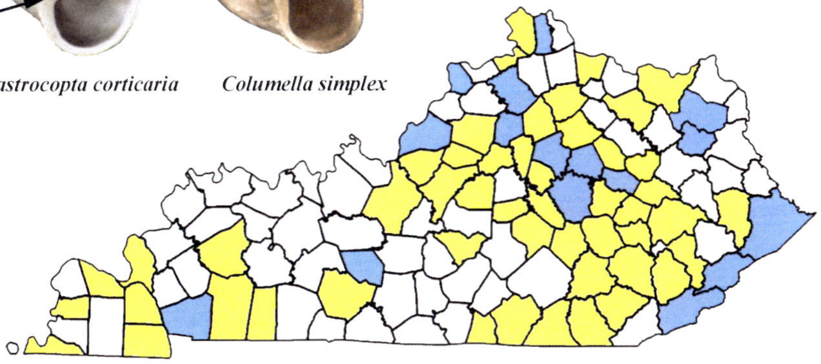

Lambda Snaggletooth

Gastrocopta holzingeri (Sterki, 1889)

Height: 1.75 mm, width 0.8 mm

Description: Pupa-shaped; lip reflected; shell with five whorls; angulo-parietal lamella is grooved down the middle (a); aperture crowded with teeth, (typically six); shells are translucent in live individuals.

Similar Species: *Gastrocopta contracta* has a notably different shape and is larger; *G. procera* is larger having one less tooth on the outer palatal wall, is brown (not white) and the angulo-parietal is forked (not grooved as in *G. holzingeri*).

Habitat: A calciphile species frequently found around talus with sparse vegetation; in Kentucky the species is found on limestone river bluffs in the thin cedar detritus accumulations over bedrock.

Status: G5/S1; State Endangered; All previous records for this tiny species in Kentucky were from Pleistocene fossil deposits found along the Ohio River; it was presumed extinct in the state until a 2009 survey at Raven Run Nature Sanctuary found extant populations on the Palisade Bluffs above the Kentucky River in Fayette County.

Specimen: West Virginia, Greenbrier County, cedar glade above Greenbrier River (author's collection).

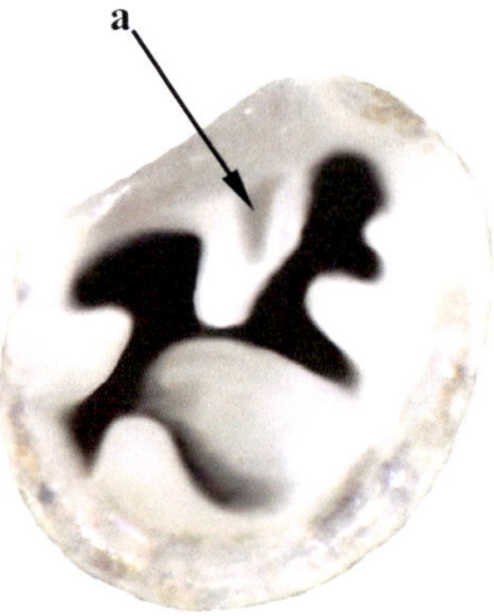

G. holzingeri

G. procera

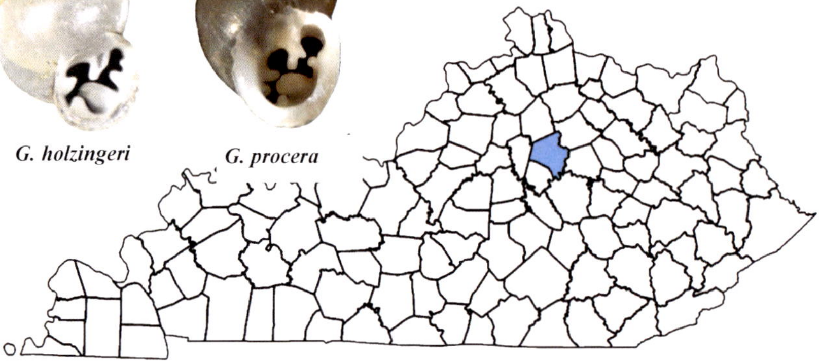

Wing Snaggletooth

Gastrocopta procera (Gould, 1840)

Height: 2.2-3 mm, width 1-1.1 mm

Description: Pupa-shaped; lip reflected; shell with 5-6 whorls; angulo-parietal lamella strongly bifid or forked at the lip (a); columellar forked or with a spur (b); lower palatal tooth deeply set in aperture; five to six teeth present in aperture; shells are somewhat translucent and brownish in live individuals and fresh shells, but become bleached and white with age.

Similar Species: *G. holzingeri* has a similar shell form but is smaller, whitish (not brown), and has a channeled, not forked angulo-parietal; *G. rogersensis* (found in Ohio near the KY border) has the same shell form as *G. procera* but has a channeled (c) not forked angulo-parietal (d) as in *G. procera*.

Habitat: An obligate calciphile on dry ground where the vegetation is sparse such as cedar glades, also occurs in sandy river floodplains scrub and forest (Nekola and Coles 2010); it is especially common in old parking lots.

Status: G5/S4; Common; in Kentucky, found from across the state.

Specimen: Illinois, Monroe County, Fults Hill Prairie Nature Reserve (Nekola and Coles image 2010).

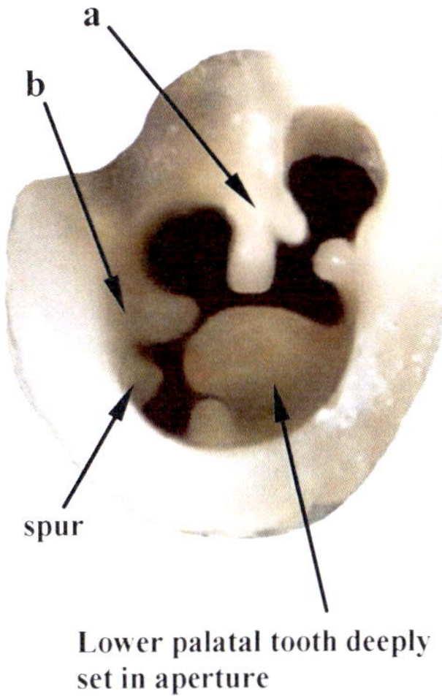

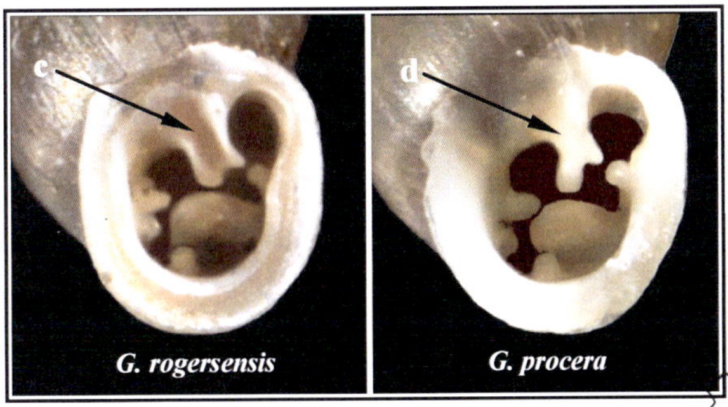

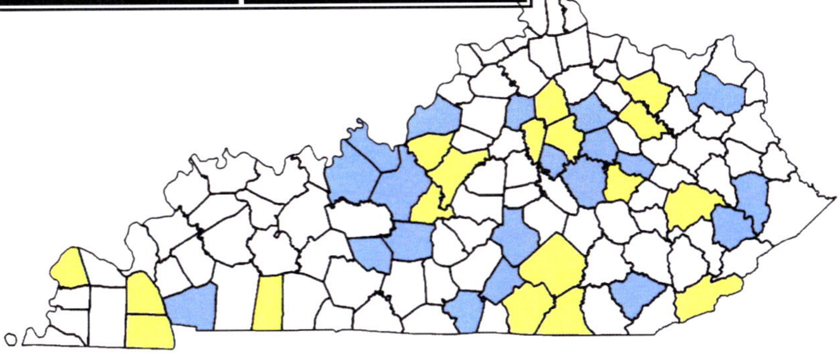

Gulf Coast Snaggletooth

Gastrocopta riparia Hubricht, 1978

Height: 2.5 mm, width 1 mm

Description: Pupa-shaped; lip reflected; shell with 6 whorls; angulo-parietal lamella forked (a); Basal palatal, lower palatal and upper palatal more or less in line (figure b); columellar without a spur or fork (c).

Similar Species: Similar in form and size to *G. procera* but distinguished primarily by the shallow depth of the lower or middle palatal lamella (*G. procera* being set deeper—see opposite page of shell comparisons) and lacks the columellar spur seen in *G. procera.*

Habitat: Found in decomposed leaf litter accumulations, often under dense shrub or vine thickets and disturbed sites, more tolerant of acidic conditions than *G. procera.*

Status: G4/S1; State Endangered; A southeastern species; likely non-native/disjunct population occurring in Harlan County around anthropogenic sites such as parking lots and along railroad beds.

Specimen: Figure (d) from Kentucky, Harlan County next to a railroad track in town of Cumberland (images and collections made by Wayne & Amy Van Devender) and figure (e) from Texas, Galveston County (ANSP 78684, Lectotype).

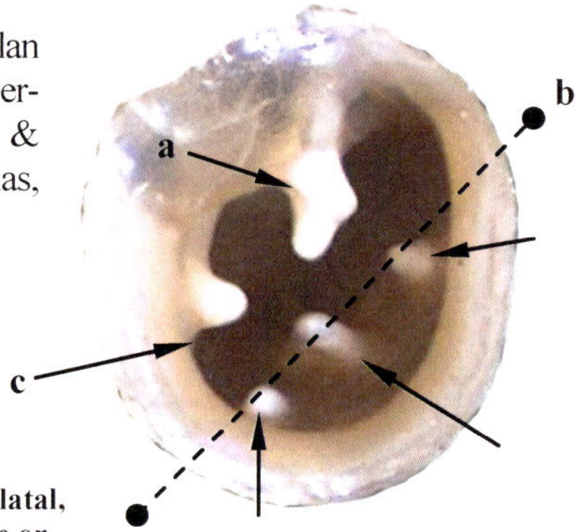

Basal palatal, lower palatal, and upper palatal more or less in line

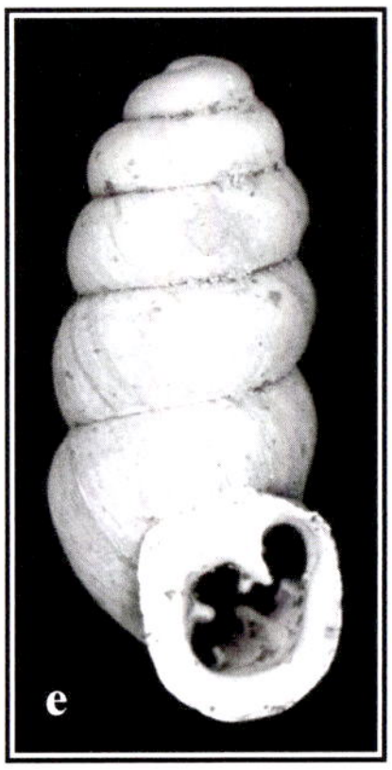

Lectotype

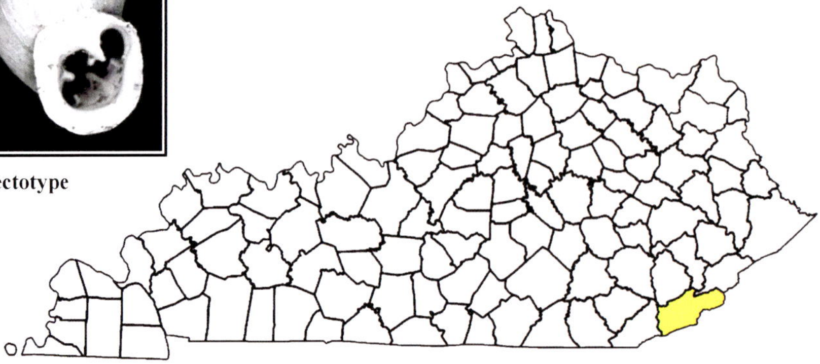

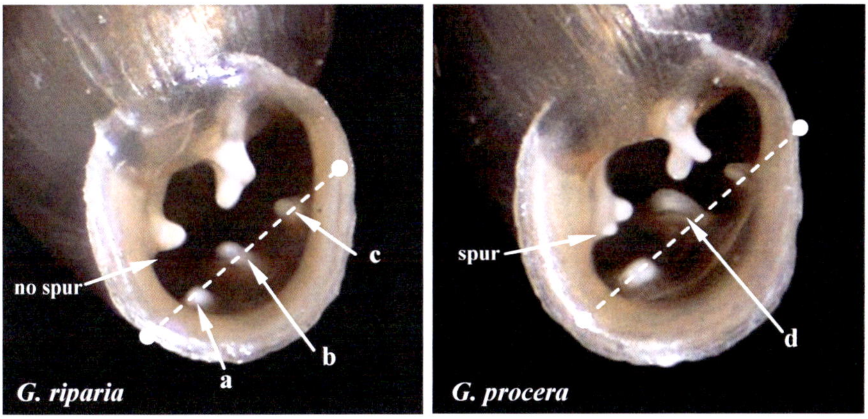

Note that the basal palatal (a), lower palatal (b) and upper palatal (c) of *G. riparia* are more or less in line, but the same teeth of *G. procera* are not, the lower palatal (d) of *G. procera* set deeper in the aperture. *G. riparia* is without a spur on the columellar while *G procera* is usually with a spur.

Cupped Vertigo

Vertigo clappi Brooks & Hunt, 1936

Height: 1.4-1.8 mm, width 0.7-0.8 mm

Description: Pupa-shaped; left side of lip reflected while the right side of the lip is simple containing a small indentation in the aperture wall (a); shell with 5.5 whorls; 6 well-developed teeth, the face of the upper palatal (b) is parallel to the front of the shell; in aperture (side) view the face of the upper palatal is seen, not the edge; shells are somewhat translucent in live individuals and fresh shells but become bleached and white with age; irregular, blunt striae.

Similar Species: *Vertigo milium* is around the same size but has a strongly curved lower palatal; *V. ovata* is larger and the lower and upper palatal lamella does not turn sideways as in *V. clappi*.

Habitat: Found in well decomposed leaf litter, fine soil on shaded boulders, talus, ledges and bases of forested bedrock outcrops and can be locally abundant (Nekola pers. comm. 2010).

Status: G1/S1; State Endangered; Extremely rare across its entire known range.

Specimen: Tennessee, Monroe County, Tellico Gorge (Nekola and Coles image 2010).

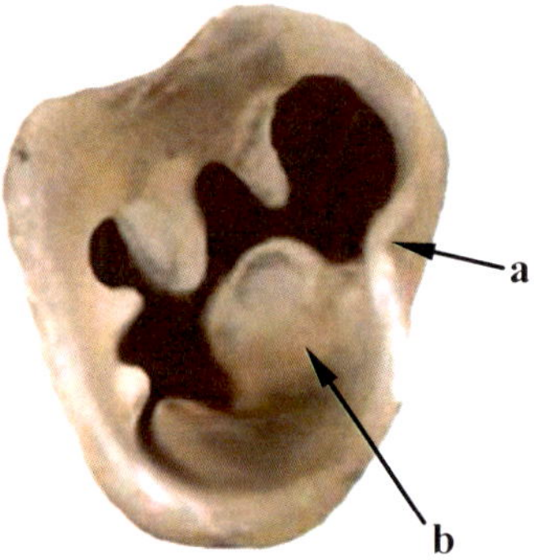

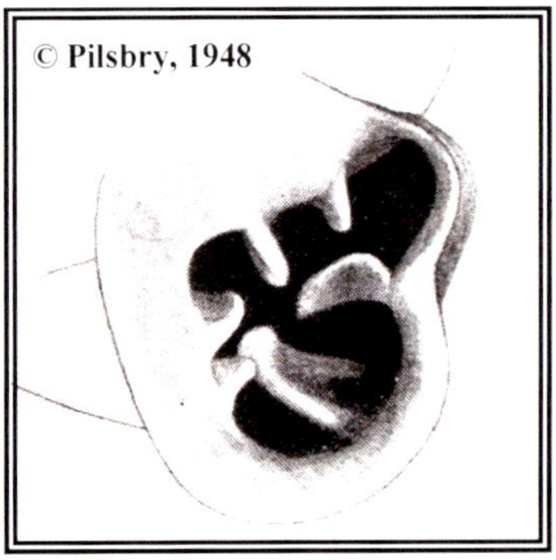

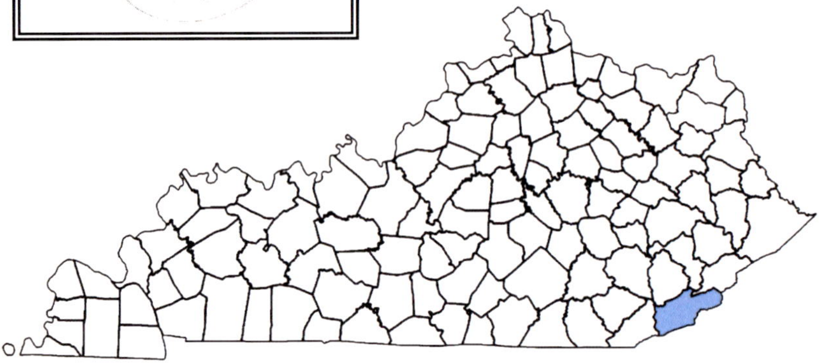

Blade Vertigo

Vertigo milium (Gould, 1840)

Height: 1.3-1.8 mm, width 0.8-1.0 mm

Description: Pupa-shaped; small indentation in the outer aperture wall; shell with 4.5-5 whorls; usually with 6 teeth that crowd the aperture, one of the palatal teeth (a) especially long with a wide entering base; shell surface glossy, weakly striate; shells are somewhat translucent in live individuals and fresh shells, but become bleached white with age.

Similar Species: The inner teeth are relatively long and fill the aperture; the long, curved lower palatal and the broad inward-sloping columellar plate of *V. milium* are very different from the peg shaped columellar of *V. ovata*. This species is unique in the genus by possessing a curved lower palatal lamella that deeply enters the shell (Nekola et al 2018)

Habitat: Found in damp, well-developed thatch and leaf litter across a wide varicty of mesic to wet sites including rocky woodland, riparian woodland, cliffs, wet prairie, sedge meadows, roadside verges, fens, and swamps (Nekola and Coles 2010).

Status: G5/S3; Uncommon; reported from 12 counties scattered across Kentucky.

Specimen: Wisconsin, Green Lake County, Berlin Fen (Nekola and Coles image 2010).

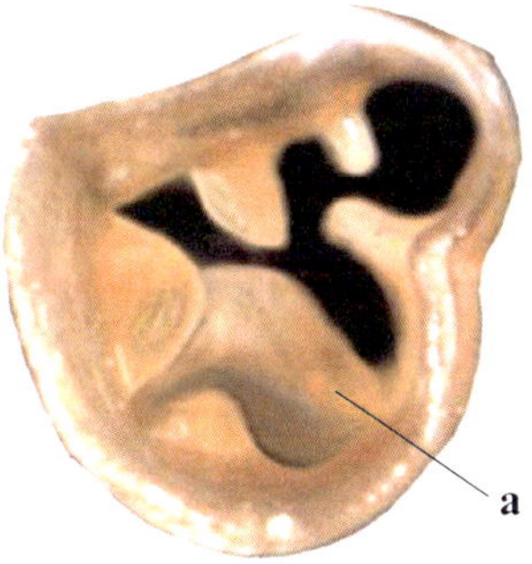

Vertigo milium

Vertigo clappi

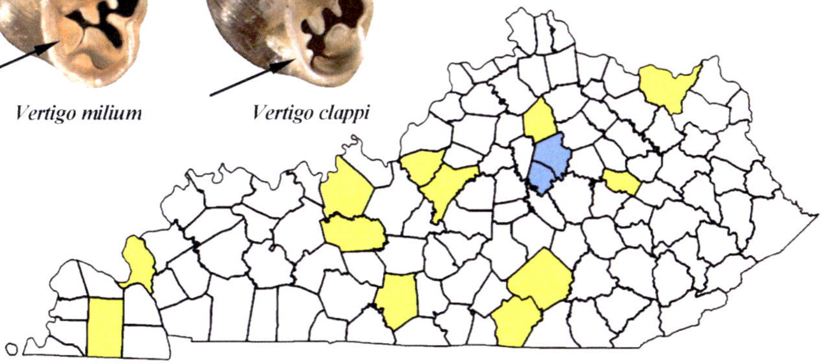

Crested Vertigo

Vertigo pygmaea (Draparnaud, 1801)

Height: 1.4-2.1 mm, width 1.0-1.1 mm

Description: Pupa-shaped; left side of lip reflected while the right side of the lip is simple containing a small indentation in the aperture wall; shell with 5 whorls; 5 peg-like teeth that are thickened and stand on a strong callus rim (a); aperture opening reduced.

Similar Species: Unlike most *Vertigo* species, *V. pygmaea* has a large crest with a white-calcareous callus on the palatal wall of the aperture; it lacks the glassy luster of *V. elatior* (now synonymized with *V. ventricosa* (Nekola and Coles 2010).

Habitat: Found in well-developed humid leaf litter and graminoid thatch in a variety of anthropogenically disturbed grassland including roadsides, old fields, yards, and abandoned quarries (Nekola and Coles 2010).

Status: G5/S2; Rare; in Kentucky reported from only two counties; Nekola et al. (2018) document that populations of this species represent Eurasian waifs brought to North America over a century ago; DNA identical to western European populations. Possibly exotic.

Specimen: Wisconsin, Manitowoc County, Kingfisher Farm (Nekola and Coles image 2010).

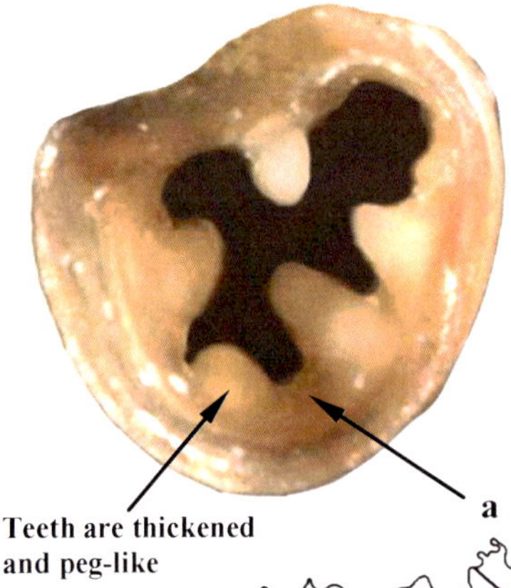

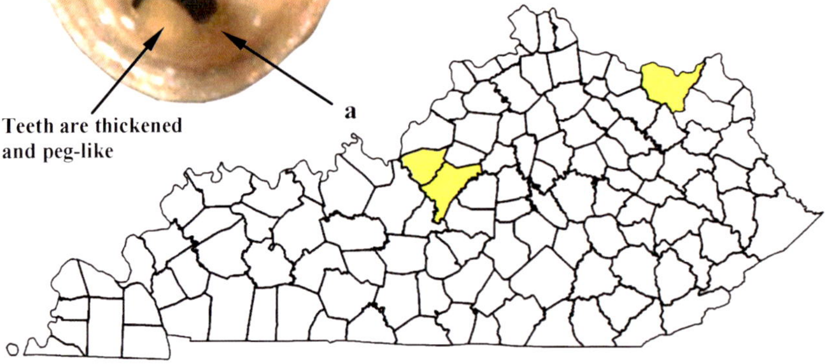

Ovate Vertigo

Vertiginidae

Vertigo ovata (Say, 1822)

Height: 1.6-2.4 mm, width 1.0-1.4 mm

Description: Pupa-shaped; left side of lip reflected & right side of the lip is simple containing a small indentation in the outer aperture wall; shell with 4.5-5 whorls; there are 6-9 teeth, four are well-developed, five are petite; together they crowd the small aperture.

Similar Species: *V. ovata* does not have a long, curved lower palatal as in *V. milium*; *V. ovata* is strongly ovate, shell volume easily 5X that of *V. milium* (Nekola and Coles 2010). Distinguished from *V. teskeyae* by its smaller volume, stronger apertural lamellae and curved columellar wall of the aperture (Nekola et. al 2018).

Habitat: Found on cattail leaves in swamps, sedge meadows, wet and mesic prairie, low calcareous meadows, river banks, lakeshores, roadside ditches, bedrock outcrops and upland forests (Nekola and Coles 2010); it can also ascend vegetation to around one meter off the ground.

Status: G5/S3; Uncommon; in Kentucky, reported from only 6 counties.

Specimen: Iowa, Dubuque County, Epworth Fen (Nekola and Coles image 2010).

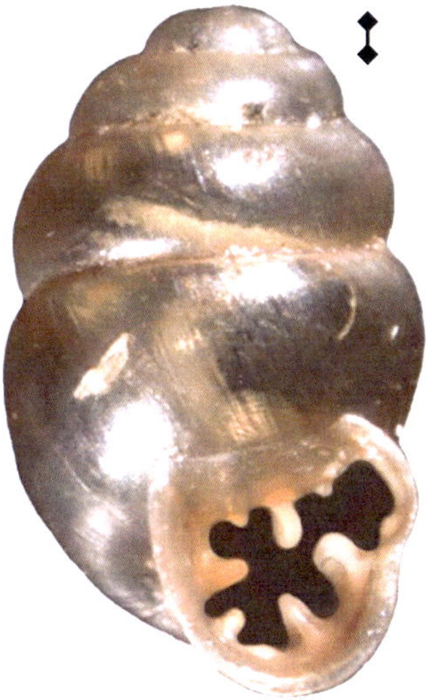

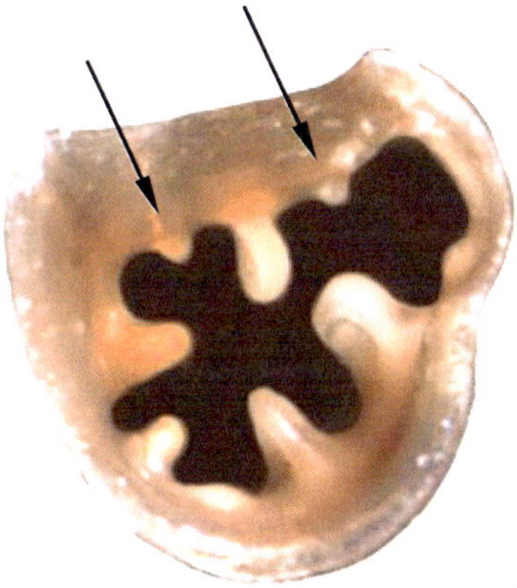

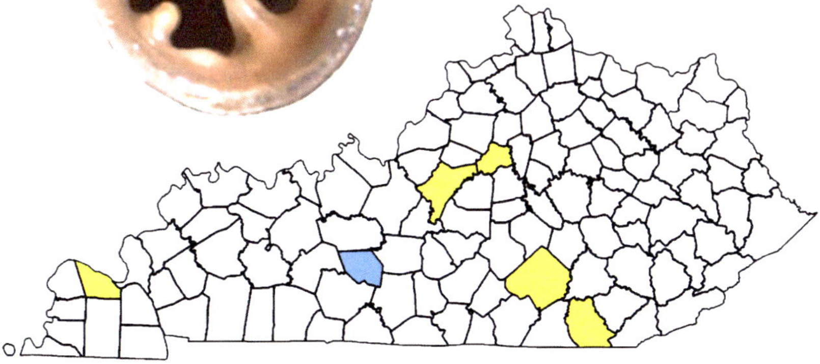

Capital Vertigo

Vertigo oscariana (Sterki, 1890)

Height:1.4-1.6 mm, width 0.85 mm

Description: Pupa-shaped; small indentation in the outer aperture wall; shell with 4.5-5 whorls; 3 medium sized teeth, columellar lamella blunt and wide (a); palatal fold (tooth) short, thick and set deeply within the aperture (b); shells are translucent in live individuals while old shells become bleached white.

Similar Species: While *Vertigo parvula* is closely similar, *V. oscariana* differs by its larger strongly ovate shell with a narrower body whorl (d) than penultimate whorl, vertical plate-like columellar lamellae, stronger sinulus, and deep depression over the palatal lamella (Nekola et. al 2018).

Habitat: Found in well decomposed accumulations of broadleaf and pine litter in mesic, wet woodlands and shaded rock outcrops; in western VA this species is found most often in dry upland pine litter (Nekola and Coles 2010).

Status: G4/S1; Rare, State Special Concern; in Kentucky this tiny *Vertigo* was reported from only one county by Hubricht (1985) but during the 2020 statewide survey two new counties were added.

Specimen: South Carolina, Berkley County, Wadboo Creek (Nekola and Coles image 2010).

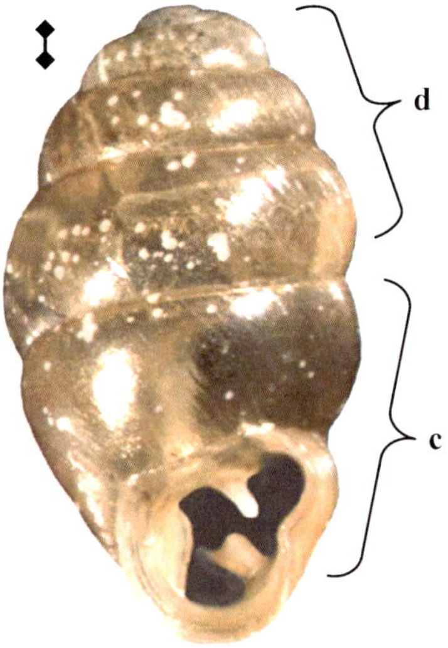

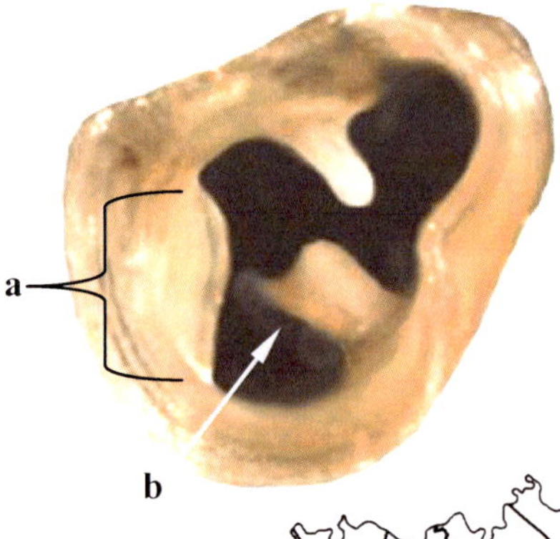

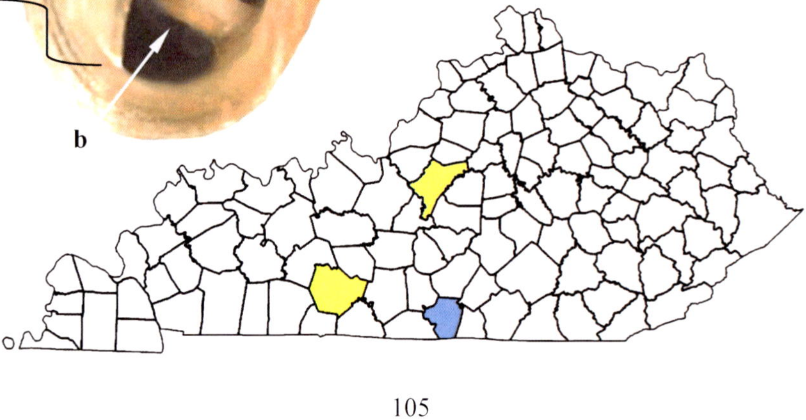

Smallmouth Vertigo

Vertigo parvula Sterki, 1890

Height: 1.4-1.6 mm, width 0.7-.0.9 mm

Description: Pupa-shaped; the lip indentation weakly developed; shell with 5 whorls; 3 rather large teeth; columellar lamella small; the shells are somewhat translucent in live individuals.

Similar Species: Most similar to *Vertigo oscariana* with three apertural lamellae. It differs from that species in its smaller cylindrical ovate shell with a weaker sinulus, peg-shaped columellar lamellae and weak depression over the palatal lamella (Nekola et al 2018). *V. tridentata* is similar in shell shape, color and apertural lamellae but differs in its much smaller size, shinier luster and lack of callus and upper palatal lamella.

Habitat: Found in accumulations of well decomposed leaf litter in base rich cove forests, rock outcrops, and talus slopes at mid to low elevations (Nekola and Coles 2010).

Status: G3/S2; Rare this tiny *Vertigo* is reported from south-central and southeastern Kentucky only.

Specimen: Tennessee, Washington County, Buffalo Mountain (Nekola and Coles image 2010).

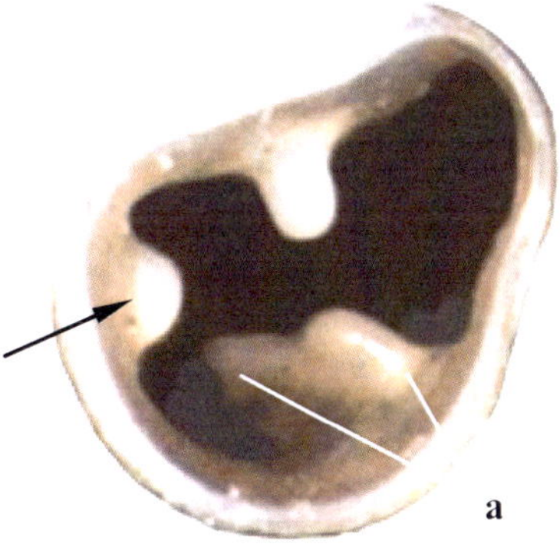

a

V. tridentata *V. parvula*

Honey Vertigo

Vertigo tridentata (Wolf, 1870)

Height: 1.5-2.4 mm, width 0.9-1.2 mm

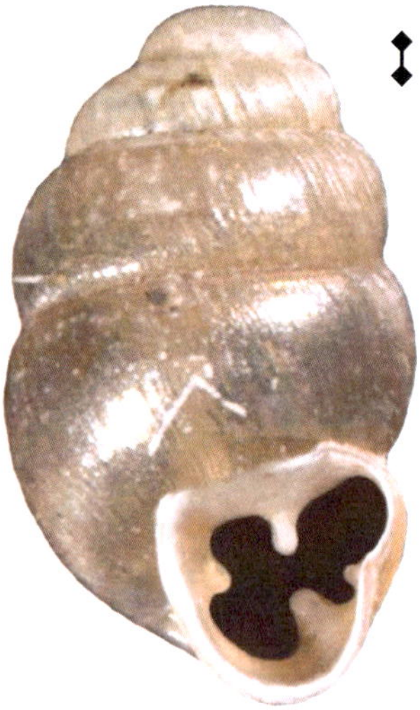

Description: Pupa-shaped, but with a taper; outer lip containing a small indentation in the outer aperture wall (a); shell with 5 whorls; 3 or 4 short teeth, aperture semicircular in shape; shells are smooth, lacking sharp striations, somewhat translucent in live individuals with a honey-yellow color shading to somewhat browner below, like most *Vertigo* species the shells becoming bleached with age.

Similar Species: The shell color of *V. tridentata* is honey-brown whereas *V. gouldii* is a dark-brown; the parietal lamella (b) is pointing at the lower palatal in *V. tridentata*, not the upper palatal as seen in *V. gouldii* (page 109), figure (a) (Nekola and Coles 2010).

Habitat: A species found climbing on herbs in the Lamiaceae family (mints) of low sunny places (Hubricht 1985), bedrock glades, in well decomposed leaf litter accumulations on shaded cliff ledges and talus (Nekola and Coles 2010).

Status: G5/S4; Fairly Common; scattered throughout central and eastern regions of the state.

Specimen: Iowa, Dubuque County, Little Maquoketa River (Nekola and Coles image 2010).

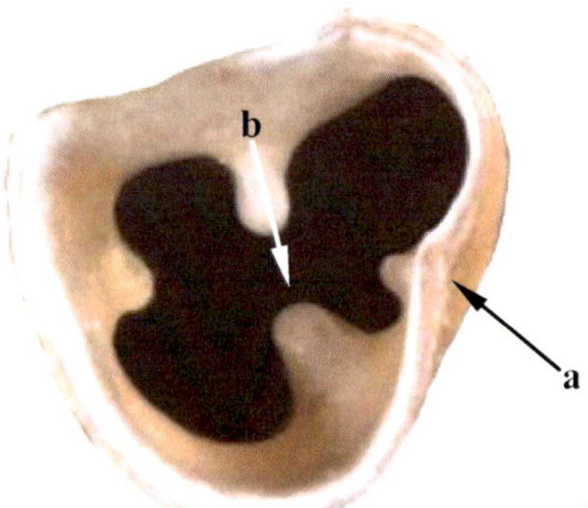

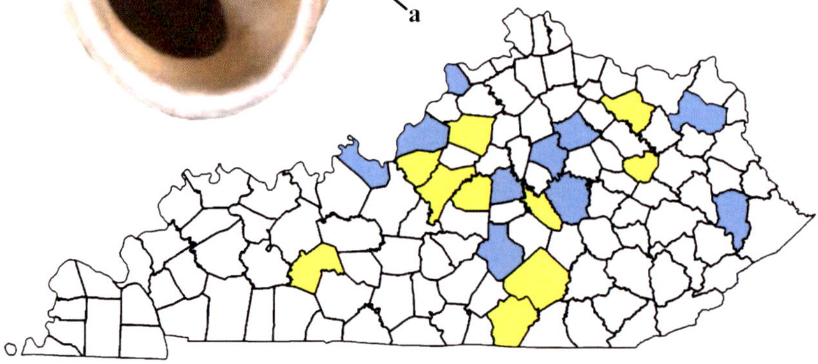

Bluff Vertigo **Vertiginidae**

Vertigo meramecensis Van Devender, 1979

Height: 1.8-2.0 mm; width 1.0-1.2 mm

Description: Pupa-shaped; left side of lip reflected while the right side of the lip is simple containing a small dent in the aperture wall; shell with 4.5-5 whorls; 4 teeth, parietal lamella and palatal folds well developed; shiny; color deep cinnamon-red; crest absent; shell striation irregular in strength and spacing; parietal lamellae pointed directly at lower palatal lamella, so that parietal, lower palatal and columellar lamellae form a cross; no *Vertigo* has a dull, conic, red brown shell (Nekola and Coles, 2010).

a

Similar Species: *Vertigo tridentata* is around the same size but has a yellow-brown ovate shell; *V. gouldii* has an ovate brown shell with a strong basal lamella; *V. oscariana* and *V. parvula* have 3 teeth but are much smaller; all other *Vertigo* species in KY have more than 4 teeth.

Habitat: A strict calciphile, *Vertigo meramecensis* is confined to mesic, vertical limestone cliff faces often profusely covered in mosses and lichens. The only eastern North American *Vertigo* species that is strictly arboreal which makes it one of the easiest *Vertigo* species to observe in the field, with adults being attached or foraging in full view on open rocks and not buried in leaf litter as is characteristic of most other species in the genus (Nekola and Coles 2010). Cliff-dwelling has also led to a unique adaptation in this animal because it will almost certainly die if it falls from its vertical habitat, individuals attach themselves to the rock face by a dry mucus thread. If dislodged, they fall only to the end of this thread, and are then able to crawl back to the cliff face (Nekola, unpublished data).

Status: G4/S1; State Endangered; this species was only recently discovered by Lori Schroeder in Bullitt County.

Specimen: Figure (a) from Iowa, Winneshiek County, North Bear Creek (Nekola and Coles image 2010) and figure (b) SEM image from Kentucky, Bullitt County, Bernheim Arboretum and Research Forest © Jeff and Lori Schroeder.

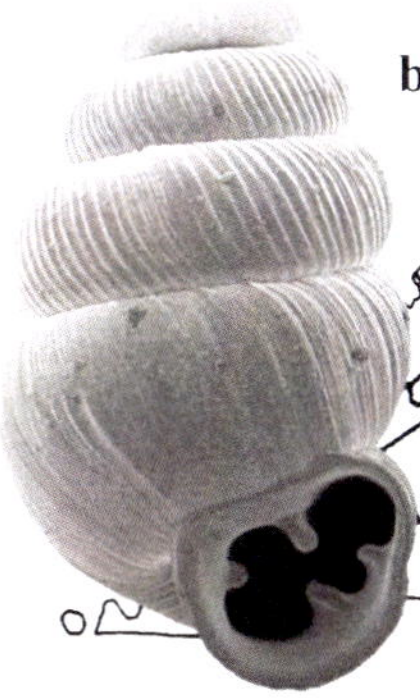

b

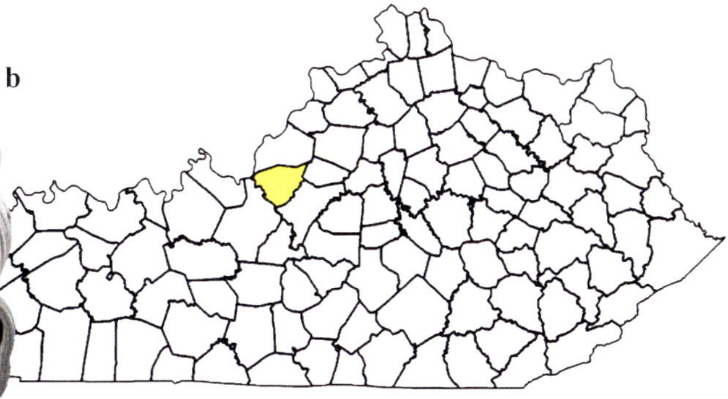

Variable Vertigo

Vertigo gouldii (A. Binney, 1843)

Height: 1.5-1.9 mm, width 0.8-1.0 mm

Description: Pupa-shaped; the indentation in the outer aperture wall is weakly developed or imperceptible; shell with 4.5-5.5 whorls; typically 4 to 5 teeth, all around the same size; parietal lamella and palatal folds well-developed; shell surface has well-developed transverse striae; shells are somewhat translucent in live individuals and fresh shells but become bleached with age;

Similar Species: *Vertigo tridentata* lacks sharp striae, has a more tapered shell, and the parietal lamella points at the lower palatal, whereas in *V. gouldii,* the parietal lamella points toward or at the upper palatal (a); *Vertigo bollesiana,* a species reported from Kentucky but not validated by actual specimens (pers. comm. Nekola 2014) are the same size as small *V. gouldii*, but differs in having a more tapered outline and stronger palatal depression.

Habitat: Limited to forested sites and is most common on wooded bedrock outcrops.

Status: G5/S5; Very Common in mostly Central and Eastern regions; one of the most common *Vertigo* in Kentucky.

Specimen: Tennessee, Monroe County, Tellico Gorge (Nekola and Coles image 2010).

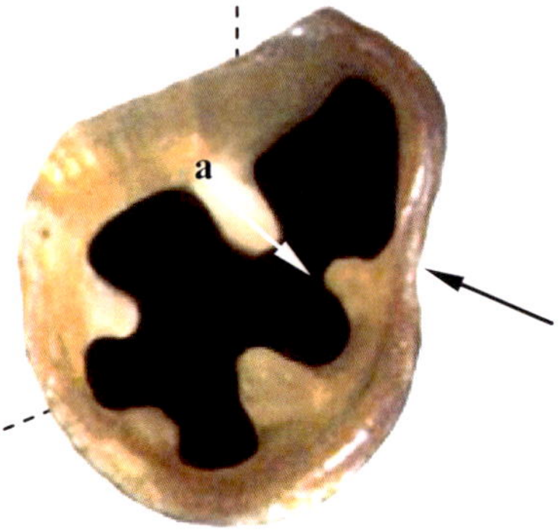

V. gouldii (short form), Warren County, KY

V. gouldii

A number of different shell forms of *V. gouldii* exist, including larger, more columnar shells; smaller, more ovate shells; and a race with coarse striation that lacks a basal lamella. Several of these forms are found in Kentucky as seen to the left.

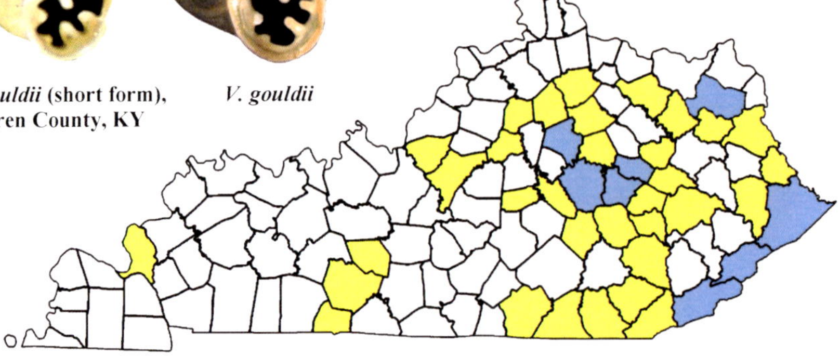

Delicate Vertigo

Vertigo bollesiana Morse, 1865

Height: 1.5 mm, width 1.1 mm

Description: Pupa-shaped; left side of lip reflected while the right side of the lip is simple containing a small dent in the aperture wall; shell with 4.5-5 whorls; 5 teeth, parietal lamella and palatal folds well developed; shell surface has poorly developed transverse striae; this species always shows a strong depression on the outside of the shell over both palatal lamellae (Nekola and Coles 2010).

Similar Species: *Vertigo gouldii* is distinguished by its more cylindrical shell with a duller luster, sharper and stronger striae and a weaker depression over the palatals (Nekola et al 2018).

Habitat: Found in leaf litter often under shrubs, on cliff face ledges and boulder tops in mesic upland forests and white cedar wetlands.

Status: G4/S1; State Endangered; this snail is reported from two counties in Kentucky (Hubricht 1985), however, records of this species in the state remain questionable and are likely that of small *V. gouldii* morphs (Nekola et al. 2009).

Specimen: Tennessee, Monroe County, Tellico Gorge (Nekola and Coles image 2010).

V. gouldii ***V. bollesiana***

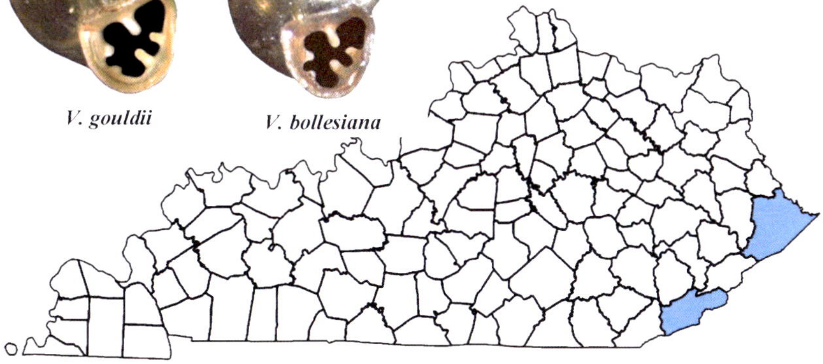

Widespread Column

Pupilla muscorum (Linne, 1758)

Height: 2.8-4 mm, width 1.7 mm

Description: Pupa-shaped; lip reflected; shell with 5.5-7.5 whorls; with or without a small angular lamella in the form of a callus (a) located in the upper right hand corner of the aperture; transverse striae fairly well developed; without spiral striae.

Similar Species: In Kentucky, other pupa-shaped shells such as *Carychium, Vertigo, Gastrocopta* are notably smaller and are not cylindrical; *Cochlicopa* species are larger and with a glass-like surface.

Habitat: Generally occurs in disturbed anthropogenic habitats such as road verges, vacant lots, abandoned quarries, old fields and concrete culverts (Hubricht 1985).

Status: G5/S2 Rare; this species is a European exotic, with at least the Iowa haplotypes identical to Moravian populations; also a number of undescribed native species, more closely related to the Eurasian *P. 'muscorum'* = exotic *P. muscorum* & native *P. hudsonianum* (in north) and *P. hebes* or *P. blandii* (in west), which have been misidentified as *P. muscorum* in North America, but none occur in eastern North America. (Nekola and Coles 2010). New State Record.

Specimen: New York, Beltrami County, Syracuse (Nekola and Coles image 2010).

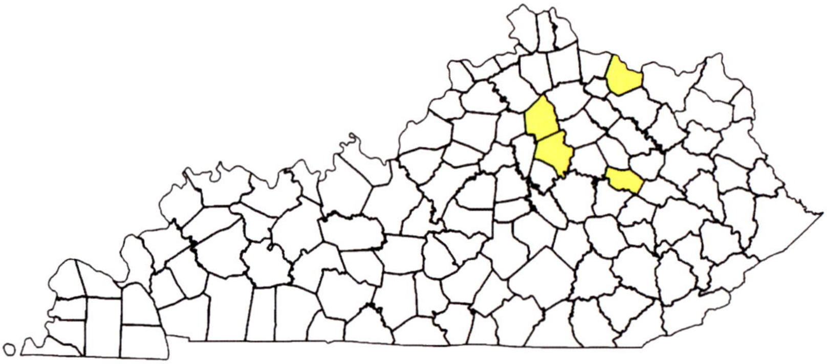

Land snails in this section have shells that are wider than tall, either heliciform or depressed heliciform and have simple lips (although a few may have slightly reflected lips). They include minute snails from 1.5 mm to snails well over 35 mm including many of the mid-size species occurring in Kentucky. Several of the smaller species in this section such as *Glyphyalinia, Helicodiscus,* and *Striatura* are in company with some of the more beautifully sculptured gastropods in North America. Many have glasslike, transparent shells while others display extraordinary micro-ornamentation. This outstanding craftsmanship, however, can only be appreciated under the strong lens.

Genera Included:
(in order of appearance in text)

Guppya

Punctum
Striatura
Hawaiia
Lucilla
Guppya
Zonitoides
Gastrodonta
Glyphyalinia
Paravitrea
Discus
Helicodiscus
Ventridens
Mesomphix
Anguispira
Haplotrema
Vitrinizonites

Dangling from a limestone rock overhang by a thin mucus thread, an aestivating juvenile Pearly Tuba, *Halotudora kusteri* from Belize is safe from most predators except perhaps small, hovering bats. Bluff Vertigo, *Vertigo meramecensis* has been documented using this protective strategy in Kentucky. Shells dangling on lines may also be empty shells in use by spiders or harvestmen (Breure 2011).

Land snails less than 3.5 mm, umbilicate (*G. sterkii* is perforate), simple lip, typically without teeth or lamellae

Land snails in this category are less than 3.5 mm in diameter, without reflected lips, and generally found in moist leaf litter and detritus. To separate these small snails, it is essential that a strong dissecting scope be employed, otherwise a correct ID will be nearly impossible. Each genus has its own unique ornamentation. These microscopic features, although sometimes difficult to see, are key for a successful identification. If you have small specimens collected from leaf litter that fit Groupings 1 through 6 below, then proceed with the following separations.

Grouping 1: shells under 1.5 mm, umbilicate, whorls tightly coiled (a), with crisscrossing lirae (b), pg. 114-119

Punctum minutissimum (very common)
Punctum vitreum
Punctum blandianum (rare)
Punctum smithi (small tooth)

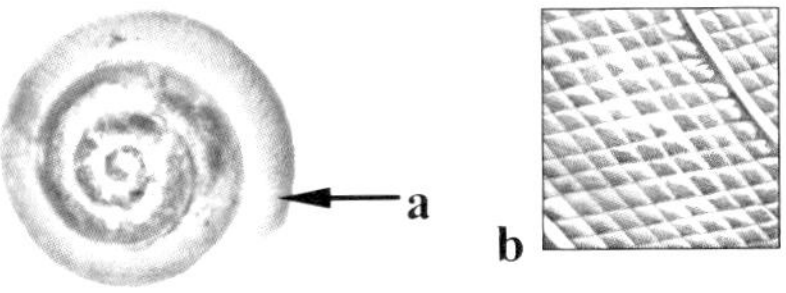

Grouping 2: shells between 1.5 mm to 3.5 mm, umbilicate, whorls loosely coiled (c), transverse ribs strongly developed (a), pg. 120-123

Pseudohyalina exigua (rare)
Striatura meridionalis (very common)
Striatura milium (rare)
Striatura ferrea (uncommon)

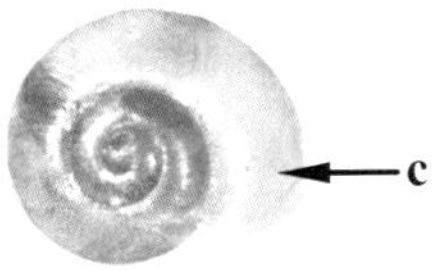

Grouping 3: shells between 1.5 mm to 2.8 mm, umbilicate, surface without crisscrossing lirae, with a dull sheen, pg. 124-126

Hawaiia minuscula
Hawaiia alachuana (uncommon)

Grouping 4: shells around 2.2 mm, umbilicate, surface without crisscrossing lirae, more or less smooth and polished, pg. 127-128

Lucilla (*Helicodiscus*) *inermis* (uncommon)
Lucilla (*Helicodiscus*) *singleyana* (uncommon)

Grouping 5: shells between 1.2 to 1.3 mm, perforate, surface with fine spiral striae, translucent and very glossy, pg. 129

Guppya sterkii (very common)

Small Spot **Punctidae**

Punctum minutissimum (I. Lea, 1841)

Diameter: 1.1-1.3 mm, height 0.75 mm

Description: Depressed heliciform; lip simple; aperture roundish, not reflected; shell with 3.5-4.5 whorls; umbilicate (variable in width); pale brown to corneous; no teeth present; fine transverse striae (fig. a) with even finer transverse and spiral lines forming a net-like feature all located between the transverse striae (fig. b), this amazing micro-feature common to all *Punctum* species can only be fully appreciated under the lens of a strong scope; fresh and live shells are thin and translucent.

Similar Species: *Punctum vitreum* has a smaller umbilicus and has more widely spaced transverse ribs; *P. blandianum* has a larger umbilicus and flatter shell, micro-surface features are the same as other *Punctum* species.

Habitat: Live animals live between the layers of moist, matted leaves of hardwood forests where leaves accumulate next to logs or near seeps.

Status: G5/S5; Common, most likely in every Kentucky county.

Specimen: Kentucky, Letcher County, Bad Branch NP (author's collection).

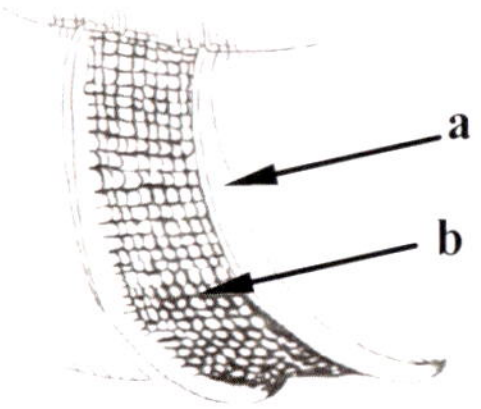

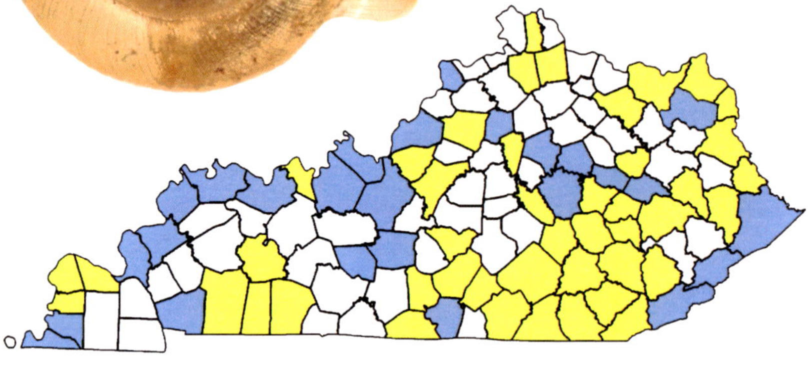

Glass Spot

Punctum vitreum (H. B. Baker, 1930)

Diameter: 1.2-1.4 mm, height 0.8 mm

Description: Depressed heliciform; lip simple; aperture roundish; shell with 4-4.5 whorls; umbilicate; corneous to colorless; no teeth present; transverse striae forming low widely-spaced and paper-thin riblets with even finer transverse lines crossed by spiral lirae (fig. a); fresh shells are thin, fragile and more or less translucent, as with most aging shells, bleaching.

Similar Species: *Punctum minutissimum* has a slightly wider umbilicus and has more tightly spaced transverse ribs; *P. blandianum* has a notably larger umbilicus but its micro-surface features are more or less the same as in *P. minutissimum*.

Habitat: Found between the moist, matted leaves around small seeps in mixed hardwoods.

Status: G5/S5; Common; understandably, this species has frequently been confused with *P. minutissimum* by past collectors.

Specimen: North Carolina, Haywood County, Purchase Knob (GSMNP collection).

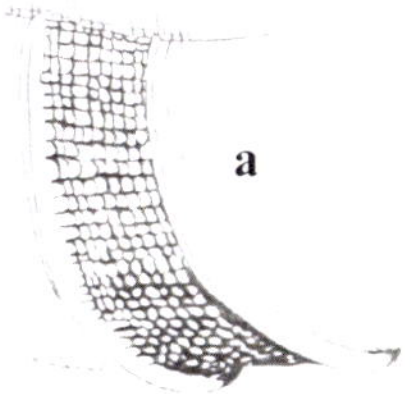

a

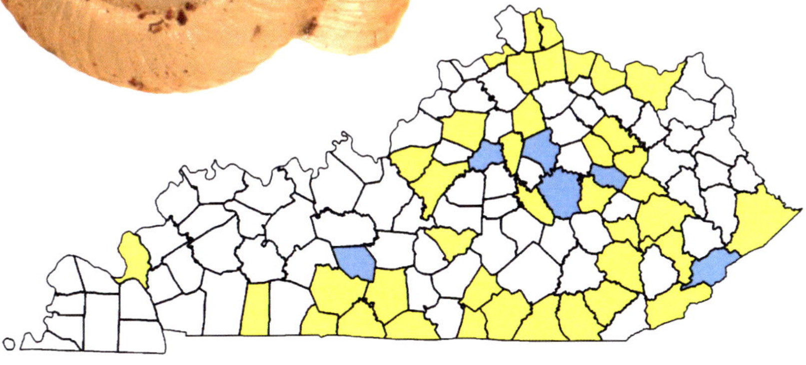

Brown Spot

Punctum blandianum (Pilsbry, 1900)

Diameter: 1.1-1.3 mm, height 0.6 mm

Description: Depressed heliciform; lip simple; aperture roundish to slightly oval shaped; shell with 4 whorls; widely umbilicate; pale brown; no teeth present; transverse striae and spiral striae are weakly-developed and crisscrossing (fig. a) but always present; fresh and live shells thin and translucent, but older shells will bleach with time.

Similar Species: *Punctum minutissimum* is the same size but has a notably smaller umbilicus and higher shell profile; *P. vitreum* has more widely spaced transverse striae than is seen in *P. blandianum,* a smaller umbilicus and a more elevated, compact shell.

Habitat: Found in deep moist pockets of leaf litter in depressions or around logs, but also between the layers of moist, matted leaves adjacent to small headwater seeps.

Status: G4/S2; Rare; this mostly southern Appalachian species is reported from scattered locations in five southeastern Kentucky counties.

Specimen: Tennessee, Rutherford County, Murfreesboro (FMNH 235261).

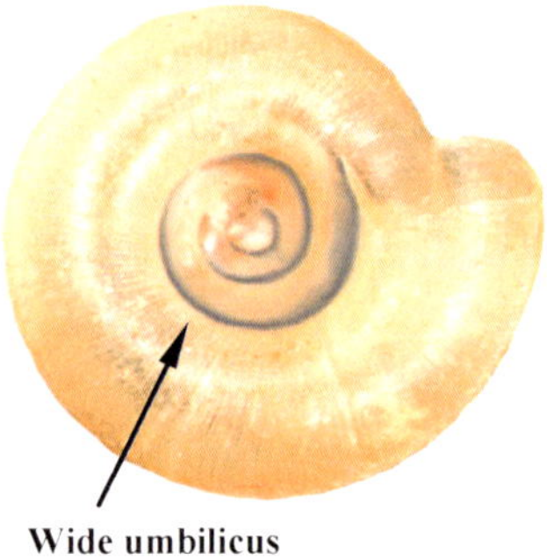

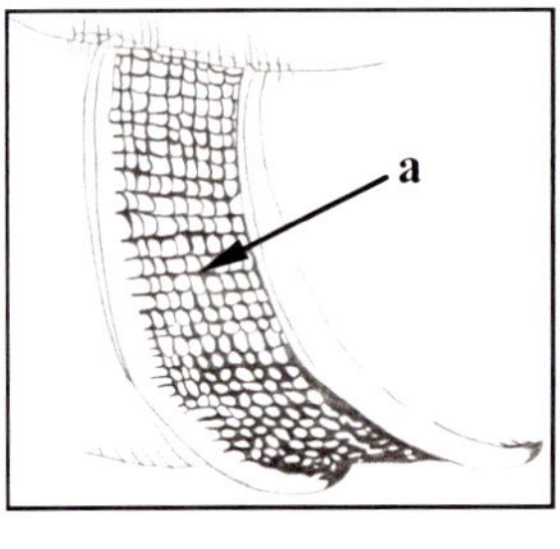

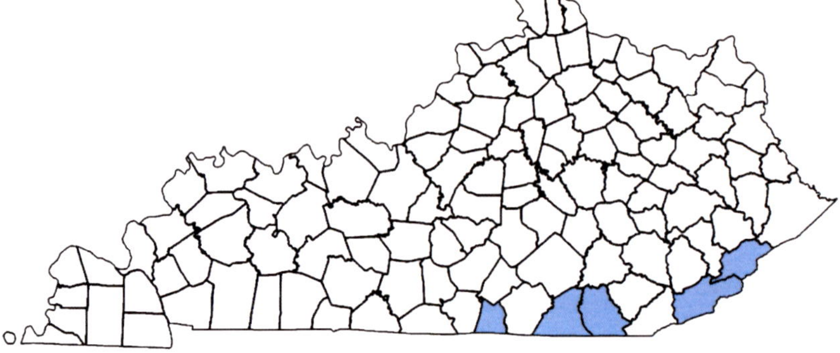

Lamellate Spot

Punctum smithi Morrison, 1935

Diameter: 1.1-1.2 mm, height .62 mm

Description: Depressed heliciform; lip simple; aperture roundish; shell with 4-4.5 whorls; umbilicate; pale tan or brown; one basal tooth (fig. a), sometimes forming a lamella (fig. b) positioned just inside the lower aperture; in some populations, lamella can be exceedingly long; transverse striae are fine and crossed with spiral striae.

Similar Species: No other small species having the shape and size of *P. smithi* has a single tooth in the aperture; other *Punctum* species have more notable micro-sculpture and are without internal armature; *Paravitrea* species typically have multiple teeth or lamellae.

Habitat: Found in deep moist pockets of leaf litter found in depressions or around rotting logs in ravines of mixed hardwood forests.

Status: G4/S3; Uncommon; found in scattered locations across central and southeastern Kentucky.

Specimen: Kentucky, Jefferson County, near McNeely Lake Dam, Okolona, (FMNH 234970)

a

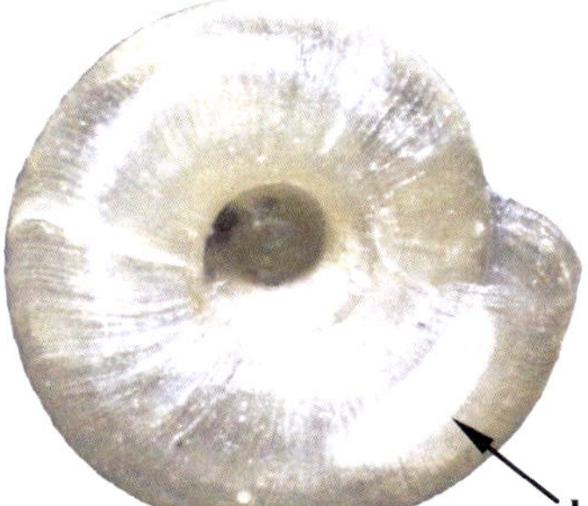

b

Black Oak Road, Mason County, West Virginia

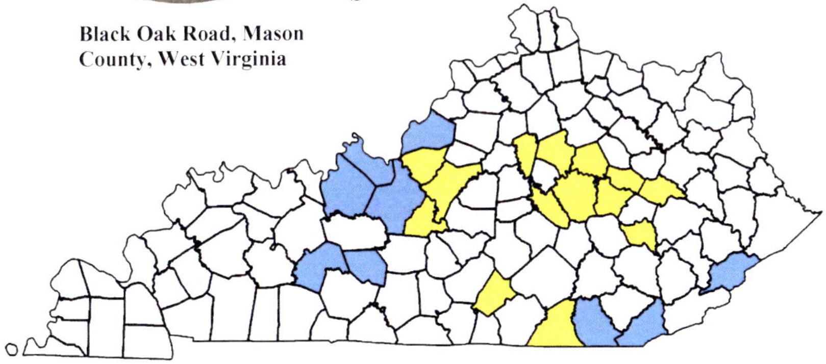

***Punctum* species Compared**

The unique crisscrossing micro-sculpture found in *Punctum* species

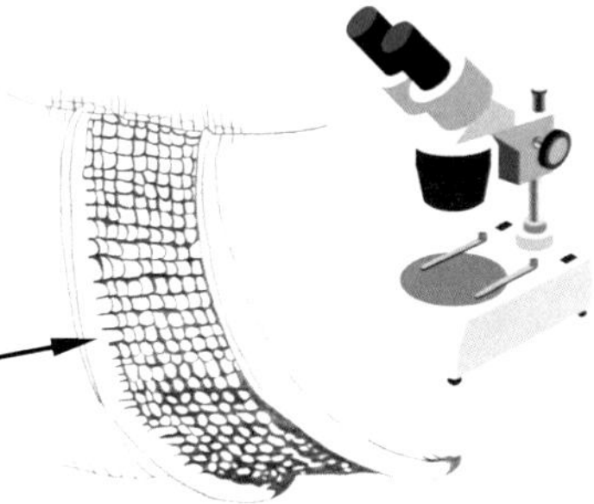

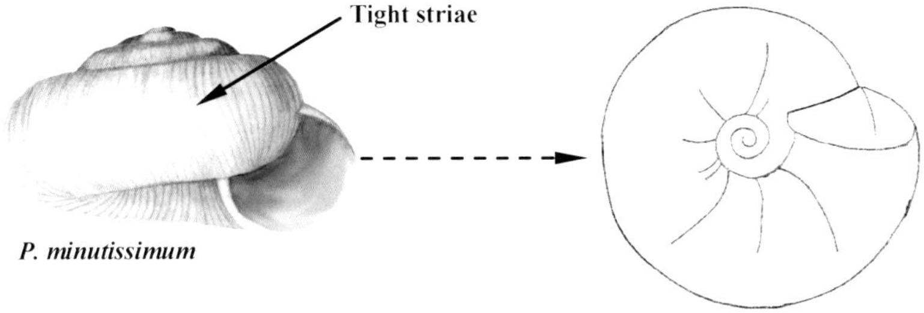

P. minutissimum

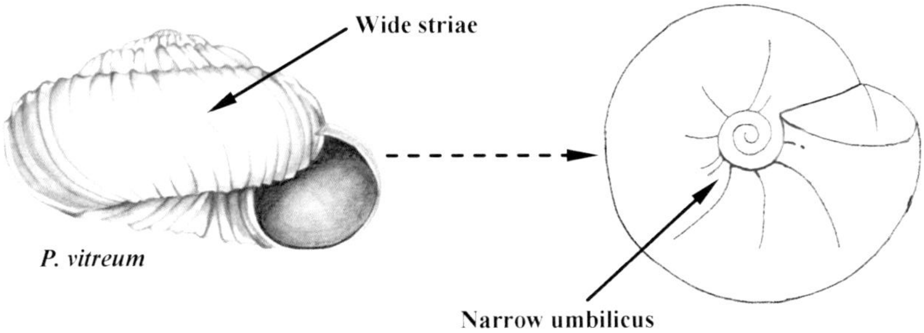

P. vitreum

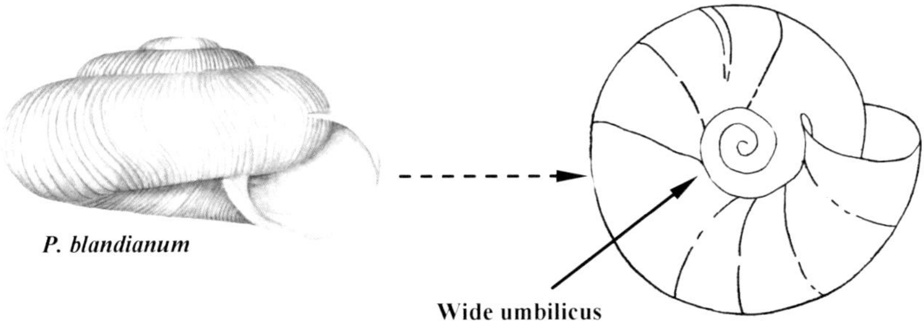

P. blandianum

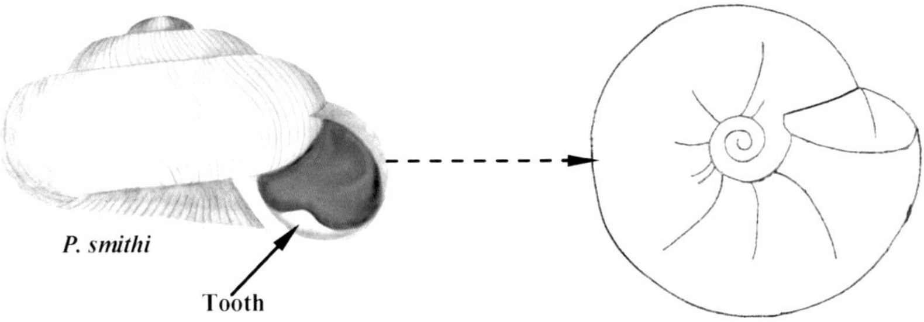

P. smithi

Ribbed Striate

Pseudohyalina exigua (Stimpson, 1847)

Diameter: 2.2-2.4 mm, height 1.25 mm

Description: Depressed heliciform; lip simple; aperture roundish; shell with 3.5 whorls; widely umbilicate; a coppery color; shell translucent in live snails and fresh dead; without teeth; transverse striae are well-developed forming minute paper-thin riblets; spiral striae also well defined but an extremely faint micro-feature of the shell surface (fig. a) below.

Similar Species: *Striatura ferrea* is larger in diameter, has a much smaller umbilicus, and has finer shell surface features.

Habitat: Recently discovered in 2020, where extant populations were found in moist leaf litter under rhododendron patches along cool headwater streams, mesic hillsides and ravines.

Status: G5/S1; State Threatened; in Kentucky, this mostly northern species is restricted to the mountain valleys of Harlan County.

Specimen: Figs. (b) from North Carolina, Swain County, Nantahala National Forest and (fig. c) from Nathaniel Mountain WMA Hampshire Co, WV (all specimens in author's collection).

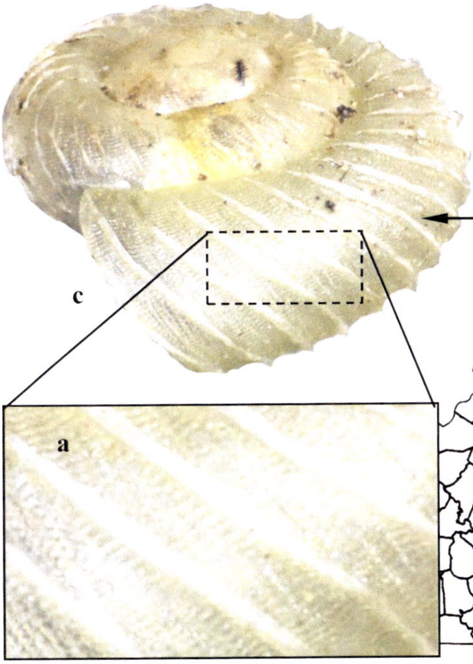

Note the raised riblets

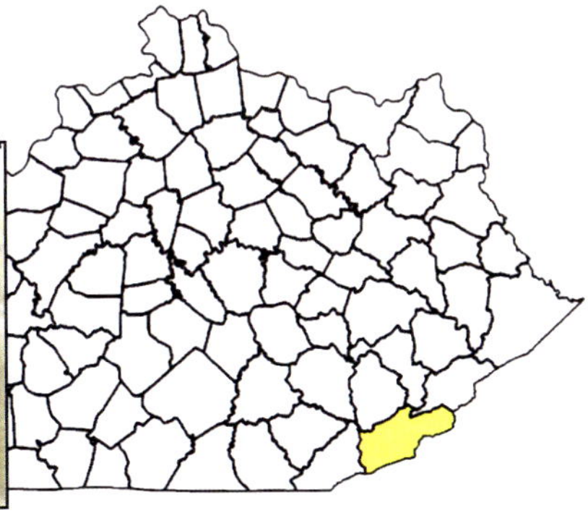

Note the spiral striae between raised lirae or riblets

Median Striate

Gastrodontidae

Striatura meridionalis Pilsbry and Ferriss, 1906

Diameter: 1.7-1.8 mm, height 1 mm

Description: Depressed heliciform; aperture roundish; lip simple; shell with 3-3.5 whorls; widely umbilicate; corneous with a greenish cast; shell translucent in live snails and fresh dead; nuclear whorl with spiral striae, (fig. a), but without transverse striae, the transverse striae developing in later whorls (fig. b) forming minute low riblets and continuing onto the base where they become somewhat weaker.

Similar Species: *Striatura ferrea* is larger in diameter, has a larger aperture and a notably smaller umbilicus.

Habitat: Found in mixed hardwood forests on hillsides and ravines living between the layers of moist leaves; also found at the base of butternut and black walnut trees.

Status: G5/S5; Common; a species that will likely be found in every Kentucky county, especially if leaf litter samples are secured during surveys.

Specimen: North Carolina, Macon County, Nantahala National Forest (author's collection).

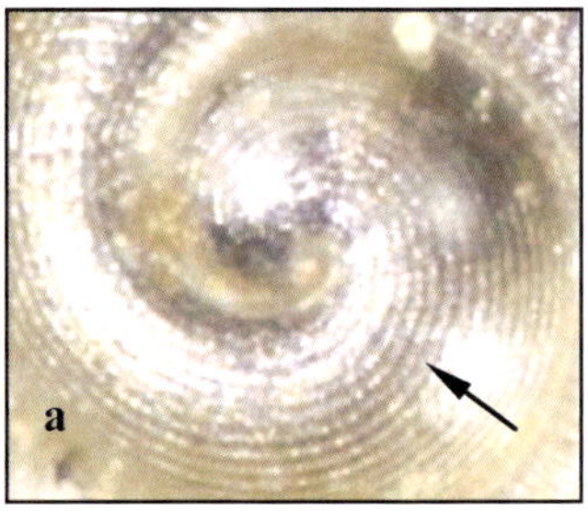

Look for spiral striae on the embryonic whorl under strong lens

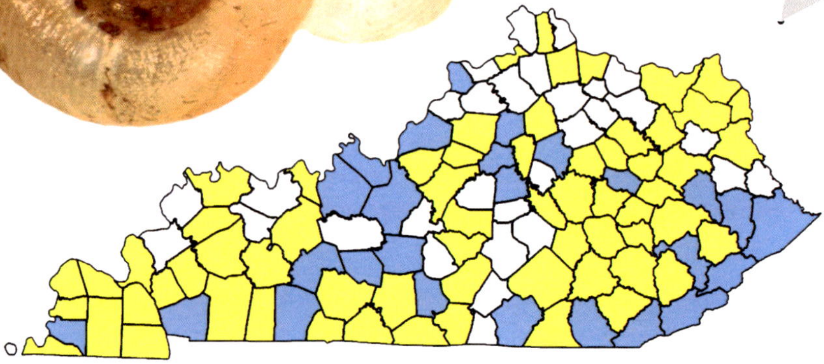

Fine-ribbed Striate

Striatura milium (Morse, 1859)

Diameter: 1.5 mm, height 0.8 mm

Description: Depressed heliciform; aperture roundish; lip simple; shell with 3-3.5 whorls; widely umbilicate; yellowish-corneous or gray; the live animal is white with dark spots on the head and tentacles; shell translucent in live snails and fresh dead; nuclear whorl smooth (fig. a) and without spiral striae; in the remaining whorls the spiral striae not prominent; fine transverse striae well-developed forming minute, low riblets; base nearly smooth.

Similar Species: *Striatura meridionalis* is larger in diameter, has spiral striae on the nuclear whorl, and has coarser riblets; *S. ferrea* is larger with a proportionally smaller umbilicus; *S. exigua* is larger with wider and taller riblets.

Habitat: In Kentucky, known from a Canada Yew patch in Middle Fork of the Red River.

Status: G5/S1; State Endangered currently known from a Canada Yew site in Wolfe County and on Black Mountain in Harlan county. *S. milium* is considered a northern relic of a much colder time as is Canada Yew.

Specimen: Iowa, Fayette County (FMNH 250966).

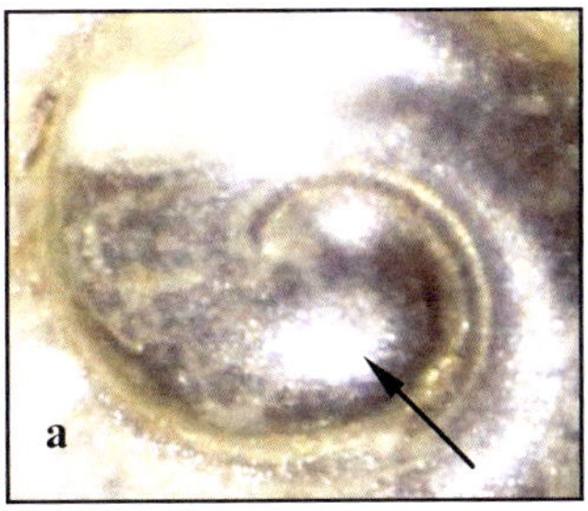

Without spiral striae on the embryonic whorl

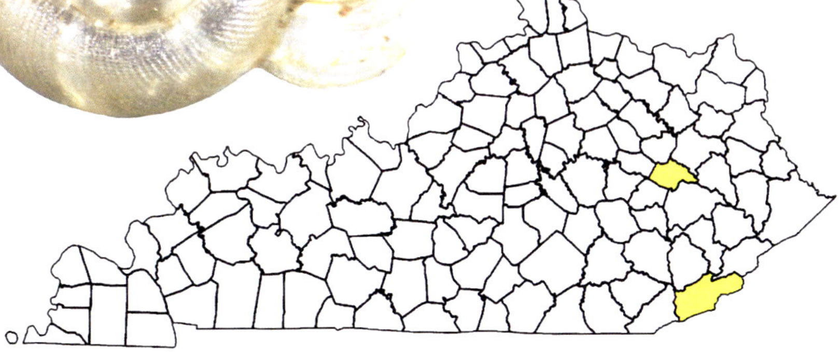

Black Striate

Gastrodontidae

Striatura ferrea Morse, 1864

Diameter: 2.5-3.4 mm, height 1.5 mm

Description: Depressed heliciform; aperture oval; lip simple; shell with 3.5-4 whorls; narrowly umbilicate; shell grayish and rather dull; transverse and spiral striae not as well developed (fig. a) as in *S. meridionalis* but are a constant and reliable diagnostic feature; without teeth; deceased and dried animal can be seen through the translucent shell (bottom view) giving it a darker appearance in some spots.

Similar Species: *Striatura meridionalis* is smaller in diameter, has a smaller aperture, and a notably larger umbilicus; *Punctum* species are smaller and have a proportionally wider umbilicus.

Habitat: Found in mixed hardwood forests on upper elevation mountainsides and in ravines living under layers of moist leaf litter.

Status: G5/S2; Rare to Uncommon; additional leaf litter sampling will no doubt add additional sites in eastern Kentucky counties.

Specimen: North Carolina, Cherokee County, Nantahala National Forest (author's collection).

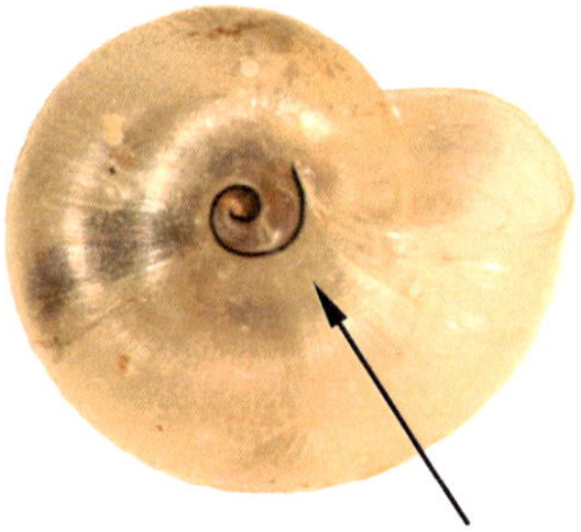

Umbilicus small

a

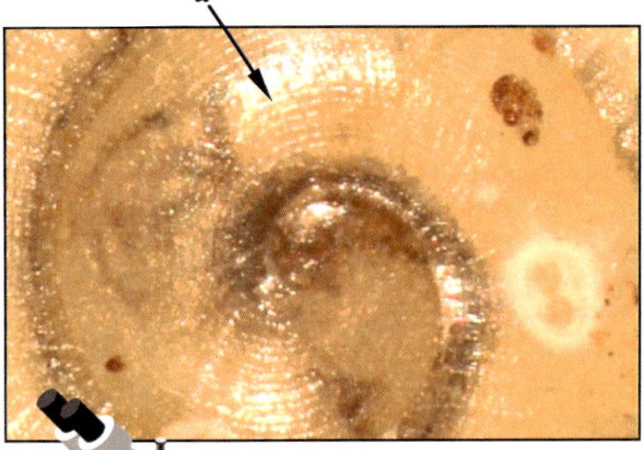

Transverse and spiral striae not as well developed (fig. a) as in *S. meridionalis* but are a constant and reliable diagnostic feature for the species

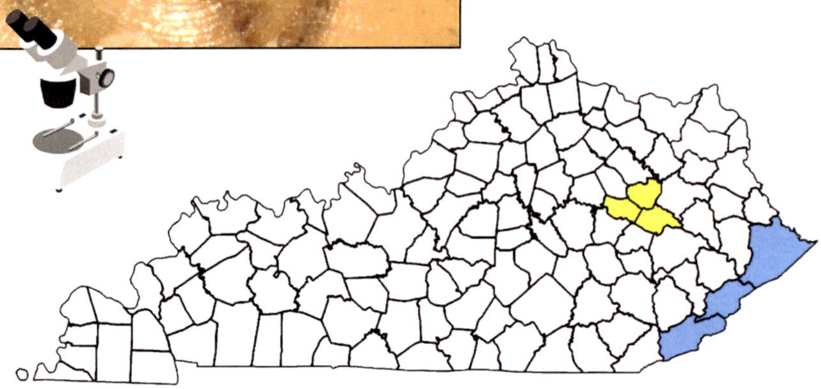

Minute Gem

Hawaiia minuscula (A. Binney, 1840)

Diameter: 2-2.8 mm, height 1.2 mm

Description: Depressed heliciform; lip simple; shell with 3.5-4.5 whorls; widely umbilicate, 0.8 mm wide; live shells thin, pale gray and translucent, dead shells quickly turning white with age; top of shell has moderately developed transverse striae that are irregularly spaced (a reliable feature for ID) and there are always faint spiral striae present under a strong lens.

Similar Species: Both *Lucilla scintilla* and *L. singleyana* have a much less distinctive sculpture and, more importantly, a smaller umbilicus; *Punctum* species have a crisscrossing sculpture; the spiral striae of *Striatura* species are more well developed.

Habitat: A snail of bare ground occurring in floodplains, meadows, roadsides, and in urban areas; the snail also occurs in the mountains in mixed hardwood forests under leaf litter and detritus.

Status: G5/S5; Common; a species that likely occurs in every Kentucky county.

Specimen: Kentucky, Letcher County, Bad Branch NP (author's collection).

Weak to moderately developed transverse striae always present on top of shell

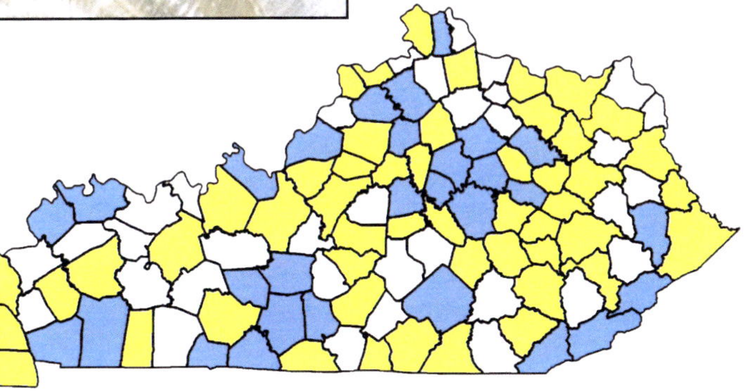

Southeastern Gem

Pristilomatidae

Hawaiia alachuana (Dall, 1885)

Diameter: 2-3 mm, height 1 mm

Description: Depressed heliciform; lip simple; shell with 4-5.5 whorls; widely umbilicate, 1.3 mm wide; live shells thin, pale gray and translucent, dead shells quickly turning white with age; transverse striae are irregularly spaced and there are always faint spiral striae under a strong lens.

Similar Species: *Hawaiia minuscula* is smaller, has a higher shell profile, and has a notably smaller umbilicus (see page 126), but this feature is highly variable in both *Hawaiia* species; *Lucilla inermis* lacks distinctive sculpture.

Habitat: A calciphile species of leaf litter and detritus and waste places.

Status: G4/S2; Rare; a species that has been confused with *H. minuscula,* a result of the two species having a wide range of umbilicus widths. Extant populations were only recently discovered in scattered locations in the state.

Specimen: West Virginia, Gilmer County, (author's collection).

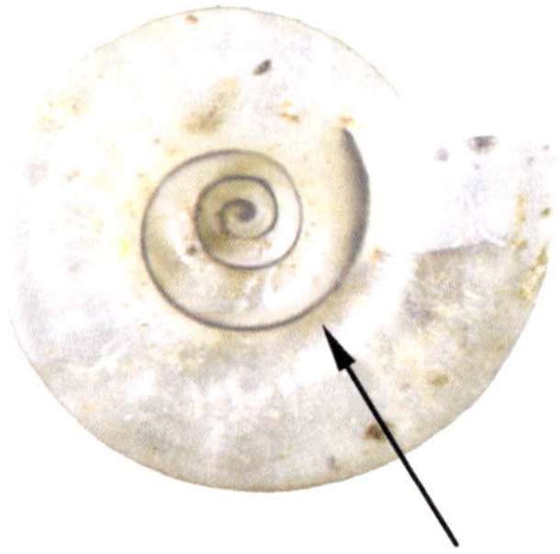

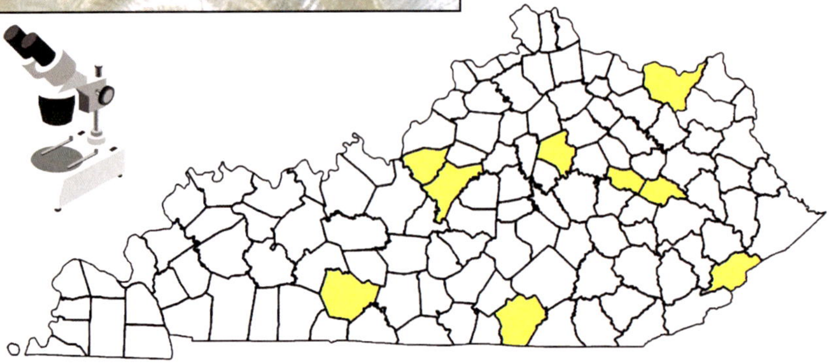

Hawaiia Shells Compared (Larry Watrous photo collection)

Minute gem, *Hawaiia minuscula* (A. Binney, 1840)

Southeastern gem, *Hawaiia alachuana* (Dall, 1885)

Smooth Coil

Lucilla singleyana (Pilsbry, 1889)

Diameter: 2.4-3 mm, height 0.9 mm

Description: Depressed heliciform; lip simple, aperture egg shaped; shell with 3-4 whorls; fresh shells subtranslucent, waxen white, shining and nearly smooth; umbilicate; no teeth present; under a strong lens slightly wrinkled by growth lines and scant traces of spiral ornamentation (the latter feature is usually present in fresh shells only); deceased and dried animal seen in frontal and bottoms views, making the shell slightly darker in places.

Similar Species: *Hawaiia* species are very similar in size and build, but have transverse striae that are notably more developed, (*Lucilla* shells are nearly smooth).

Habitat: Usually a species of open grassy areas, roadsides, along railroads, and occasionally washed into caves during rain events.

Status: G5/S3; Uncommon; likely more common than current records indicate, formerly in the genus *Helicodiscus*.

Specimen: Fig. (a) from France, (MNHN-IM-2010-13182); and (fig. b) Texas, Comal County, New Braunfels, Lecotype (ANSP 160058).

a

b

Lectotype Specimen

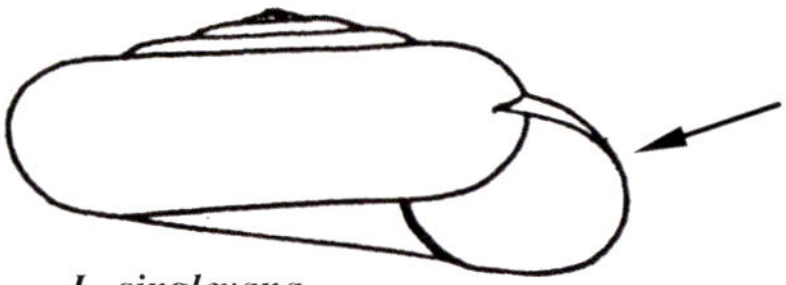

L. singleyana

© Pilsbry 1946

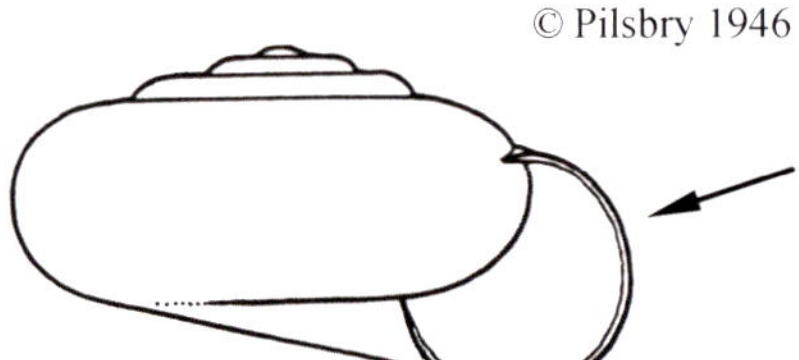

L. inermis

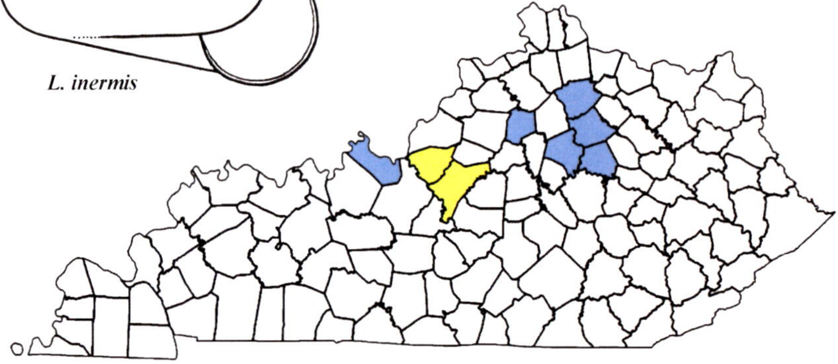

Oldfield Coil

Lucilla inermis (H.B. Baker, 1929)

Diameter: 2.2 mm, height 1.24 mm

Description: Depressed heliciform; lip simple, rounded; shell with 3-4 whorls; fresh shells are yellowish corneous with a dull sheen and are translucent, but turn white with age; umbilicate; surface weakly punctate under strong lens; no traces of spiral ornamentation.

Similar Species: *L. singleyana* displays fine microscopic spiral lines, is more notably compressed, and has an oval-shaped aperture (see Pilsbry's illustrations on page 127); be forewarned, separating *Lucilla* species is no joyride and can lead to "SBC", a debilitating mental condition known as "Shell Bewilderment Syndrome" for which there is no cure!

Habitat: Usually a species of open grassy areas, meadows, old fields and roadsides.

Status: G4/S2; Rare; likely more common than current records indicate and future leaf litter samples should find more; formerly called *Helicodiscus inermis*.

Specimen: Figs. (a) from Tennessee, Franklin County, Dry Cave (author's collection); (fig. b) *L. inermis* from Texas, Val Verde County, Pecos River and US Route 90 (FMNH 240318) and (fig. c) *L. inermis* from Dove, Tennessee, Marion County, Dove, Holotype Specimen (ANSP 147186).

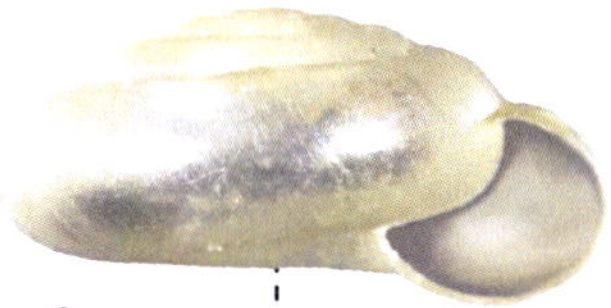

a

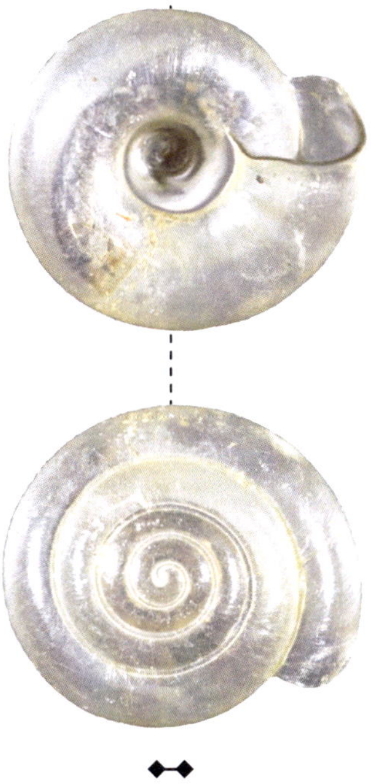

b

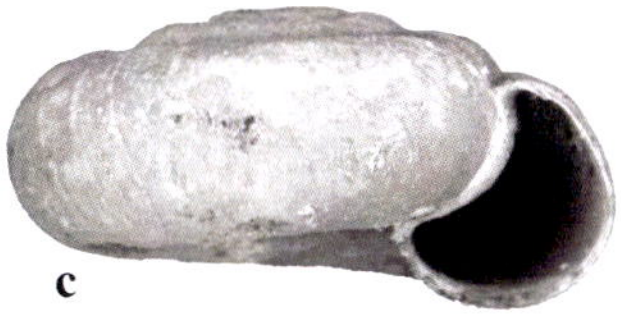

c

Holotype Specimen

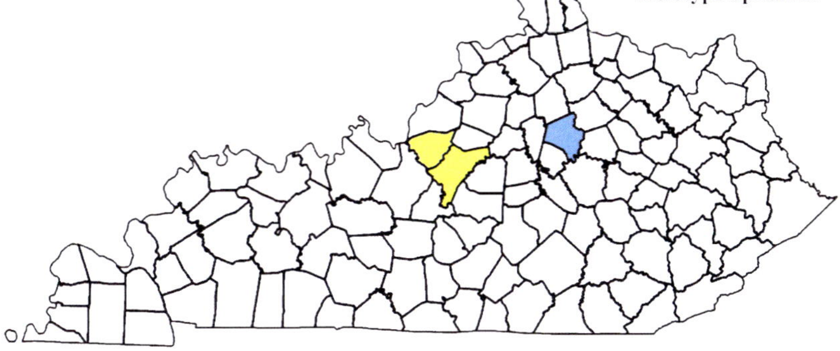

Brilliant Granule — Euconulidae

Guppya sterkii (Dall, 1888)

Diameter: 1.2-1.3 mm, height 0.75 mm

Description: Depressed heliciform; lip simple; shell with 3.5-4 whorls; minutely perforate, the umbilicus not completely closed (fig. a); shell yellowish and highly translucent in live snails and fresh dead shells; transverse striae weak (strong light and scope required here); spiral striae always present (fig. b), but sometimes a hard feature to detect, best seen around the first whorls but also on the base; tilting the shell under the scope slightly with good lighting should help bring out these hidden features.

Similar Species: This species differs from other minute snails by having a nearly closed umbilicus (fig. a), most other tiny snails this size being umbilicate.

Habitat: Found in mixed hardwood forests at all elevations; often found together with *Punctum* and *Striatura*.

Status: G5/S5; Common; one of the most frequent tiny land snail species found in leaf litter collections and it likely occurs in every Kentucky county.

Specimen: North Carolina, Swain County, Nantahala National Forest (author's collection).

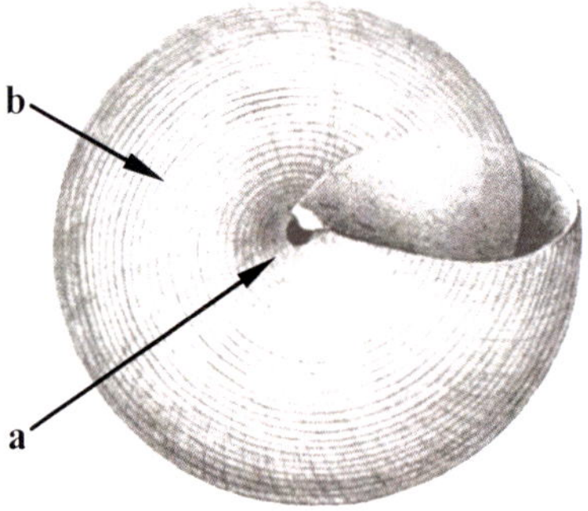

© Pilsbry, 1946

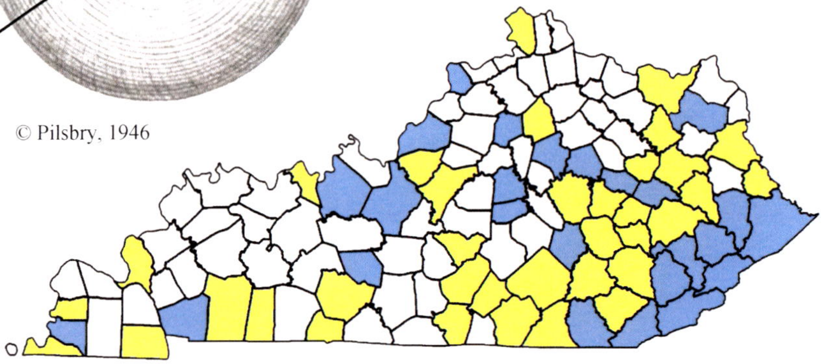

Shells 5-8.4 mm, umbilicate or perforate, simple lip

Zonitoides species are small colonial snails often associated with rotting hardwood in advance stages of decay, either standing or downed trees. They are sometimes found in large numbers (I have found as many as 50 on one log). Key features to this group of gastropods are the open umbilicus and absence of any teeth in all stages of growth. They are occasionally confused with *Glyphyalinia* and *Paravitrea* species but, in general, *Zonitoides* species are larger, have much thicker shells, and are less finely sculptured. Pay close attention to any spiral ornamentation or the absence of it under the lens. At times, this micro-feature is hard to see, depending on the lighting source and angle of the shell. It is best viewed on the top surface and will take some experience to key in on this finer feature. Slightly tilting the shell (while under the scope) will sometimes help reveal this micro-feature; wetting the surface of the shell will also aid in sculpture enhancement, especially if the shells are bleached or badly weathered. There are four species reported from the state.

Grouping 1: shells 5-8.4 mm, umbilicate (a), whorls moderately to loosely coiled (b), spiral striae usually a weak feature or absent, pg. 131-134

**Zonitoides elliotti* (uncommon)
**Zonitoides arboreus*
**Zonitoides lateumbilicatus*
**Zonitoides limatulus*

Gastrodonta species differ from *Zonitoides* in being more tightly coiled (c), perforate (fig. d), having coarse ribbing (fig. f), and are always found with well-developed teeth (fig. e). Two species are reported from Kentucky.

Grouping 2: shells 6-7 mm, perforate (d), with tightly coiled whorls (c), internal teeth (e) and strongly developed ribbing on top of the shells (f), pg. 135-136

**Gastrodonta interna*
**Gastrodonta fonticula* (rare)

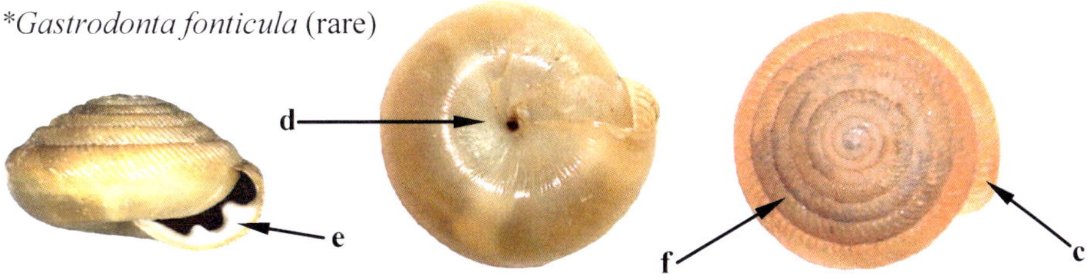

Green Gloss

Gastrodontidae

Zonitoides elliotti (Redfield, 1856)

Diameter: 7.5-8.4 mm, height 4– 4.9 mm

Description: Depressed heliciform; lip simple and slightly reflected, behind which is a small callus or thickening of the peristome (fig. a); aperture square shaped; shell with 5-6 whorls; umbilicate; greenish-horn color and glossy; transverse striae are poorly developed; spiral striae present, but are so faint that detection is often eye straining; without teeth.

Similar Species: The slightly reflected lower lip and thicker shell of *Z. elliotti* are key characteristic features in separating it from all other snail shells found in this size class.

Habitat: Found in mixed hardwood and pine woods on hillsides and ravines; in well rotting wood that is partly embedded in the soil, or under rotting hardwood logs; the species is often found in colonies, upwards of twenty-five on one single log (pers. obs.); the species appears to indicate a rich snail fauna (pers. comm. Ron Caldwell).

Status: G4/S3; Uncommon; most records in the state for *Zonitoides elliotti* are from far southeastern Kentucky counties.

Specimen: North Carolina, Swain County, Nantahala National Forest (author's collection).

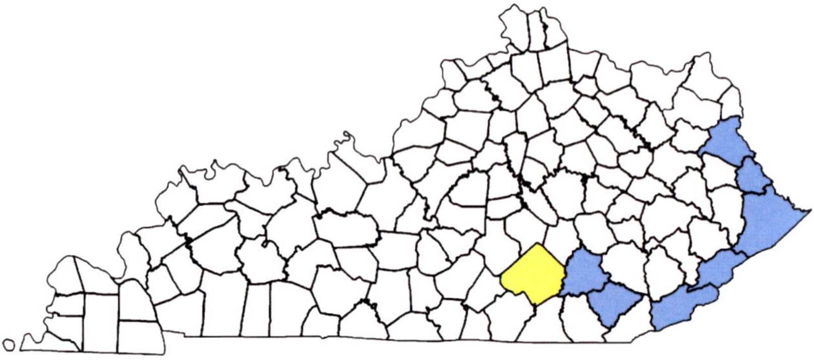

Zonitoides arboreus (Say, 1817)

Diameter: 5-6 mm, height 2.5-2.8 mm

Description: Depressed heliciform; lip simple not reflected; shell with 4.5–5 whorls; umbilicate; olive buff and glossy; transverse striae are poorly developed; spiral striae present, but a weak, eye-straining feature to see; without teeth; periphery roundish; when crawling its shell is held rather high, (fig. a).

Similar Species: *Z. elliotti* is larger, has a thicker shell and a slight reflection of the lower lip; *Z. nitidus* below (fig. c) is slightly larger and taller with a darker shell, no spiral striae, and a more roundish aperture and currently not known from Kentucky.

Habitat: A common species found on or under exfoliating bark of standing or down rotting trees in advanced stages of decay; very common on old timbers inside abandoned coal mines (MacGregor, pers. comm. 2010); usually found in small colonies.

Status: G5/S5; Common; the most common and wide ranging land snail in North America, occurring in all lower 48 states and south to Panama in Central America; in the Rockies it has been collected at 10,000 feet; it no doubt occurs in every Kentucky county.

Specimen: Figs. (b) from Kentucky, Powell County, Furnace Mountain (author's collection). (fig. c) is *Zonitoides nitidus* from Massachusetts, Suffolk County, Great Brewster Island (author's collection).

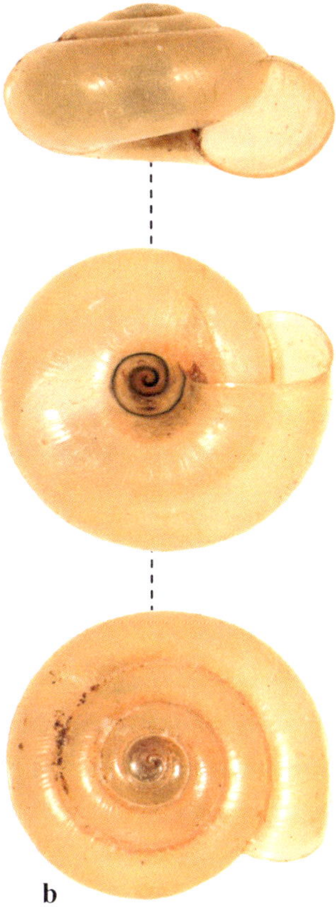

b

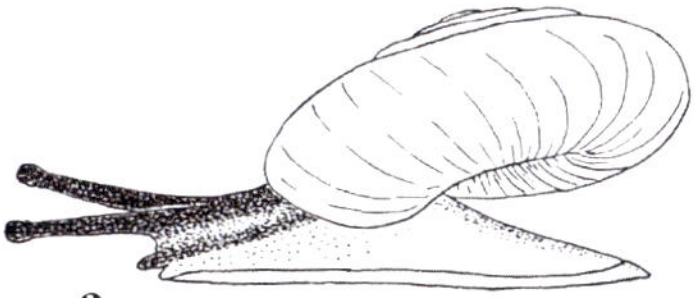

a

© Pilsbry, 1946

c

Zonitoides nitidus

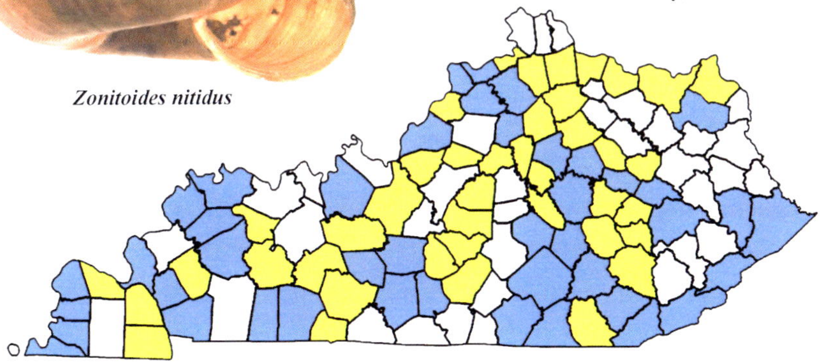

Striate Gloss Gastrodontidae

Zonitoides lateumbilicatus (Pilsbry, 1895)

Diameter: 4.2-5 mm, height 1.7-2.3 mm

Description: Depressed heliciform; lip simple, only slightly reflected; shell with 4.5 whorls; widely umbilicate; white to corneous, dull; transverse striae are well developed on the top and on the base; spiral striae present but a weak feature; without teeth in any stages of growth; shell periphery roundish.

Similar Species: *Zonitoides limatulus* has a more narrow umbilicus, is less compressed (this alone appears to be the best discriminating feature for separation) and the transverse striae are not as well-developed on its base (below illustrations from Pilsbry 1946).

Habitat: A calciphile species that is found in mixed hardwood forests on hillsides and in ravines and often close to outcropping limestone under leaf litter.

Status: G3/S4; Fairly Common; most records of this species of *Zonitoides* are from central Kentucky.

Specimen: Fig. (a) from Kentucky, Powell County, Furnace Mountain and (figs. b) from Kentucky, Bullitt County, Kochi Hill, Bernheim Arboretum and Research Forest (Image © Lori Schroeder).

a

b

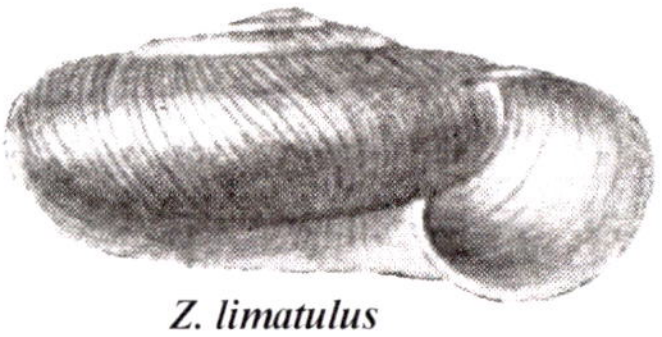

Z. limatulus

Pilsbry 1946

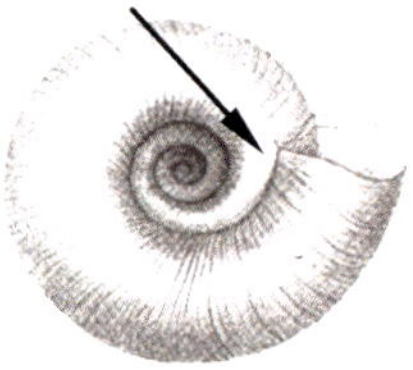

Z. lateumbilicatus

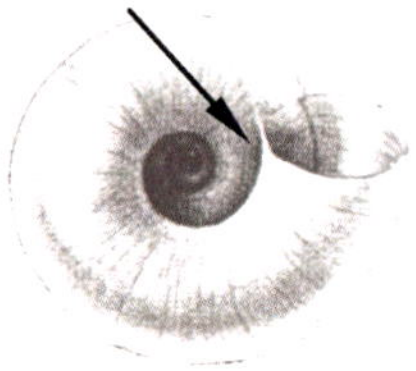

Z. limatulus

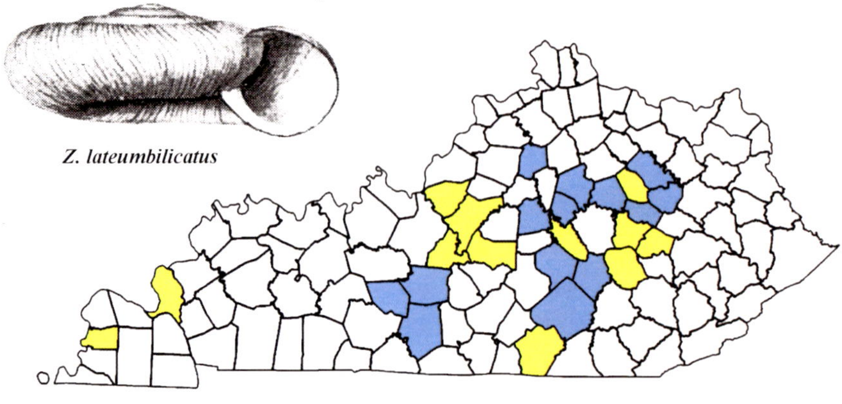

Z. lateumbilicatus

Dull Gloss

Gastrodontidae

Zonitoides limatulus (A. Binney, 1840)

Diameter: 4.3-5 mm, height 2-2.7 mm

Description: Depressed heliciform; lip simple not reflected; shell with 4.5 whorls; widely umbilicate; white to corneous, dull; transverse striae are well developed on the top (see below drawing) becoming somewhat obsolete on the base; spiral striae present but a weak feature; without teeth in all stages of growth; shell periphery round.

Similar Species: *Zonitoides lateumbilicatus* has a wider umbilicus, is more compressed (this alone appears to be the best discriminating feature for their separation) and the transverse striae are better developed on its base (see below drawings from Pilsbry 1946).

Habitat: A species of snail that prefers muddy ground of bottomland floodplains.

Status: G4/S3; Uncommon; known predominantly in central Kentucky, along the eastern knobs and several counties along the Ohio River.

Specimen: Kentucky, Edmonson County, Mammoth Cave National Park (FMNH 250897).

Pilsbry 1946

Z. limatulus

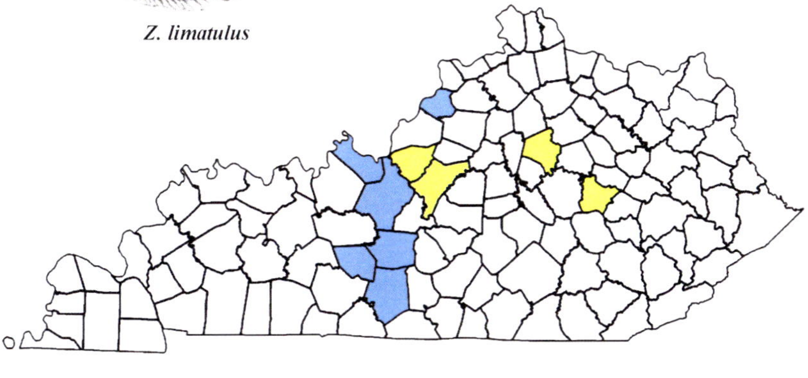

Brown Bellytooth

Gastrodontidae

Gastrodonta interna (Say, 1882)

Diameter: 6.5-7.4 mm, height 4.5-5.4 mm

Description: Heliciform; lip simple; shell with 8-9 tightly coiled whorls; perforate (fig. a), juvenile shells under 5 whorls have a proportionally larger umbilicus that gets smaller with age; cinnamon-brown; shell solid (thick) for its small size; transverse striae are distinct, rib-like and regularly spaced on top (fig. a) but much weakened on the base; two teeth are present at all stages of growth and can be seen through the bottom of live and fresh shells (fig. b).

Similar Species: *Gastrodonta fonticula* has a larger umbilicus, is smaller, and in general, has a flatter, darker shell.

Habitat: A colonial species found under leaf litter, very common under loose bark of decaying wood on hillsides and sandstone ridgetops of mixed hardwood forests.

Status: G5/S5; Common; mostly in central and eastern Kentucky.

Specimen: Kentucky, Powell County, Tunnel Ridge Road, Red River Gorge, Daniel Boone National Forest (author's collection).

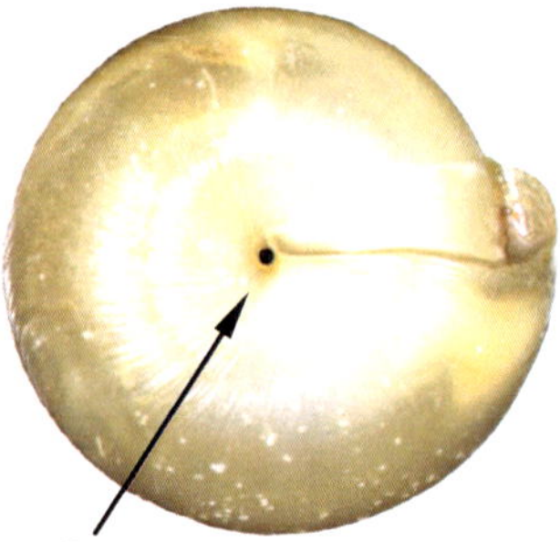

a

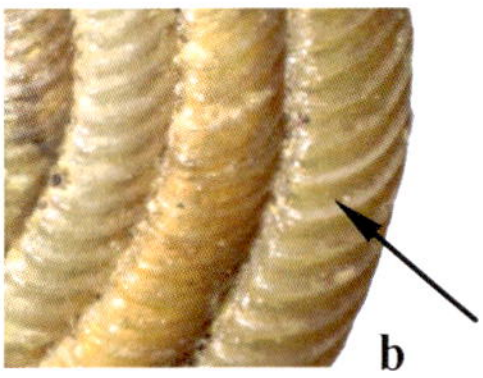

b

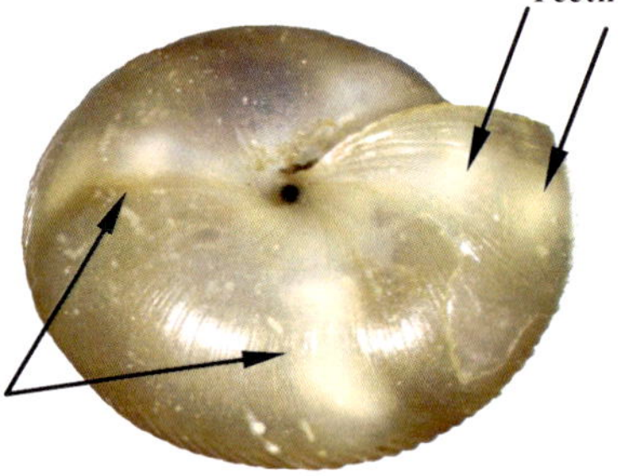

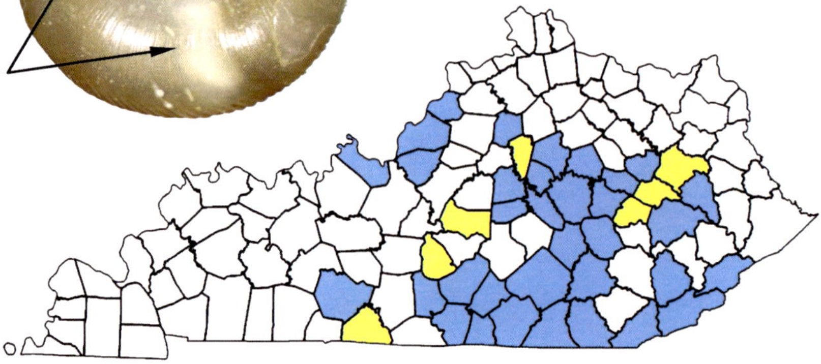

Appalachia Bellytooth

Gastrodonta fonticula (Wurtz, 1948)

Diameter: 5.7-6.9 mm, height 4.5 mm

Description: Heliciform to depressed heliciform; lip simple; shell with 7-8 tightly coiled whorls; cinnamon-brown; shell solid (thick) for its small size; umbilicate (wider than seen in *G. interna*) (fig. a); transverse striae are distinct, rib-like and regularly spaced on top but indistinct on the base; one or two pairs of teeth situated on a callous ridge which are present at all stages of growth, these teeth usually visible through the bottom of live and fresh dead shells, (fig. b).

Similar Species: *Gastrodonta interna* has the same shell form or with a slightly higher profile, this feature varying considerably, but most importantly has a smaller (pin-size) umbilicus.

Habitat: A species found under leaf litter and detritus close to rotting hardwood logs in advanced stages of decay in upland mixed hardwood locations.

Status: G3/S2; Rare; this primarily southern Appalachian species is reported from 4 counties in southeastern Kentucky.

Specimen: West Virginia, Mingo County, (author's collection).

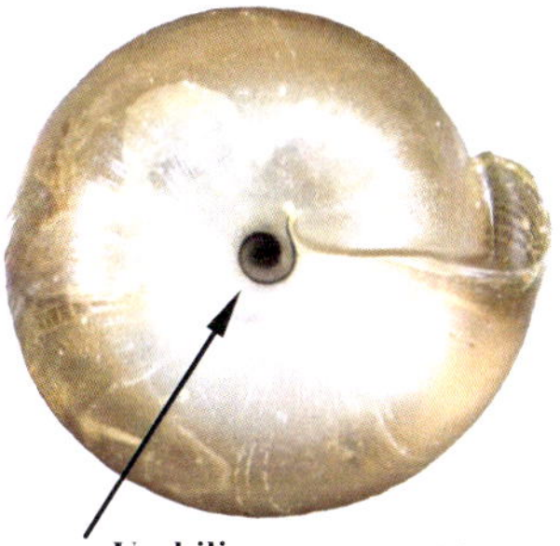

a Umbilicus more open

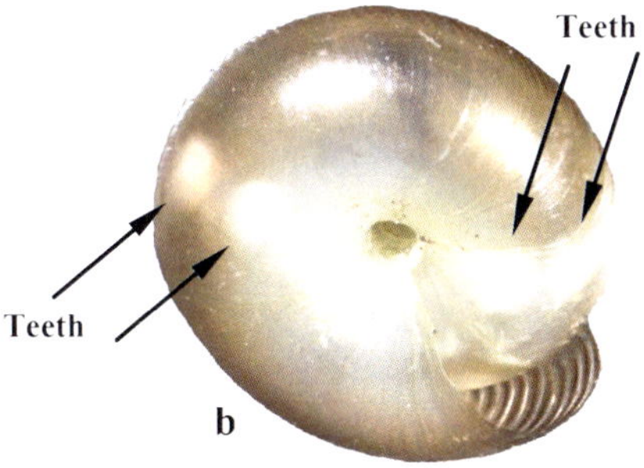

b

Shells 4-12 mm, without teeth, simple lip, loosely coiled whorls, unique indented transverse striae (*Glyphyalinia*)

Glyphyalinia species can be a troublesome group of land snails for the novice. Even for the experienced malacologist, they are challenging. Key features for this group of gastropods are the small size, the loosely-coiled whorls, the last whorl usually widely expanded (fig. a), the glossy and translucent shells and most importantly, the indented transverse striae (fig. c), which can vary from closely-spaced (fig. d) to widely spaced (fig. e). The somewhat unique indented or etched sculpture on the shell surface is the genera's signature and is easily viewed under a hand lens of 10X. Although *Paravitrea* species have indented sculpture as well, they are tightly coiled, the last whorl is not greatly expanded (fig. b) as in *Glyphyalinia* and they usually have internal armature at some stage of growth. *Glyphyalinia* species never contain internal teeth. Give special attention to any spiral sculpture under the scope. At times, this micro-feature is hard to see, depending on the light source and angle of the shell. It is best viewed on the top surface and will take some experience to key in on this fine feature. Slightly tilting the shell (while under the scope) will sometimes help reveal this micro-feature. Wetting the surface of the shell will also bring out these features, especially if the shells are bleached or badly weathered. *Glyphyalinia* snails here are arranged in three groups; **(1)** shells that are imperforate **(2)** shells that are perforate or rimate and **(3)** shells that are umbilicate.

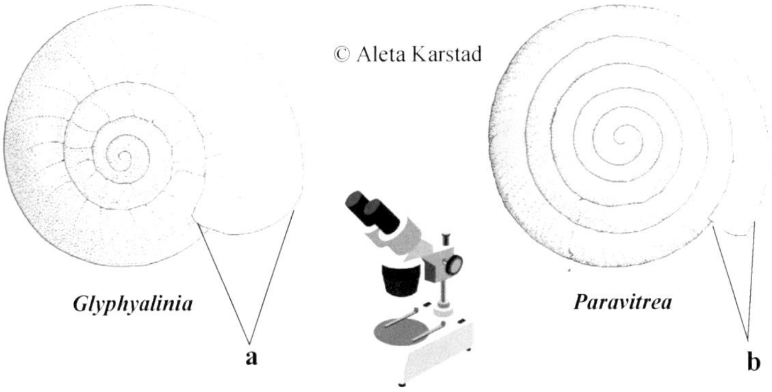

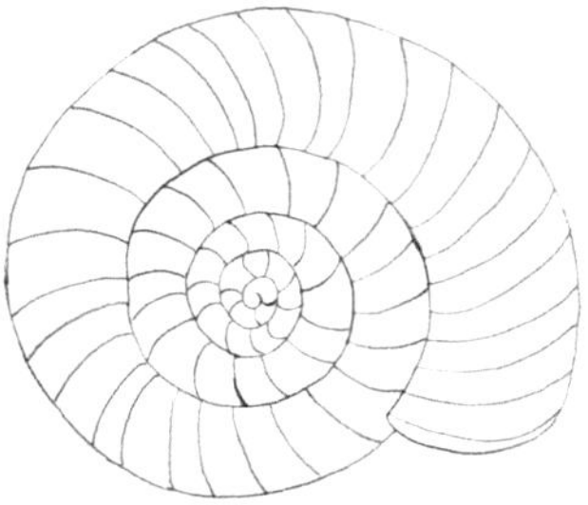

c

d

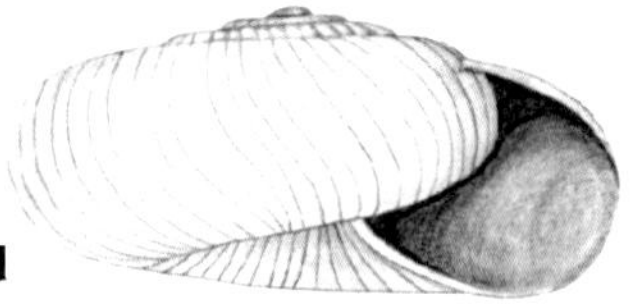

Tightly spaced traverse striae

e

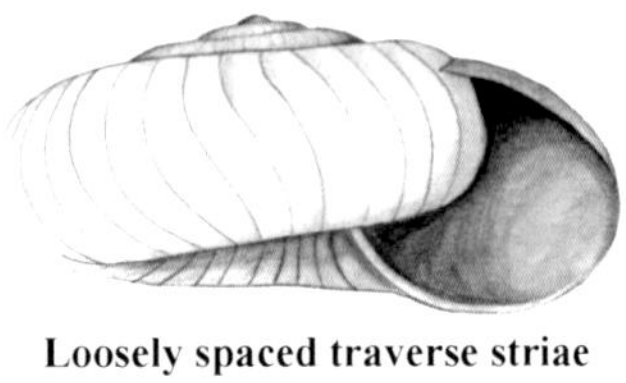

Loosely spaced traverse striae

***Glyphalinia rhoadsi* illustrating the characteristic transverse striae (indented) seen in all *Glyphyalinia* species**

Grouping 1: shells that are imperforate, pg. 139-140

**Glyphyalinia solida*
**Glyphyalinia cryptomphala*

Grouping 2: shells perforate or rimate, pg. 141-146

**Glyphyalinia indentata* (common)
**Glyphyalinia luticola* (rare)
**Glyphyalinia carolinensis* (uncommon)
**Glyphyalinia sculptilis* (uncommon)
**Glyphyalinia praecox*
**Glyphyalinia rimula*

Grouping 3: shells umbilicate, loosely spaced traverse striae, pg. 147-150

**Glyphyalinia rhoadsi*
**Glyphyalinia wheatleyi* (common)
**Glyphyalinia cumberlandiana*

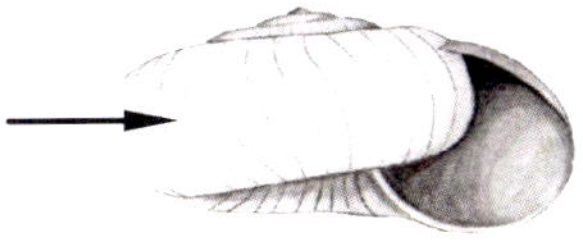

Grouping 4: shells umbilicate, tightly spaced traverse striae, pg. 151-154

**Glyphyalinia specus* (a cave species, rare)
**Glyphyalinia lewisiana* (uncommon)
**Glyphyalinia raderi* (rare)

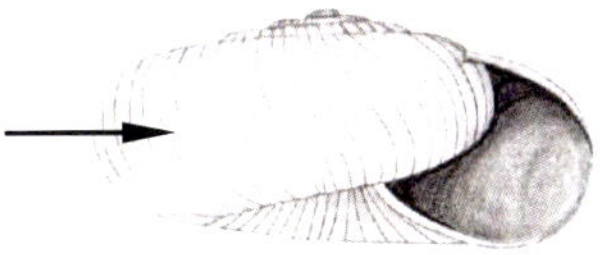

Imperforate Glyph **Gastrodontidae**

Glyphyalinia solida (H. B. Baker, 1930)

Diameter: 7.5 mm, height 4 mm

Description: Depressed heliciform; lip simple; shell with 5.25 loosely coiled whorls; shell fragile, color corneous to fulvous to almost chestnut, glossy and semi-transparent; imperforate, the umbilicus completely covered in all stages of growth (fig. a); no teeth present; indented transverse striae are well developed, closely and nearly equally spaced, continuing to the base; the spiral striae may be weakly defined but are a constant feature under a strong lens.

Similar Species: *Glyphyalinia solida* is typically larger, heavier shelled, and has a stronger sculpture than *G. cryptomphala*; other *Glyphyalinia* species are not imperforate, the umbilicus remaining at least partially open.

Habitat: A species that can be found under moist leaf litter in mixed upland hardwood forest and in talus near rock ledges.

Status: G5/S5; Common; a widespread species in central and eastern Kentucky.

Specimen: Tennessee, Marion County, Clear Spring Cave, 2 miles NE of Jasper (author's collection).

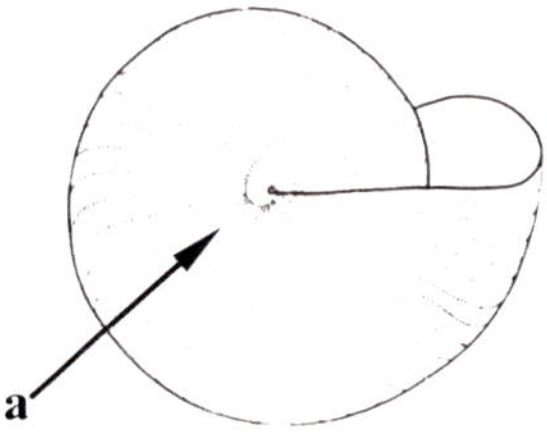

Pilsbry, 1946

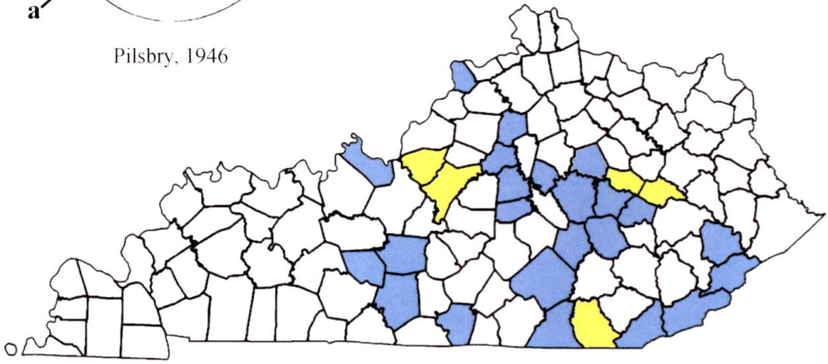

Thin Glyph

Gastrodontidae

Glyphyalinia cryptomphala

(G. H. Clapp, 1915)

Diameter: 5.1-5.9 mm, height 2-2.8 mm

Description: Depressed heliciform; lip simple; shell with 5-5.5 loosely coiled whorls; shell fragile, light horn to white, glossy and semi-transparent; imperforate, the umbilicus completely covered in all stages of growth (fig. f); no teeth present; indented transverse striae are well-developed, closely and nearly equally spaced and continue to the base; the spiral striae weakly defined but are a constant feature; deceased and dried animal can be seen through bottom view of shell.

Similar Species: *Glyphyalinia solida* is larger (by 2 to 3 mm) and has more conspicuous micro-surface features, but the two can be difficult to separate.

Habitat: This species is occasionally found in caves, but is more often found in a variety of mixed hardwood forests under moist leaf litter along river bluffs and in ravines.

Status: G5/S5; Common; a fairly widespread species in central and eastern Kentucky.

Specimen: Tennessee, Marion County, Clear Spring Cave, 2 miles NE of Jasper (author's collection).

f

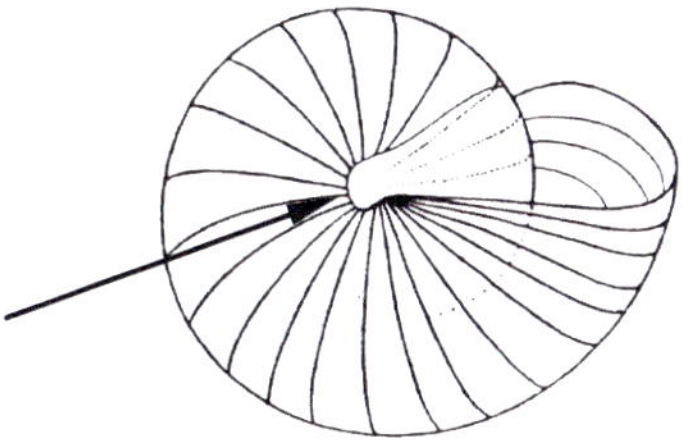

Pilsbry, 1946

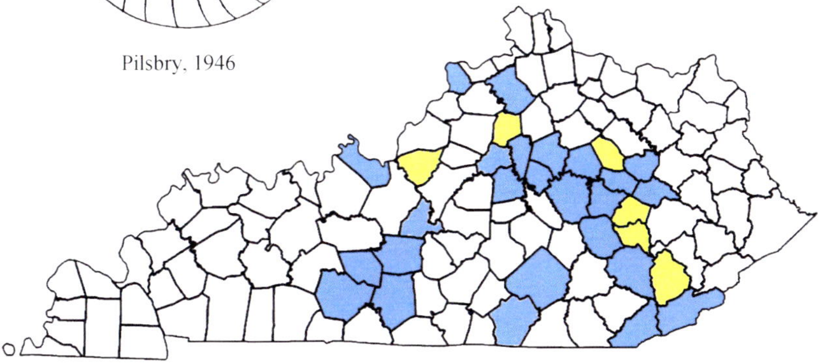

Carved Glyph

Glyphyalinia indentata (Say, 1823)

Diameter: 4.7-7.1 mm, height 3 mm

Description: Depressed heliciform; lip simple; shell with 4.5-5 loosely coiled whorls; perforate or rimate; shell fragile, pellucid, highly polished and translucent; indented transverse striae are well developed, rather widely but nearly equally spaced, about 28 on the last whorl; under a strong lens and good lighting, the spiral striae can usually be seen on some part of the last whorl, either rather weakly developed or only as traces (Pilsbry 1946); periphery well rounded; deceased and dried animal seen in all three views, making portions of the shell darker in places.

Similar Species: This species is most similar to *Glyphyalinia carolinensis* but is fully 4 to 5 mm smaller when collected from the same region; *G. rhoadsi* has a wider umbilicus and more strongly developed indented transverse striae.

Habitat: Found in a variety of mixed hardwood forests under leaf litter but also occasionally living along roadsides and in urban areas.

Status: G5/S5; Common; the most common and widespread *Glyphyalinia* in Kentucky.

Specimen: North Carolina, Macon County, Fatback Timber Sale, Nantahala National Forest (author's collection).

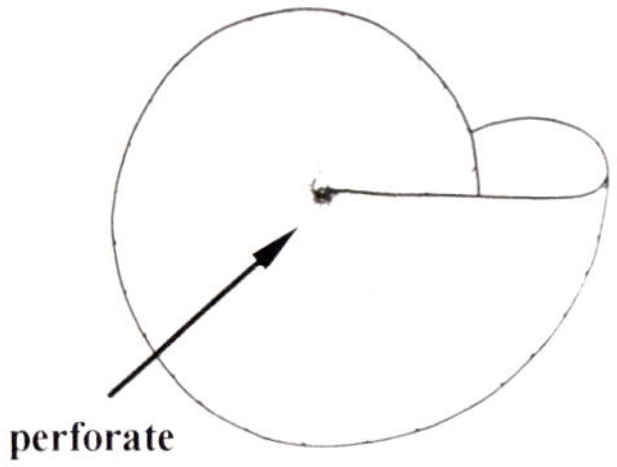

Pilsbry, 1946

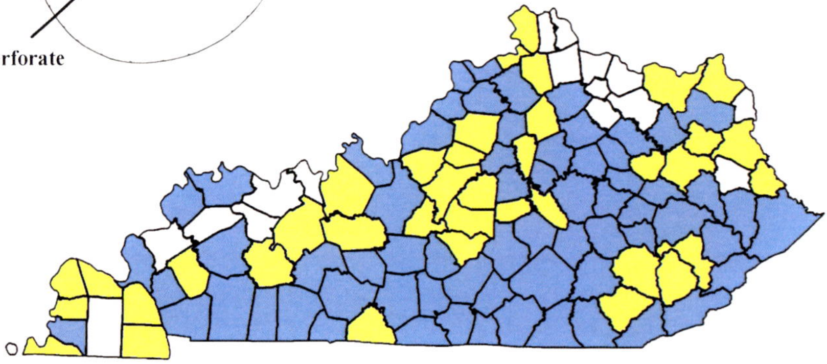

Furrowed Glyph

Gastrodontidae

Glyphyalinia luticola Hubricht, 1966

Diameter: 5.7 mm, height 2.6 mm

Description: Depressed heliciform; lip simple; shell with 4 to 4.5 loosely coiled whorls; shell fragile, coppery (when fresh), glossy, shiny, and translucent; perforate to rimate (fig. b); no teeth present; indented transverse striae are well -developed, widely but nearly equally spaced, about 20 on the last whorl; the spiral striae may be weakly defined but are a constant feature (scope required); live animal slate colored; deceased and dried animal can be seen through all three views of shell.

Similar Species: This species is most similar to *Glyphyalinia indentata* but is smaller, a coppery color, glossier and is usually found in wetter habitats (Hubricht 1966).

Habitat: Found in a variety of moist to wet habitats; found crawling on the muddy ground in wet weather in floodplain woods; also reported in waste ground in urban areas (Hubricht 1966).

Status: G4/S1; State Special Concern; currently known only from Wolfe County in a bog along the Middle Fork of the Red River and in Lyon County, Kentucky.

Specimen: Figs. (a) from Louisiana, Orleans County, New Orleans (FMNH 241570) and (fig. b) from Alabama, Baldwin County, swamp, Lillian (FMNH 147043 Holotype).

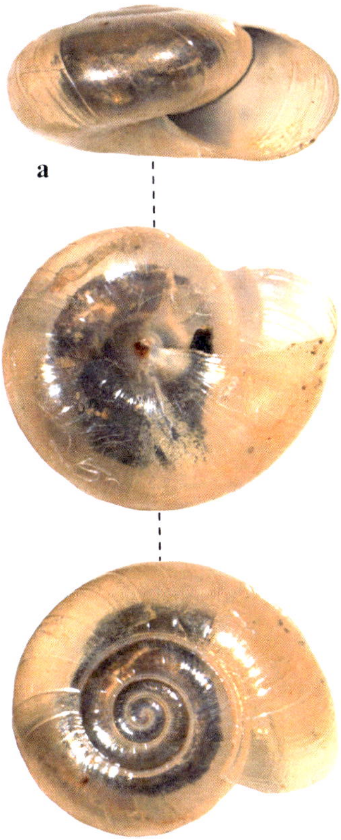

a

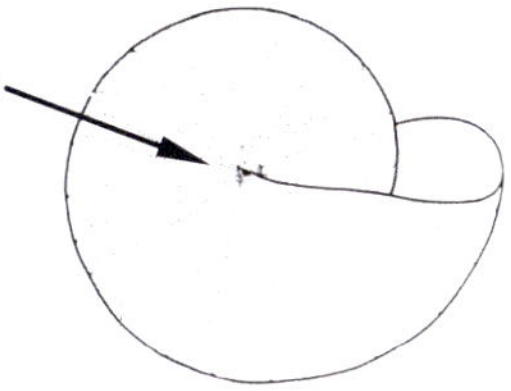

b

Pilsbry, 1946

b **Holotype**

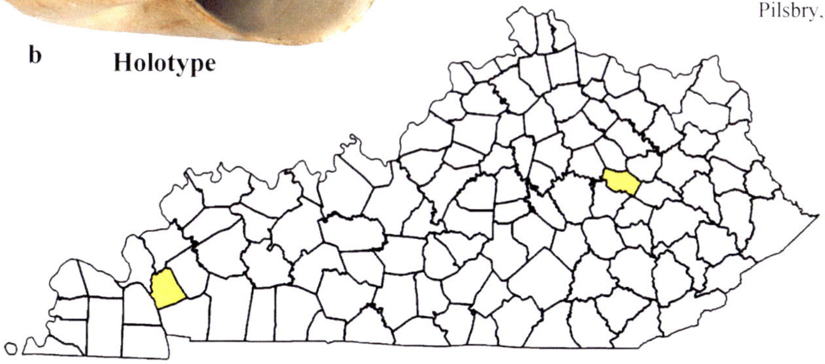

Spiral Mountain Glyph

Glyphyalinia carolinensis (Cockerell, 1890)

Diameter: 5-12 mm, height 3.4-5.5 mm

Description: Depressed heliciform; lip simple; shell with 4.5-5.5 loosely coiled whorls; perforate or rimate; shell fragile, pale-horn, shiny and semi-transparent; indented transverse striae are well developed, moderately and nearly equally spaced; a hand lens of 10X or scope will reveal numerous, closely spaced, clear-cut spiral engraved striae which, although weaker, typically continue onto the base.

Similar Species: This species is most similar to *Glyphyalinia indentata,* but is larger and has more conspicuous spiral striae; *G. praecox* is bronze-colored; *G. picea* is smaller and a species of higher elevations.

Habitat: Mixed hardwood forests under moist leaf litter along river bluffs and also a species of mountainsides.

Status: G4/S3; Uncommon; known only from southeastern Kentucky counties.

Specimen: Tennessee, Blount County, Chestnut Top Trail in the GSMNP (author's collection).

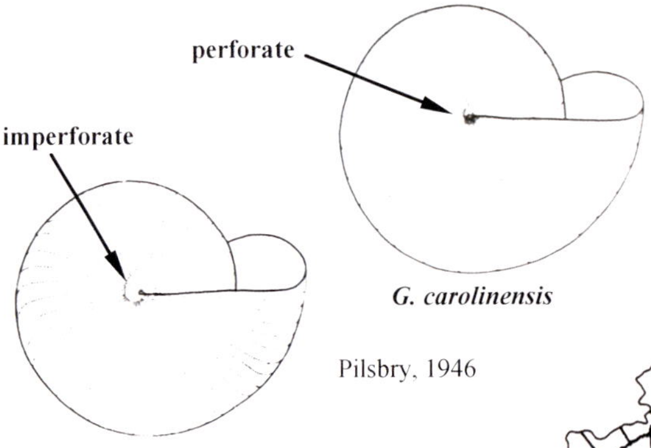

G. solida

Pilsbry, 1946

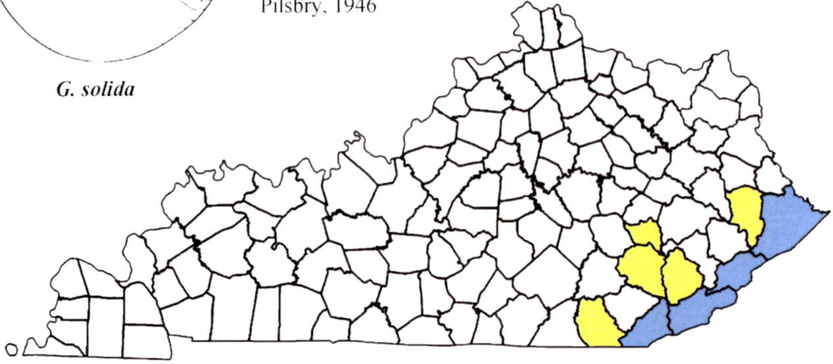

Suborb Glyph **Gastrodontidae**

Glyphyalinia sculptilis (Bland, 1858)

Diameter: 6.5-12.7 mm, height 5 mm

Description: Depressed heliciform; lip simple; shell with 7 loosely coiled whorls; shell slightly thicker than other *Glyphyalinia* species, pale horn-colored, glossy and transparent; perforate or rimate (fig. c), the minute umbilical opening is partly covered by an expansion of the columellar lip; no teeth present; indented transverse striae are well-developed, equally and closely spaced (fig. a) (about 0.5 mm apart at the periphery of the body whorl); on the middle whorl the spiral striae may be weakly defined but are a constant feature (scope required).

Similar Species: *Glyphyalinia indentata,* and *G. praecox* are smaller and the transverse striae are further apart and more irregularly spaced, (fig. b) below.

Habitat: A snail of upland mixed hardwood forests under and among moist leaf litter and detritus.

Status: G4/S3; Uncommon; reported from south-central counties in Kentucky.

Specimen: Tennessee, Grundy County, Crystal Cave, 2.5 miles north of Monteagle (author's collection).

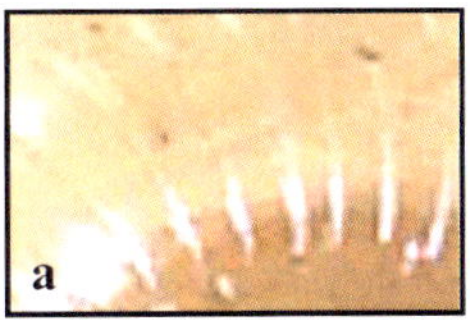
a

G. sculptilis

b

G. carolinensis

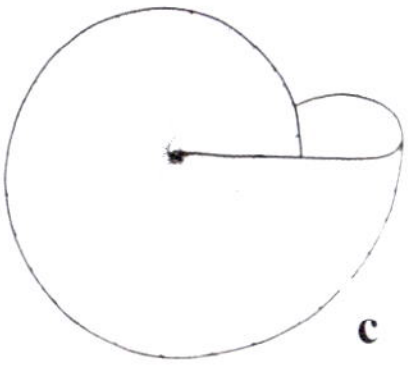
c

Pilsbry, 1946

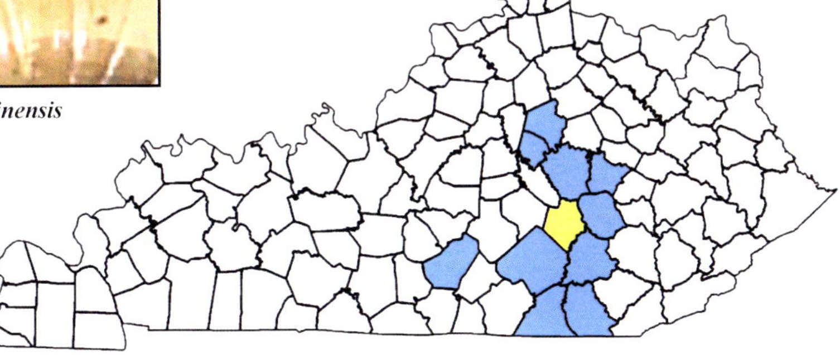

Brilliant Glyph

Gastrodontidae

Glyphyalinia praecox (H.B. Baker, 1930)

Diameter: 6.2-6.3 mm, height 2.8 mm

Description: Depressed heliciform; lip simple; shell with 4.5-5 loosely coiled whorls; perforate, the minute umbilical opening is partly covered by an expansion of the columellar lip; shell fragile, brilliant bronze-colored, glossy and transparent; indented transverse striae are well developed, moderately and nearly equally spaced and continue to the base; the spiral striae are bold and closely spaced.

Similar Species: *Glyphyalinia indentata* has a slightly higher profile, the apical whorls are less tightly coiled, the last whorl is more capacious and is corneous, not bronze colored.

Habitat: Found under moist leaf litter in floodplains and on talus slopes in mixed hardwood forests.

Status: G4/S3; Uncommon; mostly reported from central and southeastern portions of the state.

Specimen: Tennessee, Polk County, CNF (author's collection).

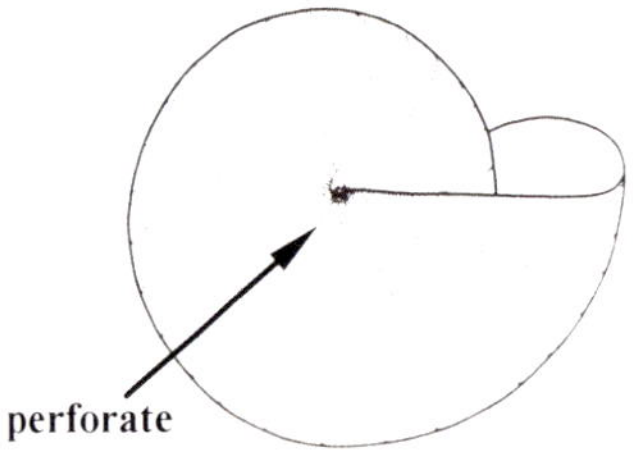

Pilsbry, 1946

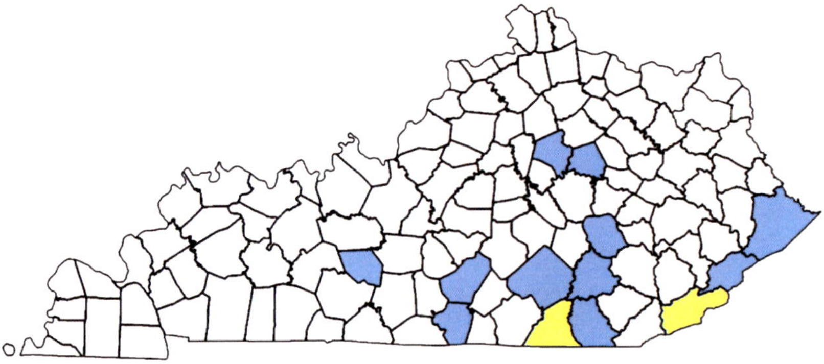

Tongued Glyph

Gastrodontidae

Glyphyalinia rimula Hubricht, 1968

Diameter: 7.7 mm, height 4 mm

Description: Depressed heliciform; lip simple; shell with 5 loosely coiled whorls; shell fragile, pale coppery (when fresh) fading with time, glossy and semi-transparent; rimate, a tongue-like callus at the columellar end which partially covers the umbilicus (fig. a); no teeth present; indented transverse striae are well developed and widely spaced; the spiral striae may be weakly defined but generally a constant feature under a strong lens; live animal pale-gray.

Similar Species: This species is most similar to *Glyphyalinia solida,* but the tongue-like callus does not completely cover the umbilicus (fig. a).

Habitat: Most often a snail of mixed wooded hillsides and ravines occurring under leaf litter; sometimes found in caves; also found in ever-increasing and unyielding kudzu tangles.

Status: G3/S3; Uncommon; most records are from southeastern Kentucky.

Specimen: Figs. (b) from Tennessee, Cannon County, in Tenpenny Cave (FMNH 156937 Holotype) and (fig. c) from Kentucky, Magoffin County, 1 mile ENE of Falcon (FMNH 241632).

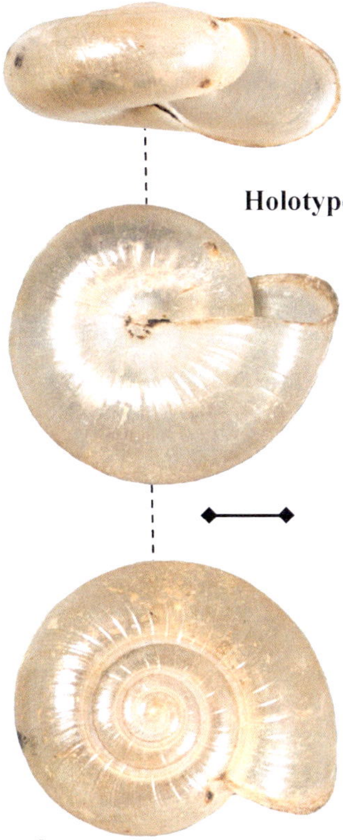

Holotype

b

c

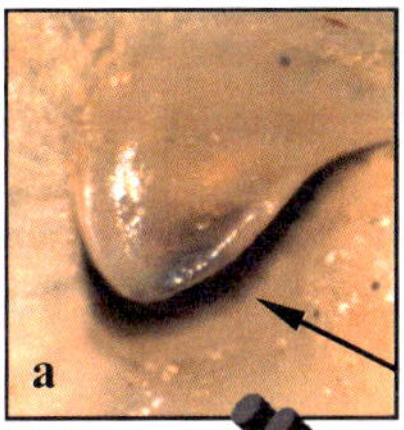

a

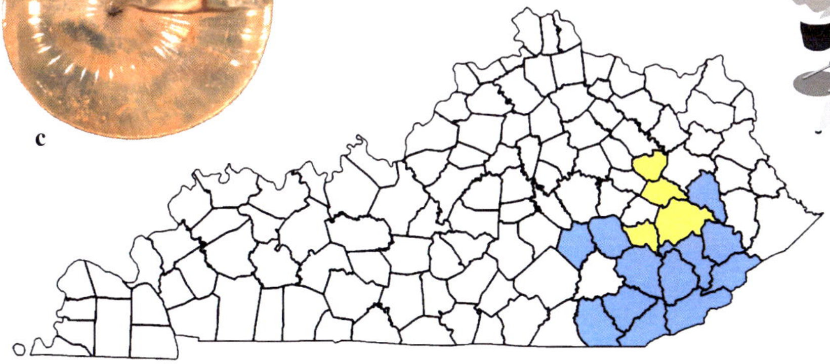

Sculpted Glyph

Gastrodontidae

Glyphyalinia rhoadsi (Pilsbry, 1899)

Diameter: 4.5-5.3 mm, height 2.5 mm

Description: Depressed heliciform; lip simple; shell with 4-5 loosely coiled whorls; umbilicate, umbilicus 1/12 the shell diameter or around 0.5 mm wide; shell fragile, corneous, glossy and semi-transparent; indented transverse striae are well developed and nearly equally spaced continuing to the base; the extremely weak spiral striae may be seen near the suture and umbilicus, but this feature is absent in some specimens (Pilsbry 1946).

Similar Species: This species is most similar to *Glyphyalinia indentata,* but is slightly smaller, has better developed transverse striae, and a notably wider umbilicus.

Habitat: Found in a variety of mixed upland hardwood forests under moist leaf litter.

Status: G5/S2; State Special Concern; all records in Kentucky are from three southeastern counties in the mountains.

Specimen: Virginia, Dickenson County, Garden Hole, rock talus above river (author's collection).

***G. rhoadsi*, Pike County, Kentucky**

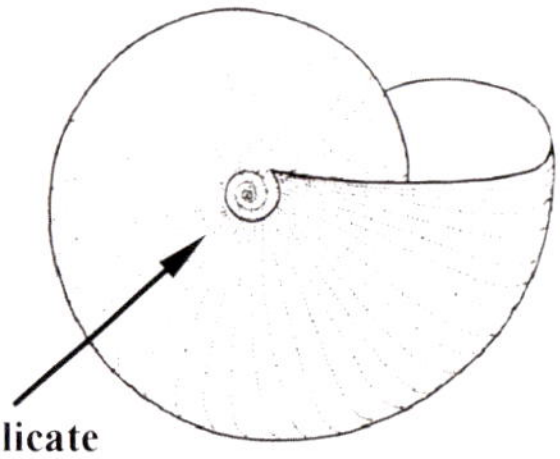

umbilicate

Pilsbry, 1946

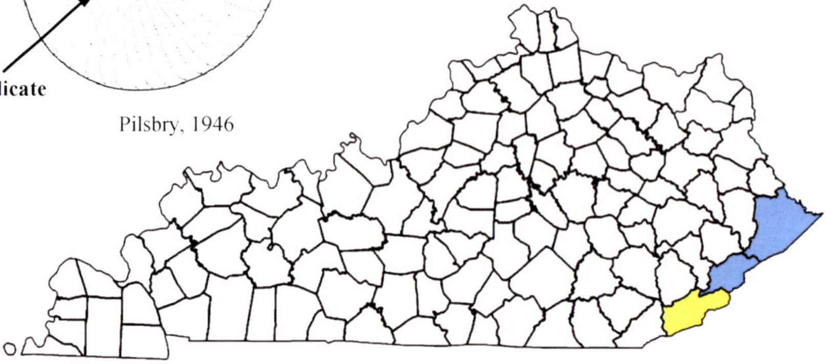

The sculpted Glyph, *Glyphyalinia rhoadsi,*
showing the extraordinary internal and external architecture of the shell

Bright Glyph

Gastrodontidae

Glyphyalinia wheatleyi (Bland, 1883)

Diameter: 5-6 mm, height 1.9 mm

Description: Depressed heliciform; lip simple; shell with 5-5.5 loosely coiled whorls; umbilicate (size varying); shell glossy and translucent; pellucid; brownish horn; indented transverse striae (radiating lines) are moderately developed and irregularly spaced; spiral striae may be absent or with only mere traces in some populations; deceased and dried animal can be seen through all three views of shell.

Similar Species: *Glyphyalinia cumberlandiana* is smaller, has a thinner, more fragile shell, and is a light horn-colored; *G. rhoadsi* has a notably smaller umbilicus, a different color, and more strongly developed transverse striae.

Habitat: Found in a wide range of habitats including moist leaf litter and deep detritus deposits located from valleys to mountaintops in mixed hardwood forests.

Status: G5/S5; Common; one of the most common *Glyphyalinia* in the state.

Specimen: Kentucky, Fayette County, Raven Run Nature Reserve (author's collection).

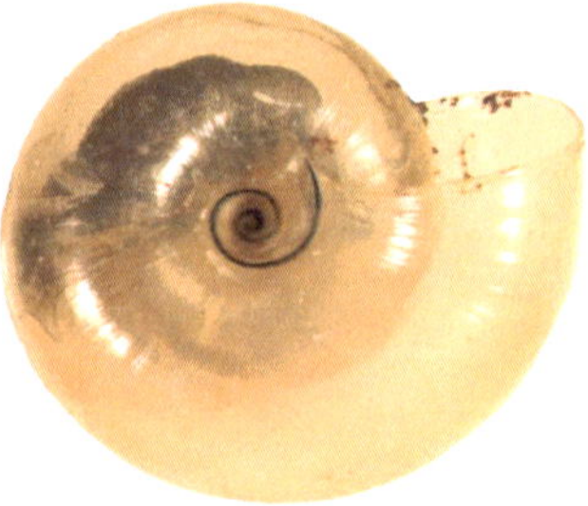

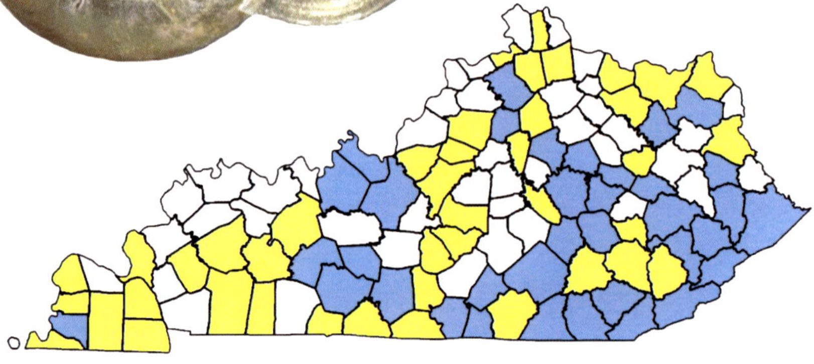

Hill Glyph

Glyphyalinia cumberlandiana

(G. H. Clapp, 1919)

Diameter: 2.5-3.5 mm, height 1.3 mm

Description: Depressed heliciform; lip simple; shell with 4-4.5 loosely coiled whorls; umbilicate; shell exceedingly fragile, glossy, light horn-colored and semi-transparent; indented transverse striae are faint and irregularly spaced; under high magnification there is merely the faintest trace of spiral sculpture (Pilsbry 1946); deceased and dried animal seen through the translucent top view of the shell, giving the first few whorls a darker color.

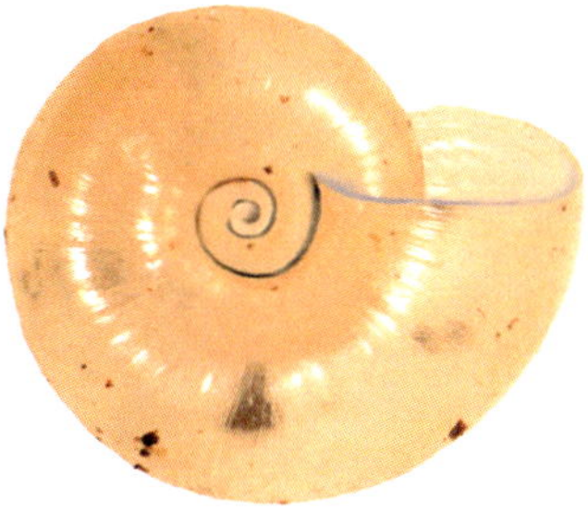

Similar Species: *Glyphyalinia wheatleyi* is similar in form, but is fully 2 mm larger.

Habitat: Reported as a calciphile of rocky limestone areas in shaded mixed hardwood forests, but also a snail of acidic habitats especially in and around rock talus.

Status: G4/S4; Fairly Common where it occurs; restricted to eastern Kentucky counties.

Specimen: Kentucky, Powell County, Furnace Mountain (author's collection).

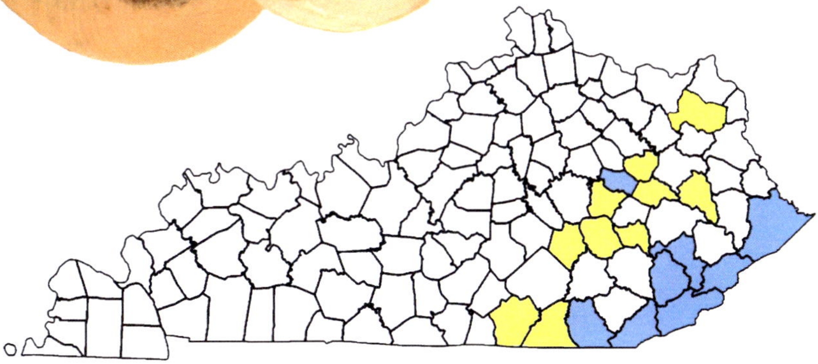

Hollow Glyph

Glyphyalinia specus Hubricht, 1965

Diameter: 4.8 mm, height 1.3 mm

Description: Depressed heliciform; lip simple; shell with 4.5 whorls, shell subhyaline, shiny; widely umbilicate; no teeth present; indented transverse striae are rather well developed and tightly spaced (fig. a); the spiral striae are only weakly defined, but often obsolete in many specimens or populations and in aged shells (dissecting scope required to view this fine feature); periphery is well rounded; the live animal is white and without functional eyes (Hubricht 1965).

Similar Species: *G. specus* is most closely related to *G. lewisiana*, differing principally in its larger size (Hubricht 1965).

Habitat: Troglobitic; known only from caves with microclimates located above flood zones where the its primary food source (cricket and bat guano) exists; also restricted to a limited number of caves and more importantly, very narrow environments within these caves.

Status: G4/S2; State Threatened; Rare

Specimen: Figs. (b) from Tennessee, White County, Ghost River Cave (author's collection) and (fig. c) from Kentucky, Barren County, Beckton Cave (FMNH 135315 Holotype).

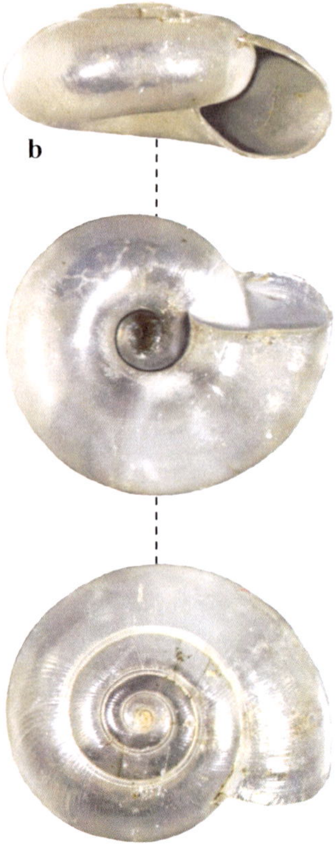

b

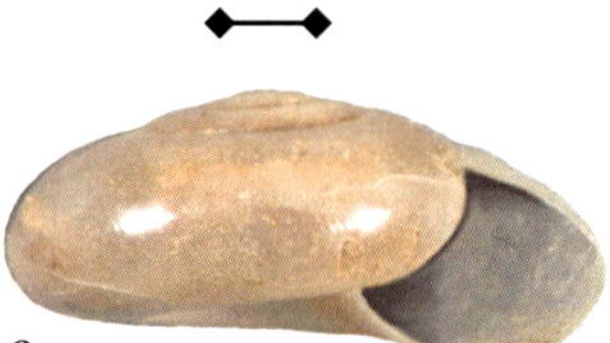

c

Holotype

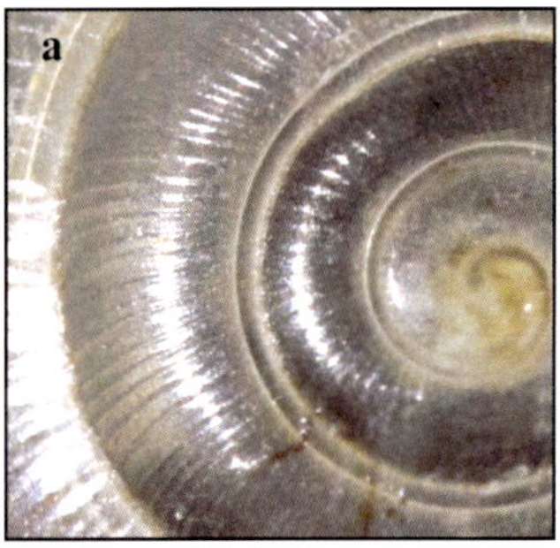

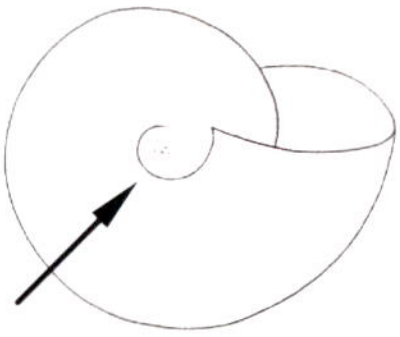

Pilsbry 1946

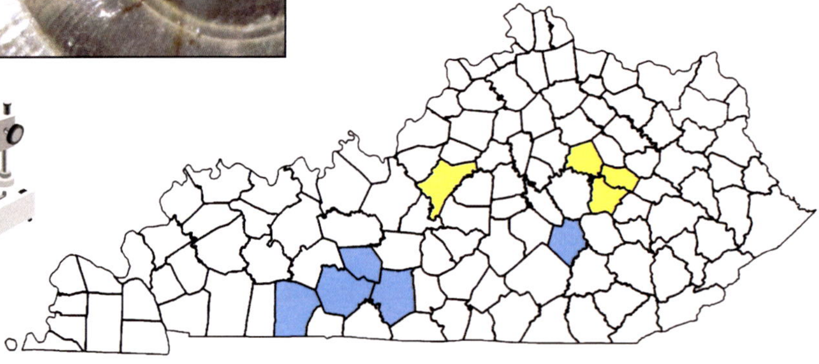

Above, *Glyphyalinia specus,* Hollow Glyph, in a cave on Furnace Mountain, Powell County, Kentucky. Caves are unique habitats unlike any found elsewhere on Earth. They contain whole ecosystems where species live, breed and die without ever seeing the light of day. These creatures who spend their entire lifetime in total darkness are called troglobitic organisms. They are usually without pigment and functional eyes. Troglobites are some of the rarest and highly specialized animals found anywhere. Not only are they restricted to the dark zones of caves **but are also restricted to exceedingly narrow microclimates within those same cave**. They are also entirely reliant on external food sources (guano) from bats and cave crickets.

For this reason, cave snails inhabit some of the smallest home ranges of any animal found in Kentucky. Because these gastropods crawl a thin line between life and death, entirely dependent on a narrow range of conditions not found in all caves, any disruptions such food shortages can affect their survival. For example, as eastern bats vanish as a result of white-nose syndrome, cave snails and other troglobitic organisms lose a food source, putting them at risk. There are currently 3 species of troglobitic land snails living in Kentucky caves and while several species are more wide ranging such as the Hollow Glyph, all cave species should be considered "Threatened" and, therefore, always taken into account during anthropogenic actions involving karst landscapes.

Pale Glyph

Gastrodontidae

Glyphyalinia lewisiana (G. H. Clapp, 1908)

Diameter: 2.8-3.5 mm, height 1.5 mm

Description: Depressed heliciform; lip simple; shell with 3.5-4 loosely coiled whorls, the last whorl greatly expanded; shell very glossy, a yellowish-white; widely umbilicate, about one-fifth to one-sixth the shell diameter; no teeth present; the delicate indented transverse striae are rather well developed and tightly spaced (fig. a) on top but mostly absent from the base; spiral striae may be weakly defined, but usually a constant feature under a strong lens; periphery is well rounded.

Similar Species: *Glyphyalinia specus* is larger by around 1.5 millimeters and is reported to be a strict obligate to limestone caves.

Habitat: A calciphile species that burrows; usually found at the base of limestone outcrops and under rocks that provide the snail ample space to crawl; occasionally carried into limestone caves during storm events.

Status: G4/S2; Rare to Uncommon; there are few records for this species in Kentucky and live animals and shells are typically scarce where the species is found.

Specimen: Kentucky, Edmonson County, Mammoth Cave NP (FMNH 240450).

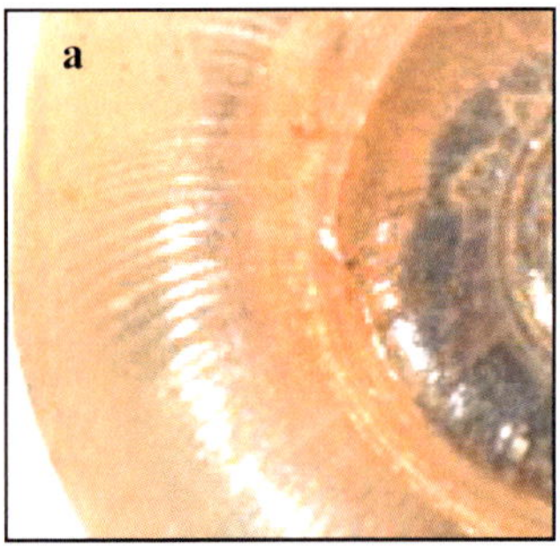

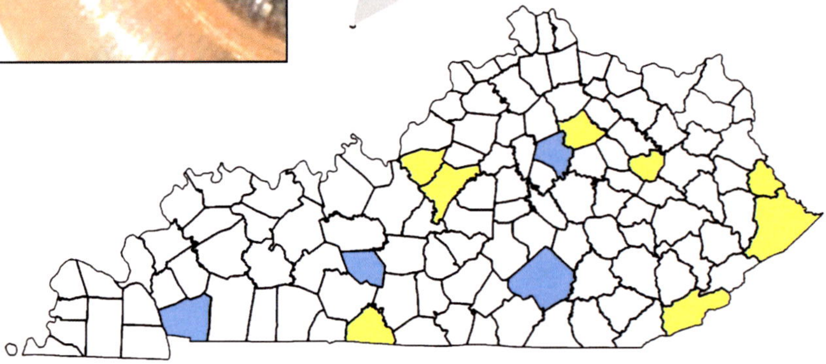

Maryland Glyph

Glyphyalinia raderi (Dall, 1898)

Diameter: 3.9-4 mm, height 1.5 mm

Description: Depressed heliciform; lip simple; shell with 4.5 loosely coiled whorls; shell pale waxen white; widely umbilicate (one of the widest of any Kentucky *Glyphyalinia*), about one-fourth the shell diameter; aperture turning downward but this feature varies among specimens; no teeth present; indented transverse striae are weakly developed, regularly and tightly spaced (fig. a); periphery rounded.

Similar Species: The aperture of *G. lewisiana* does not curve downward as it does in *G. raderi* (in frontal view), has a notably smaller umbilicus and more strongly developed transverse striae; *G. wheatleyi* is larger and has more widely spaced indented striae.

Habitat: A calciphile and possibly a burrower, usually amongst rocks and talus (Hubricht 1985).

Status: G2/S1; State Endangered; Rare; in Kentucky, known only from Carter County. Even where it occurs, few shells are ever found. Other very limited populations are reported from Pennsylvania, Virginia & Maryland.

Specimen: Virginia, Alleghany County, nine miles NNE of Covington (FMNH 240955).

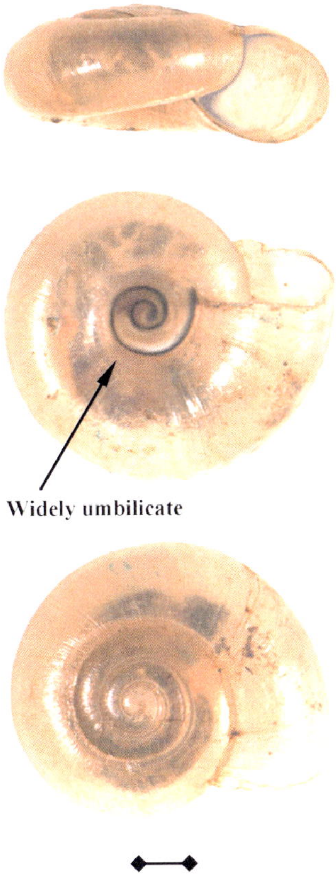

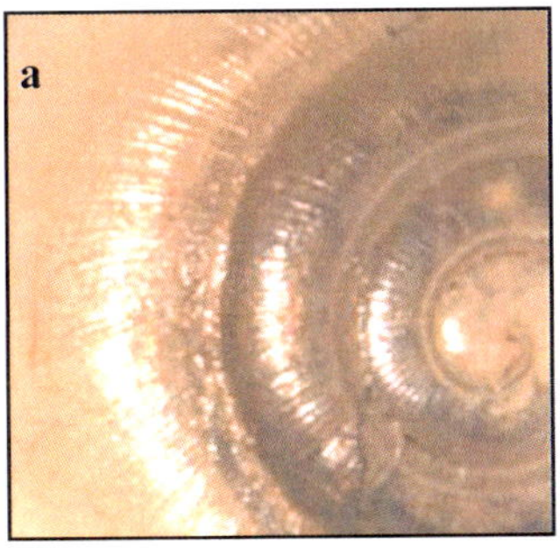

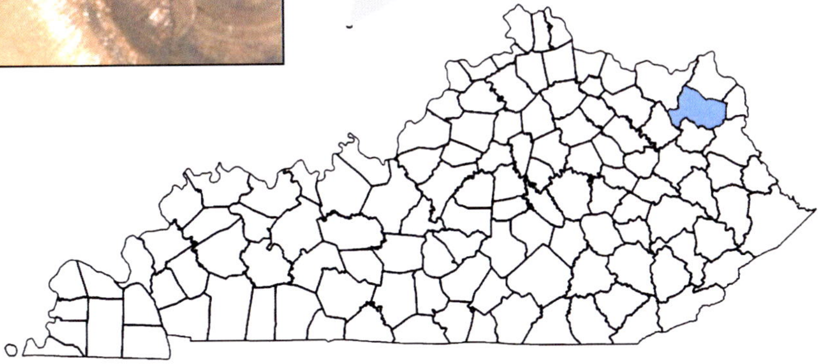

Shells 2-8 mm, teeth typically present, simple lip, the whorls are tightly coiled (*Paravitrea*), (Figs. a-i; opposite page)

Paravitrea species are undoubtedly one of the most difficult groups of land snails to differentiate. Nevertheless, most are under 6 mm (a few slightly larger), have more than 5 or 6 tightly coiled whorls, the last whorl not greatly expanded (fig. a) and most species at some stage of development possess internal armature; *Glyphyalinia* species have loosely coiled whorls, the last whorl greatly expanded (fig. b) and are without teeth. *Paravitrea* have either teeth (c and d) or lamellae (fig. e). Without a good series of *Paravitrea* shells at various stages of growth, taxon recognition is difficult. Immature specimens are essential for the identification of species in the genus. If adult *Paravitrea* species are encountered during collections, it is imperative that leaf litter samples (always a good idea anyway) also be secured at the same locations. This will likely insure that young *Paravitrea* specimens are included during the survey. Many *Paravitrea* snails contain teeth at the juvenile shell-stage that are absorbed (for the calcium content) in the adult shell-stage (see next page, figs. f-i). *Paravitrea* shells in this book are arranged in three groups: **(1)** juvenile and adult shells without teeth or lamella barriers **(2)** juvenile shells with teeth or lamella barriers but as adults are usually without these protective structures **(3)** both juvenile and adult shells contain teeth or lamella barriers.

Grouping 1: juvenile & adult shells without teeth or lamella barriers, pg. 157-159

Paravitrea capsella (common)
Paravitrea blarina (rare)
Paravitrea petrophila

Grouping 2: juvenile and adult shells with teeth or lamella, pg. 160-168

Paravitrea lapilla (rare)
Paravitrea tantilla (rare)
Paravitrea subtilis (uncommon)
Paravitrea mira (rare)
Paravitrea reesei (rare)
Paravitrea septadens (rare)
Paravitrea lamellidens (rare)
Paravitrea multidentata (most common *Paravitrea* in Kentucky)

Grouping 3: juveniles with teeth or lamella barriers, adults without, pg. 169-171

Paravitrea placentula (uncommon)
Paravitrea dentilla (rare)
Pilsbryna aff. vanattai (not a *Paravitrea* species but similar looking, very rare)

Shell Morphology & Armature of *Paravitrea* Species

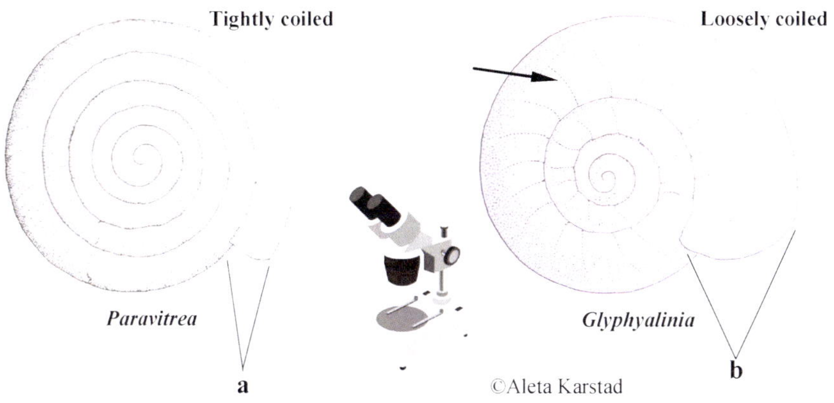

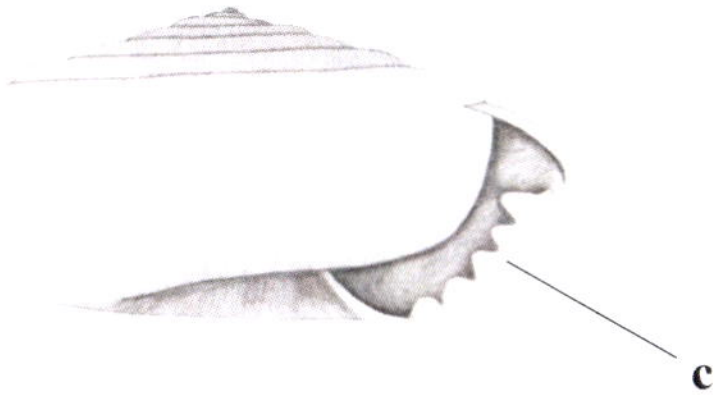

Figs. f-i are examples of *Paravitrea placentula*; the first three juvenile shells (figs. f-h) containing teeth. As the shells mature, teeth decrease in size, but in the last shell, an adult (fig. i), the teeth are completely reabsorbed for the calcium content.

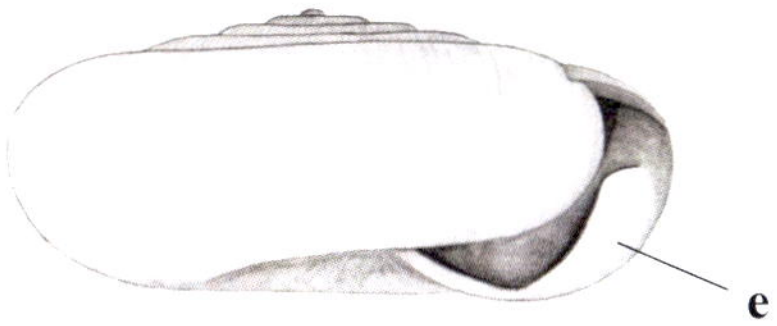

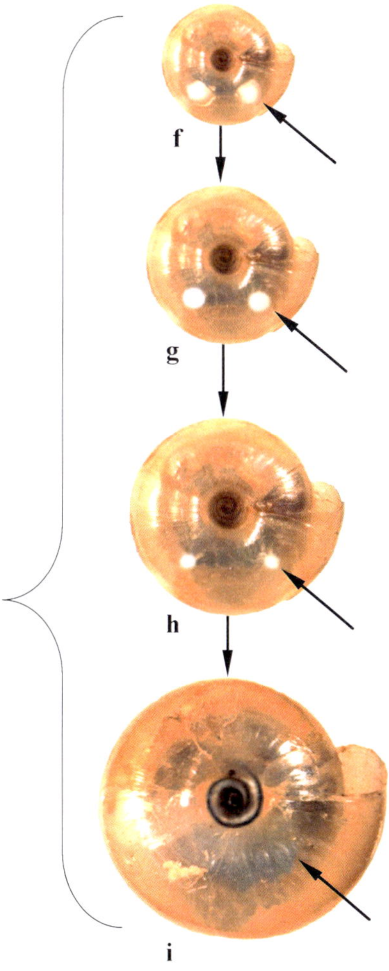

Dimple Supercoil

Paravitrea capsella (Gould, 1851)

Common name:

Diameter: 4.8-6.2 mm, height 3 mm

Description: Depressed heliciform; lip simple, roundish; shell with 6.5-7.5 tightly coiled whorls; umbilicate; aperture roundish; amber color; live animal pale-slate colored; glossy and translucent; transverse striae moderately developed on top of shell, but nearly smooth on the base; without teeth in adults; Eastern Kentucky specimens with less than 3 whorls are often seen with paired teeth (2 or 3 sets); deceased and dried animal seen through the translucent bottom and top views of the shell giving it a darker color in places (fig. a).

Similar Species: *Paravitrea placentula* is larger and has paired teeth in young shells; *P. blarina* is smaller by 2-3 mm and has a wider umbilicus (see comparison figs. b & c below).

Habitat: A habitat generalist found in a wide range of mixed hardwood forests.

Status: G4/S4; Common; one of the most frequent and widespread *Paravitrea* species in the state.

Specimen: North Carolina, Graham County, Nantahala NF (author's collection).

a

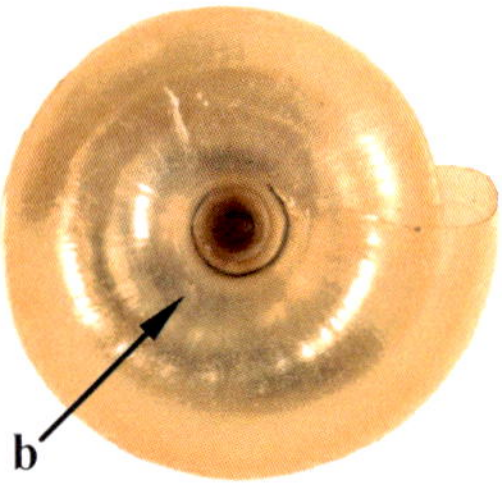

b

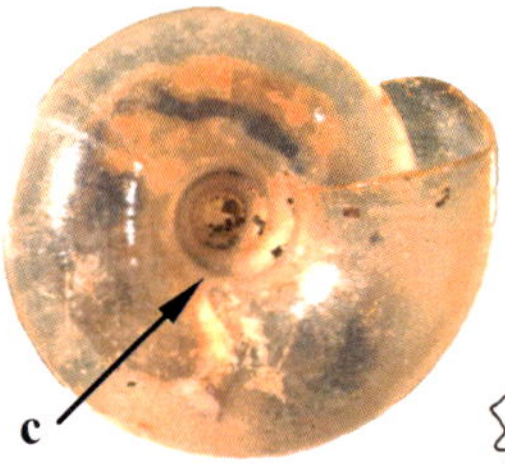

c

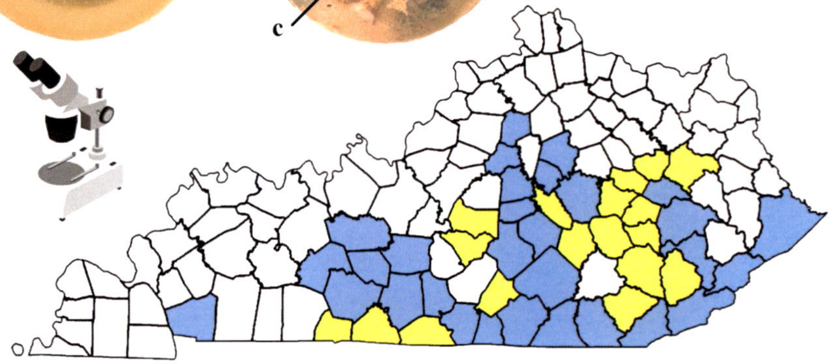

Shrew Supercoil

Pristilomatidae

Paravitrea blarina Hubricht, 1963

Common name: shrew supercoil

Diameter: 3.9 mm, height 2.2 mm

Description: Depressed heliciform; lip simple; shell with 5.5 tightly coiled whorls; shell glossy, thin and translucent in fresh and live shells; umbilicate; nuclear whorl smooth, later whorls sculptured with numerous irregularly spaced, impressed growth lines, distinct above but becoming obsolete below; there are no teeth at any stage of growth; live animal white; deceased and dried animal can be seen though bottom view of shell.

Similar Species: *Paravitrea blarina* differs from all other small (under 4 mm) *Paravitrea* species in the absence of teeth in the shell at all stages of growth (Hubricht 1985); differs from *P. capsella* in the notably wider umbilicus.

Habitat: A secretive species, found in the lower layers of leaf mold; at one location, it was found crawling about in shrew burrows under the leaves (Hubricht 1985).

Status: G3/S2; State Endangered; Rare, very little is known of this species in Kentucky, which is reported from Bell County only.

Specimen: Kentucky, Bell County, Pine Mountain Gap, .5 miles south of Pineville (FMNH 248873).

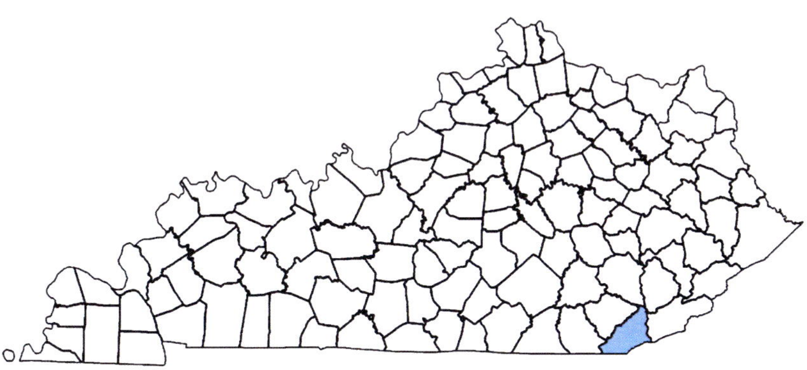

Cherokee Supercoil

Paravitrea petrophila (Bland, 1883)

Diameter: 5.7-6 mm, height 2.5-3 mm

Description: Depressed heliciform; lip simple; shell with 5.5-6 loosely coiled whorls, the last whorl notably expanded, more than is seen in other *Paravitrea* species; fresh shells are almost ivory white, live animal whitish (Pilsbry 1946); shell shiny, thin and translucent; widely umbilicate (fig. a); transverse striae (growth lines) are widely and irregularly spaced (scope required); no teeth in any stage of growth.

Similar Species: This species is often confused with *Glyphyalinia* snails but primarily differs in its tighter coils; differs from *P. blarina* mainly in its larger size (being 2-3 mm larger) and its slightly wider umbilicus; adult *G. cumberlandiana* is smaller by around 3 mm and has a thinner more fragile shell; *G. specus* is similar in size and form but is found in caves only; *P. capsella* has a smaller umbilicus and is more compact in form.

Habitat: A habitat generalist found in a wide range of mixed hardwood forests under leaf litter and detritus.

Status: G4/S3; Uncommon; restricted to 10 central and eastern Kentucky counties but likely more widespread than current records indicate.

Specimen: Tennessee, Blount County, hills above the town of Townsend (author's collection).

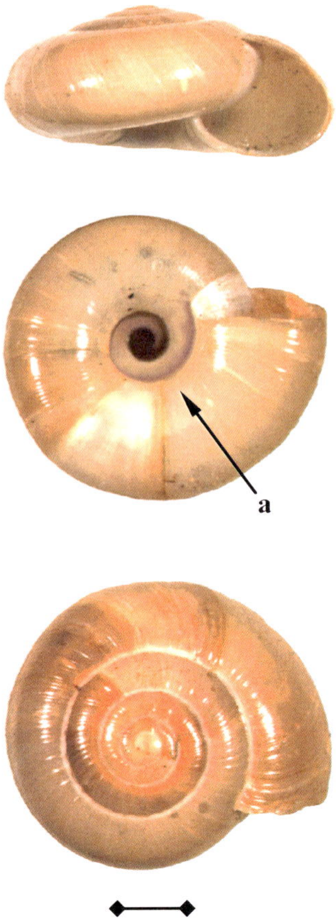

Paravitrea blarina

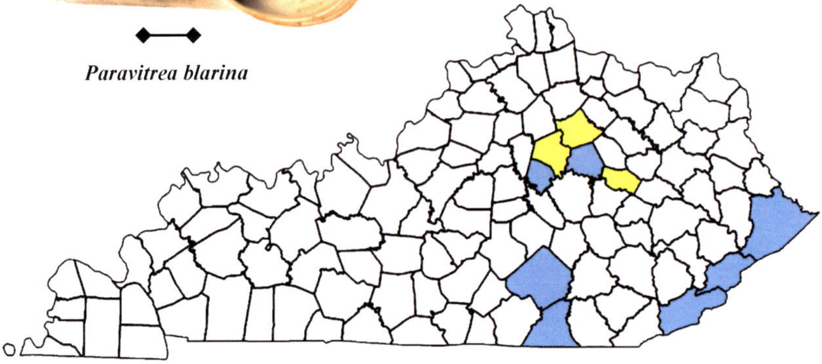

Gem Supercoil

Pristilomatidae

Paravitrea lapilla Hubricht, 1965

Diameter: 4.8 mm, height 2.2 mm

Description: Depressed heliciform; lip simple; shell with 7 whorls, end of last whorl expanding slightly; pale-buff; shell thin and translucent in fresh and live shells; umbilicate, umbilicus deep and well-like; live animal white; transverse striae are irregularly spaced, distinct above but becoming obsolete on the base; there are two or three pairs of closely set teeth , sometimes fused together, forming parallel lamellae within the last whorl usually at all stages of growth (see paratypes below).

Similar Species: *Paravitrea tantilla* is smaller, but the internal teeth are the same as in *P. lapilla*; *P. capsella* has a higher shell profile and no teeth.

Habitat: Acidic coves, rich woods under moist leaf litter.

Status: G3/S2; Rare to Uncommon; during 2019 and 2020 surveys, *P. lapilla* was discovered in 8 additional counties along the border with Tennessee.

Specimen: Figs. (a) from Tennessee, Davidson County, Todd Knob (FMNH 135313 Holotype) and Fig (b) from Kentucky, Edmonson County, Mammoth Cave National Park, Green River, Dennison Ferry (FMNH 249572 Paratypes).

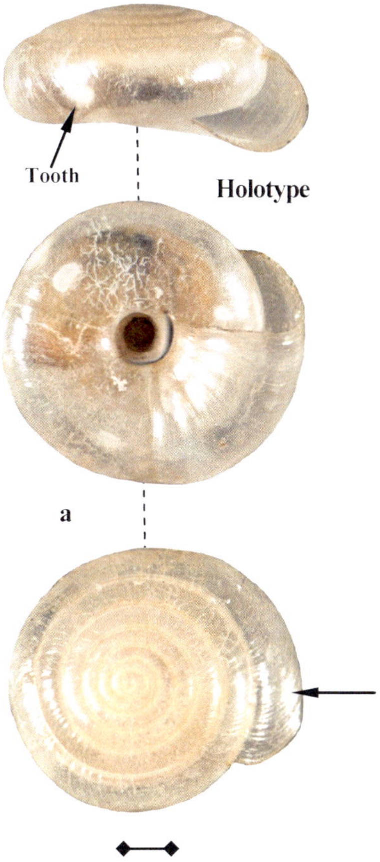

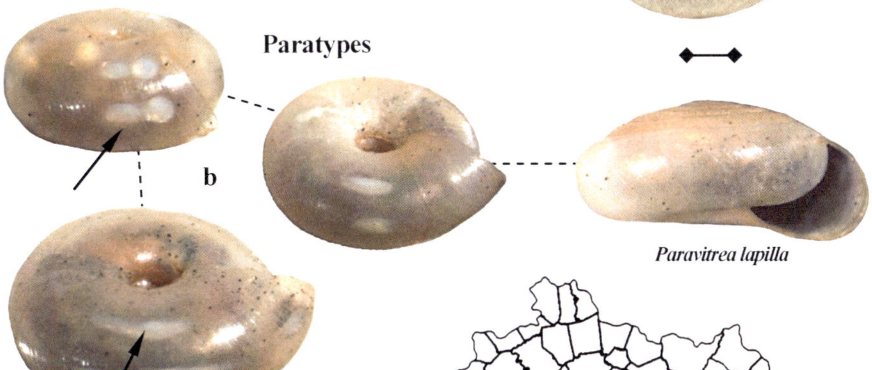

Paravitrea lapilla

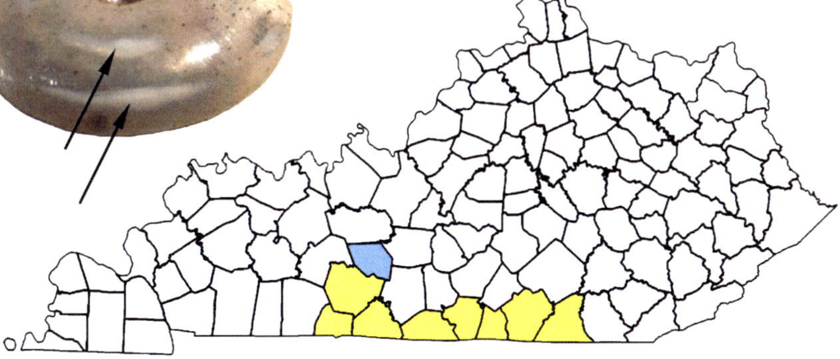

Teasing Supercoil

Paravitrea tantilla (Hubricht, 1963)

Diameter: 3.7 mm, height 2.0 mm

Description: Depressed heliciform; lip simple; shell with 6-6.5 loosely coiled whorls; shell thin and translucent in fresh and live shells; pale yellowish, subhyaline, shining; umbilicate; transverse striae are irregularly spaced; one or more pairs of rather large radial teeth that are well-spaced (fig. a) and can be seen through the bottom of both fresh juvenile and adult shells.

Similar Species: *Paravitrea placentula* is fully 3 mm larger, has paired teeth that are arranged notably closer together (fig. b) and adult *P. placentula* shells are without teeth.

Habitat: Found under moist leaf litter and detritus on hillsides in mixed hardwood forests.

Status: G3/S1; State Endangered; this land snail is reported from only one Kentucky county.

Specimen: Fig. (c) from Tennessee, Warren County, Caney Fork River (FMNH 248865) and Fig. d from Tennessee, Grundy County (FMNH 116936 Holotype).

c

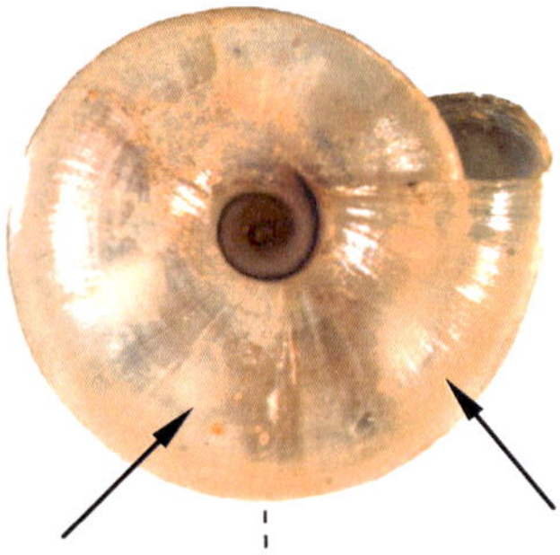

d **Holotype**

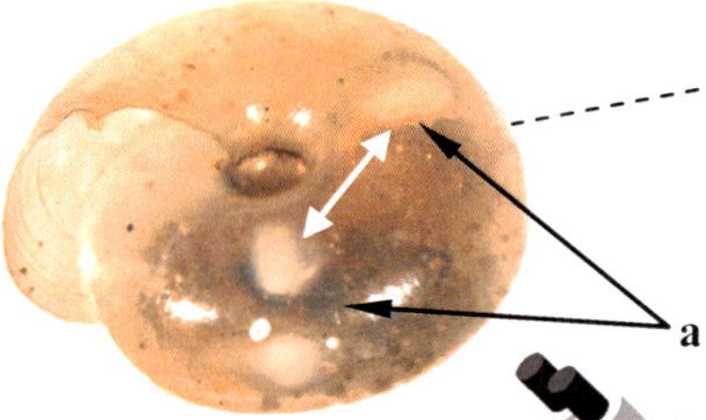

a

Paravitrea tantilla

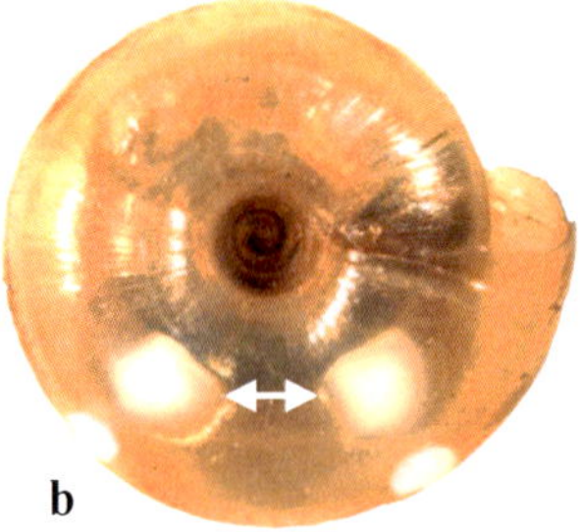

b

Juvenal *Paravitrea placentula*

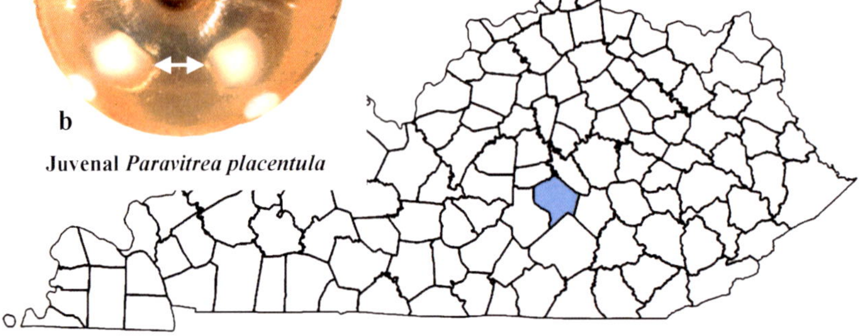

Slender Supercoil

Paravitrea subtilis Hubricht, 1978

Diameter: 2.9 mm, height 1.4 mm

Description: Depressed heliciform; lip simple; shell with 6.3 tightly coiled whorls; pale reddish-brown; glossy; umbilicate; aperture rounded; indented transverse striae are closely but somewhat irregularly spaced, becoming more widely spaced and irregular on the last whorl; 1 to 2 radial rows of 4 to 6 teeth in the last whorl, coalesced in some specimens to form a lamella (fig. a), found in both juvenile and adult shells (Hubricht 1978); *P. subtilis* is more translucent than other *Paravitrea*.

Similar Species: *Paravitrea multidentata* is most similar but has indented transverse striae that are more closely and evenly spaced on all whorls; *P. dentilla* is 3-4 mm larger and adult shells are without teeth.

Habitat: Under moist leaf litter and detritus on wooded hillsides and in ravines.

Status: G2/S2; uncommon to rare; State Special Concern; in Kentucky, the species is only very occasionally found in leaf litter collections.

Specimen: Figs. (b) from Tennessee, Macon County (FMNH 187068/1 Holotype) and (figs. c) from Kentucky, Harlan County, North side of Pine Mountain, 2 miles southeast of Bledsoe (FMNH 248709).

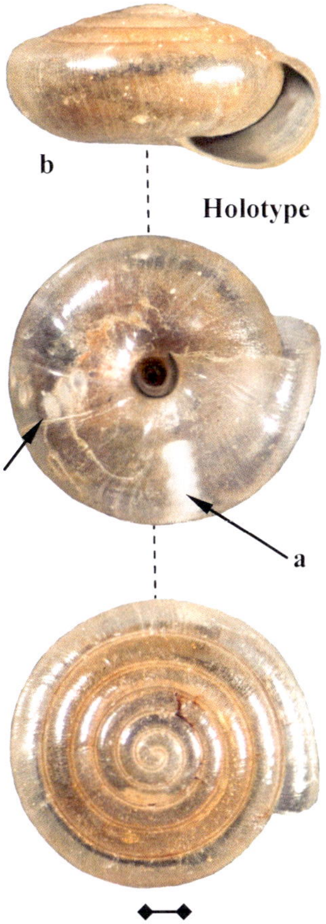

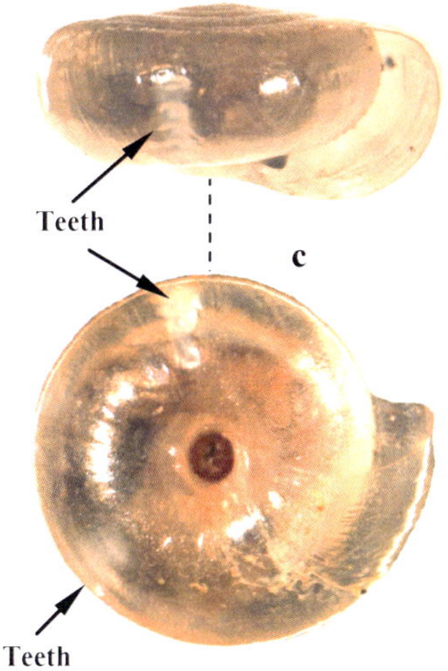

Paravitrea subtilis

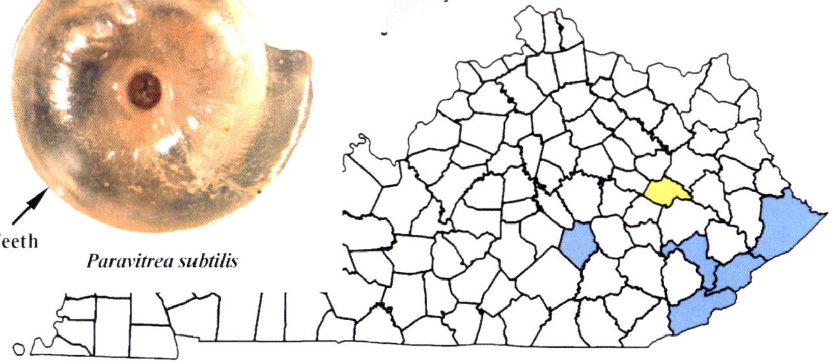

Funnel Supercoil **Pristilomatidae**

Paravitrea mira Hubricht, 1975

Diameter: 6.3 mm, , height 3.9 mm

Description: Depressed heliciform; lip simple; shell with 8.5 tightly coiled whorls; pale-amber; glossy; umbilicate; aperture lunate; irregularly spaced radial grooves above, becoming obsolete below; in the last whorl there are usually two rows of three, rather large teeth, sometimes connected by a low callus-bridge; the immature shells containing 3 teeth in each row (fig. a); the inner tooth of some adult shells is intermittently missing (Hubricht 1985).

Similar Species: *Paravitrea reesei* is smaller, has smaller teeth and three less whorls.

Habitat: Found under moist leaf litter and detritus on wooded hillsides and in ravines of mixed hardwood forests.

Status: G1/S1; State Endangered; a globally rare and restricted land snail; in Kentucky, known only from Breaks Interstate Park in Pike County and in several counties of extreme western Virginia. During a 2009 comprehensive survey of the park by Dourson and Beverly, only 5 specimens were discovered (unpublished data).

Specimen: Kentucky, Pike County, 1.7 miles west of Meta (FMNH 248934).

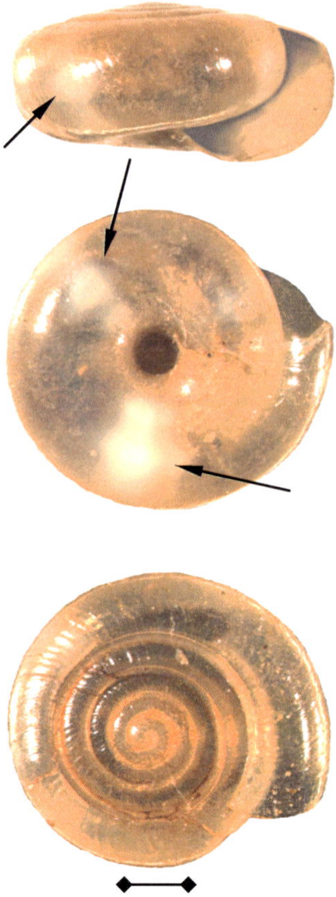

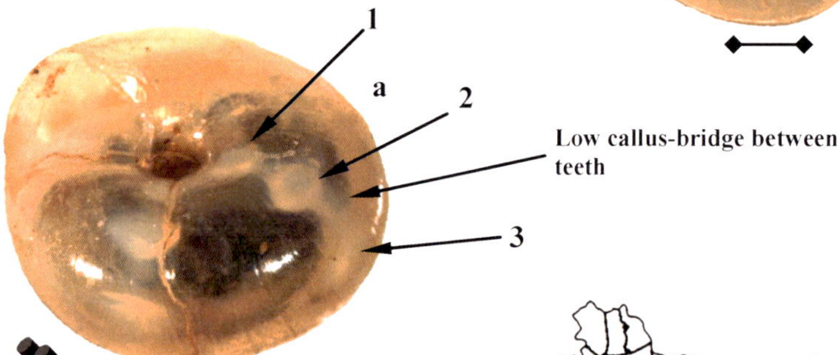

Round Supercoil

Pristilomatidae

Paravitrea reesei Morrison, 1937

Diameter: 3.5-4.7 mm, height 1.6 mm

Description: Depressed heliciform; lip simple; shell with 5-6.5 tightly coiled whorls, the last whorl not expanded; umbilicate, the umbilicus deep and well-like; shell pale-amber; glossy; aperture lunate; indented transverse striae are closely but irregularly spaced; spiral sculpture indistinct above and below; in adult shells there are rows of 3 rather large teeth (fig. a), these rows of teeth regularly spaced and visible through the base of fresh shells (Burch 1962); young shells of 2-2.5 whorls have 1 or 2 sets of paired conical teeth (fig. b), the third tooth appearing as shells mature (fig. c); teeth are usually without the callus bridge seen in *P. mira*, a species that may be found with *P. reesei*.

Similar Species: *Paravitrea capsella* is larger and is typically without teeth at all stages of growth.

Habitat: Found under moist leaf litter, detritus, and rocks on wooded hillsides, ravines, and river bluffs of mixed hardwood forests.

Status: G3/S2; State Endangered; in Kentucky, known only from Pike County.

Specimen: Virginia, Pulaski County, opposite Radford (FM 249118).

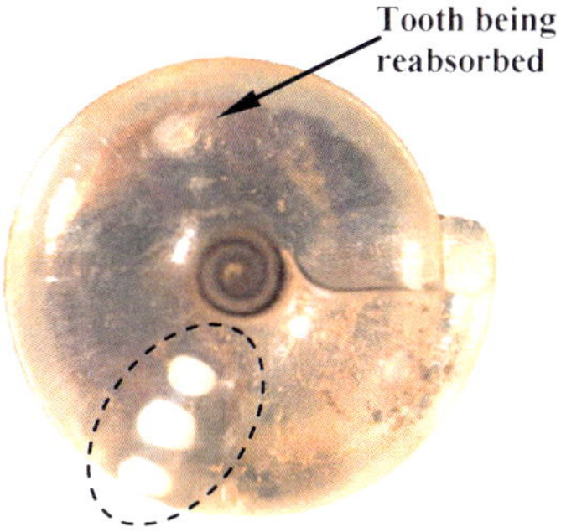

a

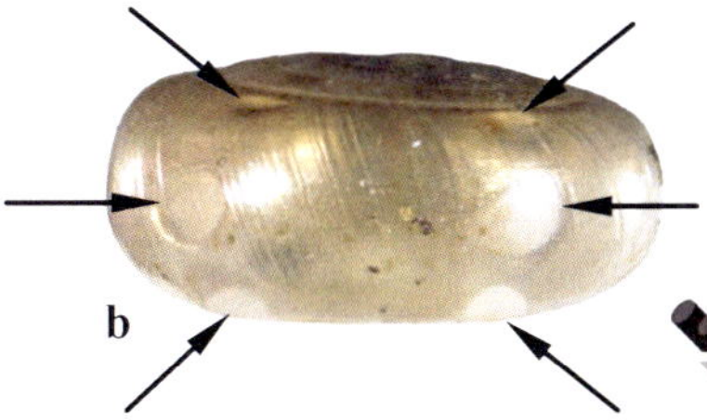

b

A slightly older shell with the top-developing third tooth, Hatfield Cemetery, Logan County, West Virginia (fig. b) and a juvenile shell showing just 2 paired teeth, the third tooth not yet developed (fig. c).

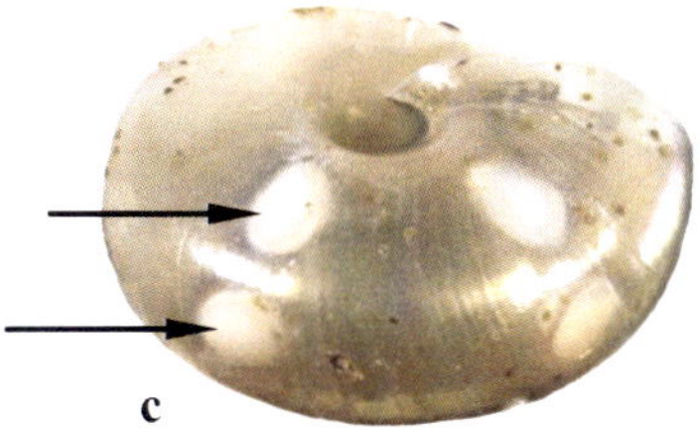

c

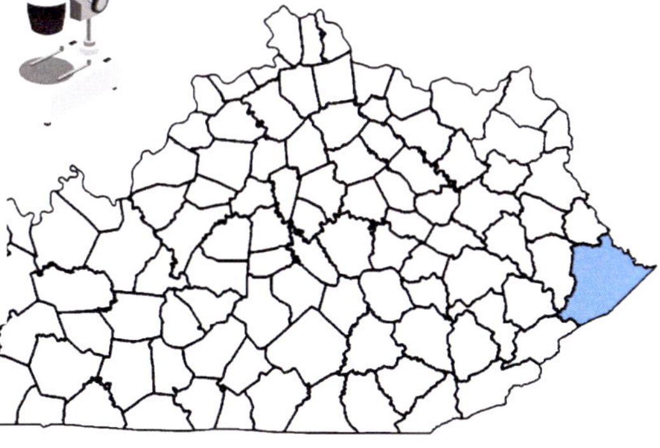

Brown Supercoil

Paravitrea septadens Hubricht, 1978

Diameter: 3.4 mm, height 1.6 mm

Description: Depressed heliciform; lip simple; shell with 6 tightly coiled whorls, the last whorl expanding slightly; shell thin, glossy and translucent in fresh and live shells; perforate and well-like; transverse striae are closely and irregularly spaced; one to two radial lamellae (figs. a & b) that can be seen through the bottom of fresh shells of juveniles and usually adults.

Similar Species: Differs from *P. multidentata* in having a more rounded periphery, fewer and weaker growth lines and lamellae instead of teeth; *P. lamellidens* is similar in size, is less rounded and has a notably smaller umbilicus.

Habitat: Found in the same habitat as *P. multidentata* and often found with it (Hubricht 1985); pockets of moist leaf litter located on rich hillsides in mixed hardwood forests.

Status: G1/S1; State Endangered; a globally rare and range-restricted land snail, reported from Dickenson and Buchanan counties, Virginia (Hubricht 1985); recently discovered (2009) in Pike County, KY at Breaks Interstate Park.

Specimen: Figs. (c) sub-adult showing internal lamellae from Virginia, Dickenson County, 1.5 miles west northwest of Bee (FMNH 249132) and figs. (d) from Virginia, Dickenson County (FMNH 171660/1 Holotype)

c

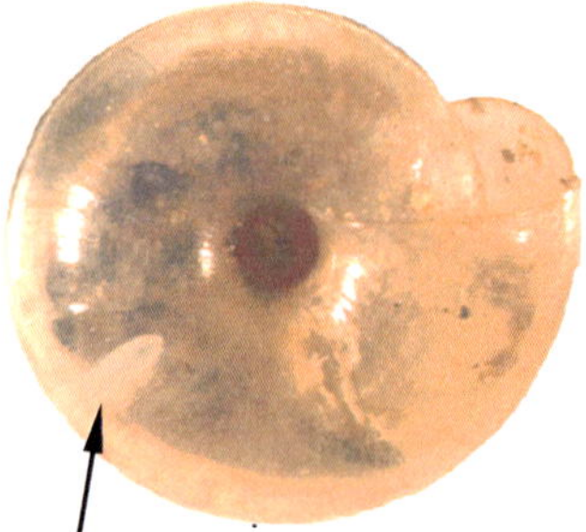

b

a

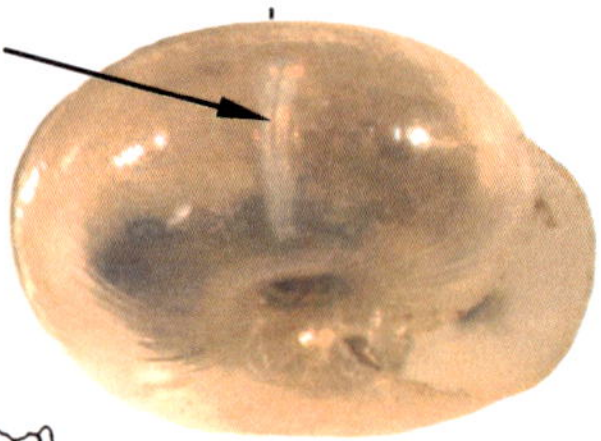

d

Holotype

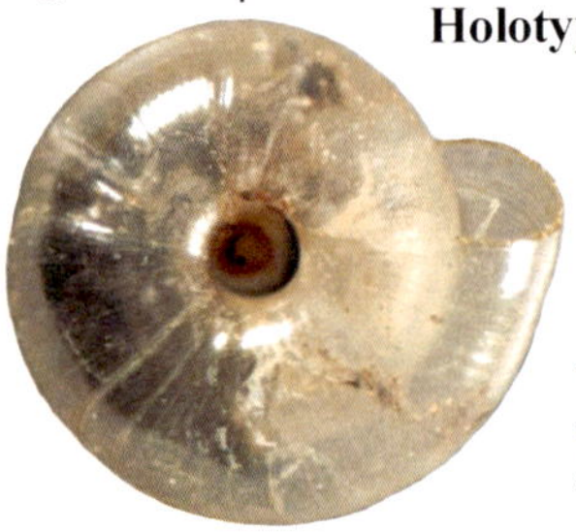

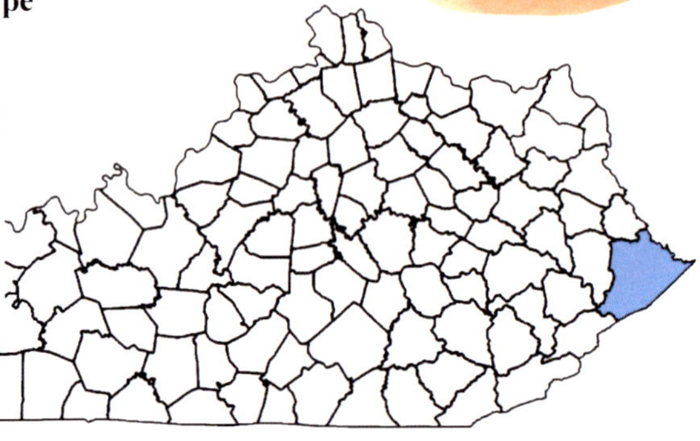

Lamellate Supercoil

Paravitrea lamellidens (Pilsbry, 1898)

Diameter: 3.5-3.8 mm, height 1.6-1.9 mm

Description: Depressed heliciform; lip simple; shell with 6.5 tightly coiled whorls, last whorl not expanding; dark reddish-brown to cinnamon buff; shell thin and translucent in fresh and live shells; perforate; transverse striae are poorly developed and weakly defined; two to three rows of obliquely radial lamellae that are closely placed and well-angled can be seen through the bottom of fresh shells of both juveniles (fig. a) and adults.

Similar Species: *Paravitrea multidentata* is similar in form but is slightly smaller, has teeth instead of lamellae and most importantly a larger umbilicus (see shell comparisons below).

Habitat: It is found in pockets of moist leaf litter, especially in rock talus located in higher elevation mixed hardwood forests; a species that burrows deep into the detritus and leaf litter.

Status: G2/S1; State Threatened; found primarily in the Great Smoky Mountains NP; in Kentucky, it was discovered at Bad Branch Nature Preserve (Dourson & Beverly 2009) and by John Slapcinsky in Pike County in 2010.

Specimen: North Carolina, Swain County, Nantahala National Forest (author's collection).

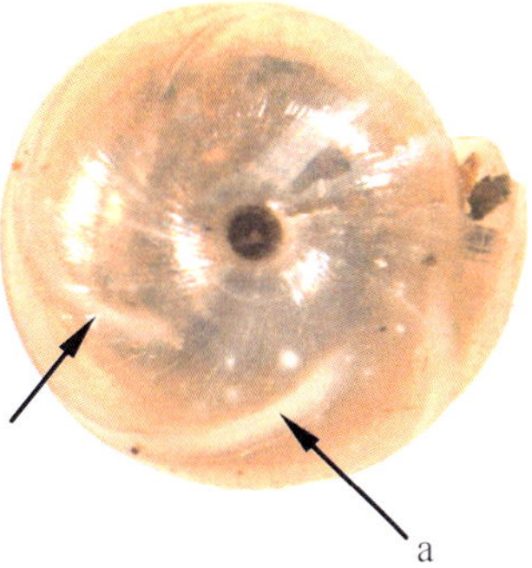

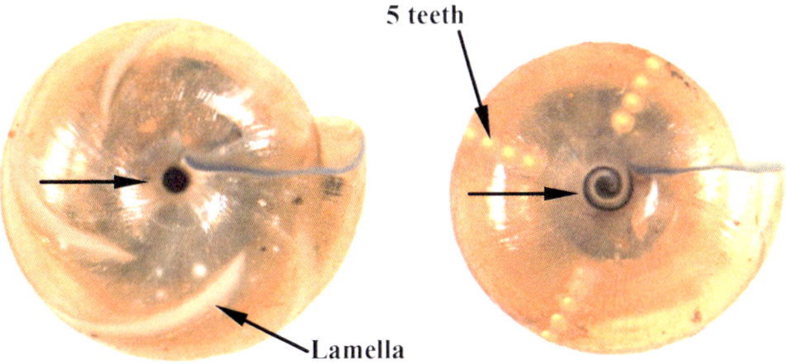

P. lamellidens (umbilicus narrow)

P. multidentata (umbilicus wider)

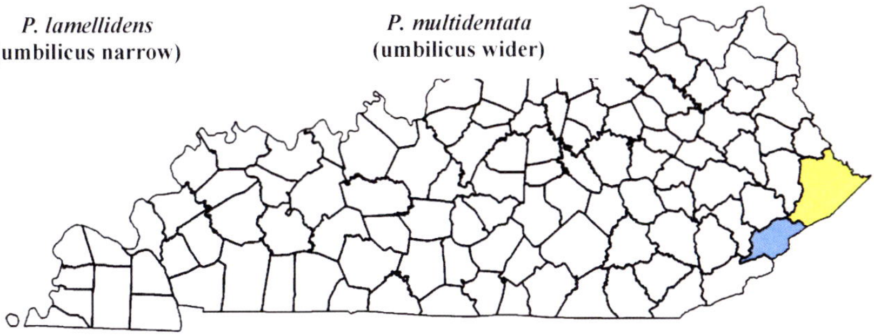

Dentate Supercoil

Pristilomatidae

Paravitrea multidentata (A. Binney, 1840)

Diameter: 2.5-3 mm, height 1.3 mm

Description: Depressed heliciform; lip simple, shell with 6 tightly coiled whorls, fully adult shells are umbilicate, while young shells are perforate; last whorl not expanding; shell thin, glossy and translucent; transverse striae are poorly developed, barely visible, very closely and evenly spaced; without spiral striae; two to four radial rows of usually five teeth that can be seen through the bottom of fresh shells of both juveniles and adults; on occasion will have rows of lamellae (fig. a); fully adult shells have a faint callus connecting the end of the lip (fig. b) and a slight reflection in the peristome (fig. c).

Similar Species: *P. lamellidens* is slightly larger, has lamellae instead of teeth, a notably smaller umbilicus, and more well developed transverse striae (see page 166).

Habitat: Found in pockets of moist leaf litter on hillsides in mixed hardwood forests; especially common in wet leaf litter surrounding small headwater seeps.

Status: G5/S5; Common; the most common and widespread *Paravitrea* species in the state.

Specimen: West Virginia, Upshur County, Stonecoal WMA (author's collection).

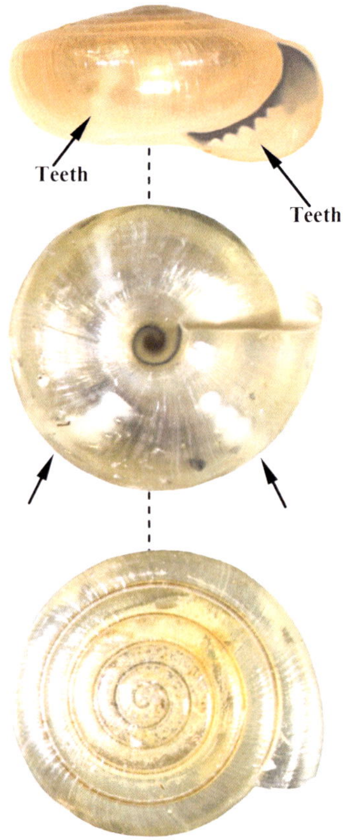

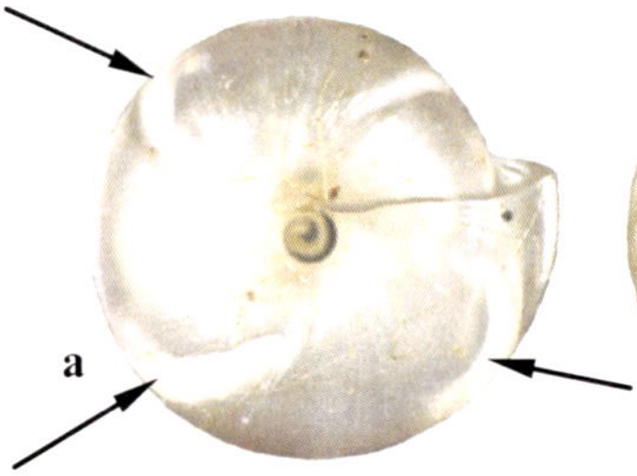

Shavers Fork,
Randolph Co, WV

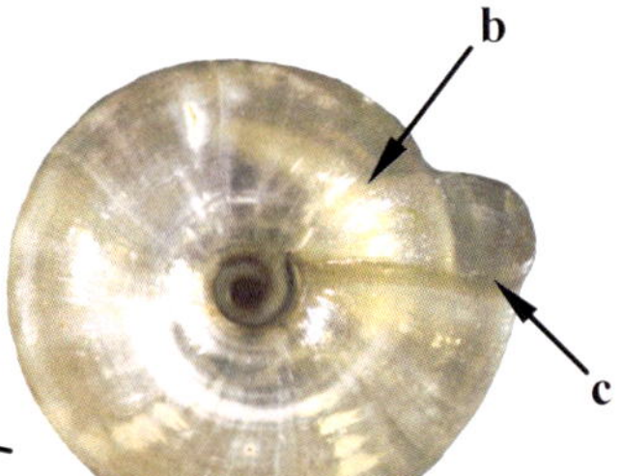

Groves Creek,
Clay County, WV

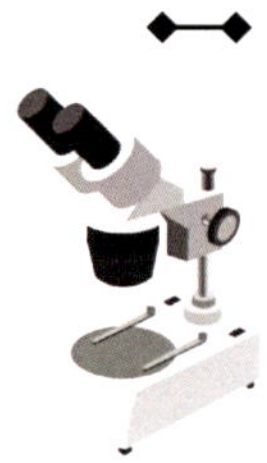

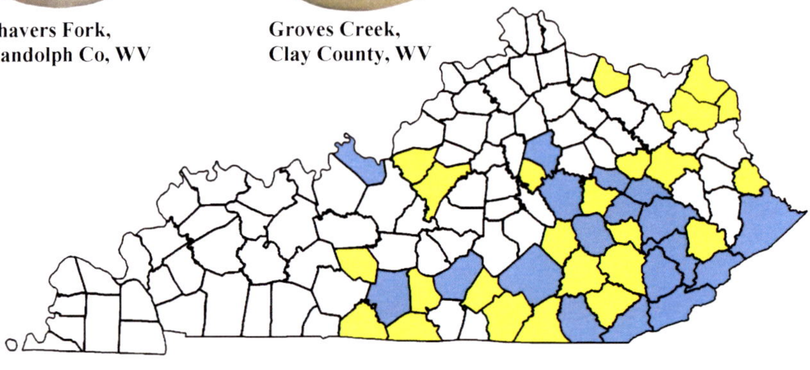

Comb Supercoil

Pristilomatidae

Paravitrea dentilla Hubricht, 1978

Diameter: 6-6.8 mm, height 3.5 mm

Description: Depressed heliciform; lip simple; shell with 7.5-8 tightly coiled whorls, the last whorl slightly expanding; shell thin, glossy and translucent in fresh and live shells; umbilicate, well-like; nuclear whorl smooth, later whorls sculptured with numerous irregularly spaced, impressed growth lines, distinct above but becoming obsolete below; in juvenile shells under 4 whorls are usually one or more radial rows of 4 or 6 small teeth within the last whorl (fig. a), the teeth are set close together, the bases usually touching; adults are without teeth.

Similar Species: *Paravitrea placentula* is larger and has paired teeth (in juvenile shells); young shells of *P. dentilla* might be confused with those of *P. multidentata*, but the surface of *P. dentilla* has irregularly and wider spaced transverse striae.

Habitat: In Kentucky, found in deep leaf litter and detritus on mesic hillsides, but also upper elevation limestone and sandstone talus.

Status: G1/S1; State Endangered; globally rare and range restricted to two counties in Virginia and known only from Bad Branch Nature Preserve and several north-facing sites on Black Mountain in Kentucky.

Specimen: Virginia, Washington County, 1 mile south of Damascus (FMNH 248919 Paratype).

Paratype

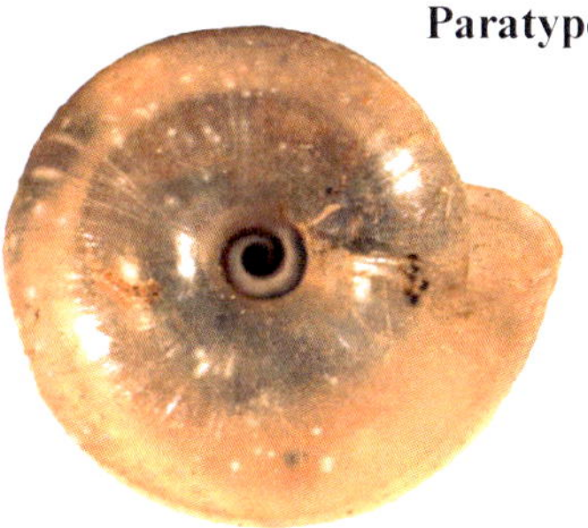

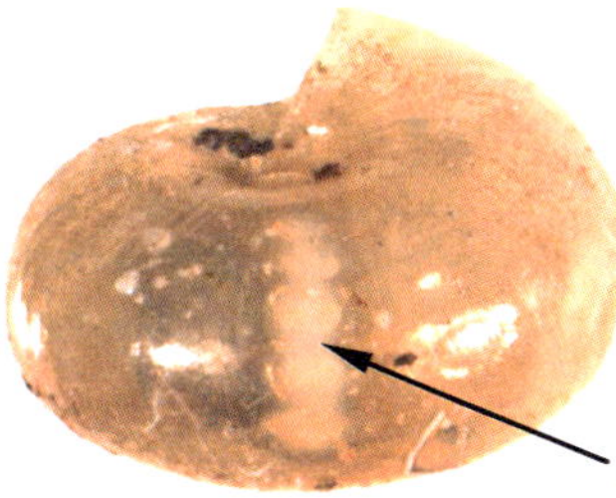

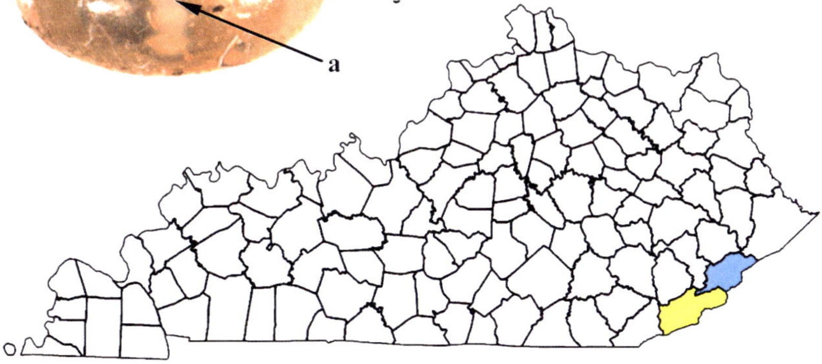

Glossy Supercoil

Pristilomatidae

Paravitrea placentula (Shuttleworth, 1852)

Diameter: 7.2-7.8 mm, height 3.6 mm

Description: Depressed heliciform; lip simple; shell with 5-7 tightly coiled whorls, the last whorl slightly expanding; corneous; shell thin, glossy and translucent in fresh and live shells; umbilicate; indented transverse striae are faint and irregular; two rows of rather large radial paired teeth that are arranged close together can be seen through the bottom of fresh and live shells of juvenile shells (fig. a); adults are without teeth.

Similar Species: *Paravitrea capsella* is smaller, has a more rounded aperture and is without teeth in both juvenile and adult shells.

Habitat: Acidic coves, rich woods, high elevation northern red oak and montane oak hickory forests.

Status: G3/S3; Uncommon; in Kentucky, known from the southeastern mountain counties.

Specimen: North Carolina, Swain County, (author's collection).

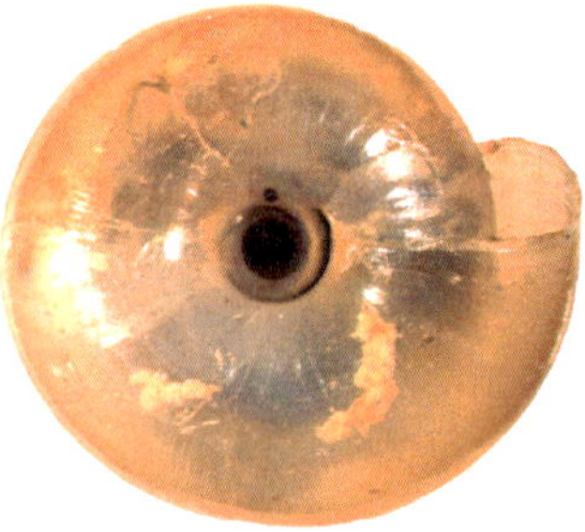

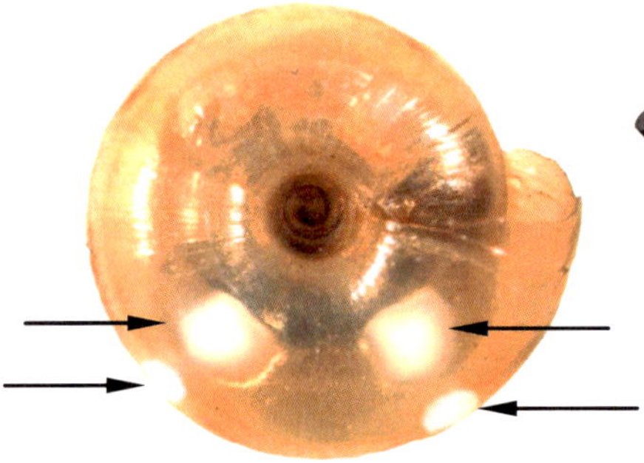

a

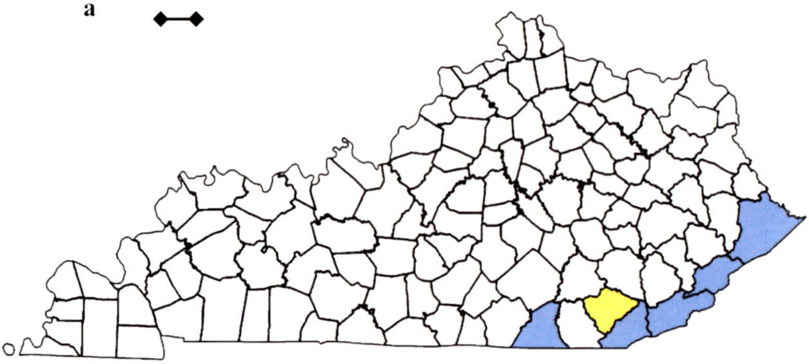

Land snails are often seen on spider webs, easily moving across the web's sticky surface without becoming entangled. Pictured above is a juvenile *Patera panselenus*. Note the spider in the background. There is some evidence that snails are attracted to the silk nets that trap condensed moisture for a source of drinking water especially during drier weather. Pictured below, the same species appears to glean water from a web. Both images from Pike County, Kentucky.

Black Mountain Bud

Oxychilidae

Pilsbryna aff. vanattai **(Pilsbry & Walker, 1902)**

Diameter: 3.8-4.4 mm, height 2 mm

Description: Depressed heliciform; lip simple; shell with 4.5-5 loosely coiled whorls, the last whorl greatly expanded; shell glossy, honey colored; umbilicate; internal lamellae or teeth present in shells under 3 whorls, in mature shells, all traces of apertural lamellae are reabsorbed; delicate indented transverse striae are irregular and a constant feature; juvenile shells are essential for conclusive identification.

Similar Species: *Glyphyalinia* species are without internal armature; *Paravitrea* have tightly coiled whorls, their last whorls generally not expanding greatly as in *Pilsbryna.*

Habitat: Rocky, wooded mountainsides, most common in wet leaf litter along streams, seeps but also found in deeper leaf litter at the base of limestone and other sedimentary rocks next to stream banks (Slapcinsky and Coles 2004).

Status: G3/S1; State Endangered; In Kentucky, this rare land snail is currently known only from the north side of Black Mountain near the summit. There is some speculation that Black Mountain populations may represent an undescribed species (per. comm. John Slapcinsky, 2000). Even though Black Mountain shells are indistinguishable from Southern Appalachians shells, Kentucky populations are disjunct by more than 100 miles, putting the species taxonomic position in question; DNA sequencing will be required to make a final determination.

Specimen: *Pilsbryna vanattai* from Tennessee, Carter County, Cherokee National Forest (author's collection).

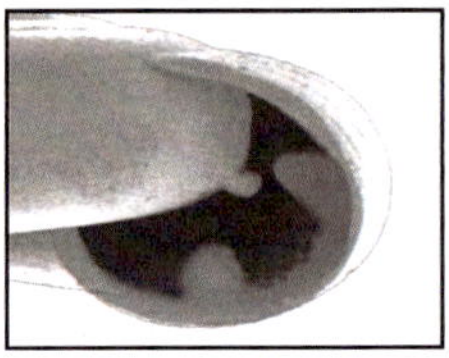

SEM image showing internal teeth by John Slapcinsky

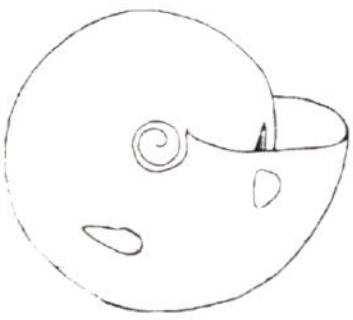

Juveniles shell showing paired Palatal teeth

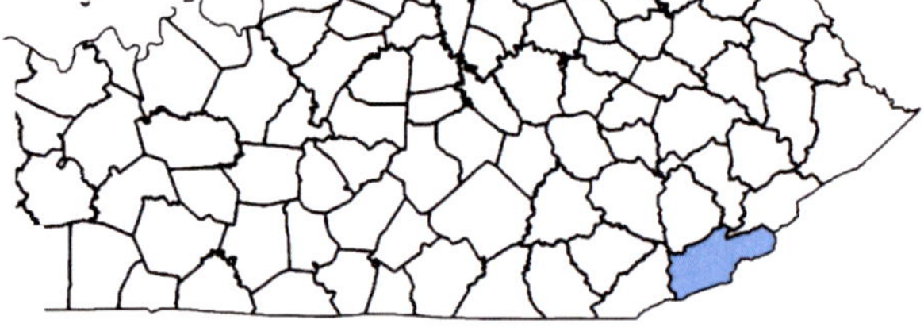

Shells 6-8 mm, simple lip, shell finely ribbed, <u>without teeth</u>, defense mucus fluorescent under UV light (*Discus*)

<u>Discus</u> species are depressed heliciform, less tightly coiled than *Helicodiscus* and do not feature color bands or blotches on shells as in *Anguispira*. <u>*Discus* snails are usually without internal teeth,</u> although *D. patulus* has a small callus tooth a short distance within its shell. *Discus* species are much larger and have an entirely different shell sculpture than *Helicodiscus*, having transverse riblets (fig. a), instead of spiral sculpture (lirae) or fringed ornamentation (fig. b). Like *Anguispira*, the defensive mucus or slime of *Discus* is fluorescent under UV light (Dourson 2013). *Discus* are generally found associated with rotting hardwood in advanced stages of decay and are sometimes found in relatively large numbers (30-50 plus).

<u>Shells 6-8 mm, ribbed usually without teeth, pg. 173-176</u>

*Discus patulus
*Discus nigrimontanus
*Discus whitneyi

Shells proportionate

Helicodiscus parallelus

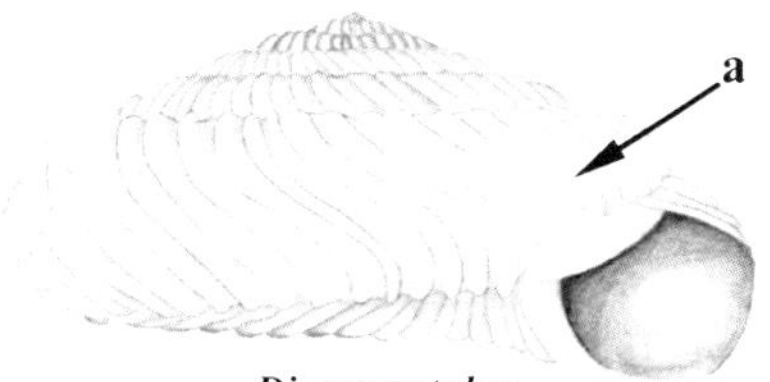

Discus patulus

Shells 3-5 mm, simple lip, raised spiral (lirae) sculpture, <u>with teeth</u>, defense mucus fluorescent under UV light (*Helicodiscus*)

<u>Helicodiscus</u> are flattened or discoidal and under a hand lens of 10X, have raised spiral lirae, fringes or hairs. <u>Most species in Kentucky have internal teeth</u> (fig. c) which can usually be seen in frontal view or through the bottom of fresh dead or live shells. Only the defense mucus is florescent, not the crawling slime. *Helicodiscus* species have some of the most extraordinary micro-sculpture and are rarely found in large numbers.

<u>Shells 3-5 mm, spiral striae strongly developed, always with teeth, pg. 177-184</u>

*Helicodiscus parallelus
*Helicodiscus notius
*Helicodiscus fimbriatus (uncommon)
*Helicodiscus hadenoecus (rare)
*Helicodiscus punctatellus (rare)
*Helicodiscus aldrichianus (rare)

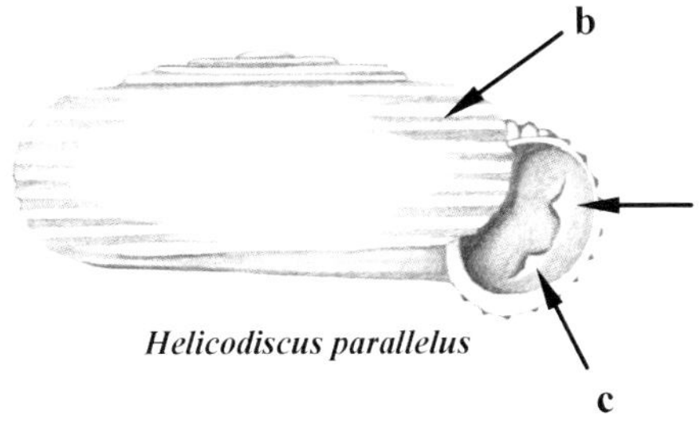

Helicodiscus parallelus

Domed Disc

Discidae

Discus patulus (Deshayes, 1830)

Diameter: 7-8 mm, height 3.3-4 mm

Description: Depressed heliciform; lip simple; aperture roundish to oval shape; shell with 5.5 whorls; widely umbilicate; there is a small rounded tubercle or callous tooth a short distance within the aperture (fig. a) but this may be a wanting feature in some specimens; transverse striae are well developed, rib-like (3-4 ribs per mm) and are distinct on top, sides and base of shell; periphery rounded.

Similar Species: Other *Discus* species in Kentucky are without an internal parietal tooth; *D. nigrimontanus* has a smaller aperture, is flatter in profile, and has an angular periphery; juvenile *D. patulus* under 3 whorls will also have an angular periphery.

Habitat: A species of upper slopes in mixed hardwood under leaf litter; it is most common on or under the exfoliating bark of rotting hardwood logs in advanced stages of decay.

Status: G5/S5; Common; the most common *Discus* species in Kentucky and likely found in every single county.

Specimen: North Carolina, Macon County, Nantahala National Forest (author's collection).

A freak of nature, *D. patulus* from the Cherokee National Forest, Polk County, Tennessee (above); all other *D. patulus* found with it had the typical flattened profile

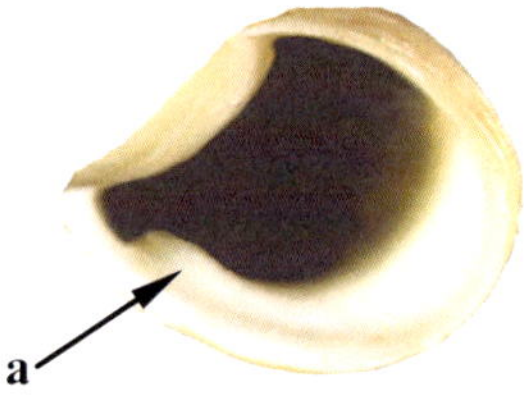

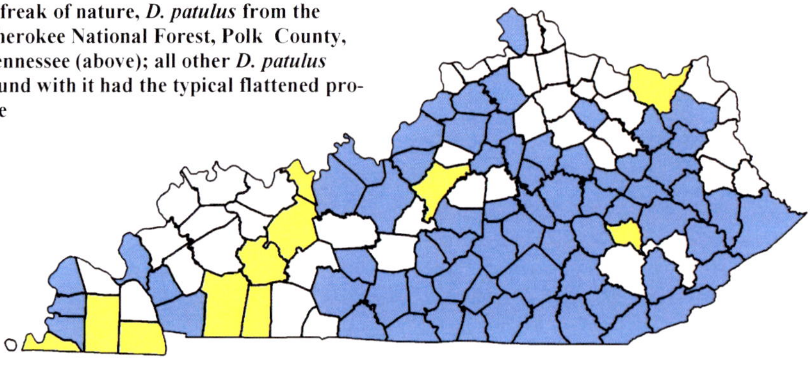

Black Mountain Disc **Discidae**

Discus nigrimontanus (Pilsbry, 1924)

Diameter: 7.4 mm, height 2.4 mm

Description: Depressed heliciform (more compressed than other *Discus* species in Kentucky); shell with 5.25 whorls; widely umbilicate; striae are well-developed and rib-like; periphery angular; the microscopic crisscross sculpture of the embryonic whorl is weak and wanting; no internal columellar tubercle as seen in *D. patulus*.

Similar Species: *D. patulus* has a rounded periphery, small rounded tubercle (fig. a), coarser riblets and a smaller more well-like umbilicus.

Habitat: A species of upland rocky talus slopes in mixed hardwood forests where leaf litter is sparse; common in the lower layers of deep limestone talus along the base of outcropped limestone on Pine Mountain in Harlan County, Ky. (MacGregor pers. comm., 2010).

Status: G4/S2; Rare; The least common of the *Discus* in Kentucky; known from only a few southeastern counties; a species more southern in distribution having disjunct populations occurring in southern Missouri and northern Arkansas.

Specimen: Tennessee, Blount County, hills above the town of Townsend (author's collection).

D. patulus

D. whitneyi

© Pilsbry, 1948

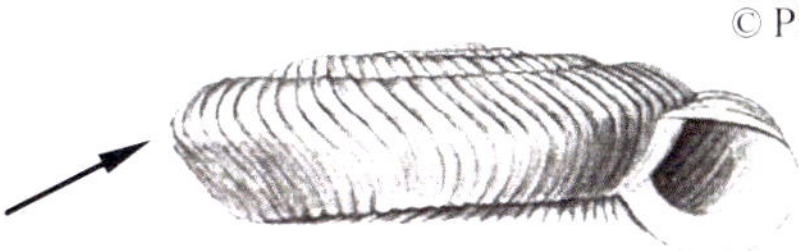

D. nigrimontanus

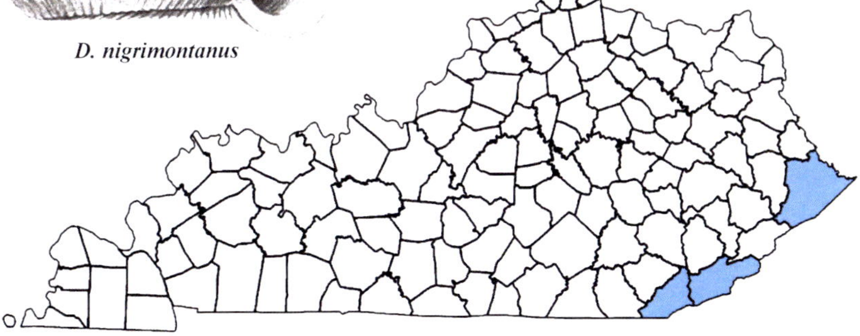

Forest Disc

Discus whitneyi (Newcomb, 1864)

Diameter: 5-7 mm, height 3 mm

Description: Depressed heliciform; lip simple; shell thin with 3.5-4.5 whorls; umbilicate; no teeth present; transverse striae are well developed, rib-like and nearly as well defined on base of shell around the umbilicus; periphery usually well rounded.

Similar Species: *Discus patulus* is slightly larger, has a thicker shell, egg shaped aperture, and has a small callous tooth; *D. nigrimontanus* (page 174) has a smaller aperture, is less compact, being flatter in profile, and has an angular periphery (not rounded).

Habitat: A species of rather low damp soils around open grassy places, roadsides, and edges of wetlands; also a frequent snail in untidy urban areas.

Status: G5/S4; Relatively Common; this mostly northern species is known from scattered counties across Kentucky; previously known as *D. cronkhitei,* but Roth (1988) demonstrated that *D. whitneyi* is a senior synonym of *D. cronkhitei.*

Specimen: Kentucky, Estill County, along railroad tracks in the town of Irvine (author's collection).

Note the course transverse riblets typical of *Discus* species, which are best observed on top of shell

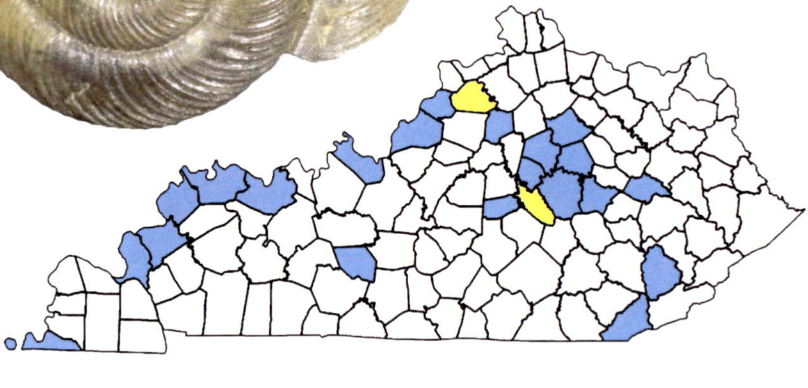

Hunters of Snail Flesh

An iridescent, snail-hunting *Cychrine* beetle dines on the flesh of an immature highlands slitmouth, *Stenotrema altispira,* Roan Mountain, Mitchell County, North Carolina.

Helicodiscus parallelus (Say, 1817)

Diameter: 3.2-3.5 mm, height 1.2 mm

Description: Depressed heliciform-discoidal; lip simple; shell with 4-4.5 whorls; widely umbilicate; paired teeth present usually deep within the aperture, can be seen in fresh shells (fig. a); spiral lirae are well developed except on the embryonic whorl where there is only a hint of spiral striae (fig. b); periphery rounded.

Similar Species: *Helicodiscus notius* is reported to be larger by one whorl, having a slightly wider umbilicus and the spiral striae on the embryonic whorl are more strongly developed, these characters however being highly variable and questionable attributes.

Habitat: Found in floodplains and upland mixed hardwoods under leaf litter and around rock structures; also in urban areas under wood debris and in vacant lots; Hubricht never found it in caves.

Status: G5/S4; Fairly Common; a widespread resident of Kentucky that appears less often than *Helicodiscus notius;* like most *Helicodiscus* species, it is rarely found in numbers.

Specimen: Kentucky, Powell County, Star Gap, Red River Gorge (author's collection).

a

Shavers Fork, Randolph County, West Virginia

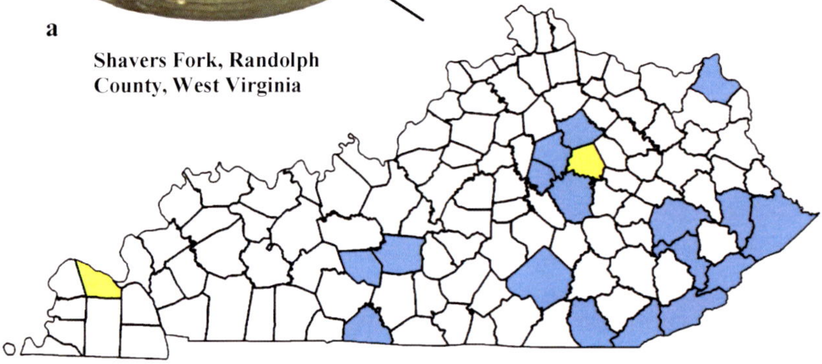

Tight Coil

Helicodiscus notius Hubricht, 1962

Diameter: 3.6 mm, height 1.04 mm

Description: Depressed heliciform-discoidal; lip simple; shell with 5-5.5 whorls; widely umbilicate; within the last whorl of juvenile and adult shells there are 2 to 3 pairs of conical teeth which can be seen through the bottom of fresh shells; the raised spiral lirae are well developed including the ones on the embryonic whorl (fig. a).

Similar Species: *Helicodiscus notius* appears to differ little from *H. parallelus* given the wide range variations in shell morphology and there have been doubts as to its validity, nevertheless, Hubricht's separating distinctions included *H notius* having better developed spiral lirae on the embryonic whorl (fig. b), this distinction being highly variable, having one more whorl and a species of drier sites; *Helicodiscus notius specus*, (a troglobitic cave snail) reported from Edmonson County, Kentucky was recently (2019) synonymized with *H. notius* by Rudiger Bieler (MolluscaBase.org).

Habitat: A species found on forested hillsides, in ravines and occasionally living in caves; usually found in drier and higher habitats than *H. parallelus*.

Status: G5/S5; Common; this widespread species likely occurs in every Kentucky county.

Specimen: Tennessee, Warren County, Storage Cave (author's collection).

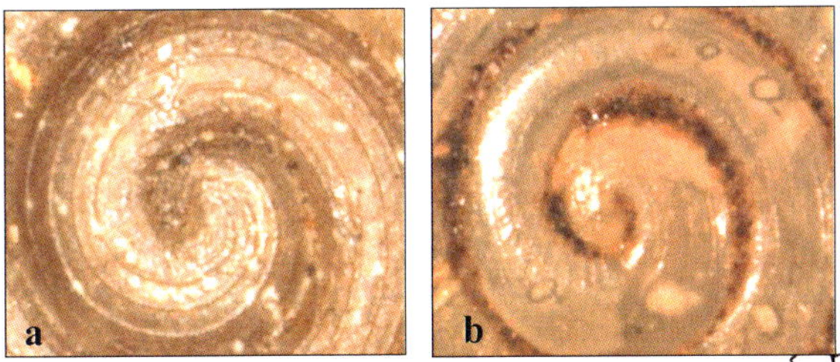

Helicodiscus notius *Helicodiscus parallelus*

Top views of embryonic whorls

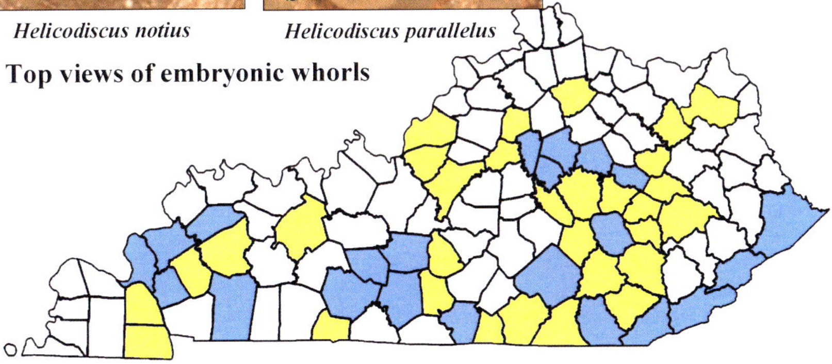

Fringed Coil

Helicodiscus fimbriatus Wetherby, 1881

Diameter: 4.5- 5 mm, height 1.5 mm

Description: Depressed heliciform; lip simple; shell with 5 whorls; widely umbilicate; paired teeth present sometimes up to six pairs that can be seen through the bottom of fresh and live shells; spiral lirae are unequal, some developed into conspicuously fringed features, but is frequently without this amazing ornamentation in old shells. Fringes, ribs and transverse striae found on land snails act like small water reservoirs, that hold and uniformly transfer (wick) water across the entire shell surface (Dourson pers. obs.). Although only speculation, these water retention features may help facilitate snail movement across drier surfaces and or help keep the snail cooler.

Similar Species: Most similar to *Helicodiscus* Species (undetermined) but has paired peg-like teeth (*Helicodiscus* Species undetermined has cupped teeth that are alternating); *H. hadenoecus* is smaller, has shorter fringes which are often lost in older shells, has one less whorl, no internal teeth at any stage of growth and lives mostly in caves.

Habitat: Found in leaf litter and under flat rocks in mixed hardwoods on hillsides.

Status: G4/S2; Uncommon to Rare; in Kentucky, reported from only four counties in the southeastern mountains of the state; the species is more common in the states of Tennessee, Alabama, and Georgia.

Specimen: Tennessee, Monroe County, (author's collection).

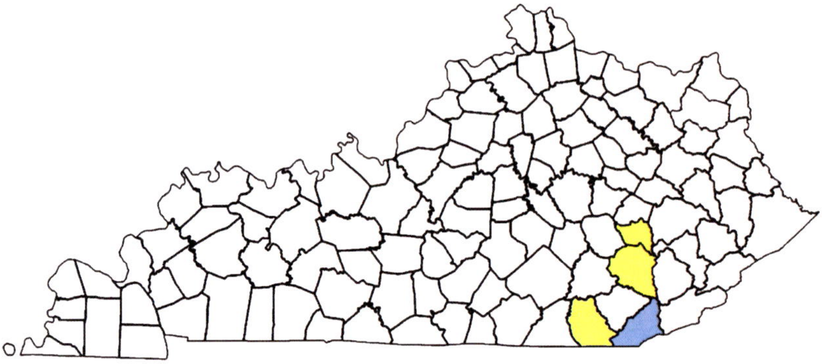

Note the spiral fringes
Helicodiscus fimbriatus
Note the internal teeth

Cricket Coil

Helicodiscus hadenoecus Hubricht, 1962

Diameter: 2.7 mm, height 1.2 mm

Description: Depressed heliciform; shell with 4 whorls; umbilicate; no internal teeth present at any stage of growth; sutures well impressed; embryonic whorl smooth, later whorls with fine revolving epidermal fringes on entire shell surface (below picture); fringes of dead and weathered shells are usually lost and show no sculpture other than weak growth lirae (Hubricht 1962).

Similar Species: *H. fimbriatus* is larger by 2 mm and has notably longer fringes; dead shells of *H. hadenoecus* can resemble those of *Hawaiia minuscula; .Helicodiscus barri,* a species found in Tennessee caves is reported to be its closest relative and is larger by around 1 mm (around 3.9 mm).

Habitat: Found in deep cave systems where it has been observed feeding on cave cricket guano but also known from leaf litter around logs and at the foot of river bluffs in 10 cm of soil or deeper (Hotopp 2012).

Status: G3/S2; State Threatened; found in just a few central Kentucky cave systems in three counties; the species is also reported from Tennessee and Alabama.

Specimen: Alabama, Marshall County, Cave Mountain Cave.5 miles southwest of Guntersville Dam (FMNH 239125).

Young shell enlarged and showing low fringes

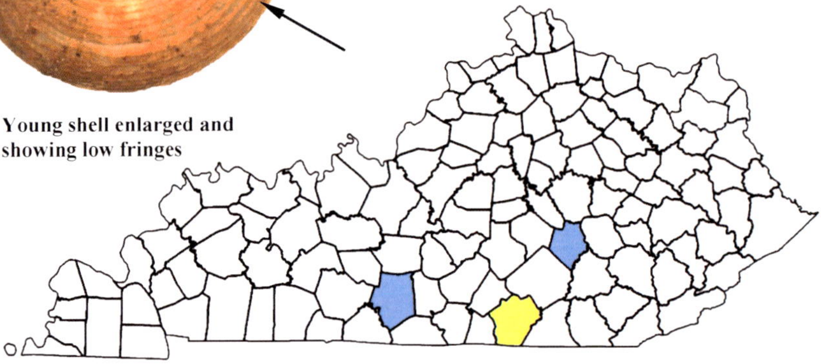

Punctate Coil

Helicodiscus punctatellus Morrison, 1942

Diameter: 3.4 mm, height 1.8 mm

Description: Depressed heliciform; lip simple but completely mature shells will have a slightly reflected lip; shell with 4.5 whorls; umbilicate; no teeth present at any stage of growth; spiral puncta (figs. a & b) (not the spiral lirae see in other *Helicodiscus*), are a well-developed features on the shell surface.

Similar Species: No other *Helicodiscus* in Kentucky has the spiral puncta feature; *Lucilla inermis* is smaller and without spiral puncta.

Habitat: In Kentucky, known only from dead shells found in several caves and one location in Tennessee; recent collections by Jerry Lewis in 2005 from caves located around Somerset suggest that extant populations still exist. Although Hubricht considered this species a troglobite, *H. punctatellus* has also been found in surface habitats and, therefore, considered a troglophilic species (able to live in both surface and cave habitats).

Status: G1/S1; State Endangered; Very Rare; very little is known of this rare snail; known only from dead but recent shells, live animals have not been observed; considered one of Kentucky's rarest gastropods.

Specimen: Kentucky, Edmonson County, Mammoth Cave National Park, Whites Cave (FMNH 239124).

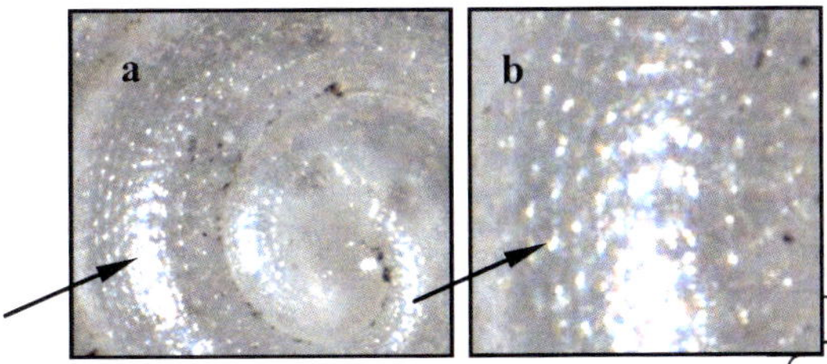

Note the spiral puncta instead of spiral striae

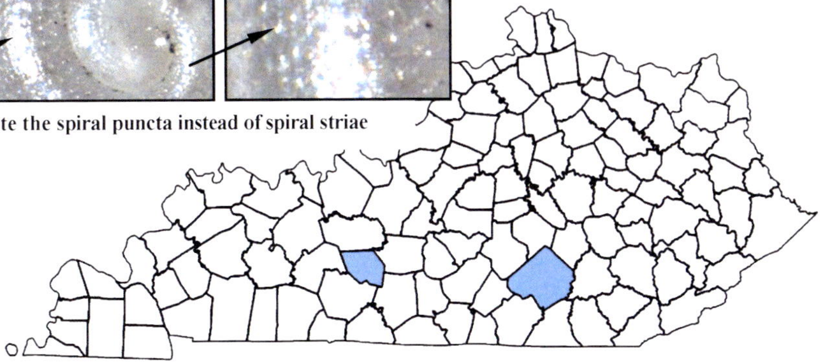

Burrowing Coil

Helicodiscus aldrichianus (Clapp, 1907)

Diameter: 1.9-2 mm, height 1 mm

Description: Depressed heliciform; lip simple; shell with 4.5-5 whorls; shell pale greenish-white, glossy and transparent in live and fresh dead shells; shells becoming bleached with age (fig. c); umbilicate; very faint growth lines but otherwise nearly smooth; in the last whorl there are usually three to four elongated teeth on the palatal wall (fig. a) and 1-2 basal snake-like lamellae (fig. b) running parallel with the spiral; periphery well rounded.

Similar Species: *Helicodiscus parallelus* have knob-like teeth, well developed spiral lirae, are larger and have a thicker shell; the internal armature in *Helicodiscus aldrichianus* is unique among Kentucky land snails.

Habitat: A burrowing species in chert rubble, undersides of stones and from deep pockets of leaf litter.

Status: G3/S1; State Endangered; Rare; this remarkable species was only recently discovered in Kentucky in Wayne County from around and in sinkholes. Originally placed in the genus *Clappiella* (Pilsbry, 1946).

Specimen: Tennessee, Warren County (FLMNH 408151 all images by Wayne Van Devender).

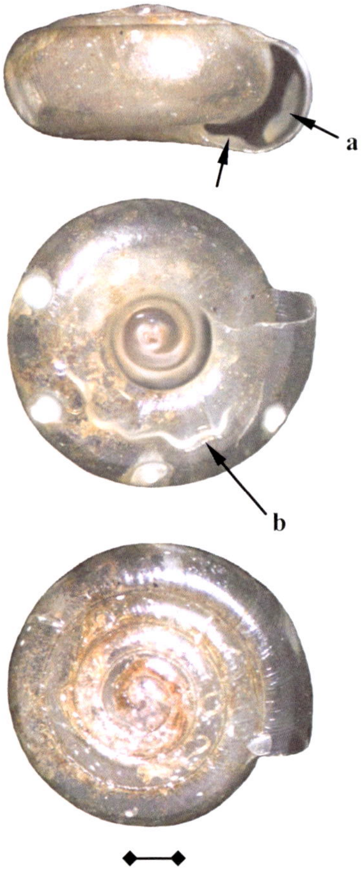

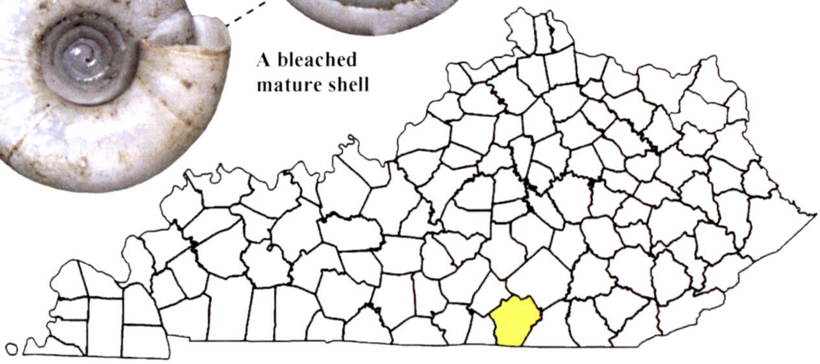

Helicodiscus aldrichianus
2 mm
Note the elongated teeth and lamellae seen through these young transparent shells

Shells 6-15 mm, with or without teeth, simple lip (*Ventridens*)

Ventridens superficially look like miniature *Mesomphix*, but a closer look reveals that *Ventridens* have tightly-coiled shells (fig. a), the last whorl not as widely expanded as seen in *Mesomphix*. *Ventridens* are also more compact in form and average 2-3 more whorls than *Mesomphix*. *Ventridens* often have teeth or lamellae (fig. b) while *Mesomphix* are without these defensive features. Like *Mesomphix* however, *Ventridens* have a thickening or thin callus, usually a whitish color, just inside the aperture's bottom (fig. c). The shell is perforate (never completely closed) to umbilicate and in many *Ventridens* the umbilicus is proportionately wider in the youngest shells.

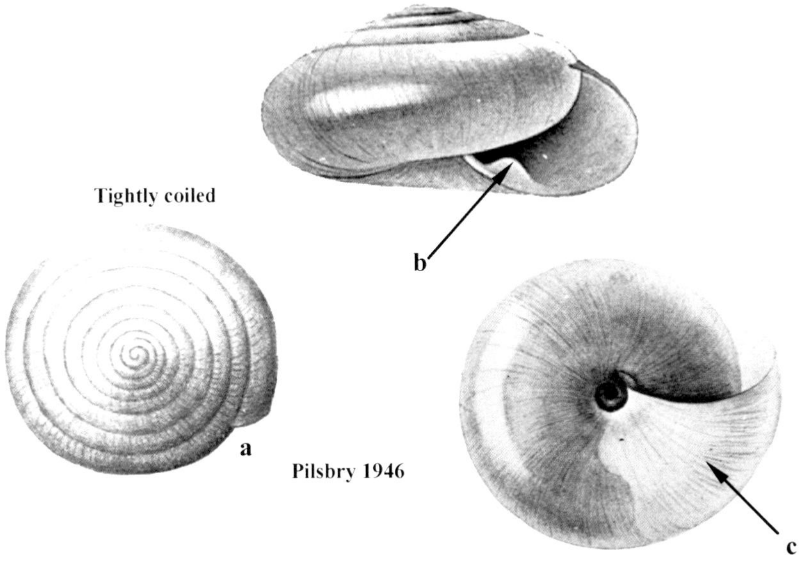

The bidentate species of the genus *Ventridens* are characterized by the presence of 2 apertural lamellae in the pre-adult shells (Hubricht 1964). *Ventridens* in this group can be divided readily into two basic groups by the color of the live animal and the presence or absence of internal armature. In the ***V. gularis*** group, the live animal is dark, olive or bluish-gray (opposite page). In the ***V. pilsbryi*** group, the live animal is pale with a hint of yellow and perhaps some grayish flecking along the back; opposite page (Hubricht 1964). Multiple species of the *V. gularis* group are frequently found with species of the *V. pilsbryi* group. Species of the *V. gularis* group have not been found living together. Species of the *V. pilsbryi* group have not been found together with the exception of *V. lasmodon* (these observations of co-existence are according to Hubricht 1964). In the third and final group of *Ventridens*, the adult shells are without teeth. It is important that the color of live animals be noted in the field (before preservation) as well as the presence of any teeth in shells collected.

Grouping 1: *Ventridens gularis* group, pg. 187-189: Live animal dark, olive or bluish-gray. Two or more apertural lamellae in the juvenile shell stage (see below). One, both or all of these lamellae may be reduced or wanting in mature shells of this group and shells rarely have more than 6.5 whorls. The aperture is proportionally larger than in the *V. pilsbryi* group of shells.

**Ventridens suppressus*
**Ventridens virginicus*
**Ventridens gularis*

Juvenile shell

Animal dark

Grouping 2: *Ventridens pilsbryi* group, pg. 190-194: Animal pale, (may have a hint of yellow) with some grayish flecking along the back and sides. Like the *V. gularis* group, there are two or more apertural lamellae in the juvenile shell stage and one, both or all of these lamellae may be reduced or wanting in mature shells of this group. The diameter of the whorls usually increases more slowly and the whorls are more flattened than seen in the *V. gularis* group. *V. pilsbryi* shells may have as many as 9 whorls.

**Ventridens pilsbryi*
**Ventridens lasmodon* (rare)
**Ventridens lawae* (rare)
**Ventridens theloides*
**Ventridens collisella*

Animal light

Grouping 3: Other *Ventridens*, no teeth in the adult shells, pg. 195-199

**Ventridens demissus*
**Ventridens intertextus*
**Ventridens percallosus*
**Ventridens cf. acerra* (rare)
**Ventridens ligera*

Without internal armature in adult shells

Flat Dome

Ventridens suppressus (Say, 1829)

Diameter: 5.8-7 mm, height 3.5-4 mm

Description: Depressed heliciform; lip simple; shell with 6 moderately coiled whorls; perforate to umbilicate; live animal dark, olive or bluish-gray; shell thin, pale horn, glossy and semi-transparent; transverse striae (sometimes called growth wrinkles) are poorly developed; spiral striae present but a weak feature; in early shell stages multiple teeth or lamella are present but in fully mature shells 6–6.5 mm with more than 6 whorls, there is usually only one basal lamella (figs. a & b), or very occasionally, no lamella at all; a member of the *V. gularis* group (Hubricht 1964).

Similar Species: *Ventridens virginicus* is smaller and contains an outer wall tooth in adult shells and a slightly larger umbilicus.

Habitat: Found in mixed hardwood forests on hillsides and in ravines under leaf litter, often times in rocky areas.

Status: G5/S2; Rare to Uncommon; most populations of this species found in southeastern Kentucky on Pine and Black Mountain.

Specimen: Virginia, Craig County, Millers Cove (author's collection).

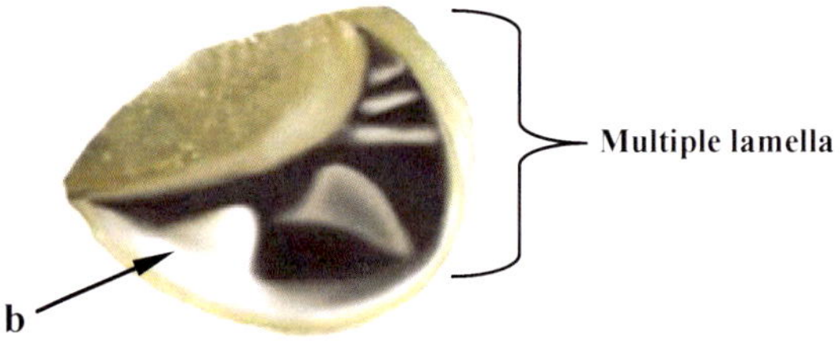

Juvenile shell showing multiple lamellae and non-forked basal tooth

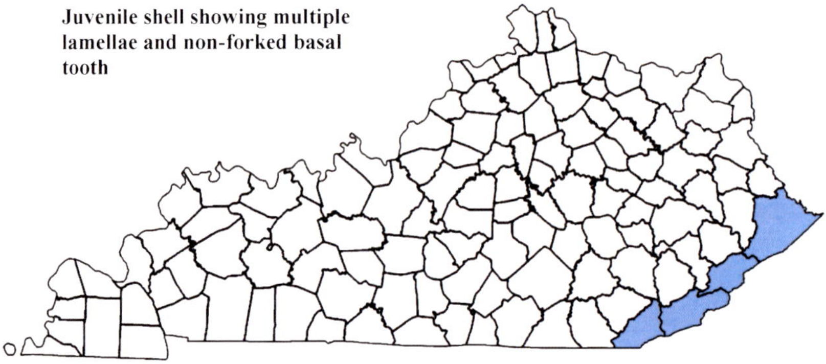

Split-tooth Dome

Ventridens virginicus (Vanatta, 1936)

Diameter: 5.4-6.4 mm, height 3.4-4 mm

Description: Depressed heliciform; lip simple; shell with 6 whorls; perforate to umbilicate; live animal dark, olive or bluish-gray; shell glossy and semi-transparent; transverse striae are poorly developed; spiral striae present but a weak feature; aperture with basal lamellae in all stages of growth, in old, adult shells the outer-wall lamellae is often absorbed; in the juvenile stage (occasionally adults) there is a distinctive double columellar lamella (fig. a); a member of the *V. gularis* group.

Similar Species: *Ventridens suppressus* is slightly larger, contains no outer wall lamellae in adult shells and its juvenile shell has a single lamella (not forked as in *V. virginicus*).

Habitat: In Kentucky the species was discovered on a dry limestone outcrops (glade) at Carter Cave State Park.

Status: G4/S1; State Endangered; currently known only from Carter Caves State Park, Carter County.

Specimen: West Virginia, Mineral County, Patterson Creek Road (author's collection).

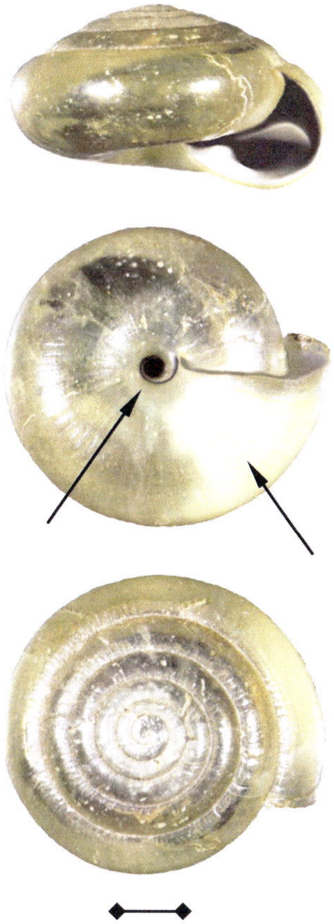

Juvenile shell showing multiple lamellae and forked basal tooth (Fig. a)

North Fork Mountain, Pendleton County, West Virginia (FM 246975)

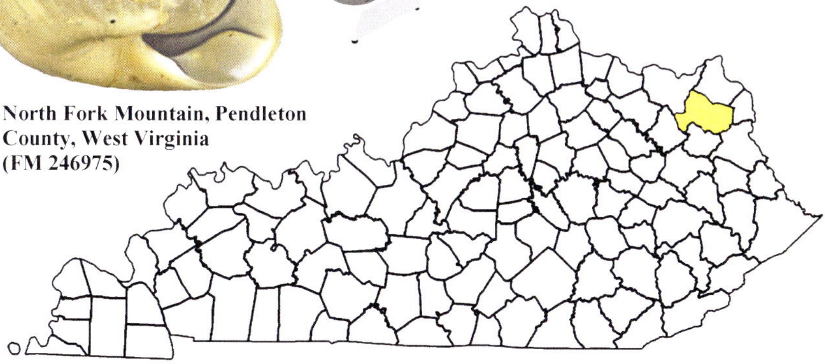

Throaty Dome

Ventridens gularis (Say, 1822)

Diameter: 7.5-9 mm, height 5-6.3 mm

Description: Depressed heliciform to heliciform, young shells are more depressed; lip simple; shell with 6-7 tightly coiled whorls; perforate, usually widest in young shells (fig. a) but in some adults, the perforation is as wide as half a millimeter; live animal nearly black (an important feature); shell translucent and glossy; transverse striae are distinct on top, but weaker on the sides and bottom; spiral striae are indistinct; within the aperture of both young and adult shells there are two lamellae which can be seen through the bottom of fresh shells (fig. b); a member of the *V. gularis* group.

Similar Species: Both *V. suppressus* and *V. virginicus* are smaller with a notably wider umbilicus; *Ventridens pilsbryi* has 7-8 whorls, a flatter shell and the live animal is pale yellow.

Habitat: Found on wooded hillsides, ravines, under leaf litter and around logs.

Status: G5/S4; Relatively Common; occurring in south central to the southeastern mountain regions of Kentucky.

Specimen: Tennessee, Monroe County, Towhee Falls Baptist Church (author's collection).

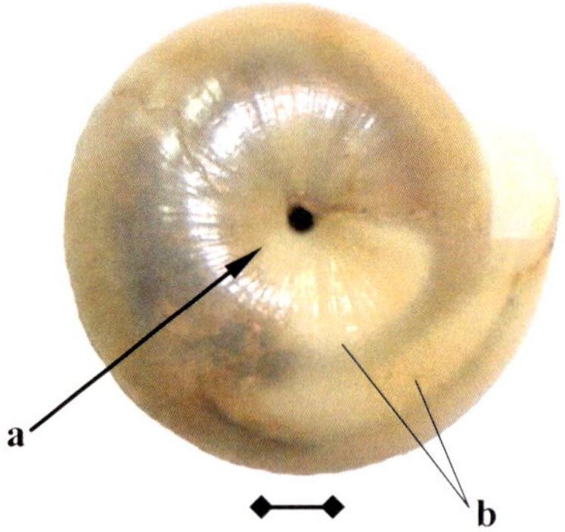

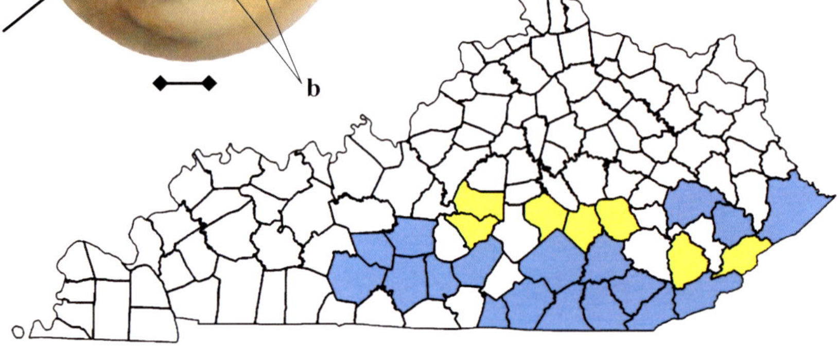

Yellow Dome

Gastrodontidae

Ventridens pilsbryi Hubricht, 1964

Diameter: 8.7-9.6 mm, height 4-5 mm

Description: Depressed heliciform; lip simple; shell with 7-8 tightly coiled whorls; perforate; live animal pale, with or without a hint of yellow; shell translucent; transverse striae (growth wrinkles) are distinct on top and somewhat weaker on the bottom; spiral striae are indistinct; within the aperture are two lamellae or teeth that are present at every stage of growth; a member of the *V. pilsbryi* group (Hubricht 1964).

Similar Species: Most similar to *V. gularis* which has a more globose shell with fewer whorls and the live animal is dark-olive or a bluish-gray (the live animal of *V. pilsbryi* is pale) .

Habitat: Found on wooded hillsides and in ravines under leaf litter and on logs, more common on limestone but also occurs on acidic sandstone sites.

Status: G5/S4; Relatively Common; occurring from central to the southeastern portion of Kentucky.

Specimen: Tennessee, Monroe County, (author's collection).

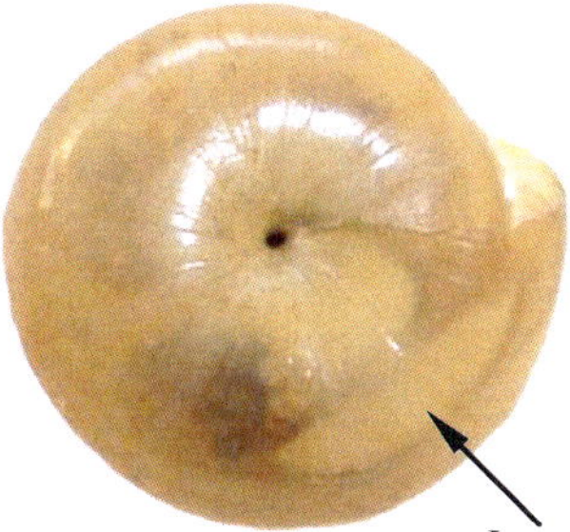

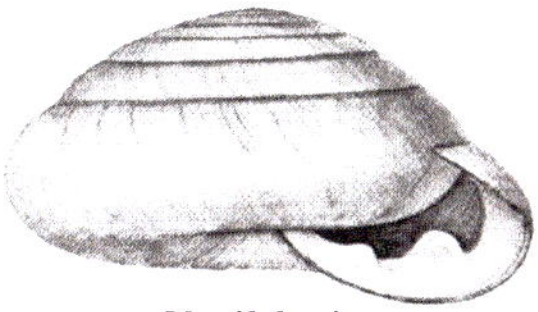

V. pilsbryi

© Pilsbry, 1946

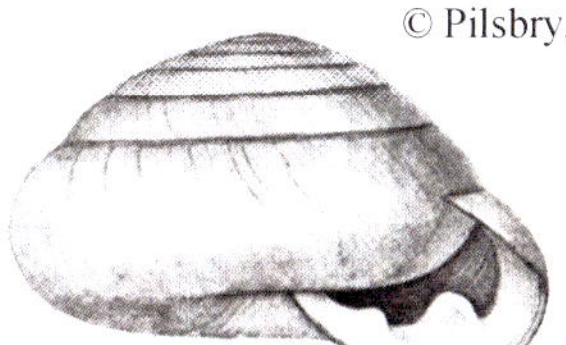

V. gularis

Young shell

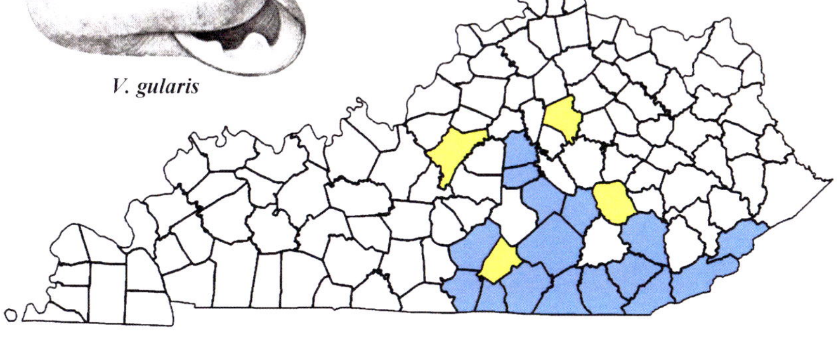

Hollow Dome

Ventridens lasmodon (Phillips, 1841)

Diameter: 7.5-7.8 mm, height 3.8 mm

Description: Depressed heliciform; lip simple; shell with 7.5-8 tightly coiled whorls; umbilicate (the widest of any *Ventridens*), the umbilicus is funnel-shaped and nearly one third the diameter of the shell; live animal pale; shell translucent, thin and glossy; finely developed transverse striae are distinct on top, but grow weaker on the sides and base of shell; spiral striae not well defined; within the aperture are two elongated lamellae (fig. a) or teeth which are present at every stage of growth, but as in *V. lawae,* sometimes out of sight in frontal view; a member of the *V. pilsbryi* group.

Similar Species: *Ventridens lawae* has a higher shell profile, a thicker shell, and a notably smaller umbilicus.

Habitat: Found on wooded hillsides and in ravines under leaf litter on both limestone and sandstone sites.

Status: G4/S3; Uncommon; only reported from 6 counties in southeastern Kentucky.

Specimen: Tennessee, Monroe County, Cherokee National Forest (author's collection).

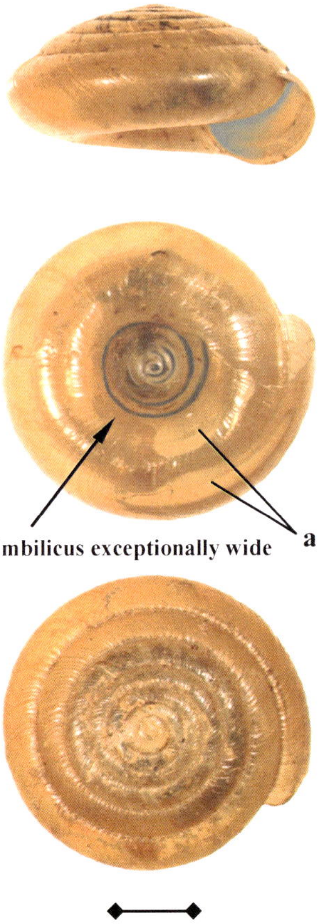

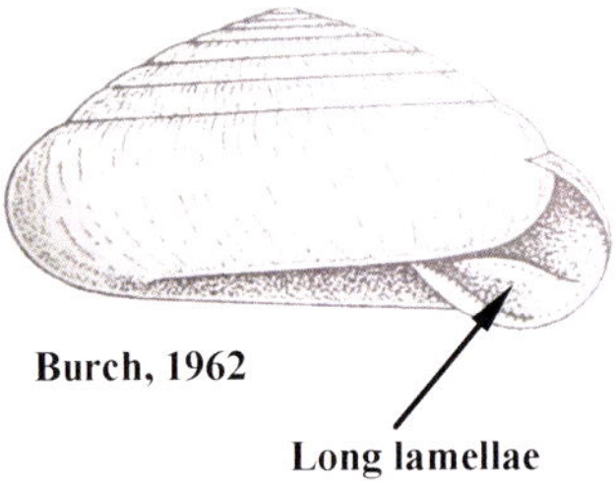

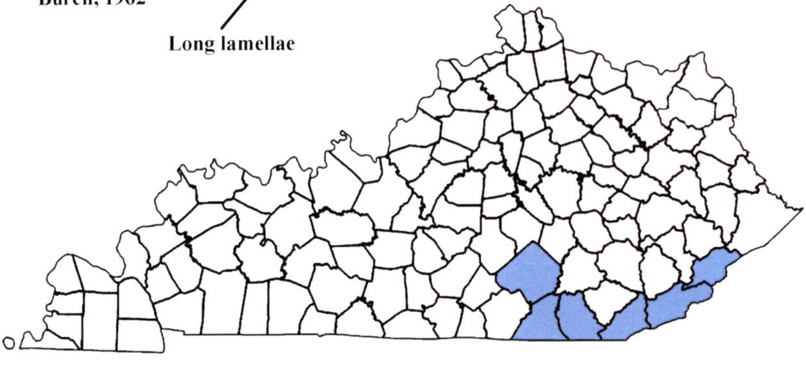

Rounded Dome

Gastrodontidae

Ventridens lawae (W. G. Binney, 1892)

Diameter: 7.8-9 mm, height 4 mm

Description: Depressed heliciform; lip simple; shell with 7.5-8.5 tightly coiled whorls; umbilicate, the umbilicus is deep with nearly parallel sides (well-like); in adults the umbilicus may vary from 0.4 mm to 2 mm in size (Hubricht 1964); live animal pale; shell light horn to yellow and glossy; finely developed transverse striae are distinct on top, but dwindle on the sides and base of shell; young shells with lamellae (fig. a); adults with or without a basal lamella; a member of the *V. pilsbryi* group.

Similar Species: *V. theloides* is similar in build, but with a smaller umbilicus and typically without adult lamellae.

Habitat: Found in mixed hardwood forests on hillsides and in ravines under leaf litter and around logs, more common on limestone, but also occurs on sandstone sites (Hubricht 1985).

Status: G3/S3S4; Uncommon; reported from scattered locations in southeastern Kentucky.

Specimen: North Carolina, Macon County, Nantahala National Forest (author's collection).

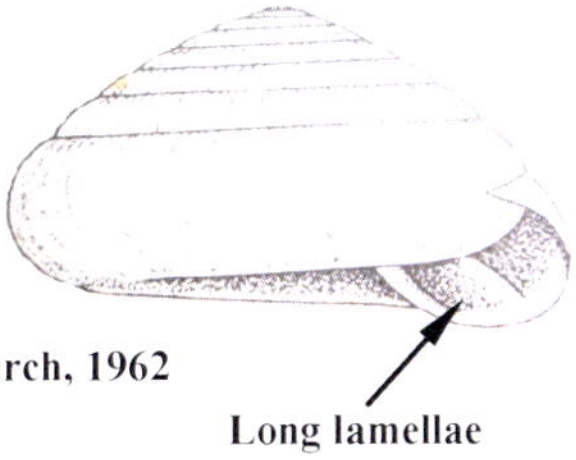

Burch, 1962

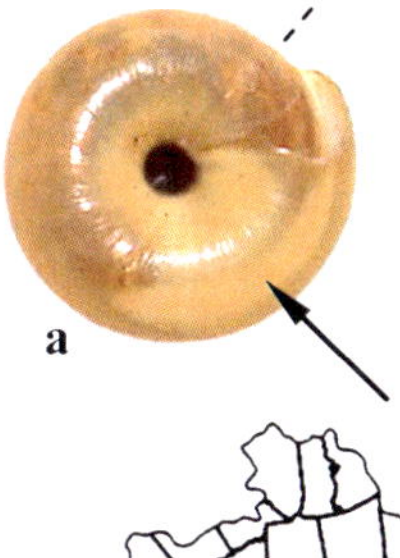

a

V. lawae, Bad Branch Nature Reserve, Letcher CO. KY

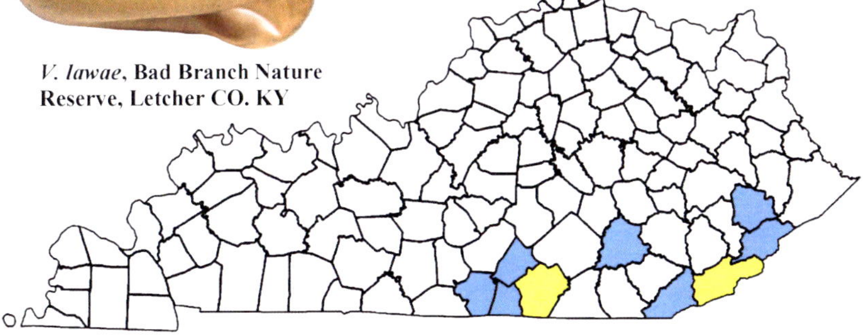

Copper Dome

Ventridens theloides (Walker & Pilsbry 1902)

Diameter: 7.5-9 mm, height 4.5 mm

Description: Heliciform and dome shaped; lip simple or with a slight basal deflection (fig. a); 7.5-9 tightly coiled whorls; perforate but open, the area around umbilicus well excavated and funnel-like (fig. b); live animal pale; shell yellowish or coppery, the base very glossy; transverse striae are moderately well developed on top, but are diminished on the base; young shells are usually with basal lamellae (fig. c), adult shells without; a member of the *V. pilsbryi* group.

Similar Species: *Ventridens gularis* has well-developed basal lamellae and a smaller umbilicus; *V. lawae* has a much wider umbilicus and is more compressed.

Habitat: Found in upland hardwood forests, in ravines, under leaf litter and around logs; generally a species of limestone but also occurs on sandstone sites.

Status: G5/S4; Uncommon; reported from scattered locations in southeastern Kentucky

Specimen: North Carolina, Swain County, Nantahala NF (author's collection).

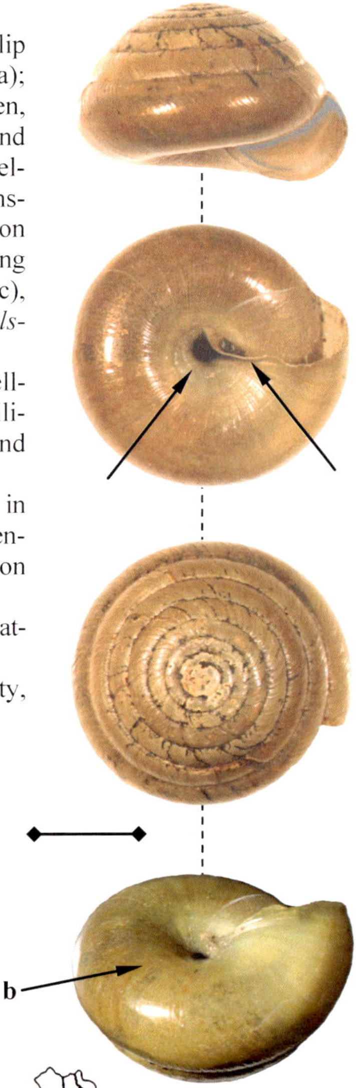

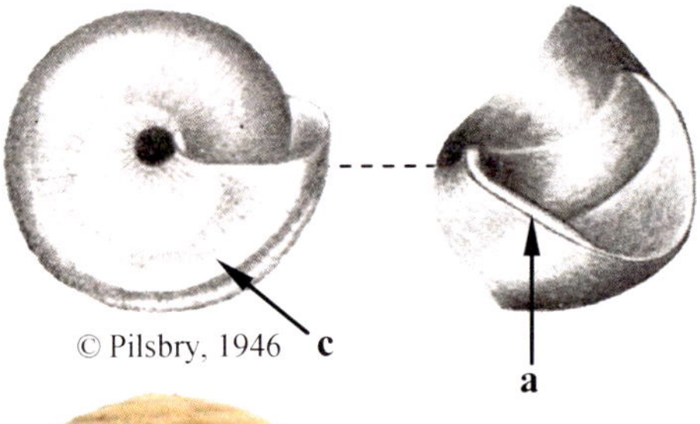

© Pilsbry, 1946

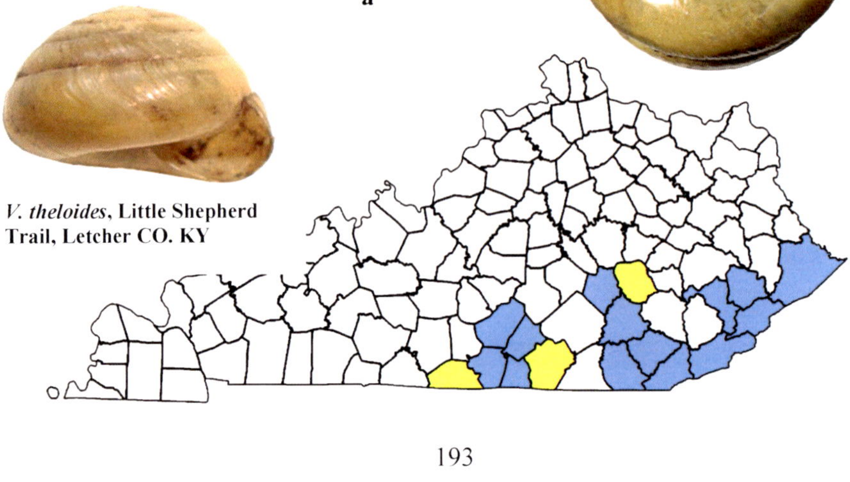

V. theloides, **Little Shepherd Trail, Letcher CO. KY**

Sculptured Dome

Gastrodontidae

Ventridens collisella (Pilsbry, 1896)

Diameter: 8.4-9.6 mm, height 6-8.1 mm

Description: Heliciform; lip simple; shell with 7.5-8 tightly coiled whorls; perforate, young shells are umbilicate, the umbilicus becoming smaller as shells mature (fig. a); live animal pale; shell pale yellowish horn color and dull glossy; transverse striae are well-defined and boldest near the sutures; without any notable spiral striae; usually with two entering lamellae in both juvenile and adult shells, becoming lower in stature with age, although some specimens will be without this armature; a member of the *V. pilsbryi* group (Hubricht 1964).

Similar Species: *Ventridens ligera* is larger and does not contain teeth at any stage of growth.

Habitat: A calciphile species occurring on wooded hillsides in hardwood forests.

Status: G4G5/S4; Relatively Common; known from scattered locations in central and southeastern Kentucky.

Specimen: Virginia, Tazewell County, (author's collection).

V. collisella, Menifee Co. KY

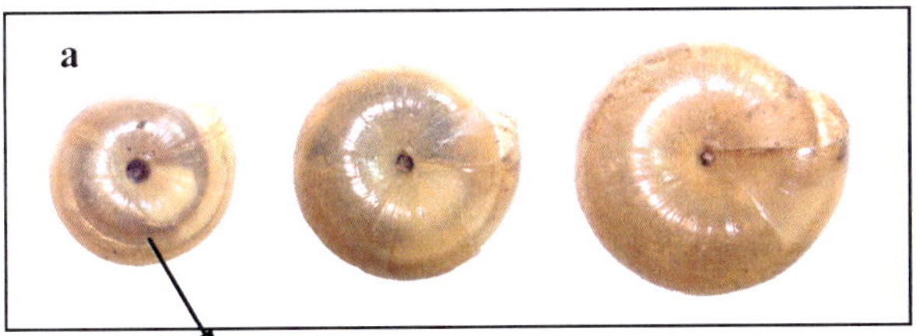

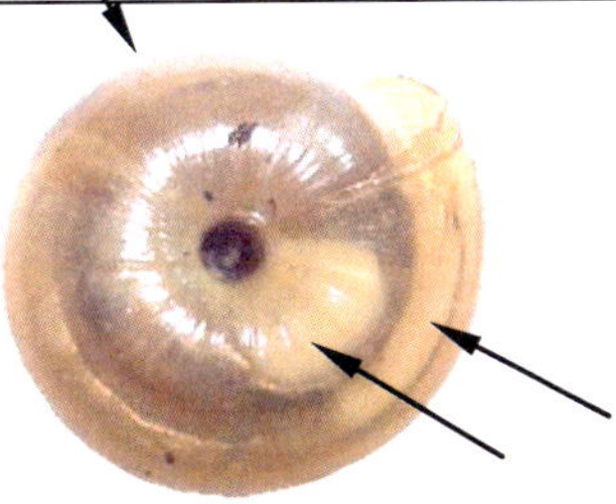

Translucent shell showing internal lamellae

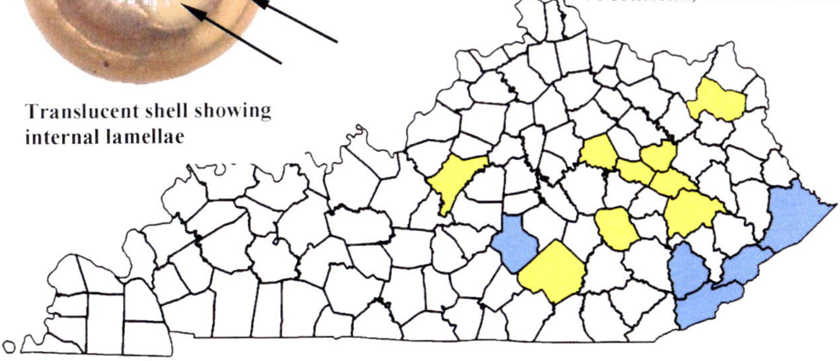

Perforate Dome

Gastrodontidae

Ventridens demissus (A. Binney, 1843)

Diameter: 7.5-11 mm, height 4.8-6.8 mm

Description: Heliciform; lip simple; shell with 6-7.5 tightly coiled whorls; perforate; shell horn-yellow and glossy; transverse striae are moderately developed on top of shell, but largely weakened on the sides and bottom; aperture typically without basal lamella in adult shells, but young shells (fig. a) are typically with these features.

Similar Species: *Ventridens ligera* is larger by a few millimeters, without teeth in juvenile shells and in general a higher shell than *V. demissus*; *V. percallosus* is only slightly smaller, has better developed transverse striae and a notable buildup of callus within the aperture (fig. a, page 197).

Habitat: Found on wooded hillsides, in ravines and floodplains under leaf litter; also a snail found frequently in urban areas.

Status: G5/S5; Common; one of the most common and widespread *Ventridens* in Kentucky, likely occurring in every county.

Specimen: Kentucky, Letcher County, Bad Branch Nature Reserve (author's collection).

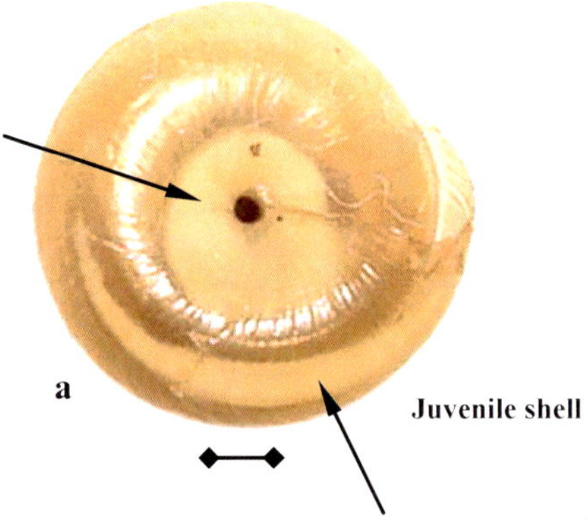

a

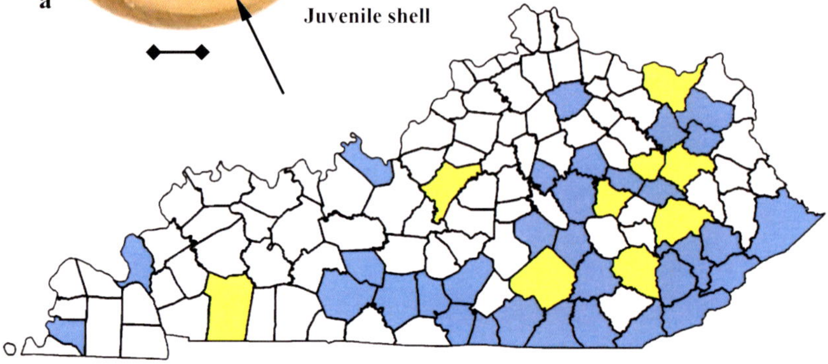

Pyramid Dome

Gastrodontidae

Ventridens intertextus (A. Binney, 1841)

Diameter: 8-20 mm, height 10-13.3 mm

Description: Heliciform; lip simple; shell with 5.5-6 moderately coiled whorls; perforate; shell yellowish-horn to olive-buff and dull; broken transverse striae (fig. a) are well-developed and are the single best feature for identifying this species; spiral striae are well developed; without teeth or basal lamellae in all stages of growth; adult shell periphery can be round or slightly angular, the shells of juveniles have a more angular periphery; shells occasionally with a faint color band.

Similar Species: *V. ligera* has a similar form, but does not have the unique broken striae found in *V. intertextus*.

Habitat: Found in mixed hardwood forests on hillsides, ravines, and acidic ridgetops, under leaf litter.

Status: G5/S5; Common; a frequent and widespread land snail in Kentucky.

Specimen: Figs. (b) from Kentucky, Powell County and (fig. c) a rounded form from West Virginia, Mineral County (author's collection).

b

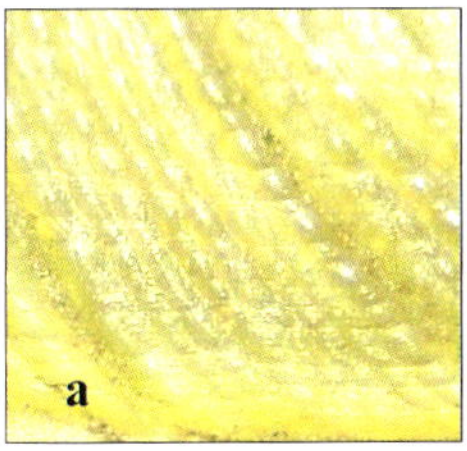

a

Young shells are more angular

c

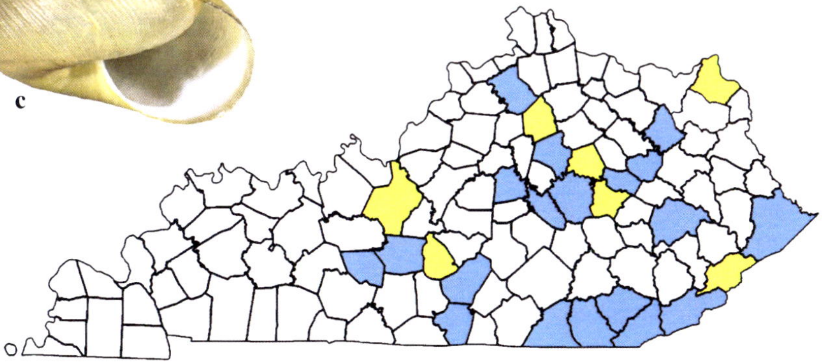

Tennessee Dome

Gastrodontidae

Ventridens percallosus (Pilsbry, 1898)

Diameter: 8.4-10 mm, height 7-7.9 mm

Description: Heliciform; lip simple; shell with 7-7.5 whorls; perforate; shell pale yellowish-horn color; transverse striae (growth wrinkles) are weakly defined on top and somewhat weaker on the sides and bottom of shell; superficial spiral striation is more or less developed on the base; notable buildup of callus within the aperture (fig. a), young shells up to 6 whorls and nearly 7 mm in size have a strong outer basal lamella and a short callus on the columella; without lamellae in adult shells.

Similar Species: *Ventridens collisella* has better developed transverse striae on the upper surface (fig. b) and usually a lamella in adult shells (fig. c); *V. ligera* is larger, has a thinner callus in its aperture and the young shells of *V. ligera* are without an outer-basal lamella.

Habitat: Found on wooded hillsides, in ravines and floodplains under leaf litter; also a snail found frequently in urban areas.

Status: G5/S1; State Special Concern; in Kentucky, known only from Todd County. More collections are needed to make final deliberation on this rare species in the state.

Specimen: Tennessee, Davidson County, Nashville, 3201 Pillsboro Road (FMNH 246994).

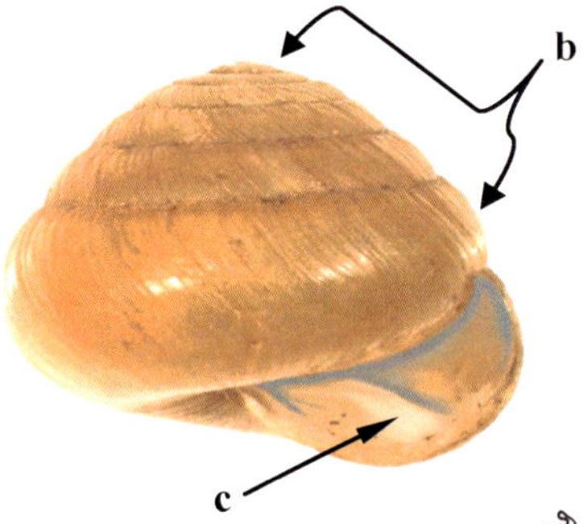

Juvenile shell of *V. percallosus*
© Pilsbry, 1946

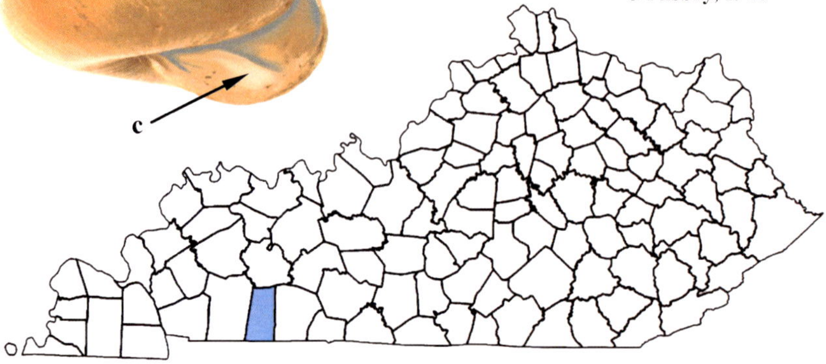

Glossy Dome

Gastrodontidae

Ventridens cf. acerra (J. Lewis, 1870)

Diameter: 12.6-18.3 mm, height 10.8-13.2 mm

Description: Heliciform; lip simple; shell with 7-8 moderately coiled whorls, the last whorl only slightly expanded; perforate; shell light yellowish-olive and very glossy; transverse striae are poorly developed, strongest on the top but fade away on the sides and base; without notable spiral striae; no teeth at any stage of growth; periphery rounded.

Similar Species: *Ventridens acerra* resembles *V. ligera* but is slightly larger, has a glossier shell surface, slightly higher shell profile (this feature variable), has the transverse striae wider on the last whorl, more widely excavated around the umbilicus and except at the columella, no internal callus.

Habitat: A habitat generalist found in a variety of mixed hardwood forests on hillsides under leaf litter usually in rocky places (Hubricht, 1985).

Status: G4/S1; State Special Concern; Specimens found in Martin County refer to *V. acerra* but additional shells are needed to make a positive ID.

Specimen: Kentucky, Martin County, Route 292 near Naugatuck, KY, above Tug Fork, GPS 37.79000-82.351940 (author's collection).

***V. acerra*, Swain County, North Carolina**

***V. acerra*, Graham County, North Carolina**

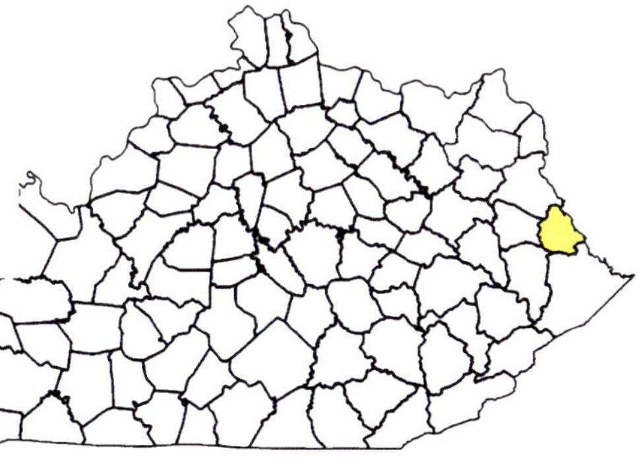

Globose Dome

Gastrodontidae

Ventridens ligera (Say, 1821)

Diameter: 11-15.6 mm, height 8-12.7 mm

Description: Heliciform, the height of this species will vary considerably; lip simple; shell with 6-7 moderately coiled whorls; perforate; shell pale yellowish-horn and somewhat glossy; transverse striae are comparatively well developed on top of shell but much less defined on the sides and bottom; aperture without a basal lamella in any stages of growth.

Similar Species: *Ventridens intertextus* has a similar form but has better developed transverse striae, a slightly larger umbilicus, often, a weakly angular periphery, and on occasions, a light color band on its periphery; *V. demissus* is smaller and young shells are armed with internal lamellae.

Habitat: Found in a variety of open, weedy and mixed hardwood forests in floodplains and other wet low-lying areas; also along roadsides.

Status: G5/S5; Common; one of the most common and widespread *Ventridens* in the state but appears to be absent from Kentucky's higher mountain ranges.

Specimen: Kentucky, Estill County, along the Kentucky River around Irvine (author's collection).

***V. ligera*, Fayette County, West Virginia**

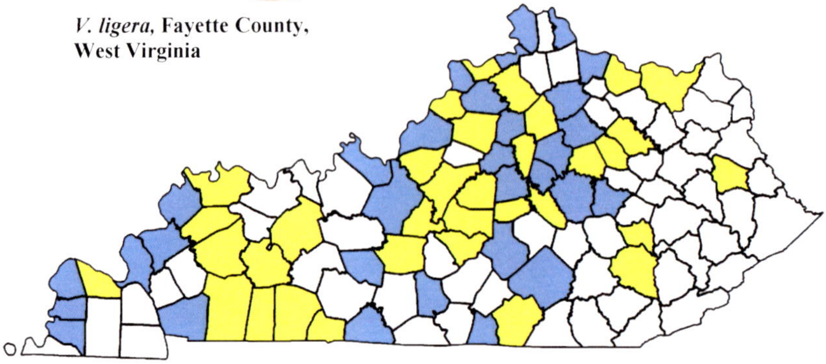

Kentucky *Ventridens* Compared (shells proportionate)

Adult Shells Usually with Teeth

V. collisella, Tazewell Co, VA

V. pilsbryi, Monroe Co, TN

V. gularis, Haywood Co, NC

V. lawae, Macon Co, NC

V. lasmodon, Monroe Co, TN

V. suppressus,
Craig Co, VA

10 mm

V. virginicus,
Mineral Co, WV

Adult Shells without Teeth

V. ligera, Powell Co, KY

V. ligera, Fayette Co, WV

V. ligera, Fayette Co, KY

V. intertextus, Taylor Co, WV

V. demissus,
Letcher Co, KY

V. theloides,
Swain Co, NC

Ventridens cf. acerra,
Martin Co, KY

10 mm

Ventridens percallosus
Davidson Co, TN

Kentucky *Ventridens* Compared (shells proportionate)

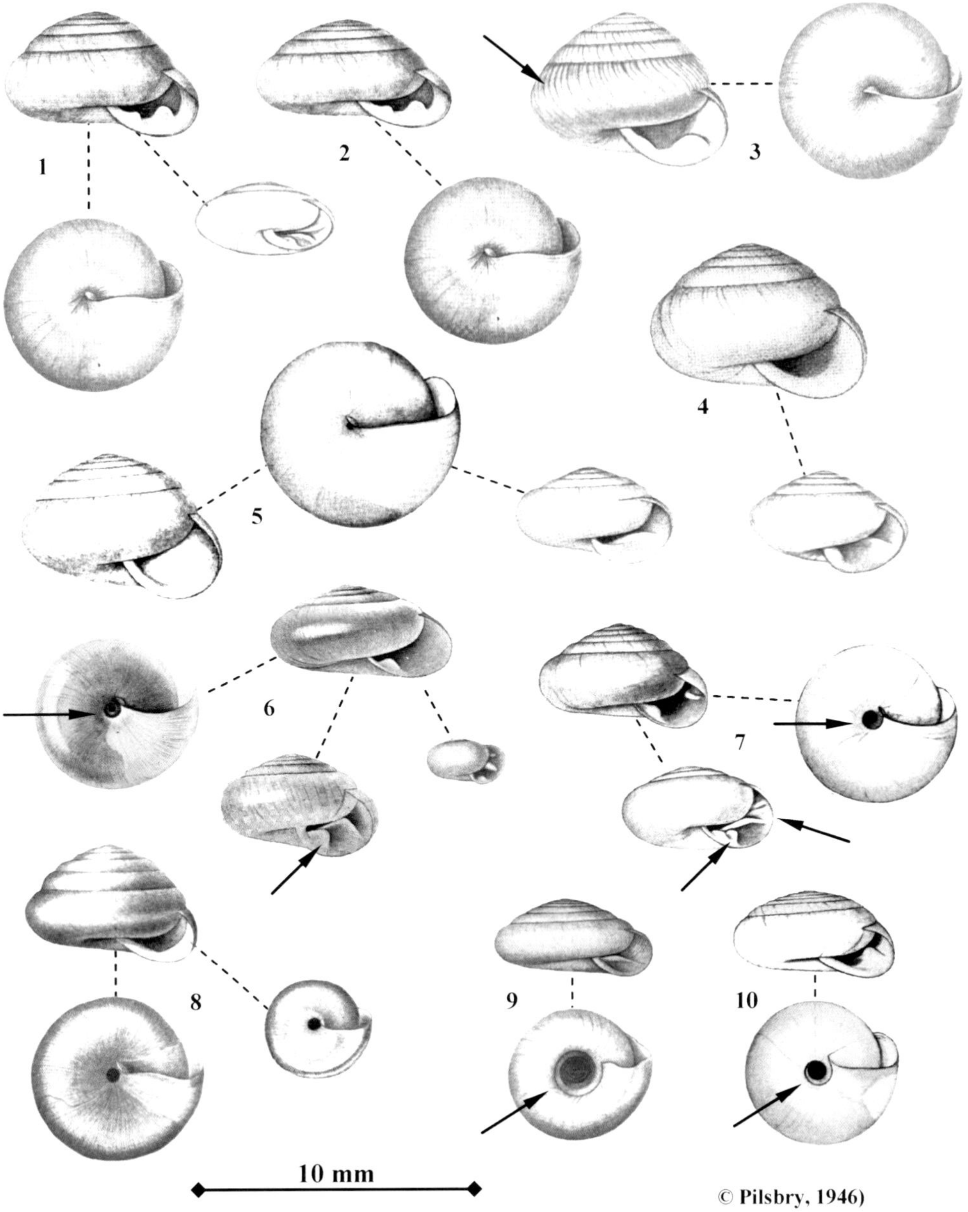

1. ***Ventridens gularis*** (common, always with basal lamellae; shown here with young shell)
2. ***Ventridens pilsbryi*** (uncommon, live animal yellow, flatter than *V. gularis*)
3. ***Ventridens collisella*** (transverse striae well developed)
4. ***Ventridens percallosus*** (notable buildup of callus inside bottom lip, shown here with young shell)
5. ***Ventridens demissus*** (transverse striae not well developed; shown here with young shell)
6. ***Ventridens suppressus*** (one basal lamella in adult shells; shown here with young shells)
7. ***Ventridens virginicus*** (two basal lamellae in adult shells, basal lamella forked in young shells)
8. ***Ventridens theloides*** (shell dome-shaped, no teeth in adult shells; shown here with young shell)
9. ***Ventridens lasmodon*** (wide open umbilicus)
10. ***Ventridens lawae*** (open umbilicus, but smaller than *V. lasmodon*)

With perfect balance, a five millimeter wide juvenile Flat Bladetooth, *Patera appressa,* crawls across a single strand of silk, demonstrating its remarkable adaptable shape-sifting foot. Red River Gorge, Powell County, Kentucky.

Shells 15-36 mm, without teeth, simple lip (*Mesomphix*)

Mesomphix species have medium to large shells (15-35 mm) that are either depressed heliciform or heliciform. The shells are loosely coiled, the last whorl expanded (fig. a), while *Ventridens* species have more tightly coiled whorls (fig. b) and are typically smaller. The shell surface can be dull to glossy (some species with a glass-like surface). Lips are not reflected and are without teeth in all stages of development. The transverse striae vary from closely spaced to indistinguishable. Several species of this genus exhibit spirally arranged papillae. The shells are umbilicate to perforate; the umbilicus never completely closed. Like *Ventridens*, most *Mesomphix* have a thickening which often forms a whitish callus just within the aperture. Two or three fresh shells of *M. cupreus,* shaken in one loosely closed hand will sound remarkably like glass. Shaking other species of snails like *Mesodon* will have a duller sound. Immature *Mesodon* species are easily confused with mature *Mesomphix* snails (see below comparison). The *Mesomphix* species will be arranged according to the absence or presence of papillae. **Group 1** shells that have spiral papillae and **Group 2** shells that are generally without any spiral papillae, although some shells in this group may occasionally have scant traces of this micro-feature (see next page)

Important Note: Immature *Mesodon* shells are often confused with immature and mature *Mesomphix* species (see below illustrations). An immature ***Mesodon*** has a more rounded non-reflected lip (only reflected when fully mature), while the mature ***Mesomphix*** lip is more oval in shape.

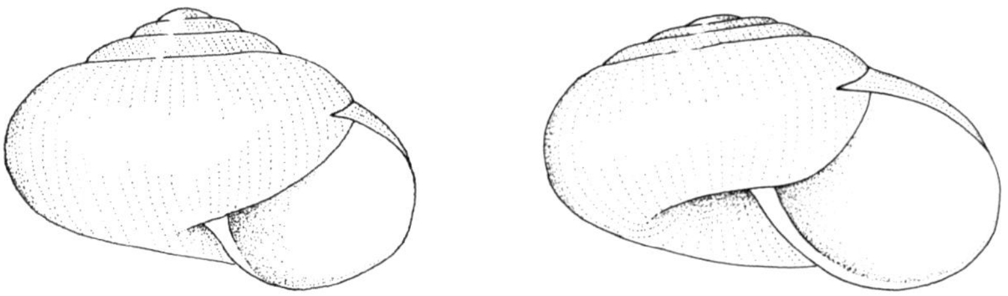

Immature *Mesodon* Mature *Mesomphix*

Grouping 1: typically with notable spiral rows of papillae, pg. 205-219

Mesomphix capnodes (rare)
Mesomphix perlaevis
Mesomphix vulgatus
Mesomphix anurus (rare)
Mesomphix inornatus
Mesomphix rugeli oxycoccus (rare)

Shell generally with notable spiral papillae

Grouping 2: typically without notable spiral rows of papillae, pg. 211-214

Mesomphix rugeli (rare)
Mesomphix cupreus
Mesomphix globosus
Mesomphix friabilis

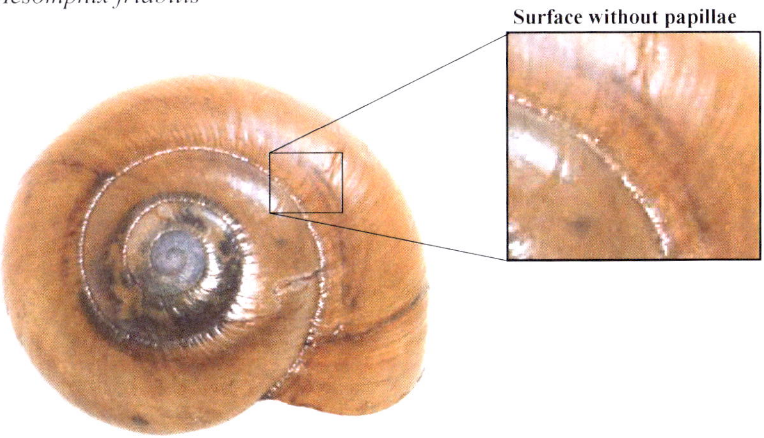

Shell generally smooth without notable spiral papillae

Dusky Button

Gastrodontidae

Mesomphix capnodes (W. G. Binney, 1857)

Diameter: 29.5-35.5 mm, height 18-26.2 mm

Description: Depressed heliciform to heliciform, in degree of elevation the shells vary widely, even within the same site, as illustrated on this page; lip simple; shell with 5 loosely coiled whorls; perforate to umbilicate; shell thin but not fragile; periostracum dark; somewhat glossy; embryonic whorl usually worn; no teeth present; with or without a thickening or thin whitish callus inside the aperture; transverse striae are poorly developed; spiral rows of papillae are present, but not well developed.

Similar Species: *Mesomphix cupreus* is typically smaller, with a slightly higher shell profile, a wider umbilicus, and very rarely displays any spiral papillae.

Habitat: A calciphile snail of upland mixed hardwood forests usually found under loose leaf litter on steep hillsides.

Status: G5/S3; Rare; currently known from only three Kentucky counties.

Specimen: All figures from Tennessee, Hamilton County, Ruby Falls (author's collection).

Umbilicus wide for a *Mesomphix*

A more elevated form of *Mesomphix capnodes*

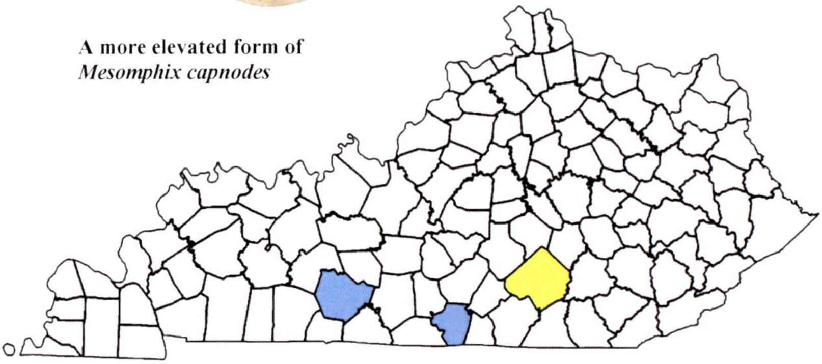

Smooth Button

Mesomphix perlaevis (Pilsbry, 1900)

Diameter: 17-21 mm, height 11-12 mm

Description: Heliciform, shell height variable, even within the same populations (see below); lip simple; shell with 5 loosely coiled whorls; perforate; shell thin, but not fragile, light olive, glossy, embryonic whorl, if not worn, with notable transverse striae (can be seen with a hand lens of 10X); with or without a thin whitish callus just inside the aperture's bottom of the last whorl (fig. a); transverse striae are strongest on the third whorl, the striations of the last whorl become notably weaker; fine spiral rows of papillae are not well developed (wanting in most places) but usually present on at least some portions of the upper shell surface.

Similar Species: *Mesomphix inornatus* is flatter in build and has well-developed spiral papillae; *M. cupreus* is larger, with a more open umbilicus and without spiral papillae.

Habitat: A snail of rich upland mixed hardwood forests usually found under and among moist leaf litters and detritus.

Status: G5/S5; Common; this is a widespread species in central and eastern Kentucky.

Specimen: Tennessee, Cocke County, Cosby Campground, GSMNP (GSMNP collection).

***M. perlaevis*, both specimens from same site, Furnace Mountain, Powell County, Kentucky**

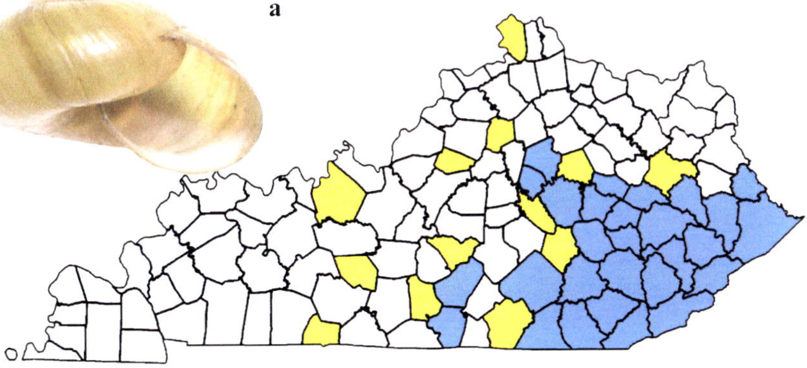

Common Button

Mesomphix vulgatus H. B. Baker, 1933

Diameter: 16-27.8 mm, height 14 mm

Description: Heliciform; lip simple; shell with 4.5-5.5 whorls; shell thin but not fragile, olive-tan with darker and lighter streaks but often uniform in color; surface dull; perforate; no teeth present but *Mesomphix* species have a thickening or thin whitish callus that is usually just inside the aperture's bottom of the last whorl; transverse striae (growth wrinkles) are well developed and closely spaced; spiral rows of papillae are well developed continuing to the lower surface.

Similar Species: *Mesomphix anurus* is smaller, having a more polished surface and the spiral rows of papillae are a wanting feature and generally absent on the base; *M. perlaevis* is smaller, glossier and has a higher shell profile.

Habitat: Found in dry habitats around limestone outcrops, hillsides, ravines and sinkholes covered in hardwood forests and cedar forests; a common species along the Kentucky River Palisades.

Status: G5/S5;Common; found throughout most of Kentucky. but absent from higher elevations.

Specimen: Fig. (a) typical form from Kentucky, McCreary County, Yahoo Falls, and (fig. b) a more elevated form from Kentucky, Fulton County, Wilson's Hill (author's collection).

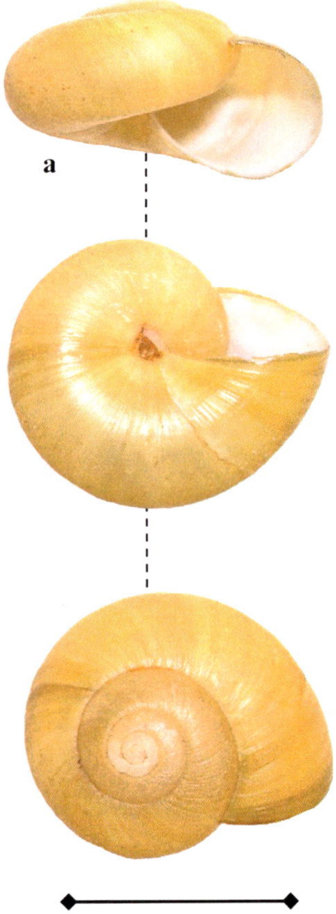

A darker, more elevated form Fulton Co, KY

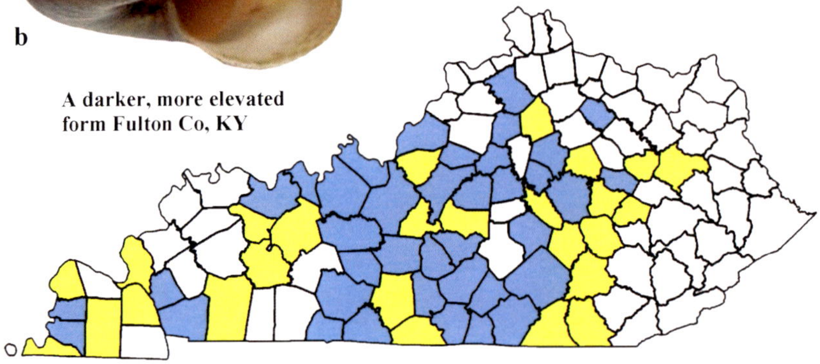

Frog Button

Gastrodontidae

Mesomphix anurus Hubricht, 1962

Diameter: 20.3 mm, height 11-12 mm

Description: Heliciform; lip simple; shell with about 5 whorls; shell thin but not fragile, pale brownish-yellow with darker and lighter streaks; surface glossy and translucent; perforate, umbilicus sealed with hardened slime; no teeth present; a thickening or thin whitish callus that is usually just inside the aperture's bottom of the last whorl (clearly seen in the specimens to right); transverse striae are well developed and closely spaced; spiral rows of papillae are poorly developed but always present, the base is smooth and shiny.

Similar Species: *Mesomphix vulgatus* is larger, has a duller surface and the spiral rows of papillae are much better developed on both the top and base; *M. inornatus* is notability flatter.

Habitat: Found in the same habitat as *Mesomphix vulgatus* but usually not found together.

Status: G4/S3; Uncommon; All records in Kentucky are from just two south-central counties the species becoming more common in Tennessee south.

Specimen: Tennessee, Campbell County, a roadside at Cove Lake State Park, (author's collection).

***Mesomphix inornatus*,**
Powell Co, KY

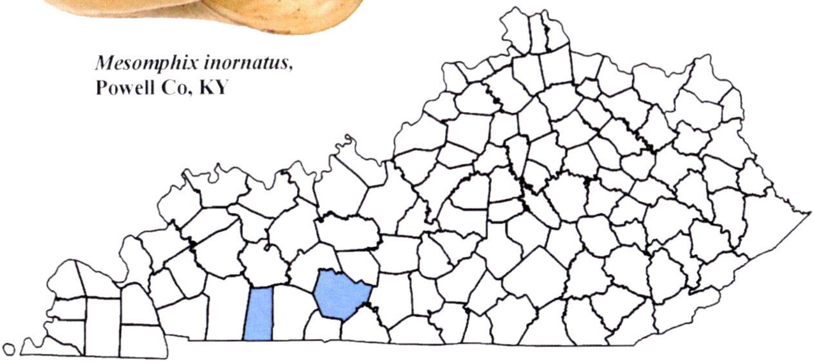

Plain Button

Mesomphix inornatus (Say, 1821)

Diameter: 16-21 mm, height 8.4-9.8 mm

Description: Depressed heliciform; lip simple; shell with 5 loosely coiled whorls; perforate, shell thin but not fragile, olive-tan, very glossy; embryonic whorl smooth (a hand lens of 10X can be used to view this feature); no teeth present; a thickening or thin whitish callus that is typically just inside the bottom of the aperture of the last whorl; transverse striae and spiral rows of papillae are present; a carnivorous species which feeds on other live snails (Dourson pers. obs.).

Similar Species: *Mesomphix perlaevis* has a higher shell profile and the transverse striae are clearly more defined on the embryonic whorl.

Habitat: A snail of rich, upland mixed hardwood forests including beech and hemlock, usually found under or in moist leaf litter and detritus, but also a species found thriving along shale banks of road-cuts.

Status: G5/S5; Common; throughout central and eastern Kentucky, becoming increasingly rare in western portions of the state.

Specimen: Kentucky, Powell County, Red River Gorge (author's collection).

Specimens of *M. inornatus*, from Warren Co, Pennsylvania are small, around 12-13 mm

***M. inornatus*, Cheat River Gorge, Snake Hill WMA, West Virginia**

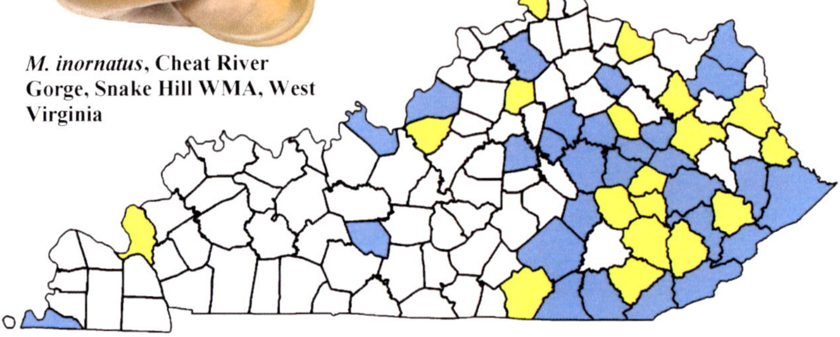

Imitator Button

Gastrodontidae

Mesomphix rugeli oxycoccus (Vanatta, 1903)

Diameter: 20-22.5 mm, height 12-15 mm

Description: Heliciform; lip simple; shell with 5.5-6 loosely coiled whorls; perforate; shell thin but not fragile, greenish-horn and glossy often with darker streaks marking former growth stoppages; occasionally has a thin whitish callus just inside the aperture; transverse striae are smooth and the shell displays finely developed spiral rows of papillae, best observed on the top last two whorls, becoming nearly obsolete on the base, or seen as mere traces; periphery well rounded.

Similar Species: *Mesomphix perlaevis* is lighter in color, has better developed transverse striae and weaker spiral rows of papillae; *M. rugeli oxycoccus* differs from *M. rugeli* (page 211) by having well-developed and crowded spiral papillae and a somewhat duller shell surface but the latter character is somewhat variable.

Habitat: Low to mid-elevation hillsides and ravines in mixed hardwood forests under moist leaf litters.

Status: Not Ranked; Rare; this interesting subspecies was first described by Vanatta in 1903 and later synonymized by Hubricht in 1985, but without explanation, consequently further DNA work is needed to make a final deliberation on its status as a species; it has been reported from southern West Virginia, southeastern Kentucky in Pike County, eastern Tennessee and western North Carolina.

Specimen: Figs. (a) from Tennessee, Carter County, Cherokee National Forest and (fig. b) from Kentucky, Pike County (author's collection).

a

b

Wrinkled Button

Mesomphix rugeli (W. G. Binney, 1879)

Diameter: 16-22.5 mm, height 13-14 mm

Description: Heliciform; lip simple; shell with 5.5-6 whorls; shell thin but not fragile, greenish -horn, glossy; perforate, no teeth present; with or without a thickening or thin whitish callus that is just inside the aperture's bottom of the last whorl; transverse striae (growth wrinkles) are smooth and the shell is usually without discernible spiral rows of papillae; *Mesomphix rugeli* is known from Harlan County only. An interesting form, *Mesomphix rugeli* form *oxycoccus* (Pilsbry 1946) occurs in Pike County on lower hillsides (under 2000 feet) in mixed hardwood forests. It differs from typical *M. rugeli* by having well developed and crowded spiral papillae.

Similar Species: *Mesomphix perlaevis* has better developed transverse striae and spiral rows of papillae; *M. friabilis* is larger, has a more expanded last whorl, and has a slightly wider umbilicus.

Habitat: A snail of higher elevation mixed hardwood forests usually found under and among moist leaf litters and detritus.

Status: G4/S2; State Threatened; in Kentucky, this rare species is known only from Black Mountain in Harlan County.

Specimen: Fig. (a) from Kentucky, Harlan County and (fig. b) from Mitchell County, North Carolina (author's collection).

a

b

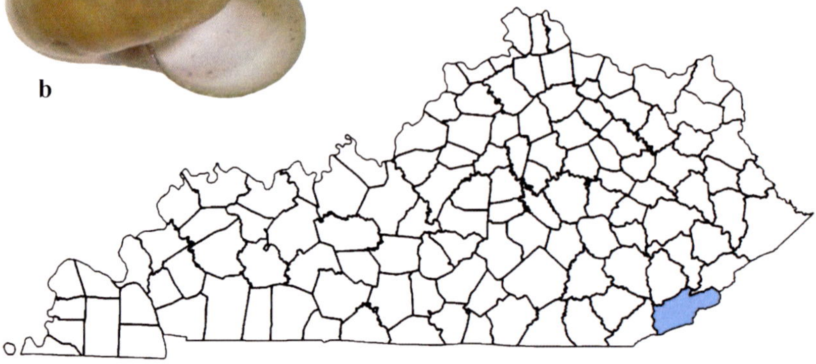

Copper Button

Gastrodontidae

Mesomphix cupreus (Rafinesque, 1831)

Diameter: 22-29.5 mm, height 12.4-15.4 mm

Description: Heliciform, but with a great deal of variation; lip simple; shell with 4.5-5 loosely coiled whorls; umbilicate, has the widest umbilicus of any *Mesomphix* species found in Kentucky; shell thin but not fragile; periostracum dark; shell glossy; embryonic whorl usually worn; a thickening or thin whitish callus just inside the aperture; transverse striae are poorly developed and rather smooth on some portions of the shell; without any detectable rows of spiral papillae; periphery well rounded.

Similar Species: *Mesomphix capnodes* is larger, has a slightly lower shell profile, a smaller umbilicus, and is always with spiral rows of papillae.

Habitat: A snail of both limestone and acidic upland mixed hardwood forests usually found under and among loose leaf litter and detritus.

Status: G5/S5; Common; the most common and large *Mesomphix* in Kentucky, likely occurring in every county.

Specimen: Tennessee, Cocke County, Cosby Campground, GSMNP (GSMNP collection).

Umbilicus wide for a *Mesomphix*

***M. cupreus*, Cocke Co, TN**

***M. cupreus*, Powell Co, KY**

***M. cupreus*, Fayette Co, KY**

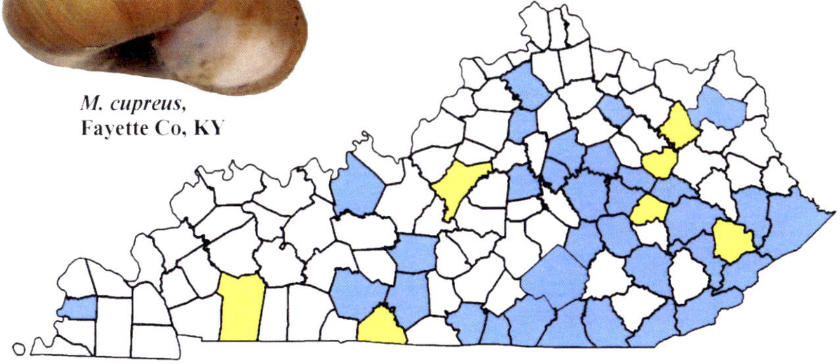

Globose Button

Gastrodontidae

Mesomphix globosus (MacMillan, 1940)

Diameter: 15.2-18.5 mm height 15-16 mm

Description: Heliciform; lip simple; shell with 5.5 whorls; shell thin but not fragile, reddish-chestnut color, glossy; perforate, no teeth present; a thickening or thin whitish callus that is typically just inside the aperture's bottom of the last whorl; transverse striae strongly developed and regularly spaced; usually without any discernible spiral rows of papillae on shell surface.

Similar Species: *M. friabilis* is larger, has less defined transverse striae and has a slightly wider umbilicus but this feature being a meager detail in their separation; may be mistaken for elevated forms of *M. capnodes* but lacks the spiral papillae, has a smaller umbilicus and is a much smaller species.

Habitat: A species of low places such as floodplains and swamps living under leaf litter and around course woody debris.

Status: G5/S3; Uncommon; Reported from mostly western counties in Kentucky with some records occurring along the Ohio River.

Specimen: Kentucky, Edmonson County, (author's collection).

A more elevated form of
Mesomphix capnodes

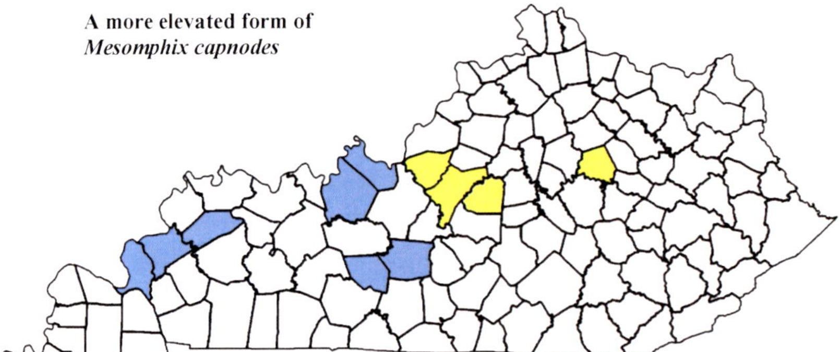

Brittle Button

Gastrodontidae

Mesomphix friabilis (W. G. Binney, 1857)

Diameter: 21-26 mm, height 12.-13 mm

Description: Heliciform; lip simple; shell with 5 whorls; shell thin but not fragile, light brownish-olive, somewhat glossy; as in most *Mesomphix* species there are random darker streaks; perforate, no teeth present; with or without a thickening or thin whitish callus that is typically just inside the aperture's bottom of the last whorl; transverse striae poorly developed; without spiral rows of papillae.

Similar Species: *Mesomphix cupreus* has a wider umbilicus (fig. a), a darker periostracum and a slightly lower shell profile; *Mesomphix rugeli* has a higher polish and is a species of upper elevation mixed hardwood forests.

Habitat: A species of low places such as floodplains living under leaf litter but also found on river bluffs and in ravines covered in mixed hardwood forests.

Status: G5/S4; uncommon; reported from only a nine counties in central Kentucky the species is likely more common than current records indicate.

Specimen: Texas, Gonzales County, Palmetto State Park (author's collection).

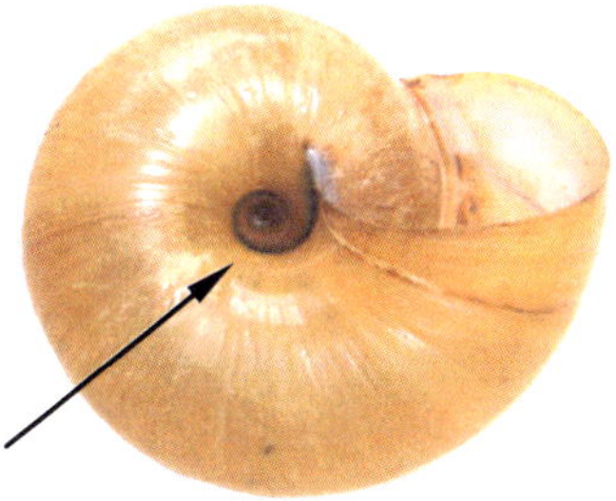

a

Mesomphix cupreus

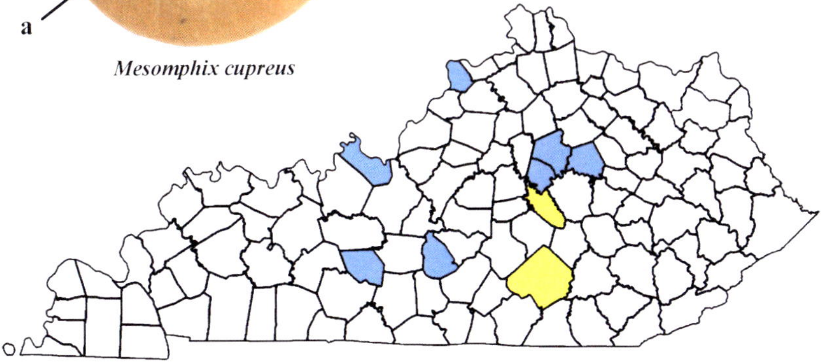

Mesomphix Shells of Kentucky Compared (proportionate)

Mesomphix with spiral papillae

30 mm

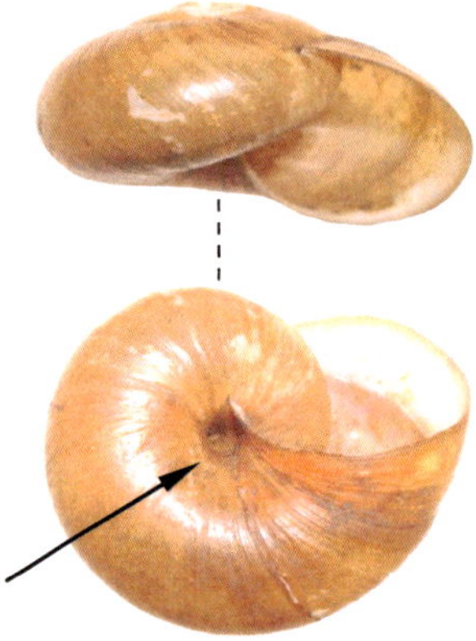

Mesomphix capnodes

Mesomphix vulgatus

Mesomphix anurus

Mesomphix rugeli oxycoccus

Mesomphix perlaevis

Mesomphix inornatus

Mesomphix without spiral papillae

30 mm

Mesomphix cupreus

Mesomphix friabilis

Mesomphix rugeli

Mesomphix globosus

A mountain tigersnail, *Anguispira jessica*, climbing a tree trunk at night. The species has been observed in excess of 20 meters high in tree tops where it is suspected to feed on slime molds (Keller & Snell 2002). Several theories have emerged explaining why the shell parades a range of contrasting color streaks. They may act as disruptive patterns, breaking up the round form, in turn making it harder to spot by predators such as birds, or, like the black and white warning colors of the skunk's bad smell, warn of distasteful defense-mucus produced by the snail under attack. Many tropical snails found living in trees are also colorfully marked (Dourson, 2009). This image was captured by back-lighting the snail's translucent shell at night, which shows the colorful alternating markings. Below the same species feeds on the sap of sugar maple from holes created by the yellow-bellied sapsucker. Other Kentucky *Anguispira* are likely more arboreal then we know. Both images taken on Roan Mountain, Mitchell County, North Carolina.

Shells 10-34 mm, without teeth, simple lip, typically with bands or streaks of color, crisscrossed protoconch sculpture, defense mucus fluorescent under UV light (*Anguispira*)

Anguispira species are among the few native land snails in Kentucky with color patterns. Interestingly, the defense mucus of this genus is fluorescent under black (UV) light for reasons that remain a mystery. In general, these medium size gastropods are most common around rotting wood and limestone cliffs and outcrops, but also found in acidic environments. Shells of *Anguispira* are usually depressed heliciform. There are crisscrossing striae (fig. a, page 217) on the embryonic (protoconch) whorl of most species of *Anguispira* (often a lost feature in worn shells), however, not all specimens within the same population exhibit this trait. The lips of *Anguispira* shells are simple in all stages of growth and there are never teeth present in the aperture. The transverse striae are more ribbed-like and easily observed with a hand lens of 10X magnification. *Anguispira* shells are always umbilicate. The periphery of the body whorl is either carinate, bluntly angular or rounded. The color patterns are usually present but this character may be poorly defined in some individuals and in shells that are badly weathered; wetting the shell will usually reveal some color. The following *Anguispira* are arranged according to shape and rib counts and are as follows: **Grouping 1:** shells globose; **Grouping 2**: shells compressed that have more than 65 ribs on last whorl; **Grouping 3**: shells compressed that have less than 65 ribs on last whorl and **Grouping 4**: shells compressed with less than 30 ribs on last whorl.

Grouping 1: shell globose, pg. 219

**Anguispira kochi*

Grouping 2: shell compressed, more than 65 ribs on the last whorl of adult shells, pg. 220-223

**Anguispira rugoderma* (rare)
**Anguispira alternata* (statewide, very common)
**Anguispira cf. jessica* (restricted to Black Mountain, rare)

Grouping 3: less than 65 ribs on the last whorl of adult shells, pg. 224-226

**Anguispira strongylodes* (restricted to extreme western Kentucky)
**Anguispira aff. strongylodes* (likely an undescribed species)

Grouping 4: less than 30ribs on the last whorl of adult shells, pg. 227
**Anguispira aff. mordax* (restricted to Whitley & Bell County, rare)

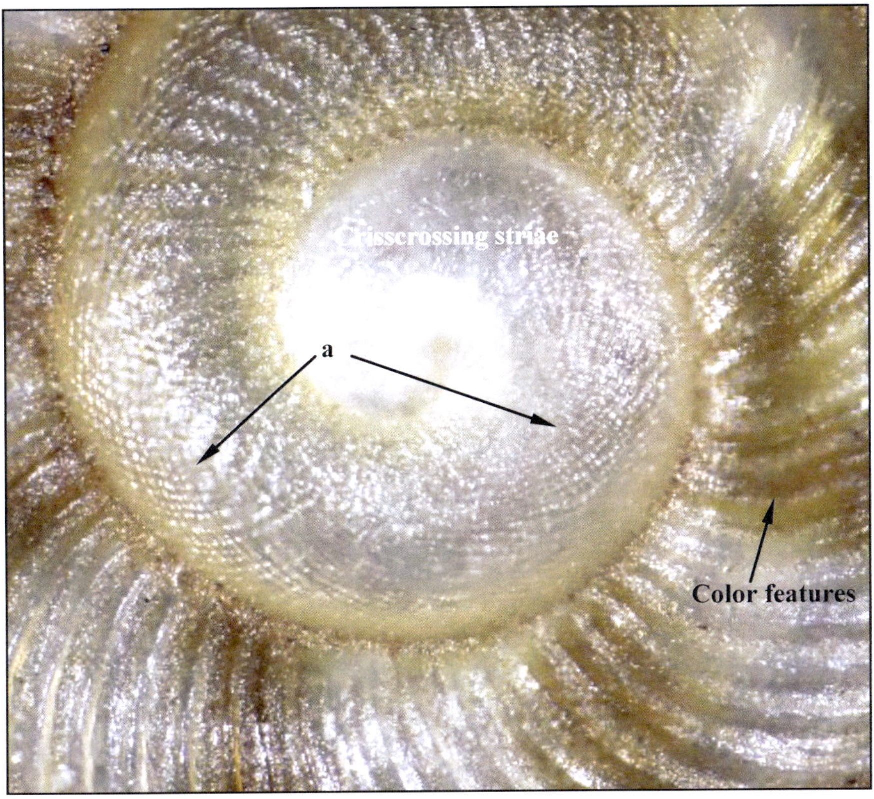
Crisscrossing striae
a
Color features

When counting ribs on the last whorl, start at (fig. b) and stop at (fig. c); this shell has around 23 ribs
5
4
3
2
1
b
c

Banded Globe

Discidae

Anguispira kochi (Pfeiffer, 1845)

Diameter:17-34 mm, height 14-25 mm

Description: Heliciform; lip simple; shell with 6 whorls; umbilicate; no teeth present, dark color bands (can be thin or wide) always present but may be a weak feature; old snails or weathered dead shells lose their brown periostracum and weather to dull white (as seen below); shell solid (thick); embryonic whorl smooth; transverse striae distinct; no spiral striae present; periphery is well rounded.

Similar Species: The look of this *Anguispira* is like no other native species in Kentucky.

Habitat: A snail of river bluffs, but also found in upland woods at or near the base of limestone outcrops, among loose leaf litter; a species indicative of quality habitat.

Status: G5/S3; State Special Concern; extant populations of *A. kochi* are becoming increasingly scarce (Dourson 2010); loss of native habitat is likely the biggest threat to this interesting land snail.

Specimen: Kentucky, Powell County, Furnace Mountain (author's collection).

Orange County, Indiana

Powell County, Kentucky

Cincinnati, Ohio

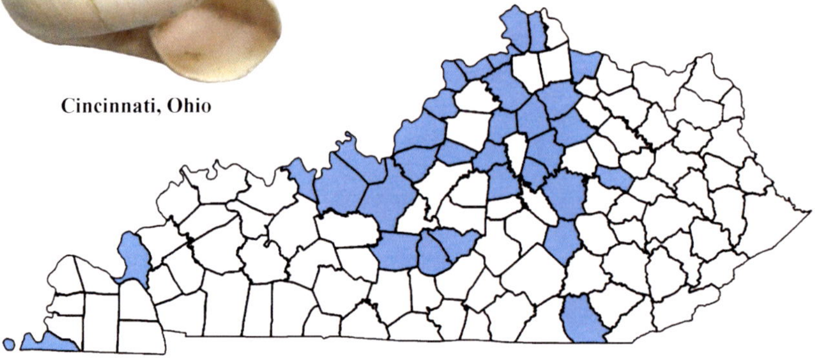

Pine Mountain Tigersnail

Anguispira rugoderma Hubricht, 1938

Diameter: 21-30 mm, height 15-16 mm

Description: Depressed heliciform; lip simple; shell with 5.5-6 whorls; widely umbilicate, the widest of the Kentucky *Anguispira*; no teeth present; greater than 80 transverse striae (ribs) are relatively well-developed; color features are faint but always present; periphery well rounded.

Similar Species: In Kentucky, *Anguispira alternata* is smaller with more finely transverse striae sculpture, more prominent color features and a proportionately smaller umbilicus; *A. mordax* is smaller, with stronger color features and more widely spaced ribs.

Habitat: A species found around large rotting hardwood logs; located on the north side of Pine Mountain; although the species is mostly associated with old growth, it also occurs in pure stands of second growth tulip popular (Kim Tarter pers. comm. 2005).

Status: G3/S3; State Threatened; this snail is endemic to Kentucky, currently known only from Clay, Harlan, Leslie and Bell counties; it is rarely found in numbers.

Specimen: Kentucky, Bell County, Pine Mountain State Park, above rock quarry next to rotting log (author's collection).

Only faint color features

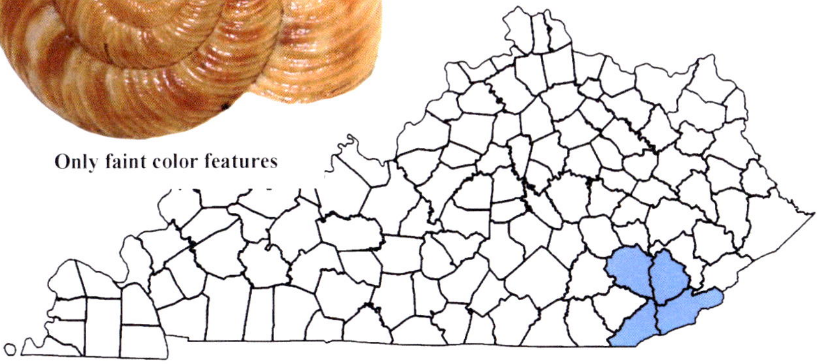

Flamed Tigersnail

Anguispira alternata (Say, 1816)

Diameter: 18-25 mm, height 10.5-14.5 mm

Description: Depressed heliciform; lip simple; shell with 5-6 whorls; umbilicate; color features usually strong on entire shell surface; in some populations there are crisscrossing striae on the embryonic whorl; no teeth present; transverse striae are closely spaced and fairly well developed on top becoming more diminished on the base, adult shells with at least 5 whorls have a count of over 80 ribs in the last whorl; most specimens have a well defined angular periphery, while some populations in the state are very weakly carinate.

Similar Species: *Anguispira mordax* is smaller having fewer than 25 ribs on the last whorl; *A. jessica* is substantially smaller, has a smaller umbilicus and a round periphery.

Habitat: A species found in a wide range of lower elevation wooded and open habitats including rocky limestone regions, glades, and along forested hillsides adjacent to rivers.

Status: G5/S5; Common; the most widespread *Anguispira* in the state.

Specimen: All figures from Kentucky, Menifee County, Red River Gorge (author's collection).

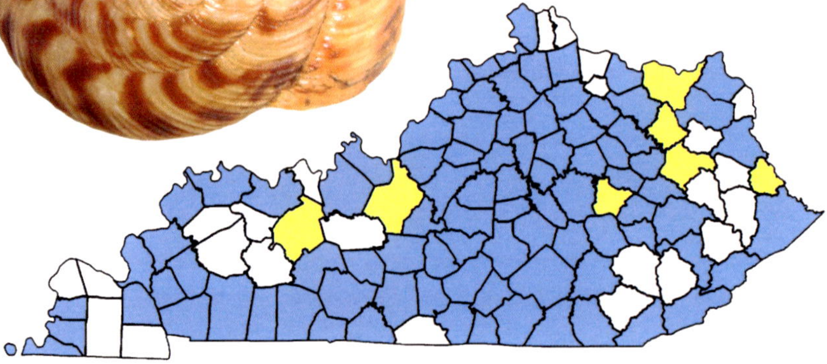

Fluorescence in Land Snails

Above image of the Flamed Tigersnail, *Anguispira alternata,* in typical defense posture; the live animal retracted inside its shell with its aperture blocked it's bubbly orange defense-mucus. Below the same specimen at night under UV light showing its fluorescence mucus. The function of the glow in land snails remains a mystery and may be nothing more than a random act of evolution. But the fluorescence mucus glowing from UV light emanating from the moon might serve as a warning sign of distasteful compounds found in the slime, repelling predators and conserving critical body moisture of the snail. Red River Gorge, Kentucky.

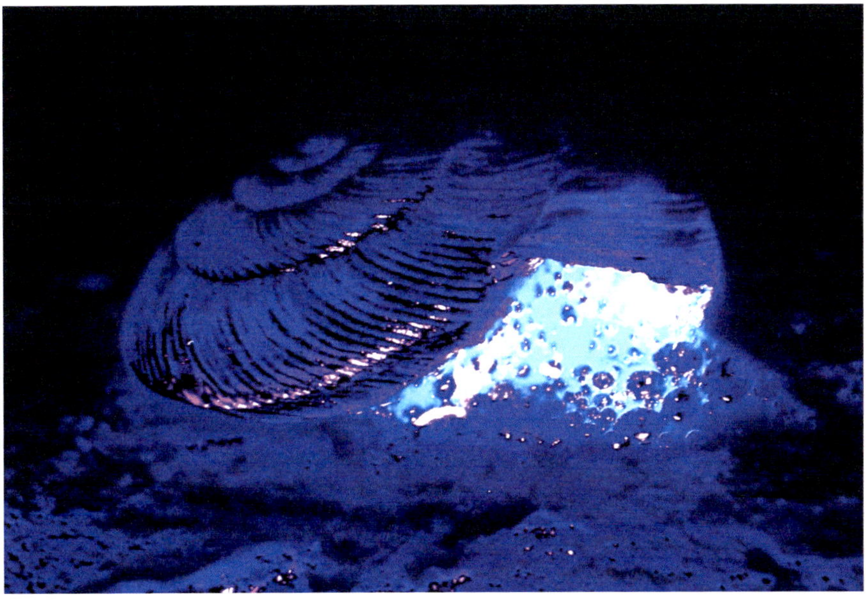

Mountain Tigersnail

Discidae

Anguispira cf. jessica Kutchka, 1938

Diameter: 15-18 mm, height 7-8 mm

Description: Depressed heliciform; lip simple; shell with 5-6 whorls; umbilicate; color features usually strong on top and sides but lacking on the bottom surface; crisscrossing striae on the embryonic whorl a strong feature; no teeth present; transverse striae are closely spaced and fairly well developed on top becoming more diminished on the base, adult shells with at least 5 whorls have a count of 65 to 80 ribs in the last whorl; a mostly rounded to bluntly periphery.

Similar Species: *Anguispira alternata* has an angular periphery-not rounded, is larger in diameter with a wider umbilicus; *A. jessica* from pine mountain genetically match *A. jessica* from TN mountains (pers. comm. John Slapcinsky, 2022).

Habitat: A species found in north facing noncalcareous soils in ravines covered by mixed hardwood forests at higher elevations.

Status: G4/S1; Rare; State Threatened; in Kentucky this small *Anguispira* is currently known only from Black Mountain.

Specimen: Figures (a & b) from Kentucky, Harlan County, north facing hillside on Black Mountain and figures (c) from Tennessee, Haywood County, Big Creek Picnic area, GSMNP (GSMNP collection).

a

b

Shell with a slightly sub-angular periphery

c

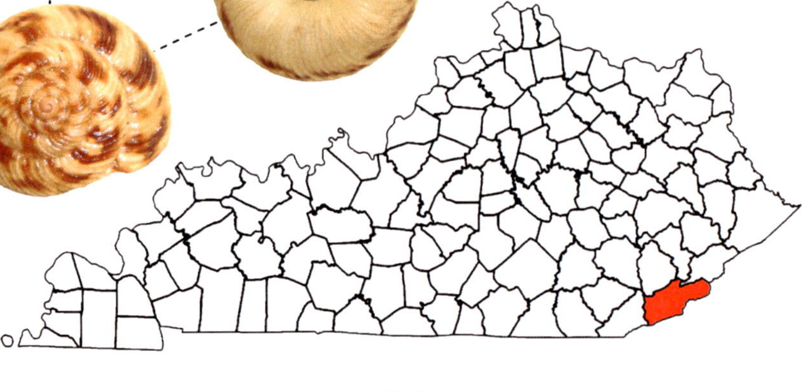

Southeastern Tigersnail

Anguispira strongylodes (Pfeiffer, 1854)

Diameter: 15.2 mm, height 8.3 mm

Description: Depressed heliciform; lip simple; shell with 5 whorls; umbilicate; color streaks and blotches a strong feature and always present; no teeth; transverse striae are well developed and rib-like, having a count of 50 to 70 in the last whorl and which are well defined onto the base of the shell; periphery roundish.

Similar Species: *Anguispira jessica* is around the same size and build but a species of the mountains while *A. strongylodes* is a lowland species, restricted to extreme western Kentucky.

Habitat: Forested river bluffs and ravines overlooking the Mississippi River; typically found around rotting logs in advanced stages of decay.

Status: G5/S1; State Threatened; very little habitat remains for this rare Kentucky snail in Carlisle and Graves County, most of that region now converted to farmland.

Specimen: Figures (a) from Kentucky, Carlisle County, Arlington, (Figure (b) from Kentucky, Graves County and Figure (c) *A. strongylodes* from Louisiana, along Red River at 84 Bridge near Natchitoches (author's collection).

a

c

b

Kentucky Tigersnail

Anguispira aff. strongylodes (Pfeiffer, 1854)

Diameter: 15-17 mm, height 10 mm

Description: Depressed heliciform; lip simple; shell with 5 whorls; umbilicate; color streaks and blotches can be either a strong or weakly developed feature but are always present; no teeth; transverse striae are well developed and rib-like, having a count of 45 to 65 in the last whorl and which are well defined to the base of the shell; periphery rounded to bluntly angular and shouldered, variable.

Similar Species: *Anguispira alternata* is larger, has a wider umbilicus and has transverse striae sculpture which is notably finer; *Anguispira aff. mordax* has more widely spaced ribs (20-30 in the last whorl) and a wider umbilicus in the last whorl; similar to *A. strongylodes* but has a more bluntly angular and shouldered periphery (*A. strongylodes* is more rounded (fig. b), although this can be a variable and somewhat unreliable feature), has a smaller umbilicus, and is generally more compressed than *A. strongylodes*. For additional comparisons see page 228.

Habitat: Limestone regions along forested hillsides usually found under the leaf litter but also among loose rock formations and limestone outcrops where it becomes increasingly abundant.

Status: Not ranked; Common; appears to be restricted to eastern portions of Kentucky. Past collectors, including myself, have called this species either *A. strongylodes* or *A. mordax,* but it likely represents an undescribed species waiting further DNA investigation.

Specimen: Figures (a) from Kentucky, Powell County, Furnace Mountain area (author's collection) and figure (b) from Kentucky, Carlisle County, Arlington, See opposite page.

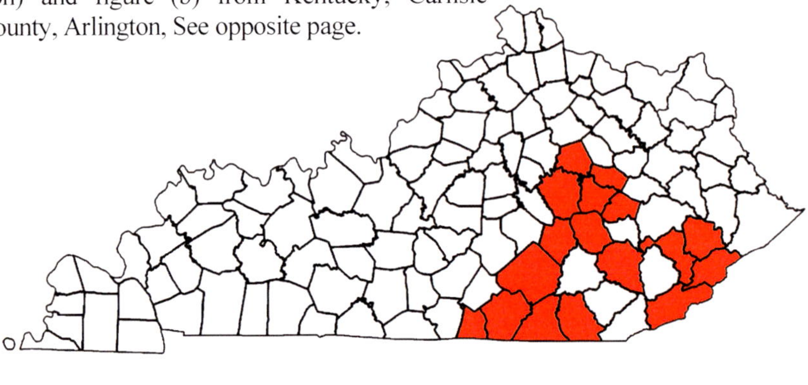

Anguispira aff. strongylodes

Anguispira aff. strongylodes
Kentucky, Wayne County, Steele Hollow WCA (author's collection). Specimens from Wayne and Clinton county are on the smaller size of this undetermined species, averaging 10 to 15 mm in diameter and 7 to 8 mm in height (author's collection).

Anguispira aff. strongylodes
A darker form from Bell County, Pine Mountain State Park, forested hillside above limestone quarry near park boundaries (author's collection).

Anguispira aff. strongylodes

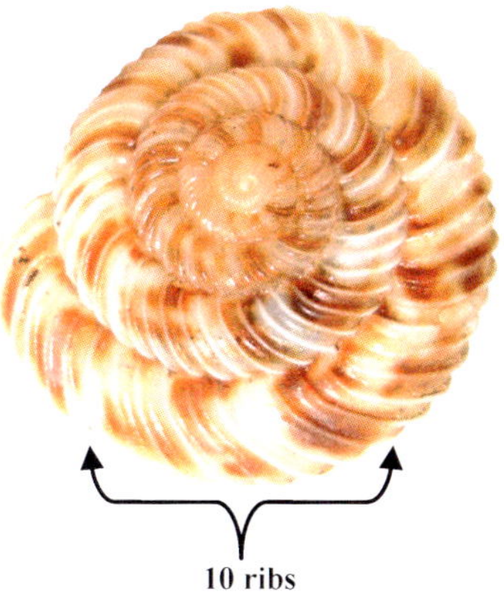

Anguispira aff. mordax

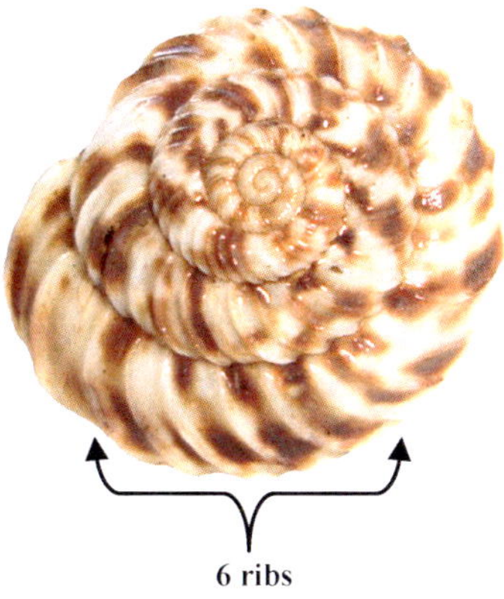

Craggy Tigersnail

Anguispira aff. mordax

Diameter: 15-18 mm, height 8-10 mm

Description: Depressed strongly heliciform; lip simple; shell with 5-6 whorls; umbilicate; color streaks can be either a strong or weakly developed feature but are always present; transverse striae are modified into distinct and very widely spaced ribs, having a count of around 20-30 ribs in the last whorl and remaining bold unto the base of the shell; periphery bluntly angular and shouldered (fig. a).

Similar Species: *Anguispira alternata* is larger in size with finer sculpture with more than 80 ribs on the last whorl; *A. strongylodes* has a higher number of ribs (55 to 65 in the last whorl) and lighter shell.

Habitat: Found in a variety of forest covered ravines and hillsides around rotting logs in advance stages of decay and at the base or inside mature sugar maples that contain heart rot.

Status; Rare, (Data deficient to rank at this time) *Anguispira aff. mordax* appears to be restricted to Bell and Whitley Counties; currently being investigated via DNA; may be a new species endemic to Kentucky.

Specimen: Kentucky, Whitley County, Pearl Siler Road, Route 1545 north side of Pine Mountain, (author's collection).

a **Darker color morphs**

Widely spaced ribs

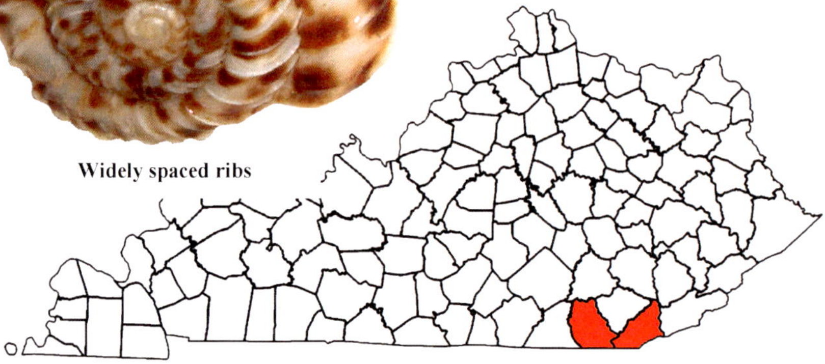

Seven Kentucky *Anguispiras* Compared (Shells Proportionate)

Anguispira kochi, Powell Co, KY

Anguispira rugoderma, Bell Co, KY

Anguispira alternata Powell, KY

30 mm

Anguispira aff. strongylodes, Powell Co, KY

Anguispira cf. jessica, Harlan Co, KY

Anguispira aff. mordax, Whitley Co, KY

Anguispira strongylodes, Carlisle County, KY

Extraordinary Feeding Behavior Observed in a Central American Land Snail (by entomologists Piotr Naskrecki and Kenji Nishida)

© Piotr Naskrecki

While using its tentacles, a wolf snail, *Pittieria aurantiaca,* politely taps the rear end of a lantern bug, *Phrictus quinquepartitus* in Costa Rica. This strange behavior appears uncharacteristic for a snail that normally hunts other gastropods for their flesh. The snail is not interested in eating the bug, but attracted to the honeydew which is ejected from the bugs abdomen. After tapping the bug's rear, the snail catches flying droplets of honeydew by forming a hood over the tip of the insect's abdomen with its head and foot. The honeydew accumulates on the snails ventral surface, and is then consumed. Why would a predaceous snail want honeydew in the first place? It appears that, in addition to the benefit of getting some high quality carbohydrates, the phloem (sap) of the tree species that the lantern bug feeds on, *Simarouba amara*, contains high levels of calcium, the mineral that snails need to build shell. Carpenter ants also want in on the action but are too short and too slow to catch the honeydew that is flying out of the lantern bug's rear end at between 0.8 and 1.7 m/sec (2.6-5.6 ft/sec.) accord-

© Piotr Naskrecki

ing to Naskrecki and Nishida. So the ants climb up the snail to harvest the sweet food from the surface of the gastropod's head. Not surprisingly, the snail quickly becomes agitated with the ants that are stepping on its eyes and chemoreceptors. Unable to rid itself of the pesky ants, the snail simply crawls away. This extraordinary land snail feeding behavior between mollusk and arthropod clearly demonstrates the wide range of calcium carbonate sources in nature and that harvesting techniques used by land snails are even more complex than we ever imagined.

© Piotr Naskrecki

Glassy Grapeskin

Gastrodontidae

Vitrinizonites latissimus (J. Lewis, 1875)

Diameter: 16.2-19.5 mm, height 8-10 mm

Description: Depressed heliciform; lip simple; shell with 2.5-3 whorls; shell paper thin, translucent and flexible with a proteinaceous coating on its outer surface and a thin calcium layer on its interior; brown to olive, yellow or sometimes green; imperforate; transverse striae (growth wrinkles) are poorly developed and mostly smooth, giving the shell a glass-like surface; live animal bluish (fig. a). If placed in ones hand, this species will bite, an interesting experience and not at all painful.

Similar Species: No other land snail in Kentucky has a flexible shell and broad shape of *V. latissimus*.

Habitat: In Kentucky, this snail is restricted to Black and Pine Mountains at elevations above 2000 feet. It is found under moist leaf litter located next to or under rotting logs in mixed hardwood, but can also be found in rather acidic conditions such as rhododendron thickets were few other snails can thrive. A study in the Great Smoky Mountains National Park found that *V. latissimus* was the most common land snail species in purple rhododendron patches (Dourson & Langdon, 2012)

Status: G4/S2; State Threatened; this rare and unique land snail is reported from Harlan and Letcher Counties only in southeastern Kentucky.

Specimen: North Carolina, Mitchell County, Roan Mountain (author's collection).

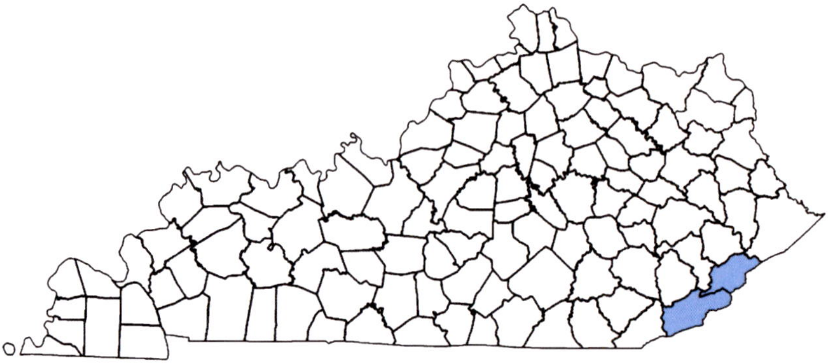

Above a Glassy Grapeskin, *Vitrinizonites latissimus,* crawling on Christmas Fern and below three Glassy Grapeskins, and a Black Mantleslug, *Pallifera hemphilli,* feeding on a fresh dead Balsam Globe, *Mesodon andrewsae*, Roan Mountain, North Carolina.

Gray-foot Lancetooth

Haplotrematidae

Haplotrema concavum (Say, 1821)

Diameter: 16-25 mm, height 5.3-9.5 mm

Description: Depressed heliciform; lip usually simple but in fully developed shells there is a slight reflection in the lower lip (fig. a); shell with 4.5-5.5 whorls; widely umbilicate; shell light greenish-yellow and glossy; transverse striae poorly developed; spiral striae present, but may be a wanting feature on some portion of the shell; live animal gray; a carnivorous species that feeds on other gastropods.

Similar Species: No other land snail of similar size has the wide umbilicus; the similar looking *Haplotrema kendeighi* of more southern mountains has a blue body whereas the animal of *H. concavum* is gray, however their shells are indistinguishable.

Habitat: Found in mixed hardwood forests at all elevations.

Status: G5/S5; Common; one of the most common and widespread land snails in Kentucky and eastern North America.

Specimen: All figures from Tennessee, Blount County, Chestnut Top Trail, GSMNP (GSMNP collection).

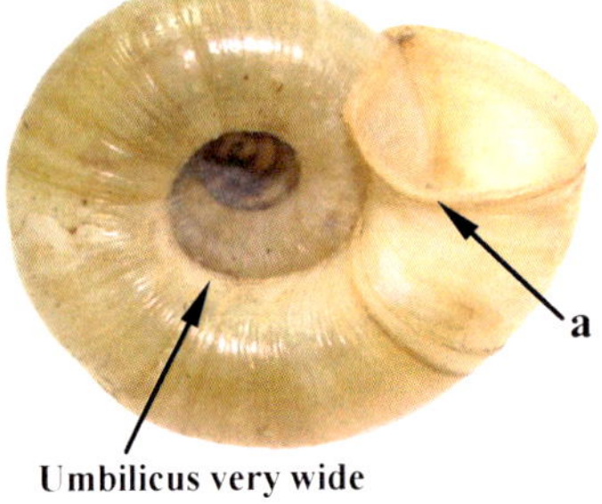

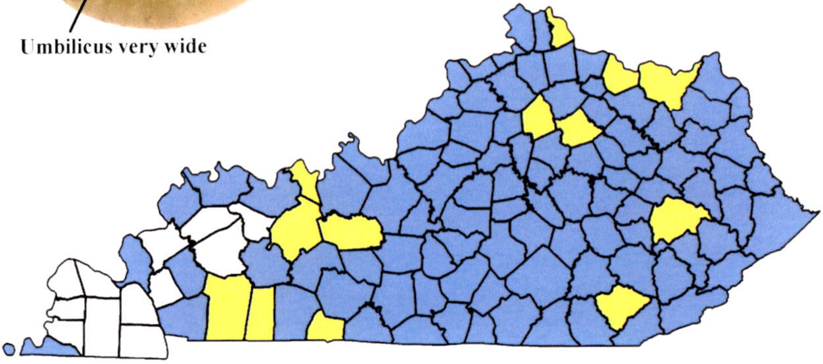

The Tiger of Land Snails, *Haplotrema concavum*

Haplotrema are the tigers of land snails, eating a variety of terrestrial gastropods including *Novisuccinea ovalis, Zonitoides arboreus, Discus catskillensis* and *Strobilops labyrinthicus* (Pearce and Gaertner 1996). *Patera, Mesomphix,* and *Mesodon* species are also taken by this formidable hunter (author's personal observations). To access the snail flesh, the species will typically enter through the aperture of the prey's shell where it slowly eats the snail alive. If the aperture is too small for entry, *Haplotrema* will drill (actually grind) through the side of the shell by using its radula (Pearce and Gaertner 1996). The nickel-sized gastropods have one of the widest umbilicus of any eastern land snail and are a common woodland species.

Above a Gray-foot Lancetooth, Newfound Gap, Great Smoky Mountains National Park and below the same species attacking and consuming a juvenile Wrinkled Button, *Mesomphix rugeli,* Mitchell County, North Carolina.

Land Snail Feeding Trails

The feeding trail of an unknown species of land snail, Cheat River Gorge, Monongalia County, West Virginia.

Land snails in this section have shells that are wider than tall, in some cases around the same, domed, or sometimes roundish in shape, (*Helicina* species being a good example of a more rounded shell). Snails in this group have simple lips, although the lips of several species such as *Strobilops* are slightly reflected. They include small snails from 1.9 mm to 8.5 mm in size. Two species *Helicina orbiculata* and *Hendersonia occulta* possess an operculum (a calcareous structure or door used to seal the aperture when the snail withdraws into its shell). *Euconulus* species look like miniature beehives. *Strobilops* shells have interesting and finely developed riblets, which is only fully appreciated under the lens of a scope.

The Genera Included:

(in order of appearance in text)

Euconulus
Hendersonia
Helicina
Strobilops

Euconulus

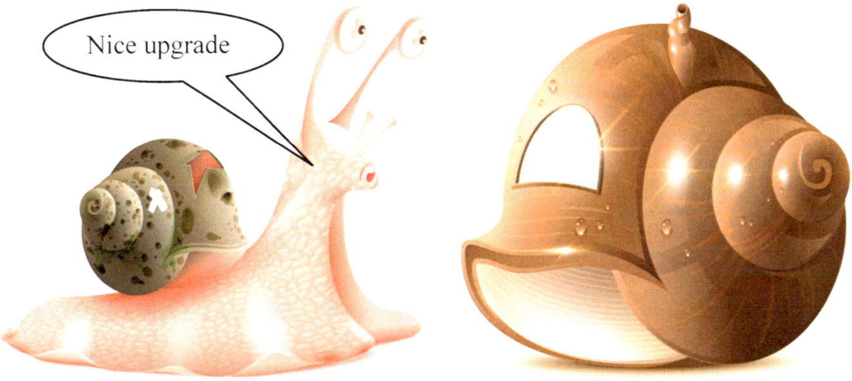

A) Shells 2.4-3.4 mm, beehive shaped, <u>no teeth</u>, simple lip, not thickened, no operculum (*Euconulus*)

<u>Euconulus</u> species are small dome or beehive shaped land snails that have many tightly coiled whorls. They are generally uncommon in leaf litter of mixed hardwood forests. They differ from *Strobilops* by having a simple not reflected lip, a smoother surface (*Strobilops* having a ribbed surface), and being more dome-shaped. In general *Euconulus* are without teeth, although *E. dentatus, E. chersinus* and *E. trochulus* have been found containing internal armature (pers. comm. Nekola, 2022). Although five species are reported from Kentucky, currently there remains much confusion among this group which can be settled by securing large sample sizes. See page 239 for further discussion and latest research on this group.

<u>Shells 2.4-3.4 mm, dome or bee-hive shape, shell surface more or less smooth (not ribbed), pg. 241-246</u>

**Euconulus fulvus egenus*
**Euconulus polygyratus*
**Euconulus chersinus*
**Euconulus trochulus*
**Euconulus dentatus*

B) Shells 6-8 mm, domed shaped or roundish, no teeth, simple lip or very slightly reflected, thickened, with an operculum in live animals, *Helicina* and *Hendersonia*

<u>Helicina</u>* <u>and</u> *<u>Hendersonia</u> species are similar in shape to ***Euconulus*** but are much larger with a thicker shell and in live specimens, containing an operculum. ***Helicina*** and ***Hendersonia*** differ from ***Strobilops*** by having a less reflected lip, a smoother surface (*Strobilops* having a ribbed surface) and averages 5-6 mm larger.

<u>Shells 6-8 mm, heliciform, transverse striae moderately developed but not ribbed, pg. 247-250</u>

** Helicina orbiculata*
** Hendersonia occulta*

8mm

All shells are proportionate

C) Shells under 3 mm, perforate to umbilicate, lip only slightly reflected, <u>teeth always present</u> (*Strobilops*)

<u>Strobilops</u> are generally uncommon gastropods in rich woods, becoming more common in glades, dry limestone outcrops and strangely enough, parking lots. They are easily separated from all snails of similar size and form by having a reflected lip; most snails of comparable size have simple lips. While *Strobilops* are dissimilar enough from *Vallonia* to make their separation trouble free, separating *Strobilops* from each other can be little more difficult. These are some of the most beautiful land snails in Kentucky.

<u>Shell 2.4-2.8 mm, lip narrowly reflected, top of shell notably ribbed, internal armature present, narrowly umbilicate, pg 251-254</u>

**Strobilops labyrinthicus*
**Strobilops cf. texasianus*
**Strobilops aeneus*

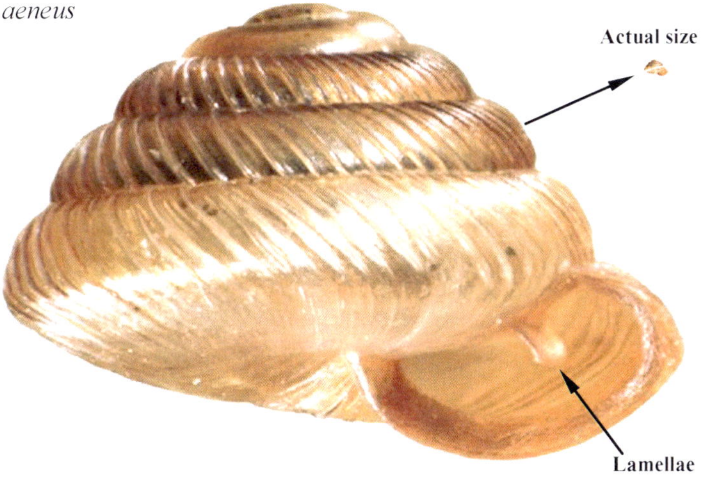

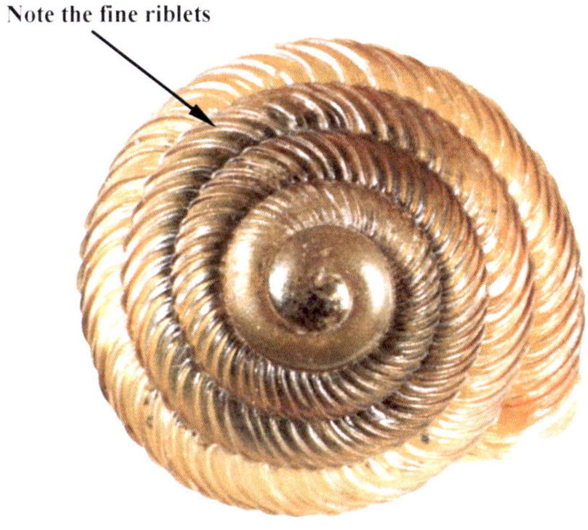

Internal lamellae seen through a fresh dead or live animal

The Euconulus Dilemma **(pers. comm. Nekola, 2022)**

The genus *Euconulus* contains no more than 20 species with half of these being limited to tropical islands in the Pacific and Indian oceans. The remaining taxa are continental where they are frequent in northern latitudes. They are present both in modern and fossil assemblages. Earlier taxonomic efforts both over-split and over-lumped actual biological species by over-emphasizing some single plastic shell traits while under-emphasizing others. A recent study published by Horsakova et al. (2020), illustrates the importance of consensus across DNA and shell/soft body features in identifying snails to species level and determining specific features that support accurate identification. While it is nearly impossible to correctly identify *Euconulus* species using a single traditionally-used single shell feature, these genetic analyses show that novel combinations of traits exist which can allow for accurate species-level identification from shells alone.

Euconulus species of Kentucky can be divided into two groups: the *Euconulus fulvus* group and the *Euconulus chersinus* group. The *E. fulvus* group is represented by two taxa: *E. fulvus egenus* occurs from the Ural Mountains in central Eurasia east across all of North America, while *E. polygyratus* is restricted to east-central North America.

The *Euconulus fulvus* found in group in Kentucky:

**Euconulus fulvus egenus*
**Euconulus polygyratus*

E. polygyratus **(clade F), cool limestone bluff, Ohio, USA, images © Michal Horsak and Veronika Horsakova**

Tightly coiled

***E. fulvus egenus*, *Salix* scrub, Western Siberia, Russia, images © Michal Horsak and Veronika Horsakova**

Loosely coiled

Euconulus fulvus consists of two subspecies, with *E. fulvus fulvus* being limited to western Eurasia. Populations from the Ural Mountains east to the Atlantic Seaboard of North America represent *Euconulus fulvus egenus*. This subspecies consists of a number of unique but still interbreeding geographic races. While considerable overlap in shell features exist, *E. fulvus fulvus* tends to have tighter coiling with a more constant rate of whorl expansion, and a grey mantle, while *Euconulus fulvus egenus* possesses more lax (looser) coiling with an initial low, then more rapid rate of whorl expansion, and mottled mantle (Horsakova et al. 2020).

Euconulus polygyratus can be distinguished from *E. fulvus* egenus by its more tightly coiled whorls with reduced coil expansion (generating 1-2 more whorls in adult shells than other members of this group; Horsakova et al. 2020). It also has a strongly lunate aperture, domed spire, reduced micro-sculpture, and lacks sillons (spiral lines cut into the shell). *Euconulus polygyratus* occurs in rich upland, often rocky forests. Range extends from Atlantic Provinces to Canada, through the Great Lakes region over to northeastern Iowa, Illinois, Ohio, Kentucky, and south along the crest of the Appalachians to western North Carolina.

The *E. chersinus* group, containing at least three species (*E. chersinus, E. dentatus, E. trochulus*), is endemic to southeastern North America. While the presence or absence of internal lamellae was thought to distinguish *E. dentatus* from *E. chersinus & E. trochulus*, preliminary genetic analysis shows that lamellae can sporadically occur across all three taxa, with their presence likely being related to high food and soil Calcium levels. Too few genetically-confirmed populations exist to determine the suite of shell features which actually allows for their accurate identification. Because of this it is impossible to list the actual shell features, ecological preferences, range and abundance for biologically-valid entities in this group.

The *Euconulus chersinus* group in Kentucky:
Euconulus chersinus (lamellae may or may not be present)
Euconulus trochulus (lamellae may or may not be present)
Euconulus dentatus (lamellae may or may not be present)

The accounts that follow only represent current conventional wisdom and as a result, the information provided should be seen as highly approximate.

Brown Hive

Euconulus fulvus egenus (Müller, 1774)

Diameter: 3.1-3.4 mm, height 2.4 mm

Description: Dome shape; lip simple; shell with 4.5-6 loosely coiled whorls; minutely perforate or nearly closed; cinnamon or dilute tawny, the summit paler; glossy; shell thin, translucent in live snails and fresh dead; transverse striae are poorly developed (dissecting scope required); spiral striae not well defined but always present; without teeth; periphery rounded or weakly angular.

Similar Species: *Euconulus chersinus* (page 243) is more finely sculptured, has a more elevated shell, is less glossy and has two more whorls.

Habitat: A Holarctic species found in mixed hardwood forests on well-shaded hillsides and ravines living under layers of moist leaf litter; also found in creek and river drift.

Status: G5/S5;tentative ranking pending further DNA investigation; all records below are also tentative, pending further research.

Specimen: Genetically confirmed shell of *Euconulus fulvus egenus* from *Salix* scrub, Western Siberia, Russia, © Michal Horsak and Veronika Horsakova (Horsakova et al 2020).

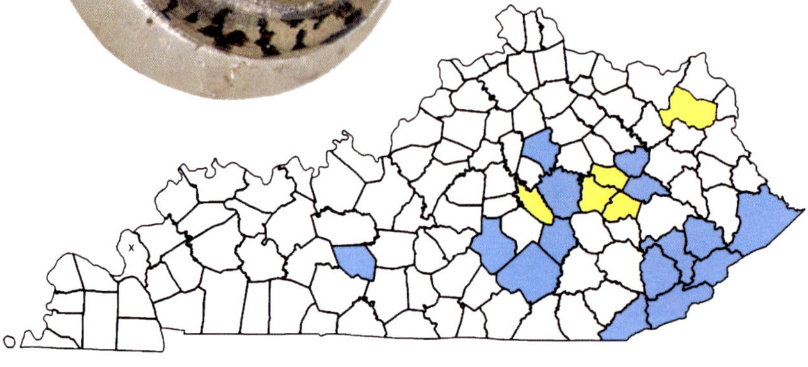

Fat Hive

Euconulidae

Euconulus polygyratus (Pilsbry, 1899)

Diameter: 2.7-3.3 mm, height 2.3-2.8 mm

Description: Dome shape; lip simple; shell with 6-7 tightly coiled whorls; scarcely perforate; aperture narrow; cinnamon-buff, somewhat dull and silky except in the middle of the base which is glossy and somewhat transparent; without teeth; in mature specimens the periphery is rounded, but in immature shells the periphery is sharply angular (Pilsbry 1946).

Similar Species: Shell much like *E. chersinus*, but with a fatter build; *Euconulus fulvus egenus* is more glossy and has 1 to 2 fewer whorls in adult shells.

Habitat: A species found in mixed hardwood forests on hillsides and ravines living under layers of moist leaf litter; in Kentucky known only from a patch of Canada Yew on a limestone talus; a Holarctic land snail.

Status: G5/S1; State Special Concern; this species was only recently discovered in a patch of Canada Yew in Red River Gorge area.

Specimen: Genetically confirmed shell of *Euconulus polygyratus* from cool limestone bluff, Ohio, images Michal Horsak and Veronika Horsakova (Horsakova et al 2020).

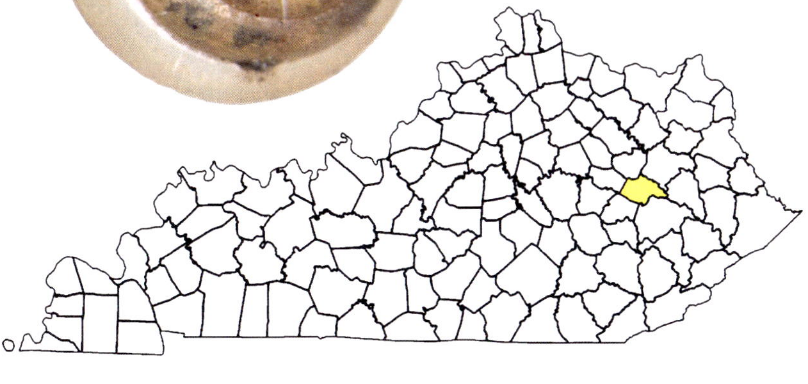

Wild Hive

Euconulidae

Euconulus chersinus (Say, 1821)

Diameter: 2.4-2.95 mm, height 2.9-3.4 mm

Description: Dome shape; lip simple; shell thin with 6-8 tightly coiled whorls; scarcely perforate; shell surface pale yellowish-white and rather dull; shells are translucent with a silky luster in live snails and fresh dead; transverse striae are poorly developed (microscope required); spiral striae not well defined, but always present (a very strong lens required); with or without teeth; body whorl slightly angular, especially in young shells which are generally more common than adults; in fully mature shells the periphery becomes rounded. Based on recent DNA research by (Horsakova et al. 2020) identifying *Euconulus* to species level should be considered tentative for now, waiting further DNA and shell morphology investigations.

Similar Species: This species differs from *E. fulvus egenus* by the more elevated spire, more numerous, narrower whorls, a narrower aperture, and in fresh specimens, by the silky luster, *E. fulvus egenus* being more glossy (Pilsbry, 1946).

Habitat: Under moist leaf litter in mixed hardwood forests.

Status: G5/S5;tentative ranking pending further DNA investigation; all records below are also tentative, pending further research.

Specimen: North Carolina, Swain County, Nantahala National Forest (author's collection).

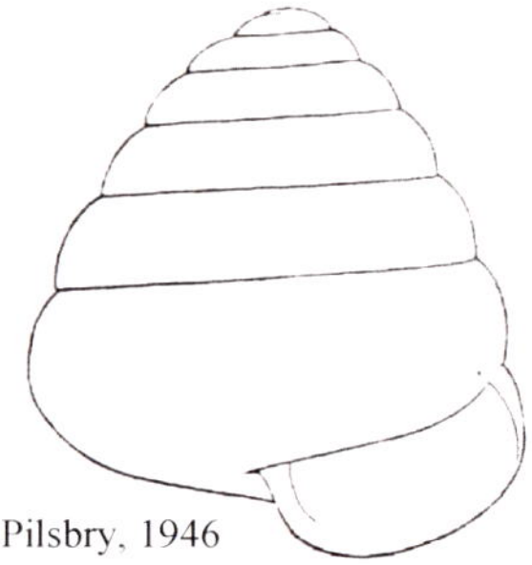

Pilsbry, 1946

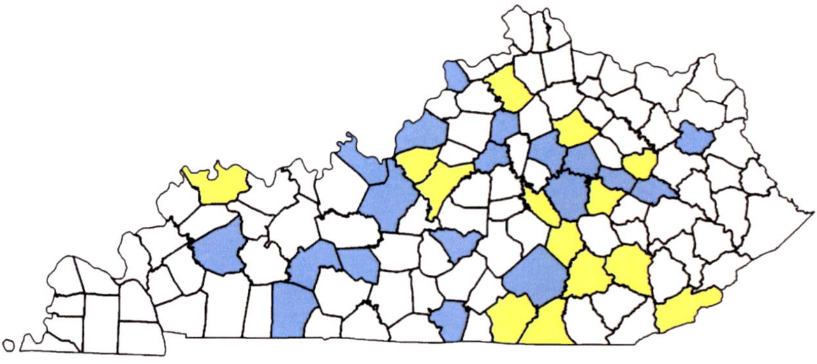

Silk Hive **Euconulidae**

Euconulus trochulus (Reinhardt, 1883)

Diameter: 2.9 mm, height 2.5 mm

Description: Dome shape; lip simple; shell with 7 tightly coiled whorls; scarcely perforate, the narrow umbilicus is completely covered by the reflected columellar margin; pale horn with a silky luster; shell thin, translucent in live snails and fresh dead; transverse striae are poorly developed (dissecting scope required); spiral striae not well defined but always present (a strong lens required); with or without teeth; in mature specimens, the periphery is slightly angular or roundish (Pilsbry 1946); as with other *Euconulus,* it is important to have a series of shells for a proper ID. Based on recent DNA research by (Horsakova et al. 2020) identifying *Euconulus* to species level should be considered tentative, waiting further DNA and shell morphology investigations.

Similar Species: Shell much like *E. chersinus*, but it has one whorl more in shells of similar size; *Euconulus fulvus* is more glossy and has 1 to 2 fewer whorls in adult shells.

Habitat: A species found in mixed hardwood forests on hillsides and ravines living under layers of moist leaf litter.

Status: G5/S5; tentative ranking pending further DNA investigation; all records below are also tentative, pending further research.

Specimen: Fig. (a) from Kentucky, Pulaski County, Meece (FMNH 252087) and (fig. b) from David Kirsh and Bill Frank photo collection at www.jaxshells.org.

a

Pilsbry, 1946

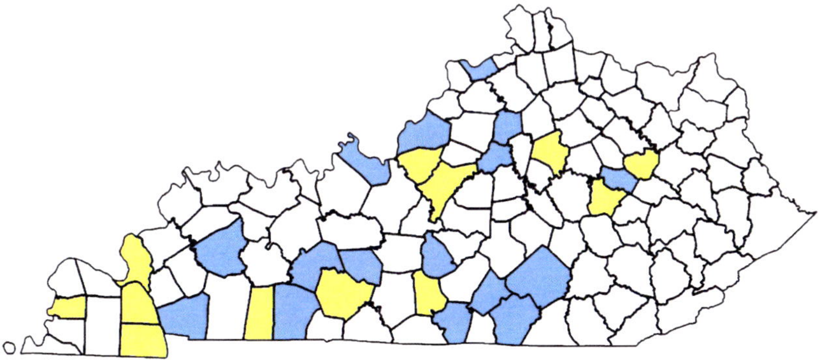

Toothed Hive

Euconulus dentatus (Sterki, 1893)

Diameter: 2.4 mm, height 2.3 mm

Description: Dome shape; lip simple; shell with 6.5 tightly coiled whorls; scarcely perforate; yellowish-white and rather dull; shell thin, translucent in live snails and fresh dead; transverse striae are poorly developed (dissecting scope required); spiral striae not well defined, but always present; teeth or lamellae in young shells (a, b & c), adults usually without teeth.

Similar Species: Most like *Euconulus chersinus* but differs in having 2 to 3 less whorls. Based on recent DNA research by (Horsakova et al. 2020) identifying *Euconulus* to species level should be considered tentative, waiting further DNA and shell morphology investigations.

Habitat: A species found in mixed hardwood forests on hillsides and ravines living under layers of leaf litter usually in drier sites than other species of *Euconulus*; it is rarely found in large numbers.

Status: G5; tentative ranking pending further DNA investigation; all records below are also tentative, pending further research.

Specimen: Kentucky, Hardin County, Spurrier (FM 251488).

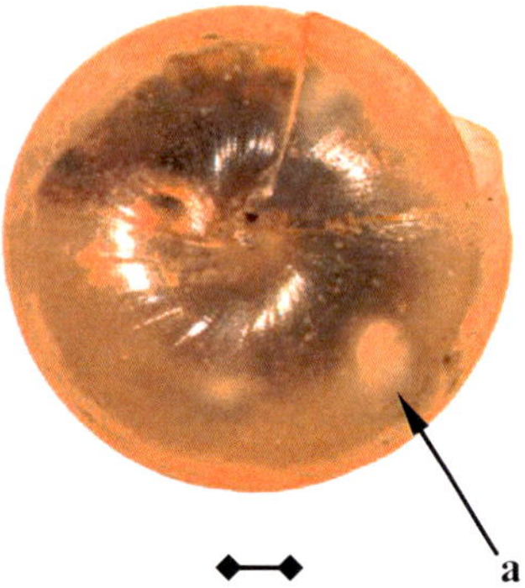

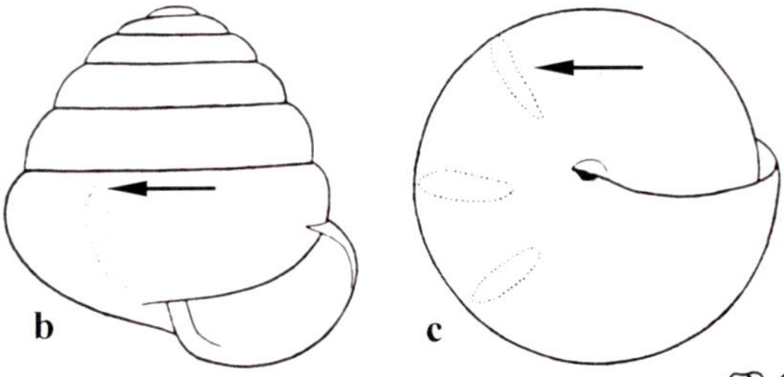

© Pilsbry, 1946

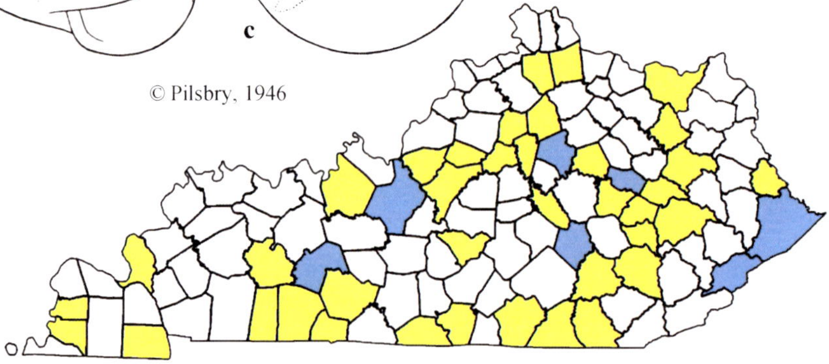

Kentucky *Euconulus* Species Compared and Sorted According to Historic Collectors L. Hubricht and H. Pilsbry (proportionate)

Euconulus fulvus **Group**
(**without** lamellae)

Euconulus fulvus egenus without teeth in young shells, Wayne County, Michigan (see page 241)

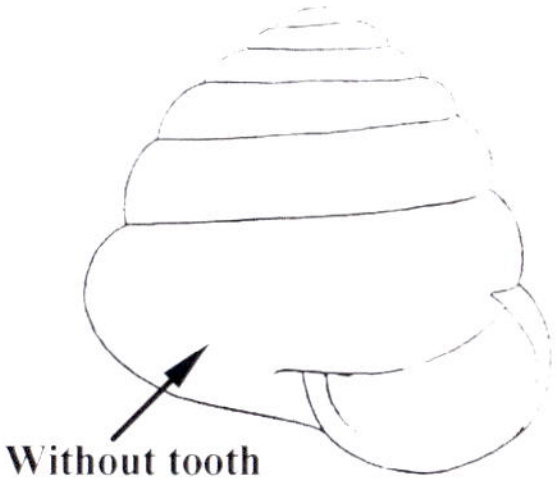

Euconulus polygyratus shells are wide (obese), or somewhat compressed for a *Euconulus*, with a narrow opening in the aperture Aroostook County, Maine (see page 242)

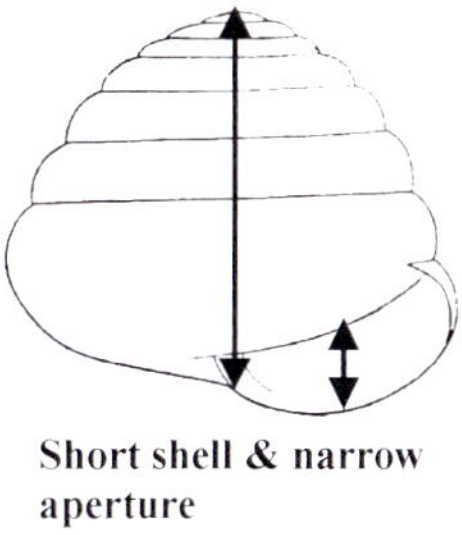

Euconulus chersinus **Group**
(with or **without** lamellae)

Euconulus chersinus adult shell tall with 7-8 whorls, Swain County, North Carolina (see page 243)

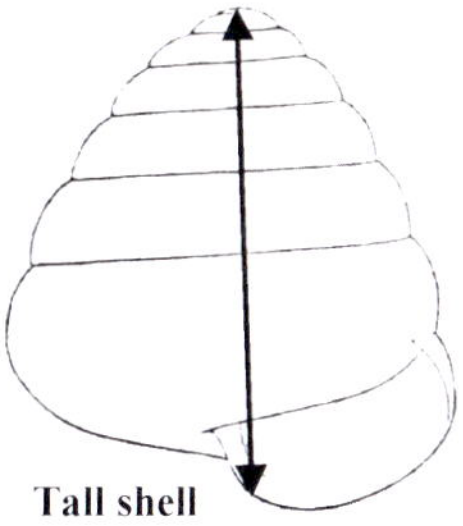

Euconulus trochulus, adult shell with a slightly angular periphery, Pulaski County, Kentucky (see page 244)

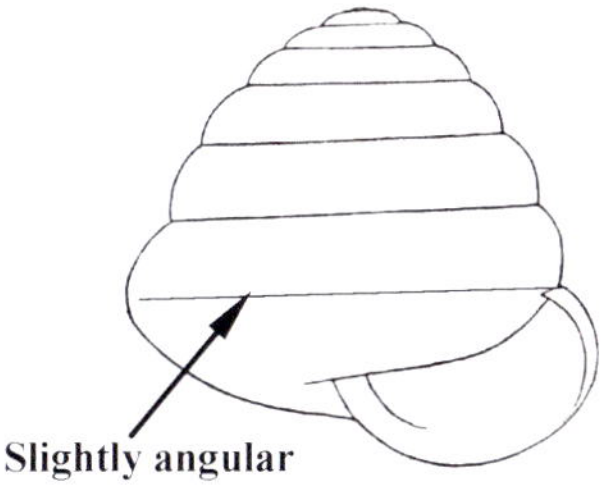

Euconulus dentatus juvenile shell and some adult shells with teeth, Hardin County, Kentucky (see page 245)

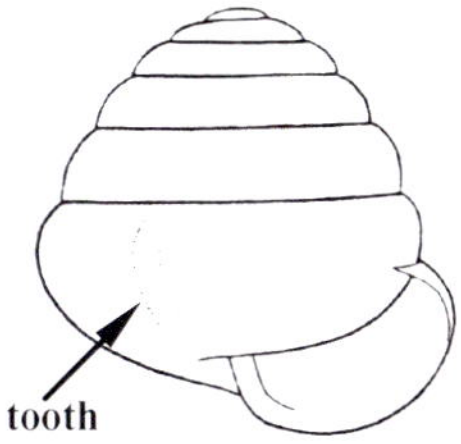

Globular Drop

Helicina orbiculata Say, 1818

Diameter: 5-8.5 mm, height 6 mm

Description: Roundish-shaped; lip slightly reflected; shell with 4.5 whorls; imperforate; operculum present; no teeth at any stage of growth; shells with brownish or sometimes white bands but many shells are often found without this feature; periphery rounded; live animal gray with a white foot; this is a primitive, largely subtropical and tropical family of land operculates (most occurring in Central and South America) that have many large and colorful members, only 2 species reaching Kentucky.

Similar Species: *Hendersonia occulta* has a slightly angular periphery typically in younger shells, no color features and a lower shell profile.

Habitat: A calciphile species found in more open habitat such as glades and limestone areas.

Status: G5/S4; Uncommon; mostly known from south central Kentucky; common on limestone outcrops in the Lake Cumberland area (MacGregor pers. comm.).

Specimen: Figs. (a) from Tennessee, Grundy County, Crystal Cave and (figs. b) pink color morph from Kentucky, Wayne County (author's collection).

a

Light color bands

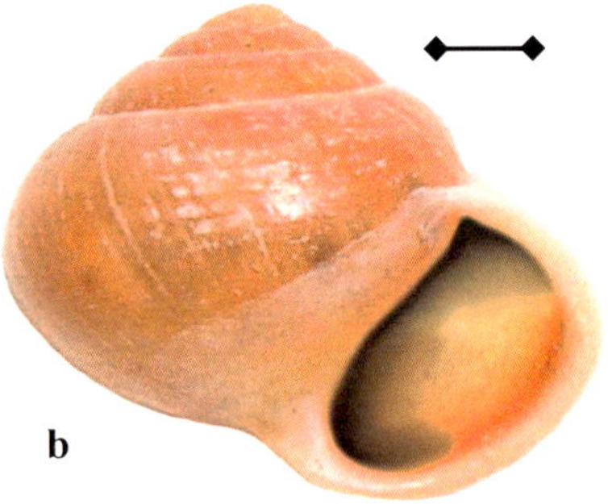

b

A pink color morph

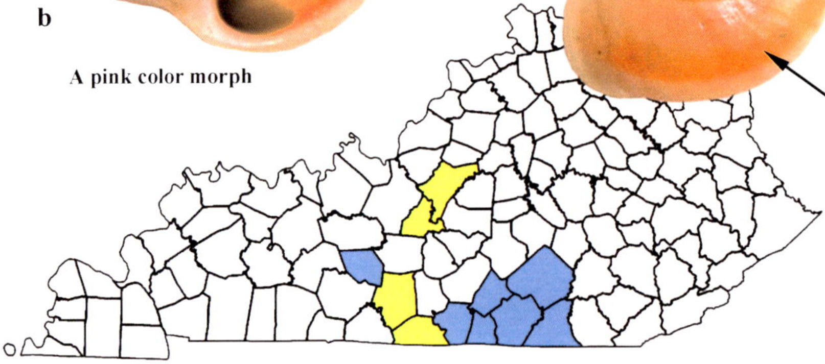

Global Distribution of Helicinidae

Helicinidae were the first gastropods to evolve to terrestrial life independently from any other known group of land snails, the Pulmonates, their closest relatives being marine and freshwater snails. Helicinidae are distinguished primarily by the presence of a calcareous operculum and one pair of antennae with eyes being at the base of these antennae. With more than 500 species the Helicinidae make up a significant portion of the molluscan fauna of the tropics both in diversity and abundance. They are fairly proportionately distributed across the tropical and subtropical zones of the New World as well as Indopacific and Pacific islands including margins of the Asian and Australian continents (Richling 2004). The islands of the Pacific and the Caribbean Sea however are the epicenters of this family with Cuba reporting more than eighty species and Jamaica around thirty-five. These are some of the most strangely-ornate and colorful gastropods on Earth. One need only study the intricate shell ornamentation of *Priotrochatella constellata* (pictured below) to fully appreciate one of nature's greatest achievements in calcium-based architecture. These island gems from Cuba are stunning examples of convoluted evolution. Interestingly, two Helicinidae species occur in Kentucky.

***Priotrochatella constellata*, Cuba**

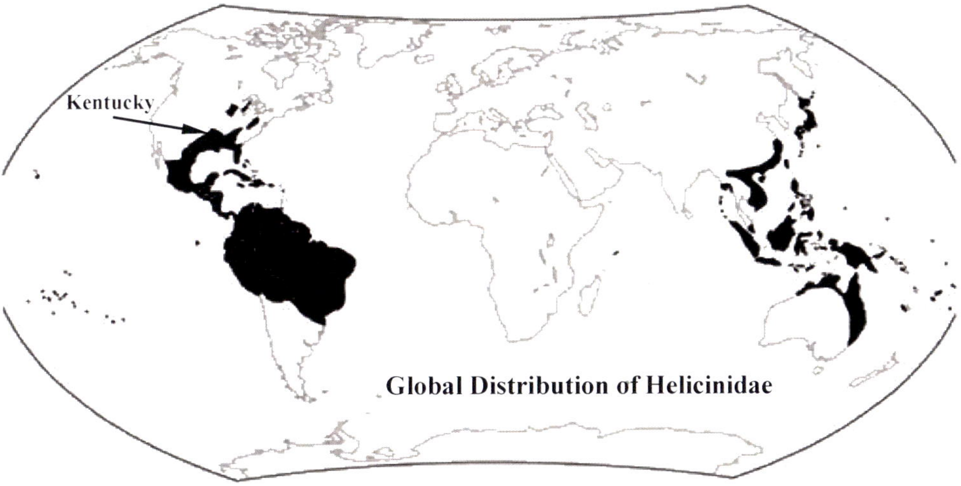

Cherrystone Drop

Hendersonia occulta (Say, 1831)

Diameter: 6-8 mm, height 4-6 mm

Description: Heliciform; lip thickened (a & b); shell with 4.5-5 whorls; imperforate; with an operculum in live individuals; transverse striae moderately developed; spiral striae are well developed and sometimes seen as fringes in young shells under 4 mm; periphery slightly angular and occasionally weakly keeled, particularly in juvenile shells; eyes are located at the base of the tentacles, (a trait common to a primitive and largely tropical family of gastropods); this species is unique among Kentucky land snails by being dioecious (having the male and females separate).

Similar Species: *Stenotrema* species are around the same size, but have a slit like aperture and are without an operculum.

Habitat: A calciphile snail of river bluffs, talus slopes, ravines and mountainsides.

Status: G4/S3; Uncommon; this species is restricted to Pine and Black Mountains in southeastern Kentucky.

Specimen: West Virginia, Randolph County, mouth of Bowden Cave (author's collection).

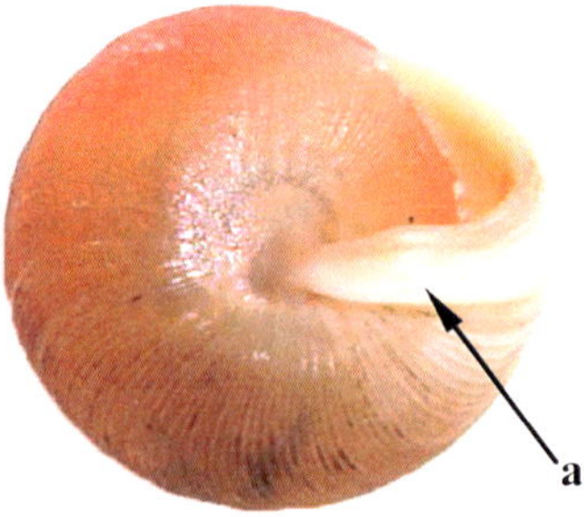

Close up of micro-sculpture

A juvenile Cherrystone Drop, *Hendersonia occulta,* illustrating stark differences between adult and young shells (common among land snails). Note the remarkable micro-sculpture and carination of shells before adulthood. Specimen from "The Trough" along the South Branch of the Potomac River, Hampshire County, West Virginia.

Maze Pinecone

Strobilops labyrinthicus (Say, 1817)

Diameter: 2.3-2.5 mm, height 1.7-1.8 mm

Description: Dome shape or beehive in form; lip only slightly reflected; shell with 5.5 tightly coiled whorls; perforate; chestnut-brown; elongated lamellae present in the aperture; within the interior are long lamellae of uneven lengths which can be seen through the bottom of live and fresh dead shells (fig. a); transverse striae are modified into well developed ribs on the top and side, becoming weaker on the base (this feature is easily viewed with a hand lens of 10X); periphery of body whorl is rounded or bluntly angular.

Similar Species: *Strobilops aeneus* has a lower shell profile, wider umbilicus, and most importantly, a more angular periphery (see page 253).

Habitat: Found in mixed hardwood forests and glade-like areas especially around loose soils of limestone substrate with red cedar trees.

Status: G5/S5; Common; this is likely the most common of the three *Strobilops* reported from Kentucky.

Specimen: Larry Watrous photo collection.

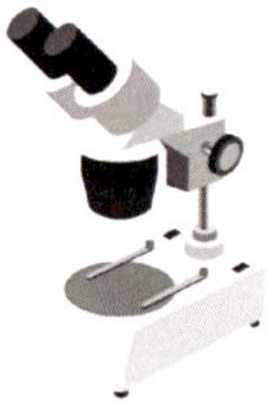

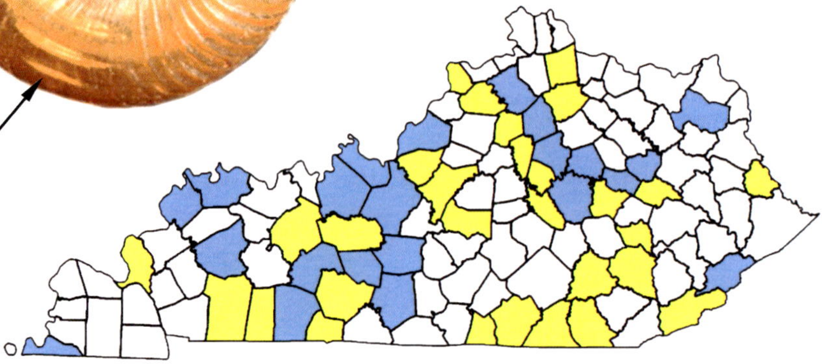

Southern pinecone

Strobilops cf. texasianus Pilsbry & Ferriss, 1906

Diameter: 2.3-2.4 mm, height 2 mm

Description: Dome shape; lip reflected; shell with 5.5 tightly coiled whorls; perforate to umbilicate; shell pale corneous; elongated lamellae present in the aperture; within the interior are long lamellae of uneven lengths, these lamellae easily seen through the bottom of live and fresh dead shells or through the cutaway top (see page 254); the transverse striae modified into strongly developed ribs, continuing boldly onto the base of the shell; periphery very slightly angular or roundish.

Similar Species: *Strobilops labyrinthicus* has a smaller umbilicus (a variable feature), weaker more closely–spaced riblets (also a variable feature), but most importantly shorter-entering parietal lamellae (see page 254 for comparisons).

Habitat: A *Strobilops* of wetter ground living among and under leaf litter of mixed hardwood forest but also on rocky limestone sites.

Status: G5/S3; Rare (?); the range of this species in Kentucky remains unclear a result of pass collections easily confused with *Strobilops labyrinthicus*.

Specimen: Larry Watrous photo collection.

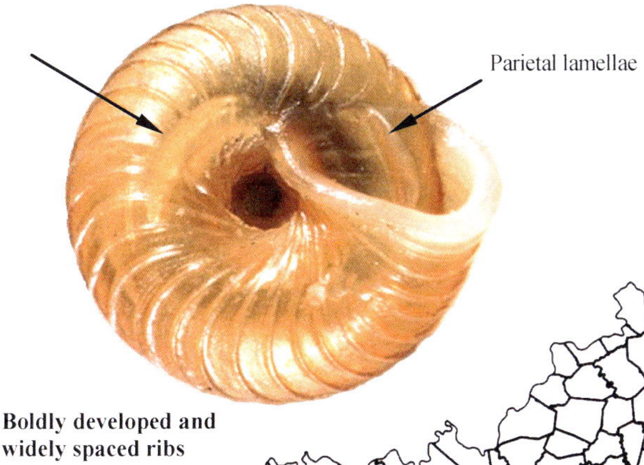

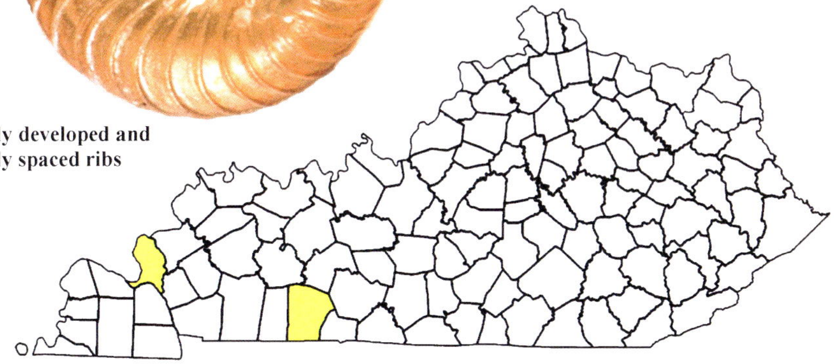

Bronze Pinecone

Strobilops aeneus Pilsbry, 1926

Diameter: 2.4-2.8 mm, height 1.5-2 mm

Description: Dome shape; lip reflected; shell with 5.5 tightly coiled whorls; perforate to umbilicate; light to dark brown; elongated lamellae present in the aperture; within the interior are long lamellae of uneven lengths which can be seen through the bottom of live and fresh dead shells or through the cutaway top; transverse striae modified into minute ribs; periphery angular.

Similar Species: *Strobilops labyrinthicus* has a slightly smaller umbilicus, higher shell profile, and a more rounded periphery (see page 254); *Euconulus* species are beehive-shaped, have smooth surfaces instead of ribs and have simple lips, not slightly reflected as in *Strobilops*.

Habitat: Found in open, glade like areas especially around limestone outcrops but also commonly found in upland woods under the bark of rotting logs often in association with *Zonitoides arboreus*.

Status: G5/S5; Common; likely occurs in every Kentucky county.

Specimen: Larry Watrous photo collection.

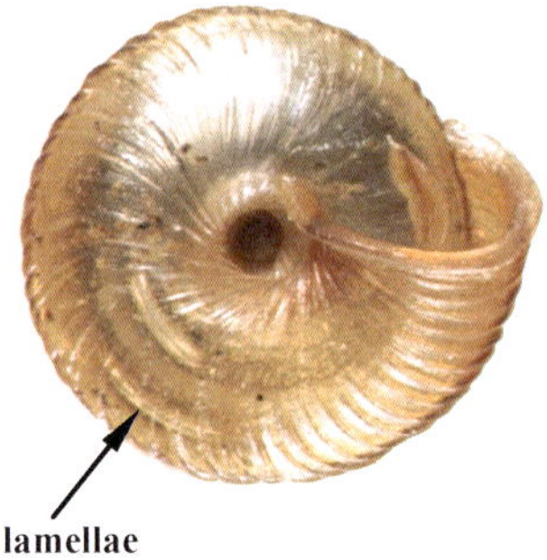

Note the fine riblets

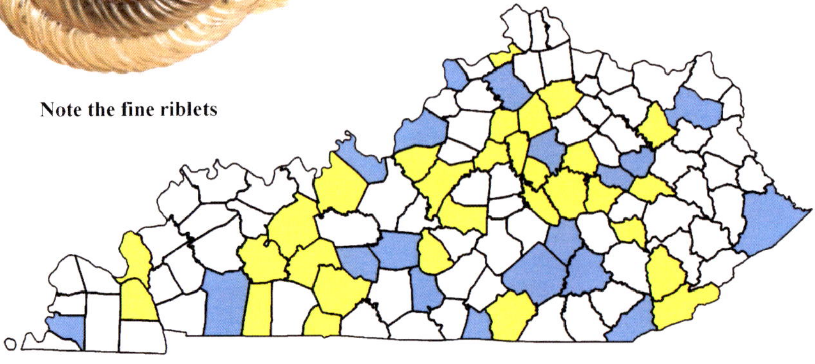

Kentucky Strobilops Compared

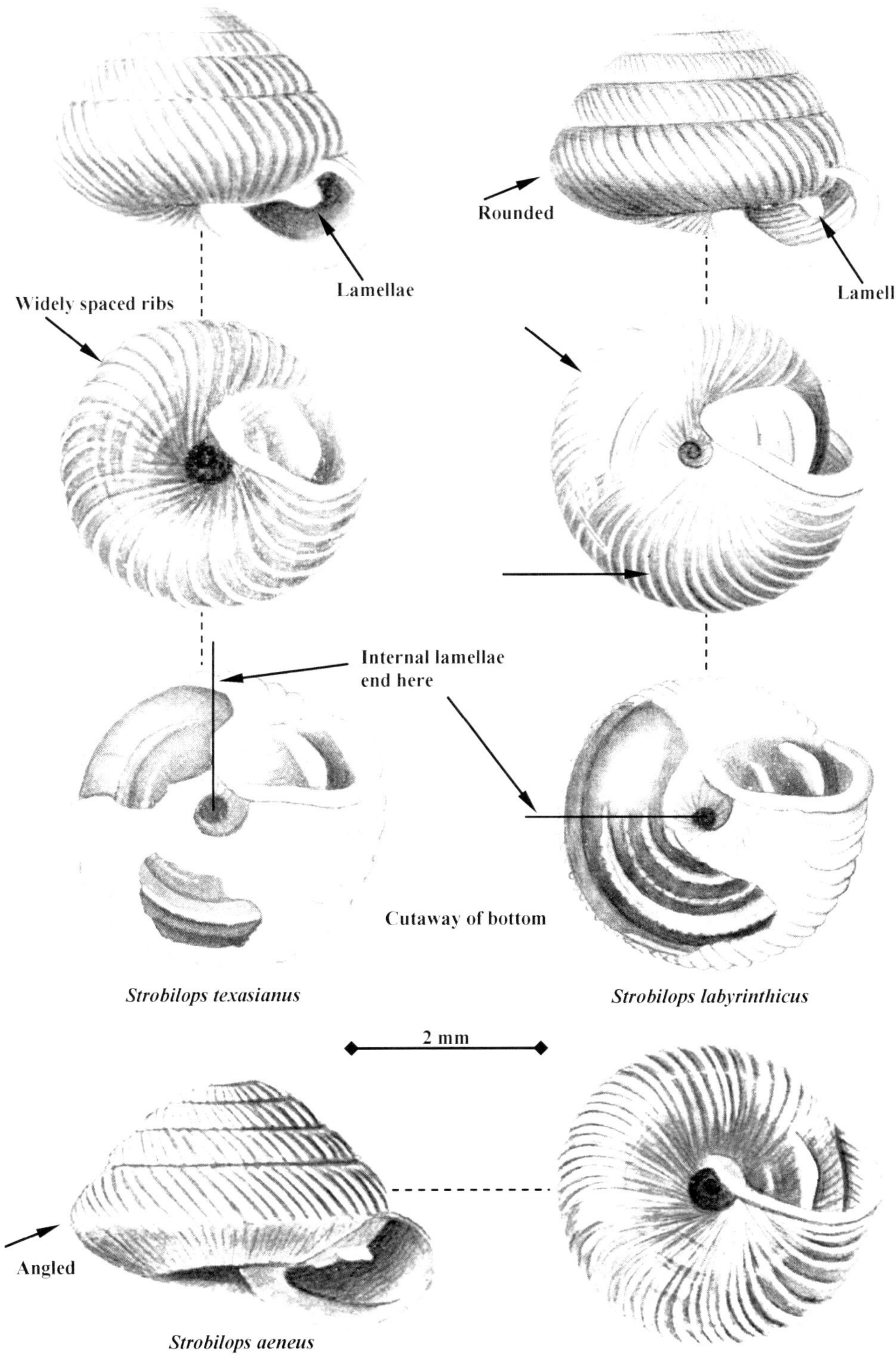

Land snails are often seen on spider webs, easily moving across the web's sticky surface without becoming entangled. Pictured above is a juvenile *Patera panselenus*. Note the spider in the background. There is some evidence that snails are attracted to the silk nets that trap condensed moisture for a source of drinking water especially during dryer weather. Pictured below, the same species appears to glean water from a web. Both images from Pike County, Kentucky.

This section includes land snails that are more or less the size and shape of pills. All species of *Stenotrema* are under 15 mm in diameter (in Kentucky, most species are between 7 and 10 mm), the majority possessing hairs (see below pictures) that cover the entire shell surface, some hairs quite long. These hairs collect forest debris, thought to help keep the snails well hidden from possible predators like birds and salamanders. The aperture of *Stenotrema* species is narrow (slit-like), denying easy access to predatory beetle larva. The long and wide parietal tooth of *Stenotrema* easily exceeds that of all other snails of similar size.

Genera Included:
(in order of appearance in text)

Stenotrema

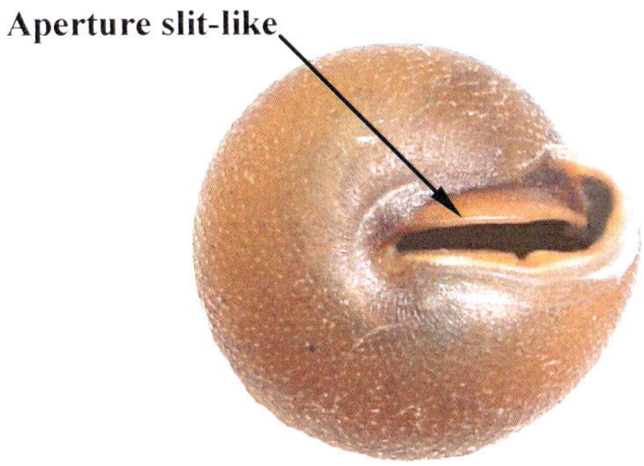

Shells 4-12 mm, pill-shaped, long parietal tooth (*Stenotrema*)

Stenotrema species are usually compact and pill-like in shape, having tightly coiled shells (a), the last whorl not expanded and most are with fine or coarse hairs. All species have a long parietal (blade-like) tooth. *Stenotrema* species are best distinguished in frontal and bottom views, top views having very little diagnostic value within the genus. Examining the aperture is key (figures e through m, page 257). The aperture opens on the underside of the shell and is slit-like and narrow. The interdenticular sinus (figures l and m, page 257) is deep in some species and shallow in others. Another important trait is the basal notch (figures g, h, i, j, and k, page 257), which can be nearly absent or a prominent feature of the lip. The fulcrum (below image), an internal structure and part of the center supporting column of the shell, can be different lengths and can be seen through the bottom of live and fresh shells, but is difficult to see in old and badly weathered shells. *Stenotrema* species differ mainly from *Euchemotrema* species by always having a completely sealed umbilicus (d), and on average are 2-3 mm smaller.

<u>*Stenotrema* species of Kentucky, pg. 259-267</u>

**Stenotrema stenotrema*
**Stenotrema macgregori*
**Stenotrema angellum*
**Stenotrema barbatum*
**Stenotrema edvardsi*
**Stenotrema hirsutum*

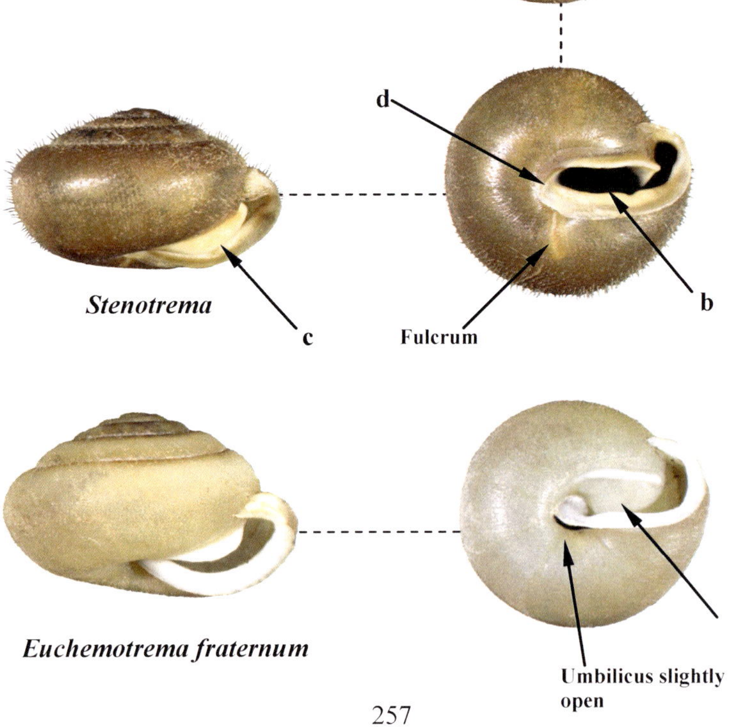

Aperture Terminology in *Stenotrema*

Bottom View

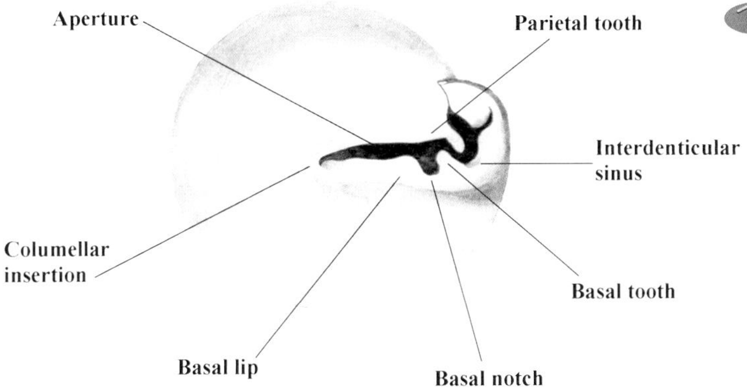

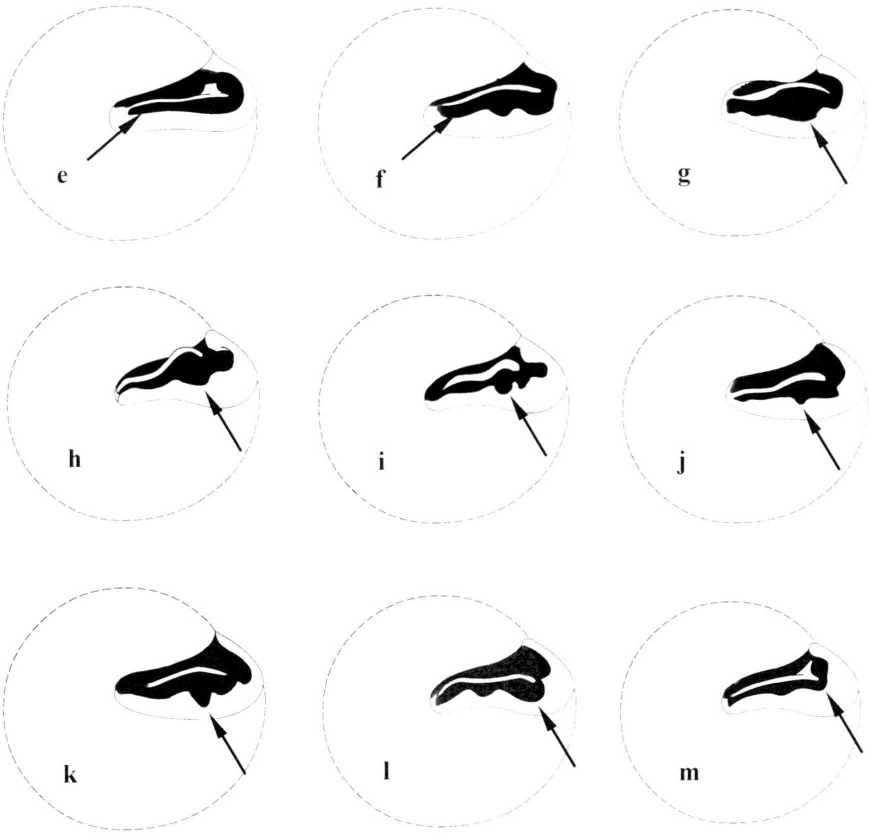

e) aperture narrow; f) aperture wide; g) basal notch wide; h) basal notch medium; i) basal notch narrow; j) basal notch shallow; k) basal notch deep; l) interdenticular sinus deep; m) interdenticular sinus shallow

Inland Slitmouth

Stenotrema stenotrema (L. Pfeiffer, 1842)

Diameter: 7.8-15 mm, height 5.3-8.6 mm

Description: Pill-shaped; aperture narrow, especially near the columellar insertion (a); shell with 5-6 whorls; imperforate; brownish-tan or cinnamon-brown; minute hairs present on entire surface, but are usually lost in aging shells; transverse striae poorly developed; basal notch is large and deep (b); interdenticular sinus deep (c); fulcrum is well developed with a convex edge (seen through an opening made on the side of the shell); periphery well rounded; this is the largest *Stenotrema* in the state.

Similar Species: *Stenotrema edvardsi* has an angular periphery; *S. macgregori* is smaller, has a more open aperture opening, but most importantly, a smaller basal notch and indistinct interdenticular sinus.

Habitat: Usually found under leaf litter in lower elevation mixed hardwood forests.

Status: G5/S5; Common; one of the most wide ranging *Stenotrema* in eastern North America and a common and widespread resident of Kentucky, especially in the central and southern portions of the state.

Specimen: Tennessee, Sevier County, Greenbrier Picnic Area, GSMNP (GSMNP collection).

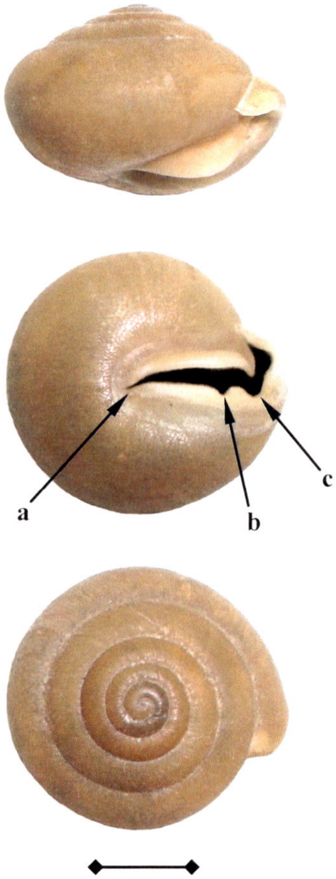

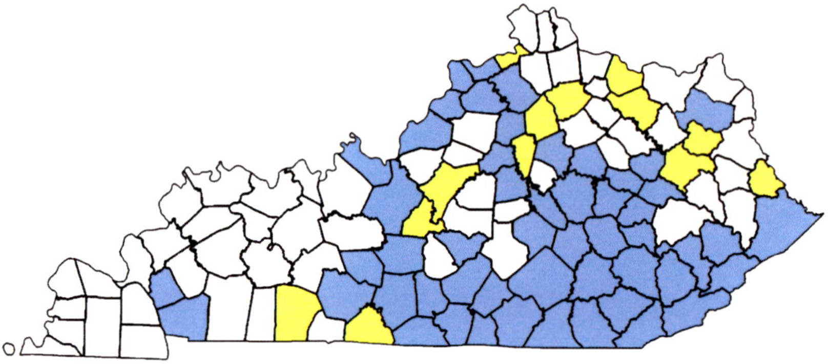

Fraudulent Slitmouth

Polygyridae

Stenotrema macgregori Dourson, 2011

Diameter: 8-10 mm, height 6 mm

Description: Pill-shaped; lip reflected; shell with 5-6 whorls; imperforate; cinnamon-buff; short stiff hairs present on entire surface, but hairs are often lost in older shells; transverse striae poorly developed; basal notch shallow and the interdenticular sinus is indistinct in most specimens; fulcrum is well developed, but short; periphery rounded.

Similar Species: *Stenotrema stenotrema* is 2-3 mm larger, has a deeper basal notch, more-closed aperture, and the interdenticular sinus is notably excavated; the two species are occasionally found together and can readily be separated by size alone; *S. angellum* is larger, more compressed with a notably shouldered periphery (a) (*S. macgregori* has a more rounded periphery), has a lighter periostracum and is a species of the Bluegrass of Kentucky.

Habitat: Found in rich hardwood forests under forest litter; apparently absent or uncommon from the drier mountain tops, dense rhododendron thickets, and Virginia pine forests of eastern KY.

Status: G3/S3; Uncommon; this recently described species is restricted to eastern Kentucky.

Specimen: Kentucky, Pike County: hillside above Russell Fork, 1.6 km SE of Elkhorn City within Breaks Interstate Park (FMNH 328317 Holotype).

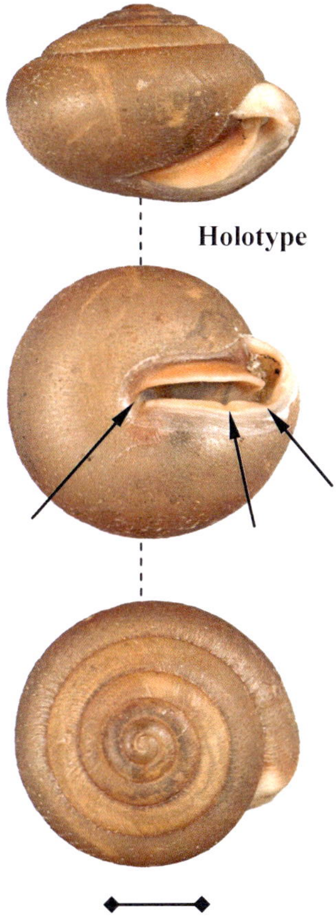

Stenotrema angellum

S. macgregori

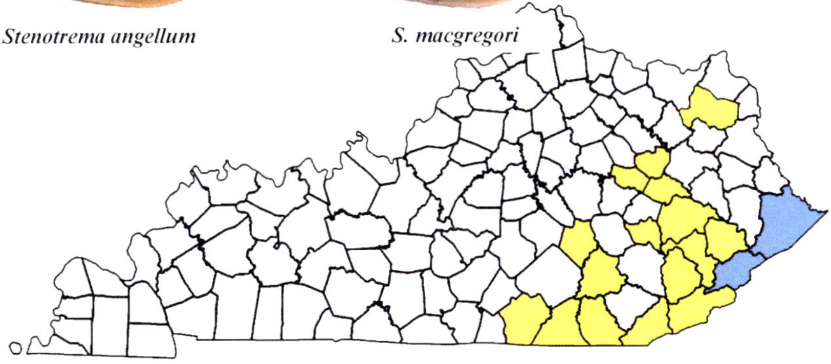

Kentucky Slitmouth

Polygyridae

Stenotrema angellum Hubricht, 1958

Diameter: 10-11 mm, height 6-6.5 mm

Description: Pill-shaped; lip reflected; shell with 5-5.5 whorls; imperforate; cameo-brown to snuff-brown; minute hairs present on entire surface but hairs are lost in older shells or when shells are cleaned; transverse striae poorly developed; basal notch is small and the interdenticular sinus is shallow and indistinct; fulcrum short; shell periphery not well rounded.

Similar Species: *Stenotrema stenotrema* averages 2-3 mm larger, has a deeper basal notch, has a more rounded periphery in side view and the interdenticular sinus is notably deeper; the two species are often found together where they can readily be separated by size; *S. angellum* refers to *S. macgregori* of the Cumberland Plateau region but is larger, more compressed with a notably shouldered periphery (*S. macgregori* has a more rounded periphery) and a darker periostracum; the two species have not been found together.

Habitat: A species of hillsides and ravines found living under leaf litter in mixed hardwood forests.

Status: G4/S4; Common; Likely the most common *Stenotrema* species in central Kentucky, not occurring much beyond the state's borders.

Specimen: Kentucky, Fayette County, Raven Run Nature Reserve (author's collection).

Stenotrema macgregori

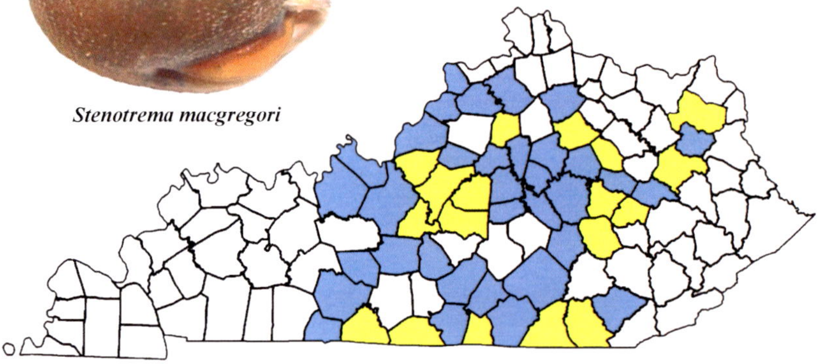

Bristled Slitmouth

Stenotrema barbatum (G. H. Clapp, 1904)

Diameter: 7.5-11 mm, height 5.5-7 mm

Description: Pill-shaped; lip reflected; shell with 5.5 whorls; imperforate; tawny-olive to light cinnamon colored; well-developed, stiff hairs (0.5 mm long) usually present on live and fresh shells, but lost with age; transverse striae poorly developed; basal notch shallow and the interdenticular sinus is indistinct; fulcrum long and well developed, fully 3 mm in length (figures a and b); shell periphery not well-rounded, with a slight shoulder.

Similar Species: Most like *S. hirsutum,* but larger and with strongly developed, stiffer hairs, 0.5 mm long; *S. edvardsi* has an angular periphery; *S. stenotrema* has a more narrow aperture opening and is less hairy.

Habitat: A snail found in mixed hardwood forests of floodplains under leaf litter and logs; sometimes an upland species.

Status: G5/S3; Common; reported from scattered locations across the state.

Specimen: West Virginia, Summer County, Bluestone River (author's collection).

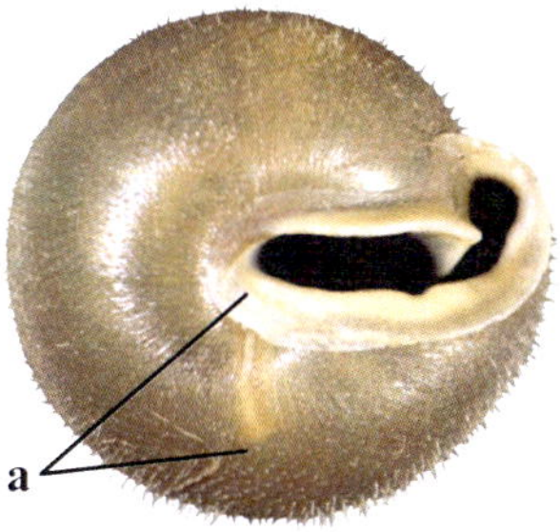

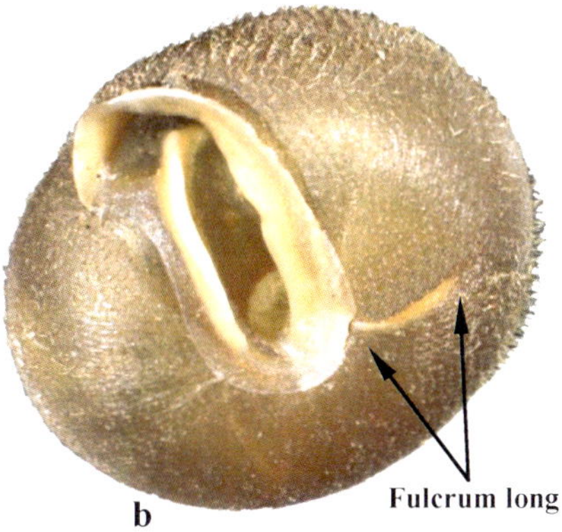

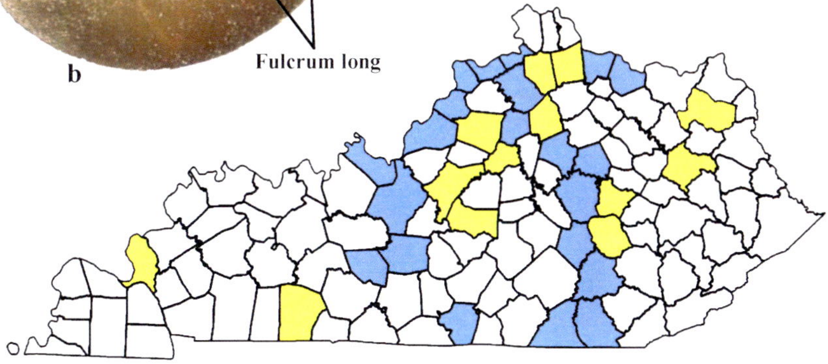

Ridge-and-Valley Slitmouth

Stenotrema edvardsi (Bland, 1856)

Diameter: 7-8 mm, height 4.8-5.4 mm

Description: Pill-shaped; shell with 5-5.5 whorls; imperforate; tawny-olive to light cinnamon colored; stiff and well-developed hairs, especially on young animals, but usually lost on aging shells (see opposite page); transverse striae poorly developed; basal notch small and shallow; interdenticular sinus indistinct; fulcrum short and moderately developed; shell periphery distinctly and always angular.

Similar Species: *Stenotrema hirsutum* is more or less the same size but differs in having a deeper interdenticular sinus, a deeper basal notch, and most notably, a boxy (not angular) periphery.

Habitat: Rocky mixed hardwood forests, around and under logs and in leaf litter; reported to be very common in hemlock dominated ravines on hillsides (Hubricht 1985).

Status: G4/S3; Fairly Common; in Kentucky, this long-haired gastropod is mostly restricted to eastern Kentucky counties.

Specimen: West Virginia, Mercer County, Camp Creek State Park (author's collection)

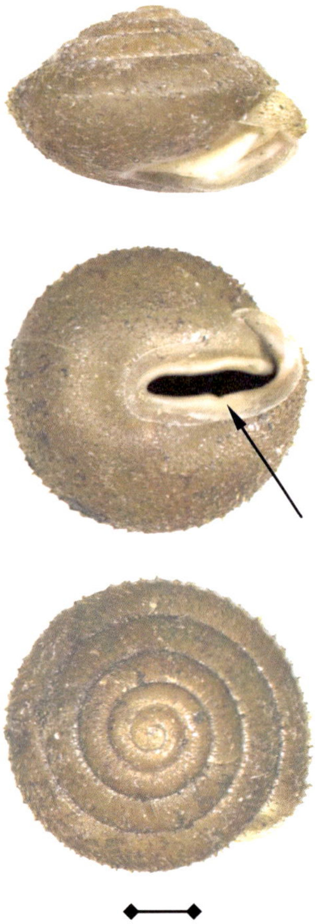

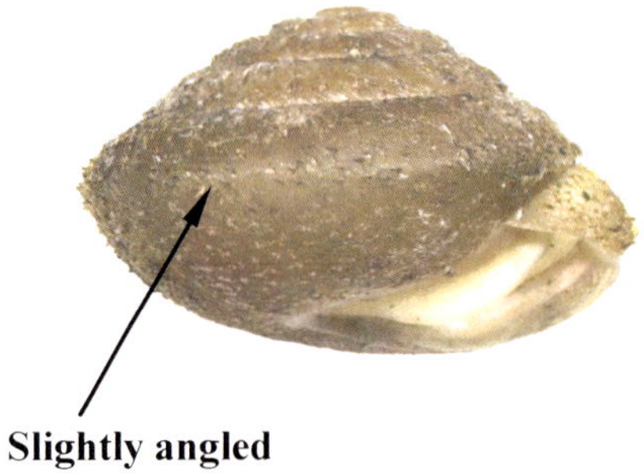

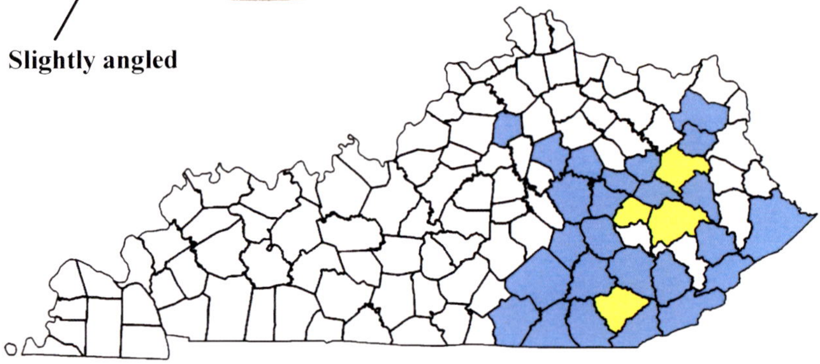

Above image of a juvenile Ridge-and-Valley slitmouth, *Stenotrema edvardsi,* crawling toward dinner; a colony of maturing slime mold growing from rotting wood. Below, observe the numerous hairs that cover the entire shell surface; the longest ones growing from the sutures. These hairs are usually lost as shells mature. Both images from Red River Gorge, Kentucky.

Hairy Slitmouth

Polygyridae

Stenotrema hirsutum (Say, 1817)

Diameter: 5-9.6 mm, height 4.3–5.7 mm

Description: Pill-shaped; aperture width generally wider than other *Stenotrema* species; shell with 5-5.5 whorls; imperforate; cinnamon-buff to tan; short, stiff hairs present on entire surface, but hairs are generally lost in older shells; transverse striae poorly developed; basal notch and the interdenticular sinus are moderately deep; the fulcrum (a) is well developed, long (around 2 mm), and can be seen through the bottom view of fresh shells; shell periphery is boxy-shaped in frontal view.

Similar Species: Most similar to *S. simile,* but with less prominent lip features, and most importantly, a much longer fulcrum than seen in *S. simile.*

Habitat: A species of dry upland mixed hardwood forests found under a variety of forest litter; also found hiding under logs and rocks and in clearings.

Status: G5/S5; Common; this widespread and frequent land snail most likely occurs in every Kentucky county.

Specimen: Kentucky, Powell County, Furnace Mountain (author's collection).

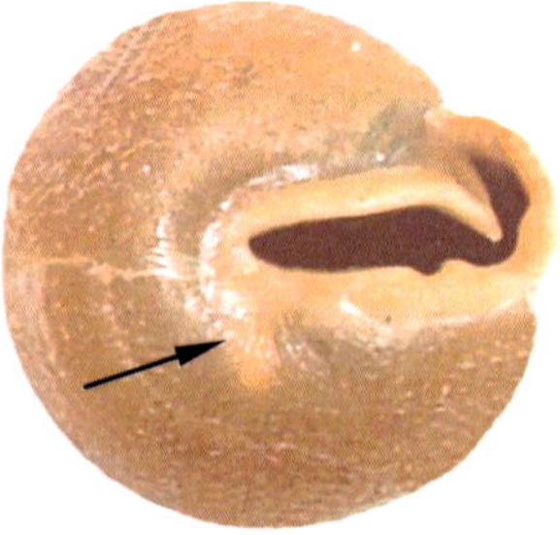

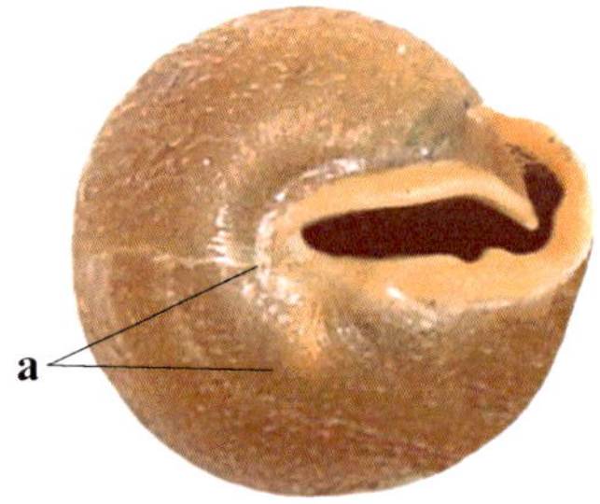

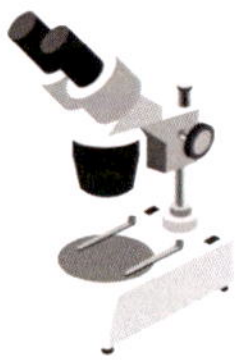

Bad Branch Nature Preserve, Letcher County, Kentucky

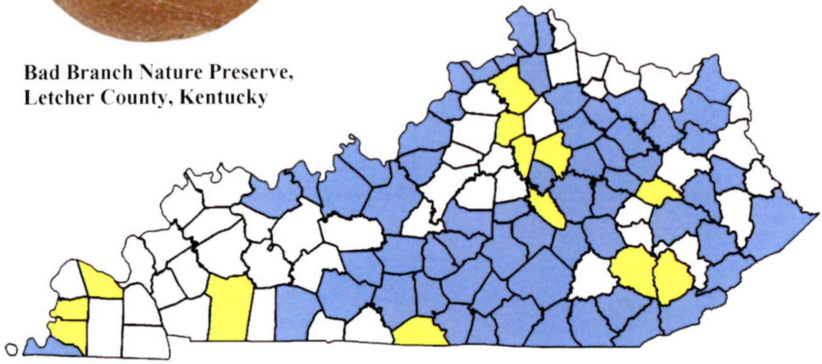

Stenotrema and *Euchemotrema* species compared (proportionate)

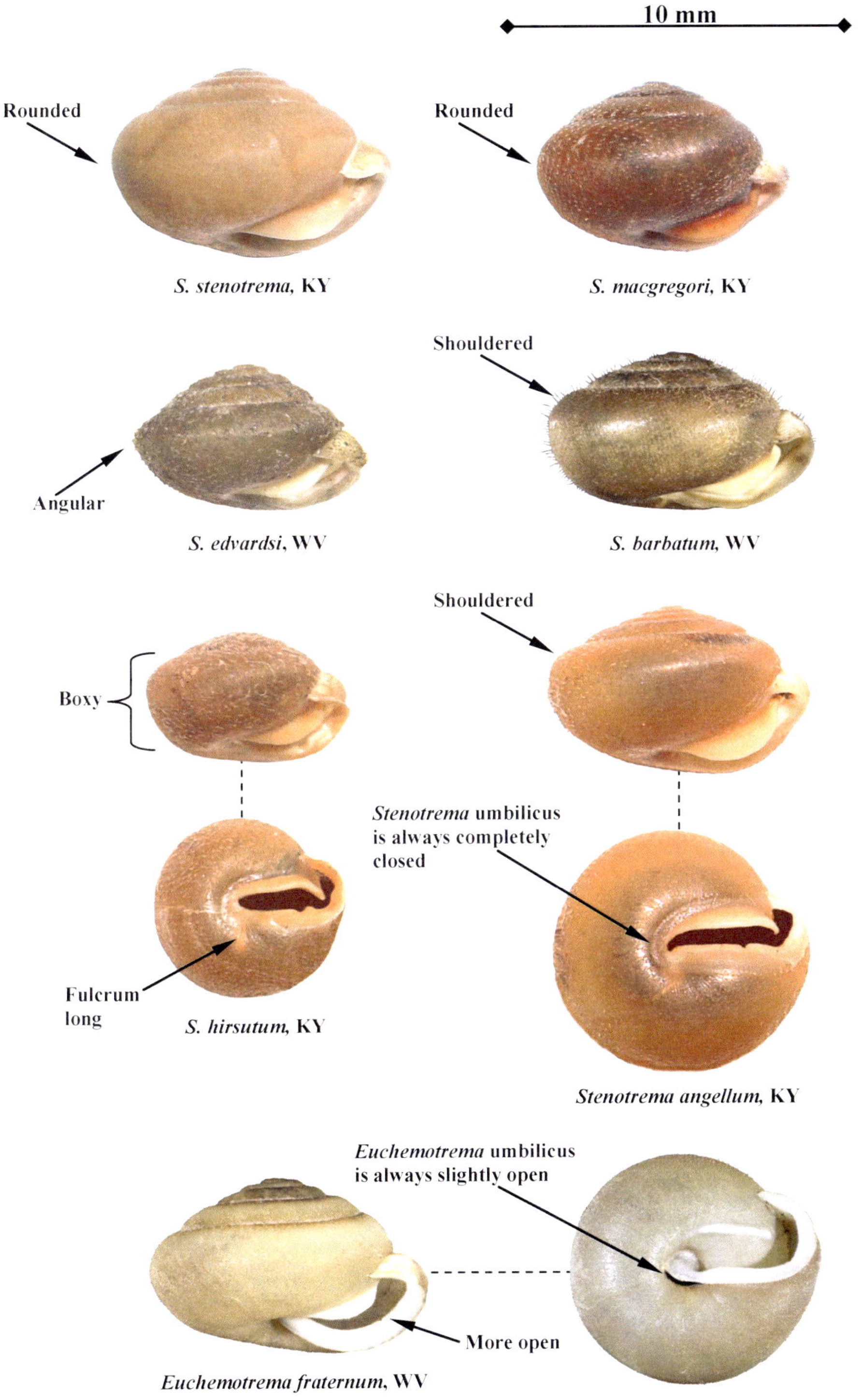

The Protective Hairs of *Stenotrema*

Above image of a Highland Slitmouth, *Stenotrema altispira,* covered in bits of soil and detritus. Below, a clean shell of same species showing the remarkable hairs that jacket the shell's exterior. The hairs collect material, thought to aide in crypsis, perhaps keeping the snail out of plain view of predators such as salamanders and birds. Both images from Roan Mountain, North Carolina.

This section includes land snails that possess shells which are wider than tall, heliciform or depressed heliciform, with or without teeth, and in most cases, have reflected lips. Shells in this group can be small, 5 mm or less, such as *Vallonia*, but most are among the largest land snails found in Kentucky, some reaching nearly 45 mm in diameter. Most are unadorned in color, although some gastropods such as *Webbhelix* and *Allogona* are rather colorful.

Genera Included:

(in order of appearance in text)

Vallonia
Daedalochila
Lobosculum
Euchemotrema
Inflectarius
Patera
Triodopsis
Xolotrema
Fumonelix
Mesodon
Neohelix
Webbhelix
Allogona
Appalachina

A Grand Globe, *Mesodon normalis,* strips rotting flesh for protein (yummy) and rasps bone for calcium from a decomposing raccoon carcass, Hot Springs, North Carolina.

Shells under 3 mm, umbilicate, reflected lip, <u>without teeth,</u> with or without paper-thin ribs (*Vallonia*)

<u>Vallonia</u>, like *Strobilops,* are generally uncommon gastropods in rich woods, becoming more common in glades, dry limestone outcrops, and abandoned parking lots. They are easily separated from all snails of similar size and form by usually having a widely reflected lip, although in several species the lip is rather narrowly reflected; with most other snails of comparable size containing simple lips.

<u>Shell 1.8-2.3 mm, lip widely reflected, without internal armature, widely umbilicate, pg. 271-276</u>

*Vallonia pulchella
*Vallonia excentrica
*Vallonia costata
*Vallonia perspectiva

Vallonia pulchella

Vallonia excentrica
Larry Watrous photo collection

Costate Vallonia

Valloniidae

Vallonia costata (Müller, 1774)

Diameter: 2.5 mm, height 1 mm

Description: Depressed heliciform; lip widely reflected, thickened within; aperture lip not continuous (a); shell with 3.5 loosely coiled whorls, the last whorl greatly expanded; widely umbilicate; gray or pale yellowish-corneous; no teeth present; shell surface with 20-35 paper-thin ribs in the last whorl (see page 218 for counting ribs); periphery is well rounded.

Similar Species: *Vallonia perspectiva* is smaller in diameter and has a smaller number of ribs in the last whorl (averaging approximately 35 or more); *V. parvula* is smaller and pale gray or nearly white.

Habitat: A Holarctic species of open habitat such as glades, grassy edges, and roadsides, also found in abundance at the edges of abandoned parking lots and around crumbling buildings.

Status: G5/S4; Uncommon; this largely northern gastropod is reported from mostly central Kentucky counties.

Specimen: Kentucky, Powell County, parking lot in the city limits of Stanton (author's collection).

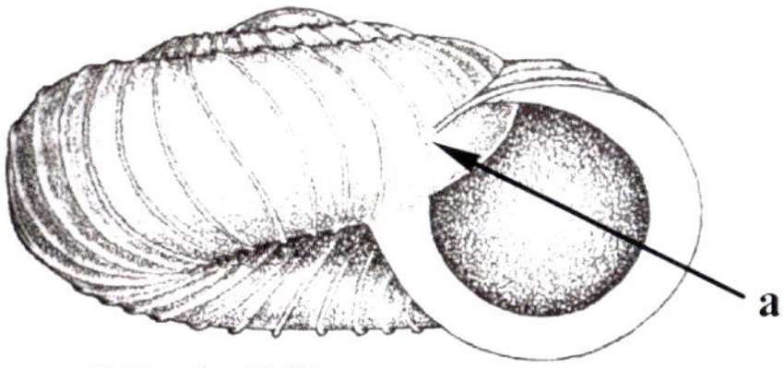

© Burch, 1962

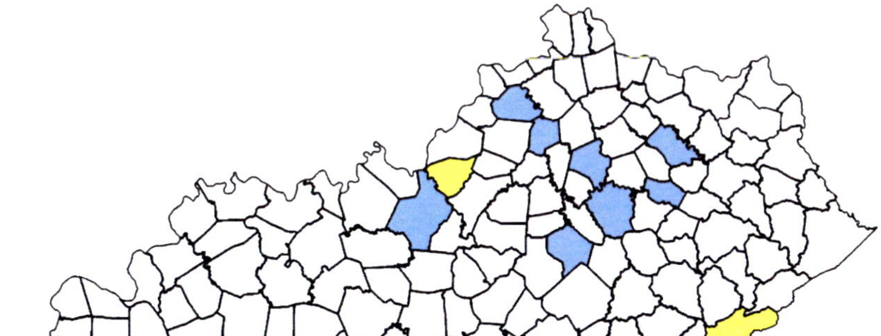

Thin-lip Vallonia

Vallonia perspectiva Sterki, 1893

Diameter: 1.7-2 mm, height 0.7 mm

Description: Depressed heliciform; lip only slightly reflected and thin, continuous across the body whorl or on occasion, growing free of it (see opposite page, fig a); shell with 3.5 loosely coiled whorls, the last whorl expanded; widely umbilicate; pale horn to colorless; no teeth present; shell surface with paper thin ribs (about 35 or more in the last whorl); periphery roundish.

Similar Species: Similar to *Vallonia costata* but with a greater number of ribs in the last whorl and a notably wider umbilicus (these two characters being somewhat variable) and smaller in size; *V. parvula* (page 379) has a strongly thickened lip and smaller umbilicus.

Habitat: A calciphile snail, found on talus slopes, dry glade-like limestone outcrops with mixed cedar trees.

Status: G4/S3; Uncommon; in Kentucky, this species is currently known from only a few central and eastern Kentucky counties.

Specimen: West Virginia, Greenbrier County, near mouth of Second Creek (author's collection).

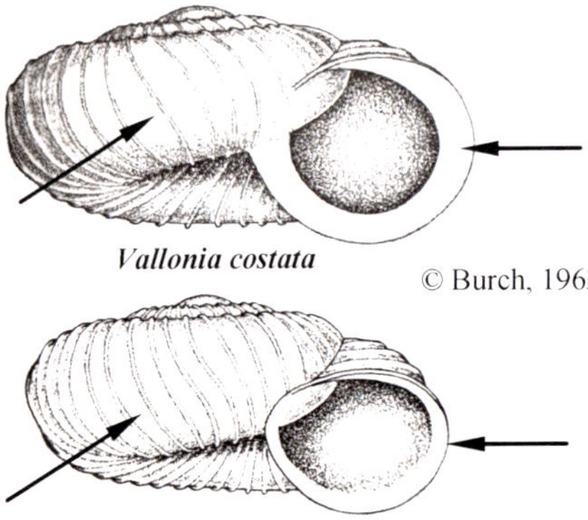

Vallonia costata

© Burch, 1962

Vallonia perspectiva

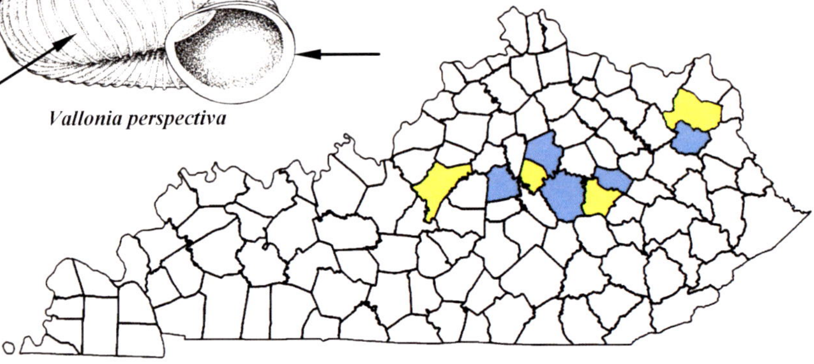

Built-in radiator — *Vallonia* snails usually live in hot dry habitats such as glades and other exposed environments. The extraordinary, paper-thin ribs on the shell surface of these little jewels are thought to act like radiators, keeping the animal inside cool. Pictured here is the Thin-lip Vallonia, *Vallonia perspectiva* from Greenbrier County, West Virginia.

Calcium Recycling and Land Snails

The skull of a bobcat doesn't go to waste as a Balsam Globe, *Mesodon andrewsae* (left) and a Mountain Tigersnail, *Anguispira jessica,* (right) glean calcium from the bone. The calcium will facilitate bodily functions, reproduction, but most importantly, shell building. Images taken on Roan Mountain, Mitchell County, North Carolina.

Shells 5-10 mm, depressed heliciform, with crowded convoluted teeth (*Daedalochila* and *Lobosculum*)

Daedalochila **and** ***Lobosculum*** are interesting species of small snails in the family Polygyridae that, in general, are most common and speciose in the southeastern USA. They are depressed heliciform, with some southern species rather remarkably flattened. The umbilicus is either rimate or widely umbilicate, tightly coiled, ranging in size from 5-10 mm. Most species, including taxa found in Kentucky, contain an elaborate arrangement of unyielding barriers or teeth within the aperture, preventing successful attacks from predatory *Cychrine* beetles. The snails are unknown from extreme eastern Kentucky and are generally uncommon in central and western Kentucky.

Shell 5-10 mm, lip moderately reflected, aperture crowded with distorted teeth, widely umbilicate to rimate, pg. 279-284

**Daedalochila fatigiata*
**Daedalochila plicata*
**Daedalochila troostiana*
**Daedalochila leporina*
**Lobosculum pustuloides*

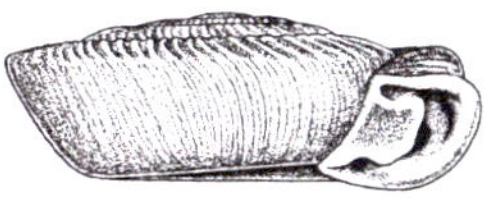

Daedalochila fatigiata

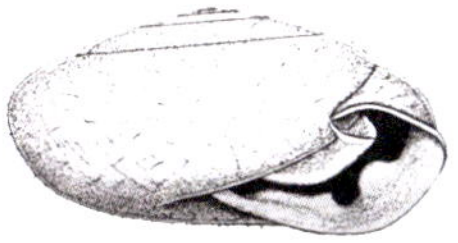

Lobosculum pustuloides

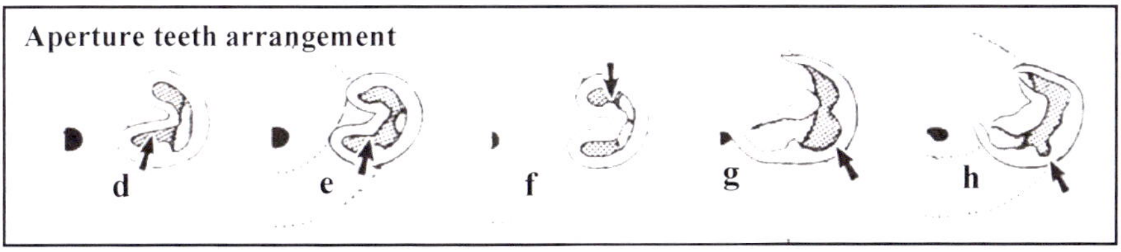

Aperture teeth arrangement; **d**) basal branch of parietal tooth without an angle; **e**) basal branch of parietal tooth with an angle; **f**) upper branch of parietal tooth with a projection; **g**) basal and palatal teeth widely separated; **h**) basal and palatal teeth narrowly separated (drawings on this page from Burch 1962).

Bluegrass liptooth, *Daedalochila fatigiata*

Bluegrass Liptooth

Daedalochila fatigiata (Say, 1829)

Diameter: 8-10.4 mm, height 2.5 mm

Description: Depressed heliciform; lip oddly reflected and deformed-looking; shell with 6 whorls; umbilicate, pale cinnamon-buff; without hairs on shell surface; transverse striae well developed and rib-like; spiral striae present; shell periphery is carinate or strongly angular; large teeth in the aperture.

Similar Species: The shell periphery of *Daedalochila troostiana* is only bluntly angular but the teeth arrangement is similar to that of *D. fatigiata.*

Habitat: A calciphile species usually found crawling on bare soils in open, sunny habitat during and after rain events (Hubricht 1985).

Status: G3/S1; State Threatened; In Kentucky this rare species is currently known from Livingston County only; the species also has a restricted range in Tennessee and Alabama.

Specimen: Tennessee, Wayne County, Clifton, along Tennessee River (author's collection).

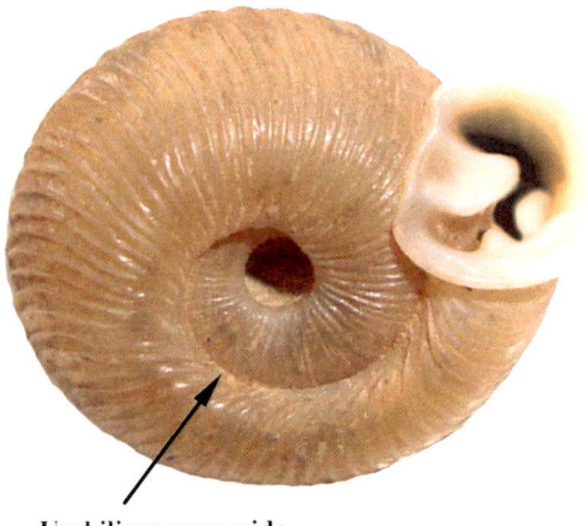

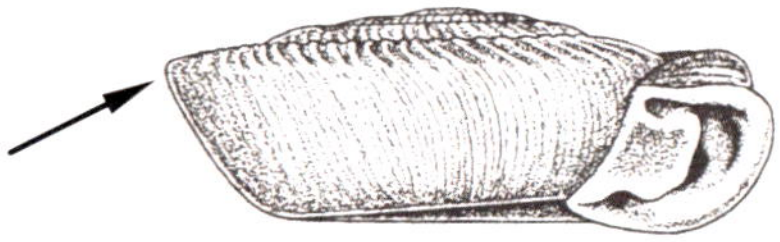

© Burch, 1962

Cumberland Liptooth

Polygyridae

Daedalochila plicata (Say, 1821)

Diameter: 5.5-5.7 mm, height 2.5 mm

Description: Depressed heliciform and compact; lip warped and deformed-looking, first 4 whorls are tightly coiled but the last whorl is greatly expanded and twisted outward; shell with 5 whorls; umbilicate, tannish-brown; without hairs on shell surface; transverse striae well developed; upper branch of parietal tooth with a small projection (see below); shell periphery squared or bluntly angular.

Similar Species: *D. fatigiata* is larger, has a flatter profile, is uniquely ribbed and the periphery is carinate to strongly angular.

Habitat: A calciphile species found in dry hardwood forests and cedar glades around limestone outcrops; usually under leaf litter.

Status: G4/S4; common; the most common and wide ranging *Daedalochila* species in Kentucky.

Specimen: Figure (a) from Kentucky, Warren County, Bowling Green (author's collection).

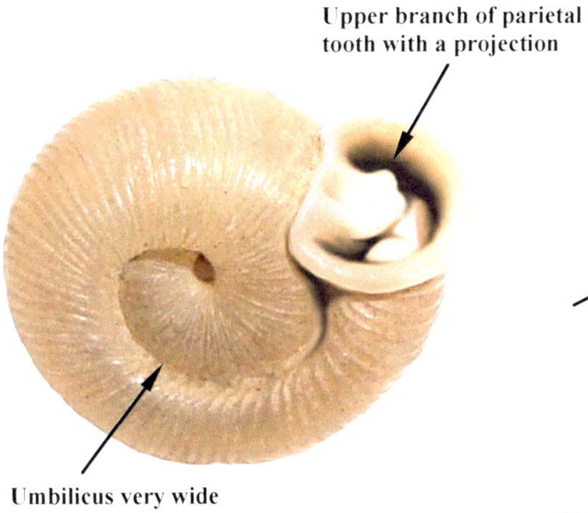

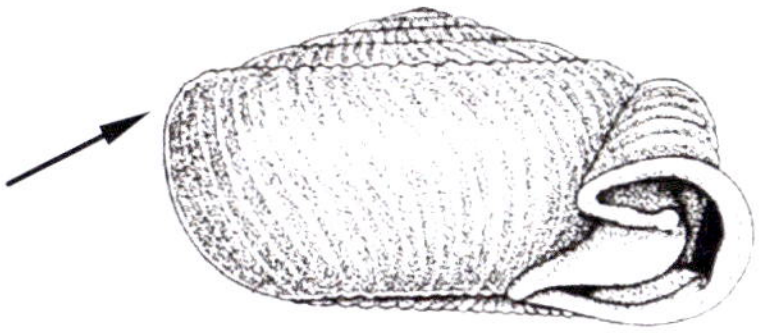

© Burch, 1962

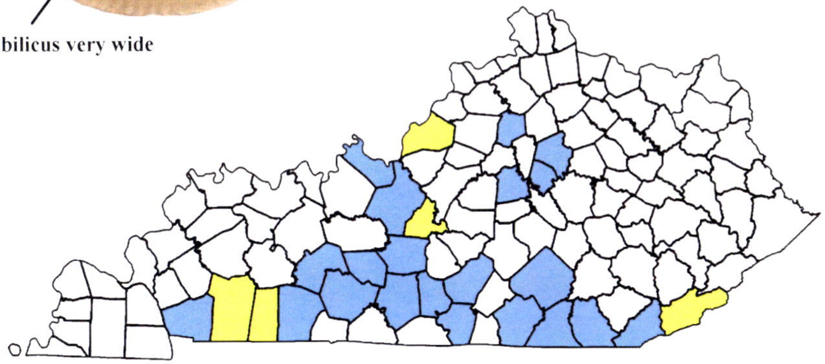

Nashville Liptooth

Daedalochila troostiana (I. Lea, 1839)

Diameter: 6.7-8.1 mm, height 3.5 mm

Description: Depressed heliciform; lip oddly reflected and deformed-looking; shell with 5-6 whorls; umbilicate, tannish-brown; without hairs on shell surface; transverse striae well developed and rib-like; upper branch of parietal tooth without a small projection; a small tooth (a) on the parietal wall a short distance within , (use an emery board to file down the outer shell surface to view this feature); shell periphery only bluntly angular.

Similar Species: *Daedalochila plicata* is slightly smaller and has a small projection on the upper branch of the parietal tooth; *Daedalochila fatigiata* is larger, has an angular periphery and does not contain the internal tooth seen in *D. troostiana.*

Habitat: A calciphile species found in dry hardwood forests and cedar glades around limestone outcrops; usually under leaf litter and other forest debris. Found in the same habitat as *D. plicata* but not found with it.

Status: G4/S3; Uncommon, reported from just two counties in south-central Kentucky.

Specimen: Kentucky, Cumberland County, 9 miles north northwest of Burkesville (FMNH 254810).

© Pilsbry, 1948

© Burch, 1962

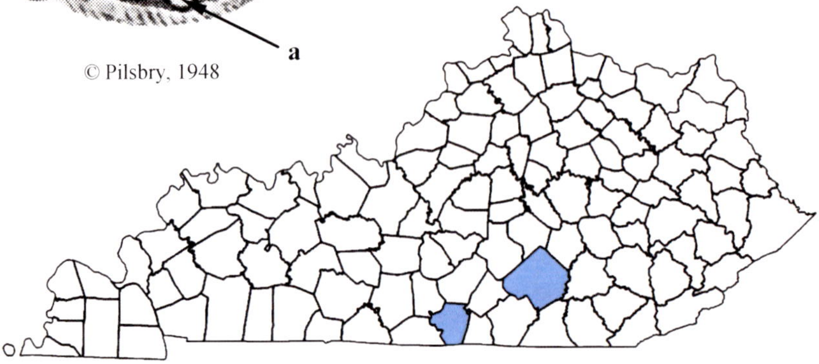

Tiny Liptooth
Lobosculum pustuloides

Gulf Coast Liptooth

Daedalochila leporina (Gould, 1848)

Diameter: 5-6.5 mm, height 2.5 mm

Description: Depressed heliciform; lip reflected; shell with 4.5-5 whorls; rimate, umbilicus nearly covered by the basal lip; pale-brown to olive-buff; upper shell surface glossy; delicate hairs on shell surface are sparse and usually lost in older shells; the basal tooth is situated on a strong ridge; shell periphery rounded.

Similar Species: *Loboscululm pustuloides* has a notably more open umbilicus; *D. plicata* is widely umbilicate, has distinctive transverse striae that are rib-like and the teeth are arranged very differently.

Habitat: A species of low, wet floodplain mixed hardwood forests, meadows and near springs.

Status: G4/G3; Uncommon; in Kentucky this mostly southern specie is reported from only four counties.

Specimen: Kentucky, Livingston County, Ferguson Creek, roadside (author's collection).

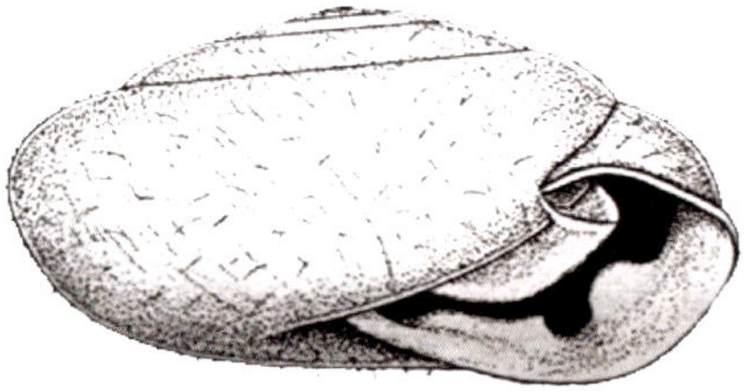

© Burch, 1962

Daedalochila leporina

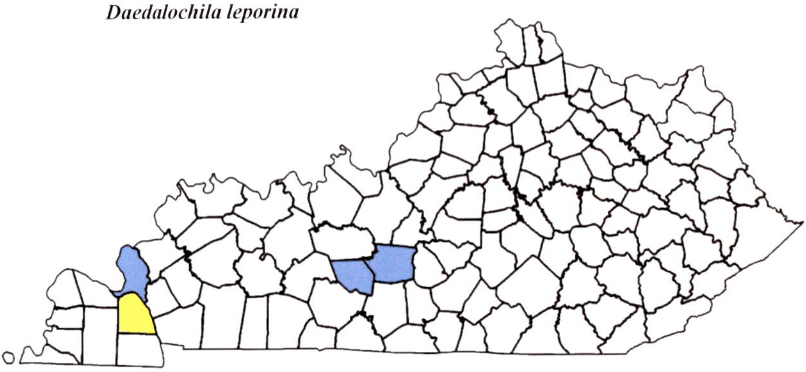

Tiny Liptooth

Polygyridae

Lobosculum pustuloides (Bland, 1858)

Diameter: 5-5.5 mm, height 2mm

Description: Depressed heliciform; lip reflected; shell with 4-5 whorls; umbilicate, umbilicus not nearly covered by the basal lip as in *Daedalochila. leporina*; orange-buff to pale horn-colored; upper shell surface dull; delicate hairs on shell surface are sparse; the basal tooth is forked forming two small knobs; shell periphery rounded.

Similar Species: *D. leporina* is larger, the top surface is glossier and the umbilicus is rimate or nearly closed; *D. plicata* is more widely umbilicate, has distinctive transverse striae that are rib-like and the teeth are arranged differently.

Habitat: A calciphile species found around logs, limestone talus and under leaf litter in mixed hardwood forests.

Status: G4/S3 Uncommon; this largely southern species is reported from only five Kentucky counties.

Specimen: Alabama (author's collection).

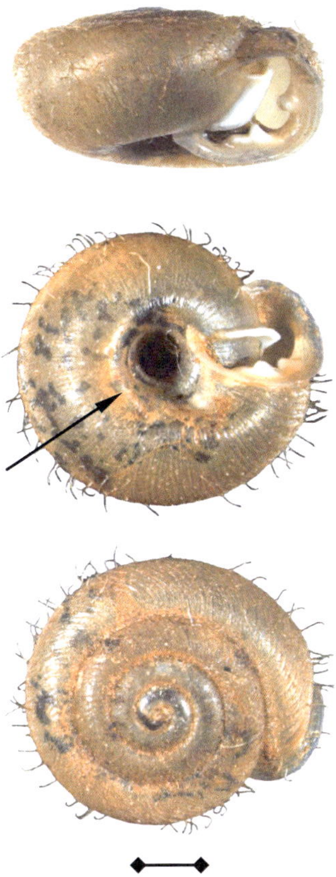

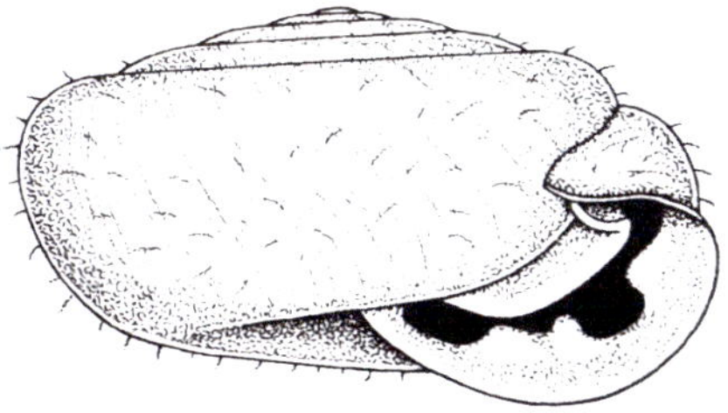

© Burch, 1962

Lobosculum pustuloides

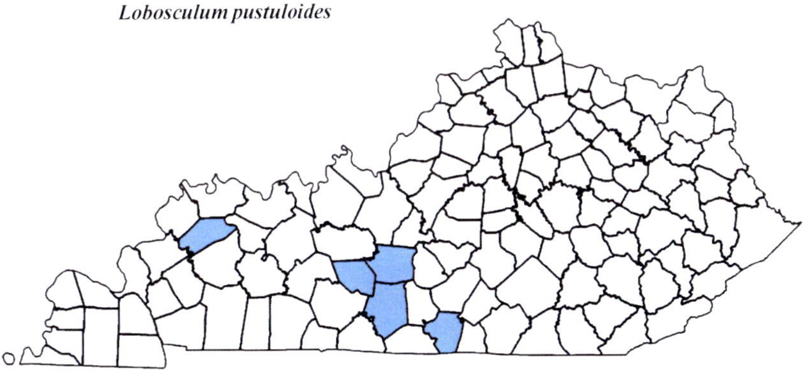

What appears to be the dappled shell of a Balsam Globe, *Mesodon andrewsae* (above and below images), is in fact, the live mottled animal seen through a very thin shell. This may be simply a consequence of life in a calcium poor environment, or perhaps, an evolutionary advantage to the snail surrounded by the usually high number of predatory shrews found on Roan Mountain. In the picture above, the patterns of the shell (a) are surprisingly similar to the patterns of the stone (b) on which the snail crawls. In the dim light of nightfall, the snail shell may be nearly impossible for a predator shrew or bird to distinguish. Roan Mountain, Mitchell County, NC.

Shells 6-11 mm, rimate, slightly reflected lip, long, blade-like parietal tooth (*Euchemotrema*)

Euchemotrema are small woodland snails that, in general, are quite common where they are found. Both species in Kentucky have an extended parietal tooth (a) and are always rimate (b), the umbilicus covered, but never completely sealed. This character may be hard to see with the naked eye, but can usually be distinguished with a hand lens of 10X.

Shell 6-11 mm, rimate, long parietal tooth, pg. 287-288

**Euchemotrema fraternum*
**Euchemotrema leaii aliciae*

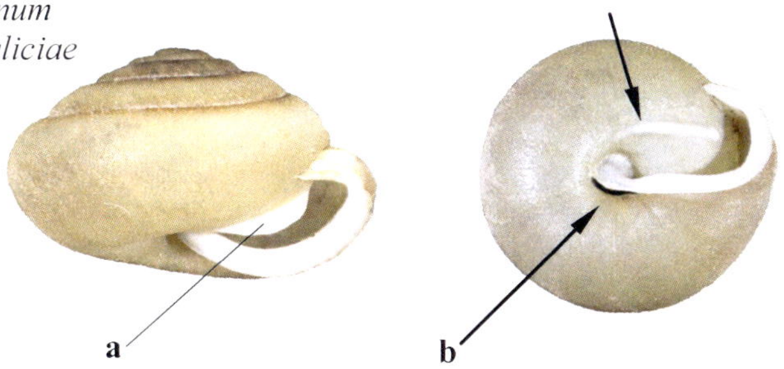

Shells 15-30 mm, rimate, reflected lip, if present the parietal tooth small and knob-like (*Mesodon*—in part)

Mesodon are conspicuous land snails of Kentucky, some species reaching 35 mm or larger in diameter. Most *Mesodon* are umbilicate with the exception of the 2 species presented here, which are always rimate (c). Both species are common in the state, and mostly a gastropod of degraded habitat or waste places. Ironically these two species were probably rare before human settlement. *Mesodon* species that are not rimate are represented elsewhere.

Shell 15-30 mm, rimate, with or without teeth, shell globose, pg. 290-291

**Mesodon clausus*
**Mesodon thyroidus*

Parietal tooth small and knob-like

Upland Pillsnail **Polygyridae**

Euchemotrema fraternum (Say, 1824)

Diameter: 7.8-11.4 mm, height 5.2-6.9 mm

Description: Heliciform; lip reflected and nearing the end a thickened deposit of calcium; shell with 5-6 tightly coiled whorls; the umbilicus is imperforate or nearly covered perforate (a); large parietal tooth that is blade like; shell dull; pale tan to cinnamon-buff; minute hairs present in last whorl but hairs are mostly lost in older shells; transverse striae poorly developed; shell periphery rounded; fulcrum rather short.

Similar Species: Similar to *Stenotrema* species but without the basal notch and interdenticular sinus; *E. leaii aliciae* has a more open umbilicus, not nearly as covered by the peristome (lip) as seen in *E. fraternum* and a long not short fulcrum.

Habitat: A species found under forest litter and around logs in upland mixed hardwood forests in both calcareous and acidic soils.

Status: G5/S5; Common; this species is widespread in Kentucky.

Specimen: West Virginia, Greenbrier County, Jct. 63 and Ft. Spring Pike (author's collection).

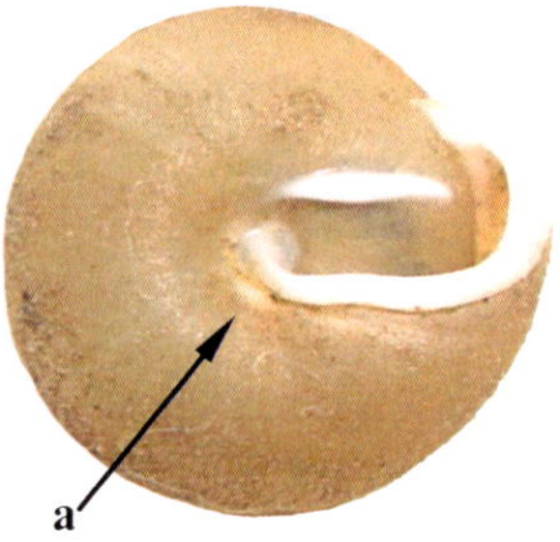

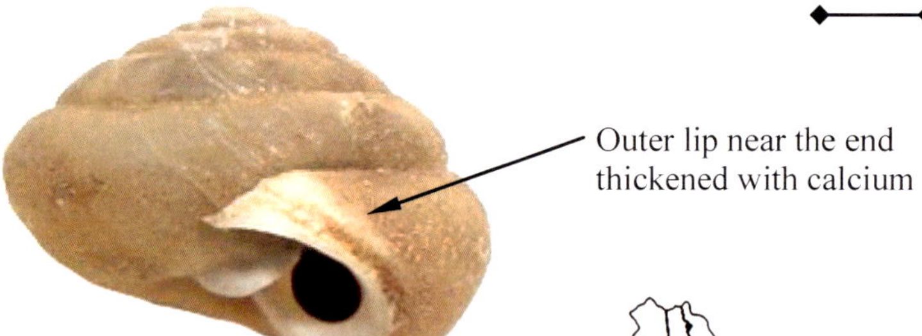

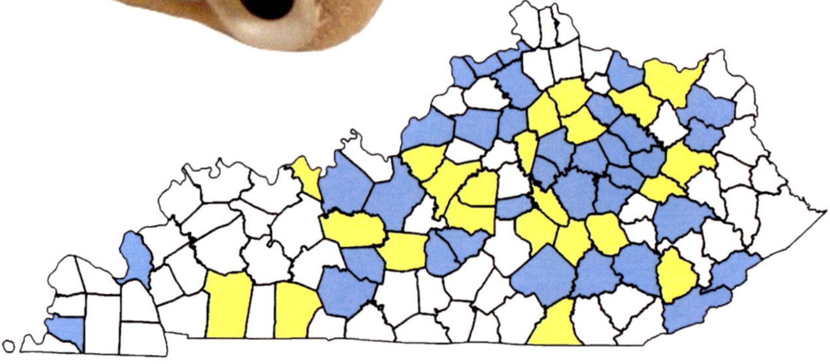

Floodplain Pillsnail

Euchemotrema leaii aliciae (Pilsbry, 1893)

Diameter: 7.3-7.8 mm, height 5.2-5.8 mm

Description: Heliciform; lip reflected; shell with 5.5-6.5 closely coiled whorls; rimate to nearly closed (a); large parietal tooth blade-like and curving toward umbilicus (b); shell somewhat glossy; tannish-olive to cinnamon-buff; minute hairs present in last whorl, but hairs are often lost in older shells; transverse striae poorly developed; shell periphery rounded; the internal Fulcrum is quite long and most easily observed in fresh shells (c and e).

Similar Species: Like *Stenotrema* species but without the basal notch and interdenticular sinus; *Euchemotrema fraternum* is larger, has a nearly closed umbilicus,, and its shell surface is a more dull glossy.

Habitat: A species of low wet areas, marshes; swamps and floodplains, but also found in meadows and along roadsides.

Status: G5/S5; Relatively Common; this species is reported from scattered locations across the state.

Specimen: Figure (d) from West Virginia, Tucker County, Canaan Valley around rocky outcrops (author's collection) and figure (f) from ©Larry Watrous.

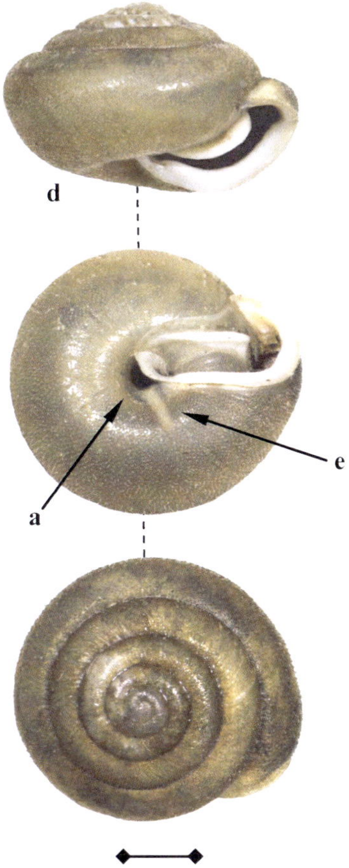

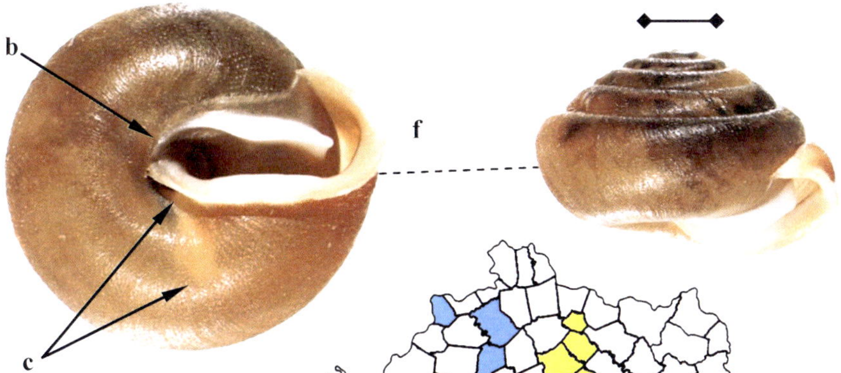

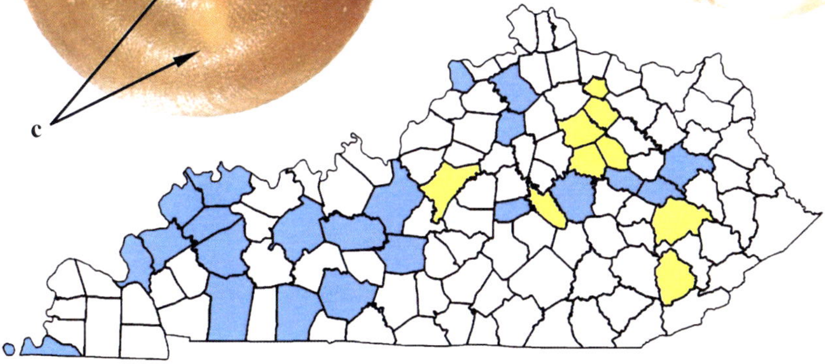

Euchemotrema and Stenotrema Compared

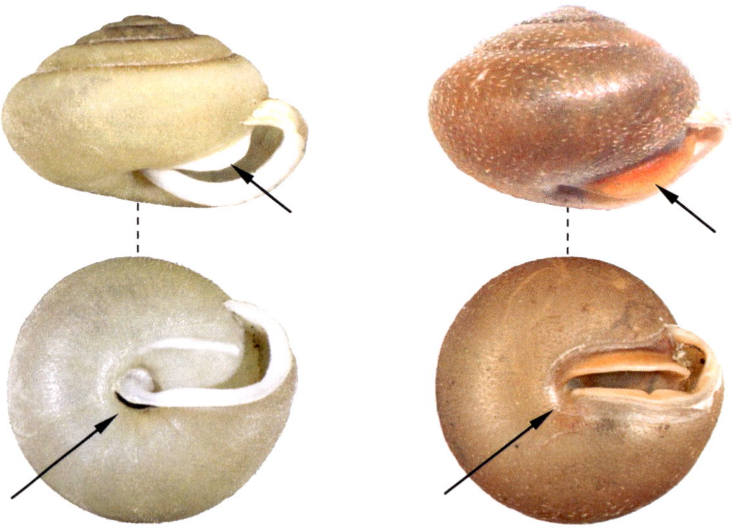

Euchemotrema ***Stenotrema***

Note that in frontal view the aperture of *Euchemotrema* is notably open and the umbilicus is rimate while the aperture in *Stenotrema* is blocked by a large parietal tooth and has a completely sealed umbilicus.

Euchemotrema and *Inflectarius* Compared

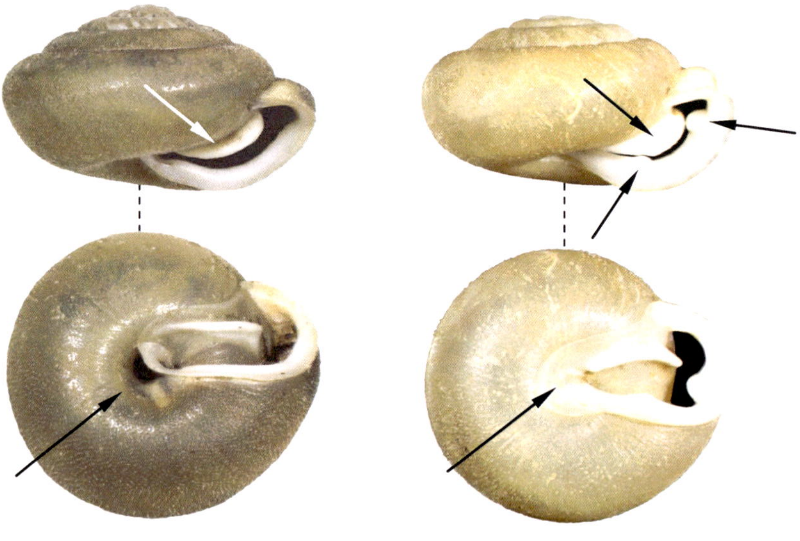

Euchemotrema ***Inflectarius***

Note in frontal view that *Euchemotrema* has a parietal tooth while *Inflectarius* has a parietal, basal, and palatal tooth. Additionally, the umbilicus of *Euchemotrema* is open while in *Inflectarius*, it is always completely closed.

Yellow Globelet

Mesodon clausus (Say, 1821)

Diameter: 15-17 mm, height 10-13 mm

Description: Heliciform; lip reflected; shell with 5-5.5 loosely coiled whorls; rimate, the umbilicus never entirely closed (a); without teeth; shell glossy and thin; pale-yellow to light tan; no hairs; transverse striae moderately developed; spiral striae present; shell periphery well-rounded.

Similar Species: *Mesodon thyroidus* is larger and has a small parietal tooth; by its color, the spiral striae, and the nearly closed umbilicus, *M. clausus* resembles *Inflectarius downieanus* (a more southern species), but in larger form.

Habitat: A calciphile species of low wet areas, marshes, swamps, and floodplains; also found along roadsides and waste places such as old parking lots, damp roadside ditches, grassy slopes, and ditches along railroad tracks; in wet weather found climbing high on herbaceous plants.

Status: G5/S5; Uncommon; a widespread and common species in Kentucky.

Specimen: Tennessee, Blount County, Cades Cove, GSMNP (GSMNP collection).

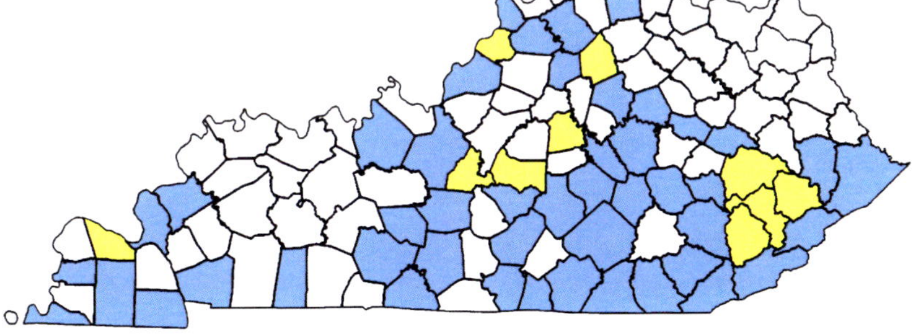

White-lip Globe **Polygyridae**

Mesodon thyroidus (Say,1816)

Diameter: 15-30 mm, height 11-18 mm

Description: Heliciform; lip reflected; shell with 5-5.5 loosely coiled whorls; umbilicus always rimate (a), never completely closed; small parietal tooth usually present; ivory yellow, pale yellowish green to brown; no hairs; transverse and minute spiral striae ever-present; shell periphery well rounded.

Similar Species: *Mesodon clausus* is smaller and is without a parietal tooth; the umbilicus of *M. normalis* is always completely closed; *M. zaletus* has a thicker shell and a completely closed umbilicus.

Habitat: This species can be exceptionally common in bamboo slicks (river cane), young floodplain forests, meadows, and waste places such as grassy roadsides; it is not a species of rich woods and old growth forests.

Status: G5/S5; Common; reported from every Kentucky county; interestingly this species was likely extremely rare before forests were cleared as it is hard to fine in mature intact woodlands.

Specimen: West Virginia, Logan County, (author's collection).

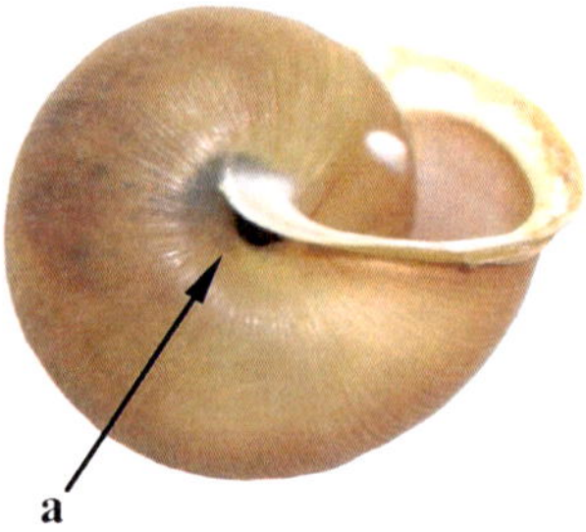

Furnace Mountain, Powell County, Kentucky

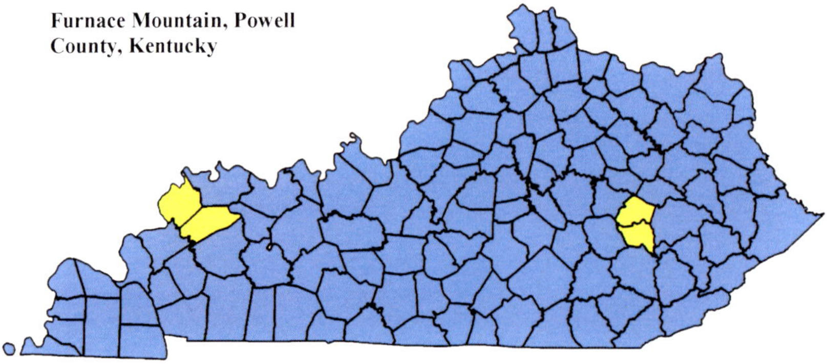

Shells under 7-15 mm, <u>imperforate</u>, reflected lip, teeth always present (*Inflectarius*)

<u>Inflectarius</u> species are under 15 mm, imperforate with the lip less widely reflected than in *Mesodon*. The two *Inflectarius* in Kentucky have tightly coiled shells, are imperforate, and contain three teeth (a), the parietal being the largest. They are best distinguished in frontal view and by examining the aperture for a small basal and palatal tooth. *Inflectarius* are sometimes confused with *Stenotrema*, but a closer examination of the aperture opening reveals their differences. Similar to *Euchemotrema*, but *Inflectarius* has a basal and palatal tooth and are imperforate, not rimate as in *Euchemotrema*. Juvenile *Inflectarius* shells have periostracal processes or fringes (see below), but these interesting shell features are usually lost in aging shells.

<u>Shell 7-16 mm, always with a parietal, basal and palatal tooth, pg. 293-295</u>

**Inflectarius inflectus*
**Inflectarius rugeli*

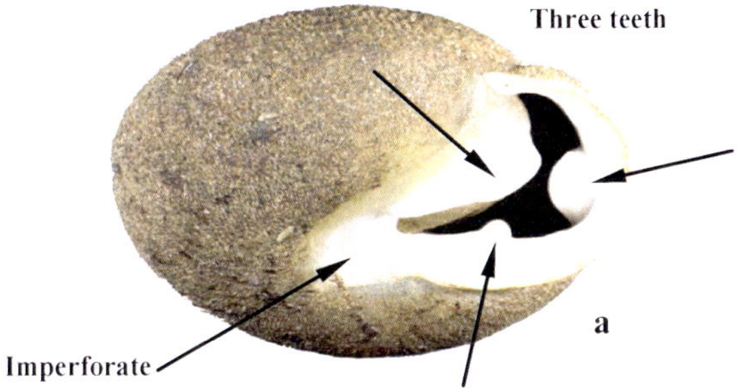

A juvenile *Inflectarius rugeli* shell enlarged (around 3 mm in actual size), illustrating the periostracal processes or fringes seen on young shells under 5 mm, but often lost in older shells. Forest litter often sticks to these fringes and is believed to help camouflage the gastropod from potential predators such as birds and salamanders.

Actual size

Juvenile *Inflectarius rugeli*

Shagreen

Inflectarius inflectus (Say, 1821)

Diameter: 7.5-13.8 mm, height 5.4-8.1 mm

Description: Depressed heliciform; lip reflected; shell with 4.5-5.5 whorls; imperforate; large parietal tooth present; basal tooth small; palatal tooth small and only slightly recessed in aperture; cream buff colored to light yellowish-horn; shell surface with short periostracal processes and fine transverse striae or wrinkles; periphery rounded.

Similar Species: *Inflectarius rugeli* is slightly larger and the palatal tooth is larger and more deeply positioned in aperture; intermediate forms between the two species are not uncommon; *Euchemotrema* species have a rimate umbilicus (a); *Stenotrema* have a slit-like opening (b) see below.

Habitat: This is a land snail of roadside ditches, railroad right-of-ways, grassy areas, and shale banks of road cuts; not a species of mature woods.

Status: G5/S5; Common; this is the most widespread of the two species of *Inflectarius* in the state.

Specimen: Kentucky, Hickman County, Obion Creek WMA, Travis Slough (author's collection).

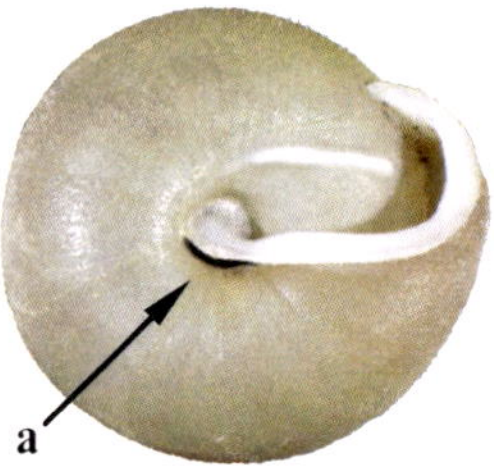

Euchemotrema

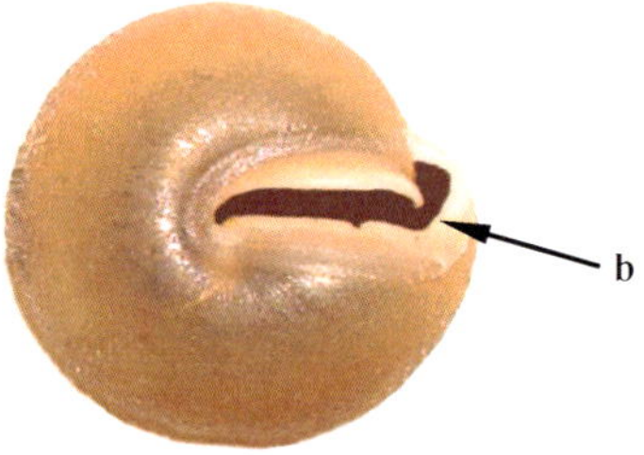

Stenotrema

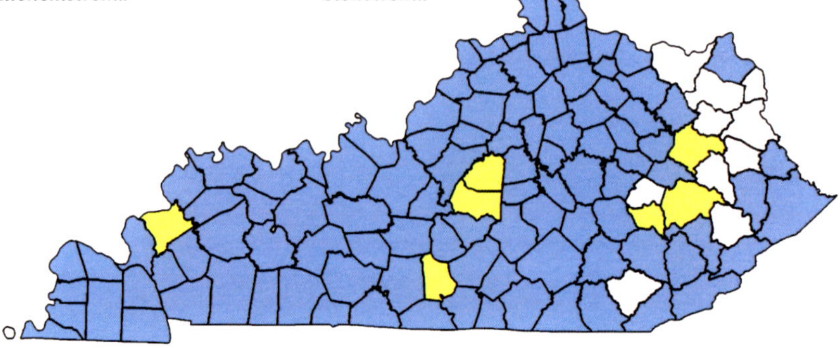

Deep-tooth shagreen

Polygyridae

Inflectarius rugeli (Shuttleworth, 1852)

Diameter: 7.8-16.4 mm, height 6.4-9.3 mm

Description: Depressed heliciform; lip reflected; shell with 5.5 whorls; imperforate; large parietal tooth present; basal tooth small; palatal tooth large and deeply positioned in aperture (a) and in bottom view somewhat hidden; waxy horn colored; shell surface with short periostracal processes that may at times appear as scales especially in older shells.

Similar Species: *Inflectarius inflectus* is slightly smaller, has a smaller palatal tooth which is less deeply set in aperture (b); *I. rugeli* is thought to be a species complex, containing multiple species showing little difference in shell morphology except for size differences.

Habitat: A habitat generalist of mixed hardwood forests, found under leaf litter and logs and along forested road edges; a species of mature woods.

Status: G5/S5; Common; a widespread and common species in Kentucky.

Specimen: Tennessee, Monroe County, Hawk Knob Bottom, Cherokee NF (author's collection).

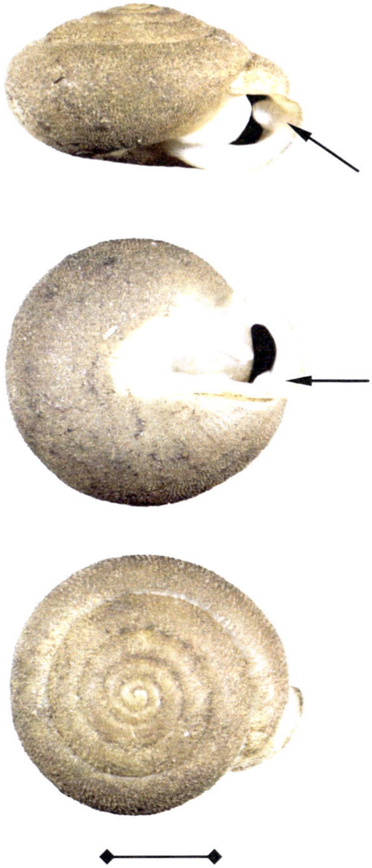

I. rugeli *I. inflectus*

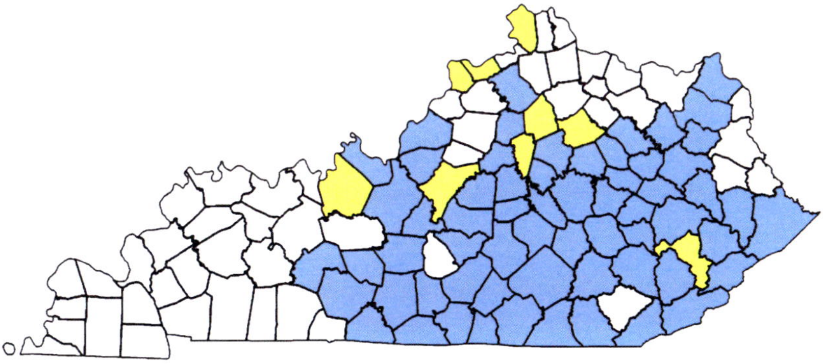

Brown Globelet

Polygyridae

Inflectarius kalmianus (Hubricht, 1965)

Diameter: 9.2 mm, height: 6.2 mm

Description: Heliciform; lip reflected and thickened; shell with 4.8 whorls; shell solid (thick) for its small size; imperforate; no teeth present; tawny-olive; no hairs; transverse and minute incised spiral striae that are strongest on the base; embryonic whorl nearly smooth; shell periphery well rounded.

Similar Species: *Mesodon clausus* is more globose, notably larger and is rimate (not imperforate as in *I. kalmianus*; *I. rugeli* and *I. inflectus* both have large and well defined teeth in the aperture.

Habitat: Open glade-like habitats are preferred by this small species; usually in open, grassy meadows and but also common along roadsides where it can be found in large numbers.

Status: G3/S3; Uncommon; In Kentucky an uncommon but widespread snail reported from south-central and western counties.

Specimen: Kentucky, Pulaski County, (author's collection).

Shells proportionate

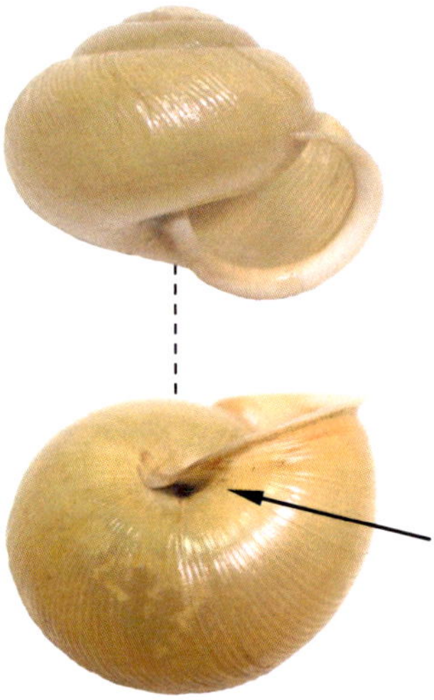

Mesodon clausus

Inflectarius kalmianus

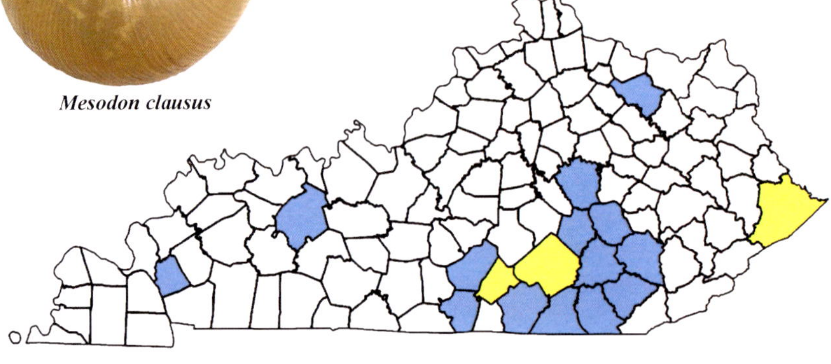

Above a newly hatched Grand Globe, *Mesodon normalis,* feeding on the unopened flower spikes of a Dwarf rattlesnake plantain orchid, *Goodyera repens,* Nolichucky River Gorge, Unicoi County, Tennessee.

Shells 15-24 mm, imperforate, with long, low basal tooth, adult shells with reflected lip (*Patera*)

Patera species are usually depressed heliciform with the oddball exception of *Patera pennsylvanica* (opposite page). This species defies the standard by being heliciform, looking like a small *Mesodon* and lacking any teeth. All adult *Patera* shells exhibit reflected lips, and in Kentucky, display a large parietal tooth, except for *P. pennsylvanica.* While the long, low basal tooth (a) is a wanting feature on many shells, it is an important diagnostic mark for the group as a whole and should not be disregarded. There are either spirally arranged papillae (b) or spiral striae (c). These diagnostic features may be an obscure attribute on aged shells that have worn surfaces. An area that generally remains protected from abrasions is located just behind the reflected lip of mature shells (d). All *Patera* are imperforate. Immature *Patera* shells will not have reflected lips, and are sometimes confused with young *Mesomphix* shells. The following *Patera* are arranged according to these features and are as follows: **Group 1** shells that are heliciform, **Group 2** shells that are depressed heliciform and have spiral papillae and **Group 3** shells that have spiral striae in the form of incised lines.

Group 1: shells that are heliciform, pg. 298

**Patera pennsylvanica*

Group 2: shells with spiral papillae, pg. 299-301

**Patera appressa*
**Patera aff. appressa* (undetermined)
**Patera estillensis* (rare)

Group 3: shells with spiral striae, pg. 302-304

**Patera laevior*
**Patera panselenus*

Spiral papillae

Spiral striae

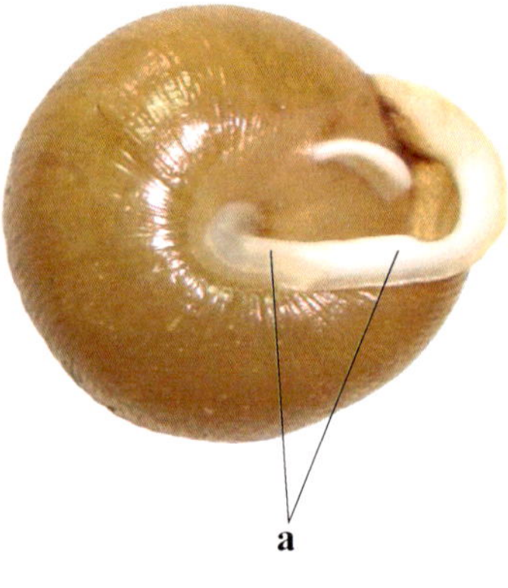

Proud Globelet

Polygyridae

Patera pennsylvanica (Greene, 1827)

Diameter: 15-20 mm, height 10.2-15 mm

Description: Heliciform; upper portion of lip points abruptly downward (a) and is only slightly reflected; right side of aperture square-like, not rounded as in other *Patera* species; shell with 5.5-6 whorls; imperforate; no teeth present; light tan to yellowish-olive; no hairs in adult shells; transverse striae well-developed; spiral striae always present and a strong feature; shell periphery rounded; an unusual *Patera* given that most species in this genus are depressed heliciform.

Similar Species: *Mesodon thyroidus* has a more rounded aperture and slightly open umbilicus.

Habitat: Open road cuts and grassy areas along abandoned railroad beds and limestone slopes; fairly common in mature sugar maple woods (pers. comm. John MacGregor).

Status: G3/S3; Uncommon; this species has a limited distribution in north central Kentucky becoming more common in states to the north.

Specimen: Kentucky, Henry County, KY River WMA, Raccoon Branch Tract (author's collection).

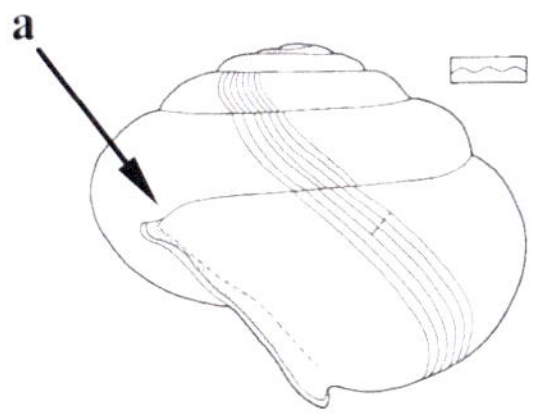

© Emberton, 1998

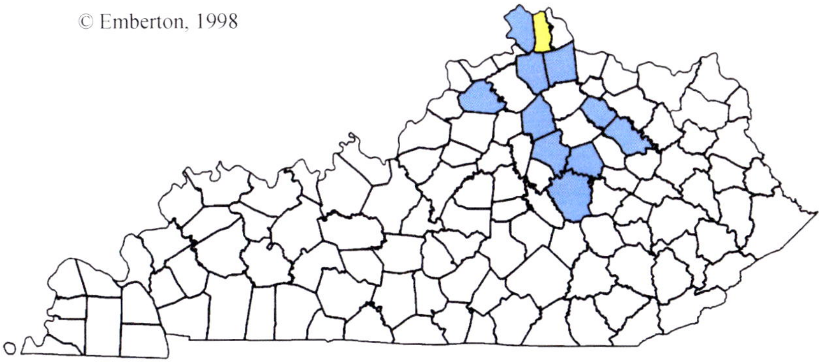

Flat Bladetooth

Patera appressa (Say, 1821)

Diameter: 13-19.5 mm, height 7-9 mm

Description: Depressed heliciform; lip reflected; shell with 4.5-5 whorls; imperforate; large parietal tooth present with a wide base (a) and a long, low basal tooth, which at times may be poorly defined (b); cinnamon-buff to brownish -horn; shell surface somewhat glossy; adult shells without hairs; transverse striae are well developed on top and sides, but weakly defined on the base; spirally arranged papillae typically present; shell periphery either rounded or in some populations very slightly angular but never carinate.

Similar Species: *Patera panselenus* has strongly developed spiral striae (not papillae); *P. laevior* is of the same size and build, but has weakly developed striae (not papillae).

Habitat: Limestone and sandstone rock outcrops and cave entrances.

Status: G4/S4; Common; *Patera appressa* is reported from scattered locations across south central and eastern Kentucky being replaced by *P. laevior* to the west of this region.

Specimen: Kentucky, Powell County, Furnace Mountain (all in author's collection) .

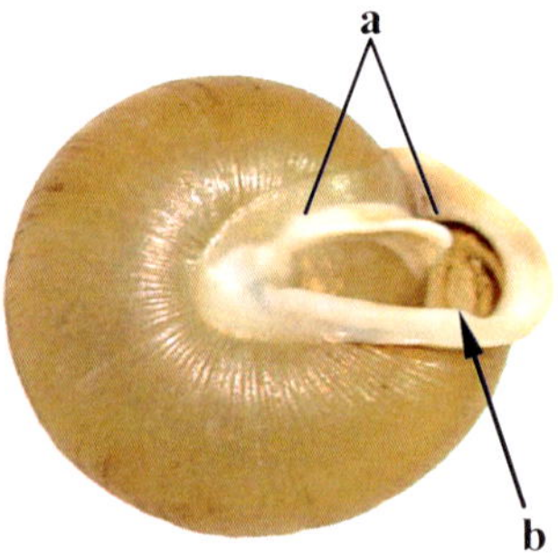

***Patera appressa*, Powell Co, KY**

Shells proportionate

***Patera aff. appressa*, Whitley Co, KY**

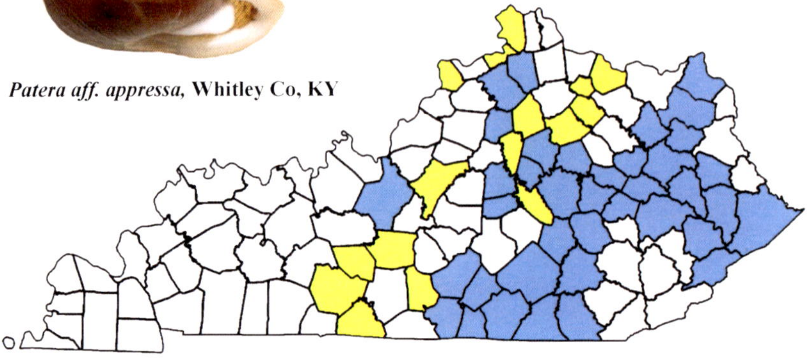

Pine Mountain Bladetooth **Polygyridae**

Patera aff. appressa (Say, 1821)

Diameter: 12-14 mm, height 6-7 mm

Description: Depressed heliciform; lip reflected; shell with 4.5-5 whorls; imperforate; large parietal tooth present with a wide base and a long, low basal tooth, which at times may be poorly defined; brownish-horn; shell surface somewhat glossy; adult shells without hairs; transverse striae are course and well developed on top and sides, but weakly defined on the base; spirally arranged papillae typically present on base but becoming fused together on top of shell forming spiral ridges or lirae (a); shell periphery either rounded or very slightly angular but never carinate as in *Patera estillensis*.

Similar Species: *Patera appressa* is typically larger by 4 to 5 mm and has spirally arranged papillae on top of the shell, not the spirally arranged and raised lirae seen in *Patera aff. appressa.*

Habitat: Hillsides and ravines of Pine and Black Mountain.

Status: Not ranked; Uncommon; this undetermined species appears to be restricted to Pine and Black Mountains and awaits further scrutiny.

Specimen: Kentucky, Whitley County, Pearl Siler Road, north side of Pine Mountain (author's collection) .

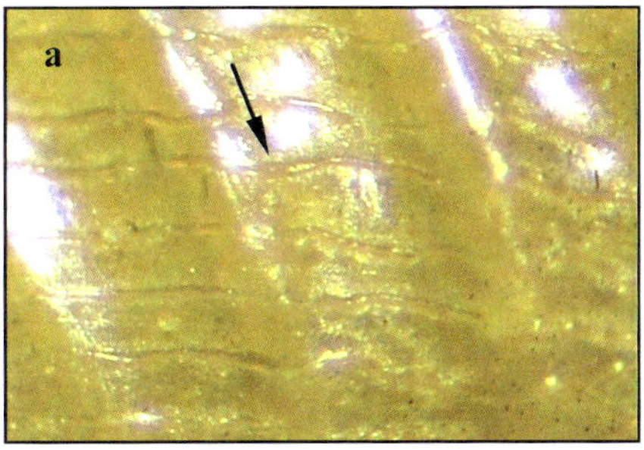

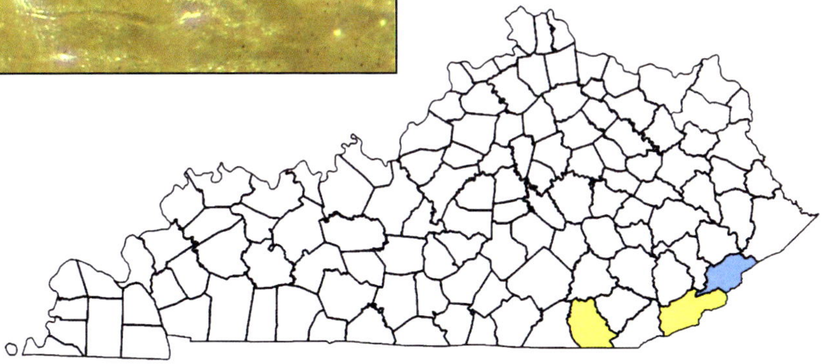

Carinate Bladetooth

Patera estillensis Dourson, 2011

Diameter: 15-20 mm, height 7-8 mm

Description: Depressed heliciform; lip reflected; shell with 4.5-5 whorls; imperforate; large parietal tooth present; basal tooth small and sometimes poorly defined but a constant feature; shell surface somewhat glossy; no hairs at any stage of growth; transverse striae well developed on top of shell but weakly defined on the base; spiral papillae are always present but are sometimes only faintly visible; shell periphery carinate to strongly angular the entire length of the shell (a key feature).

Similar Species: *Patera laevior* has spiral striae; *P. appressa* has a rounded periphery.

Habitat: Found among rocks at the base of limestone cliffs or in small crevices within the cliff face, live individuals can be found around the entrances of small caves during July and August.

Status: G1/S1; State Threatened; Endemic to Kentucky; while shells are generally common, live snails are much rarer; known only from a few narrow ridge systems and high knobs above Red Lick Creek in Estill County and one small corner of Jackson County.

Specimen: All figures from Kentucky, Estill County, Red Lick Creek area (author's collection).

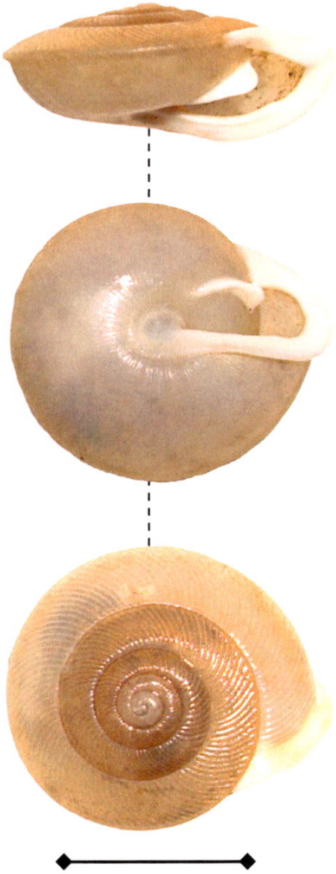

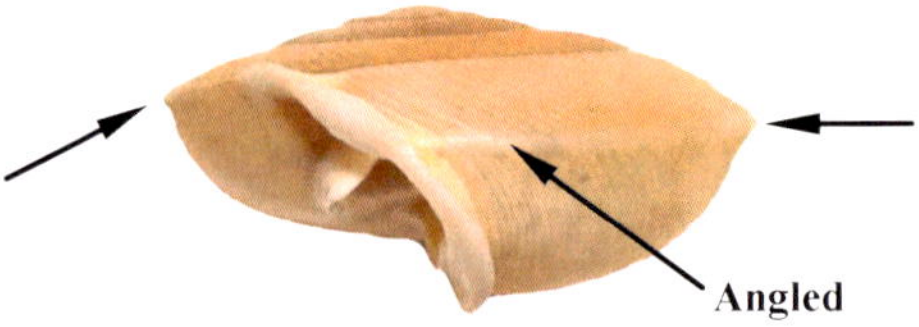

Patera estillensis

Patera appressa

Holotype

Kentucky, Estill County: Happy Top Mountain above Red Lick Creek, 1.5 km SE of Jinks (Daniel Boone National Forest) (FMNH 328316 Holotype).

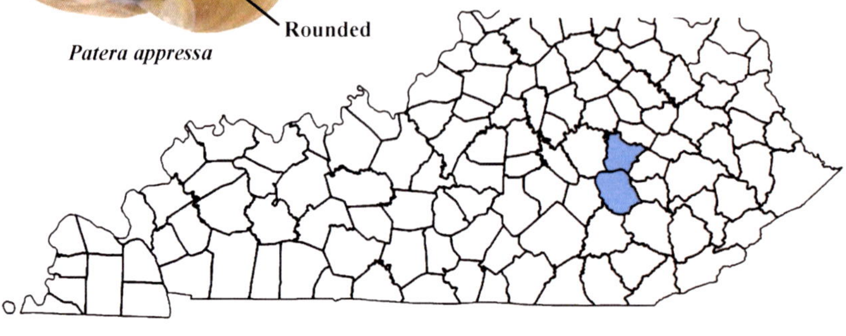

Smooth Bladetooth **Polygyridae**

Patera laevior Hubricht, 1968

Diameter: 16-18 mm, height 8-9 mm

Description: Depressed heliciform; lip reflected; shell with 4.5-5 whorls; imperforate; large parietal tooth present; long, low basal tooth, which at times, may be poorly defined (a); cinnamon buff; shell surface somewhat glossy; no hairs in adult shells; transverse striae well developed on top of shell but weakly defined on the base; thin spiral lines on the upper surface and on the base, but are sometimes only faintly visible; shell periphery rounded.

Similar Species: *Patera appressa* has moderately to well developed spiral papillae; *Patera panselenus* is flatter, glossier, and has stronger developed spiral striae.

Habitat: Mixed hardwood forests, rocky areas, both sandstone and limestone clifflines and steep wooded hillside; common along railroad beds and in old quarries (MacGregor pers. comm.).

Status: G4/S5; Common; Replaces *Patera appressa* in central and western portions of Kentucky.

Specimen: Kentucky, Franklin County; intersection of Routes 1900 and 1262, base of limestone cliffs (all figures from author's collection).

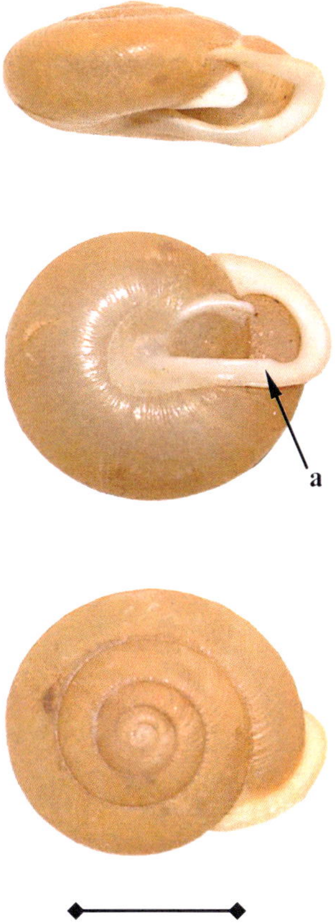

A population of *P. laevior* at Mammoth Cave NP, Edmonson County, Kentucky have excessively thickened peristome for unknown reasons,

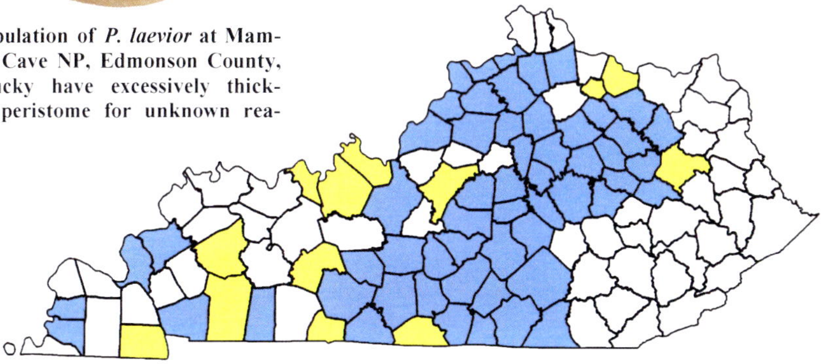

West Virginia Bladetooth — Polygyridae

Patera panselenus (Hubricht, 1976)

Diameter: 18-20 mm, height 8.3 mm

Description: Depressed heliciform; lip reflected; shell with 5 whorls; imperforate; parietal tooth large with a wide base; long, low basal tooth, which at times, may be poorly defined 9 (a); shell pale brownish, glossy; no hairs in adult shells; through a hand lens of 10X the transverse striae appear closely spaced, but weakly developed; spiral striae are a strong and reliable feature; periphery roundish.

Similar Species: *Patera appressa* has spirally arranged papillae, bolder and more widely spaced transverse striae (see page 304); the two species occasionally occur together;

Habitat: Shale banks adjacent to small streams; under rocks and logs in mixed hardwoods forests, also around moist limestone and sandstone rock faces.

Status: G3/S1; State Threatened; *Patera panselenus* is restricted to extreme southeastern Kentucky.

Specimen: West Virginia, Kanawha County, WV near the town of Charleston (all figures in author's collection).

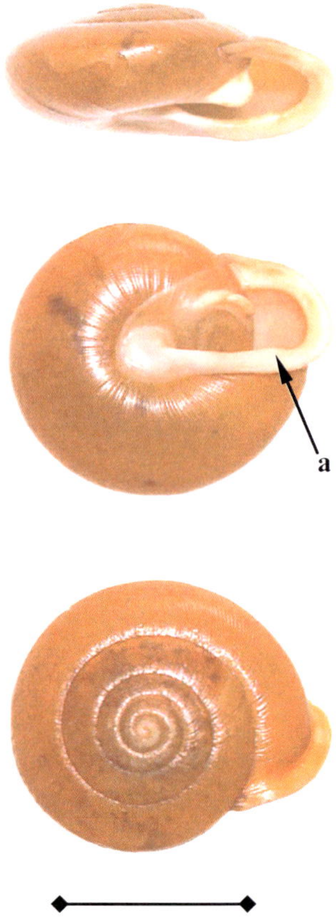

***P. panselenus*, Greenbrier County, West Virginia**

***P. appressa*, Powell County, Kentucky**

Above, top views showing the finer, lower, and more closely spaced transverse striae and glossier shell of *Patera panselenus* and the coarser, more widely spaced transverse striae and duller shell of *Patera appressa*. Both specimens from the same location from Breaks Interstate Park, Pike County, Kentucky.

Shell 9-22 mm, 3 teeth, <u>umbilicate</u> (*Triodopsis*)

Most ***Triodopsis*** are depressed heliciform, having widely reflected lips at maturity and all species are umbilicate. Shells typically contain three teeth, the parietal tooth being the largest. In some species, the basal and palatal teeth may be wanting features as in some populations of *T. tennesseensis*. <u>It is important to determine if the parietal tooth points above or below the palatal tooth (figures a, b and c).</u> Always use the bottom view to determine this attribute. The transverse striations are usually a relatively strong character. The *Triodopsis* are grouped as follows: **Group 1** parietal tooth points well above the palatal tooth, **Group 2** parietal tooth points at or above the palatal tooth and **Group 3** parietal tooth points at or below palatal tooth.

<u>Group 1: umbilicate, parietal tooth points well above palatal tooth, pg. 306-308</u>

**Triodopsis vulgata*
**Triodopsis hopetonensis*
**Triodopsis claibornensis* (rare)

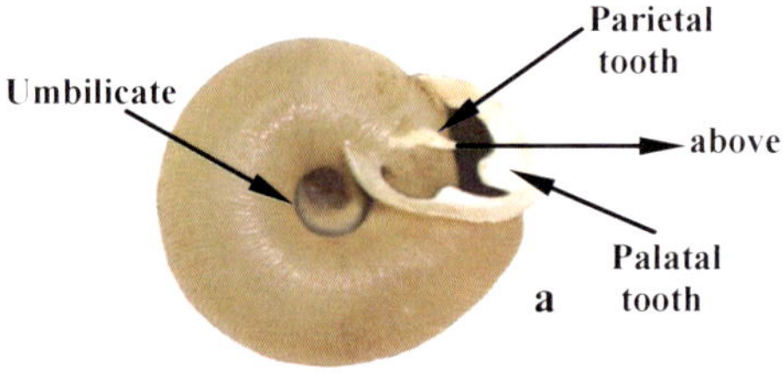

<u>Group 2: umbilicate, parietal tooth points at or above palatal tooth, pg. 309-311</u>

**Triodopsis anteridon*
**Triodopsis aff. anteridon* (rare)
**Triodopsis discoidea*

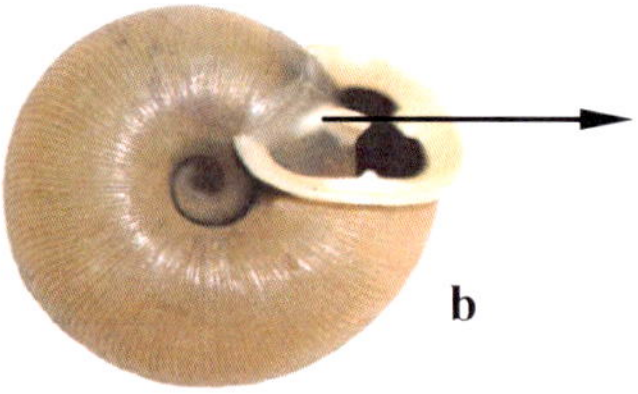

<u>Group 3: umbilicate, parietal tooth points at or below palatal tooth, pg. 312-316</u>

**Triodopsis complanata*
**Triodopsis tennesseensis*
**Triodopsis aff. tennesseensis* (rare)
**Triodopsis tridentata*

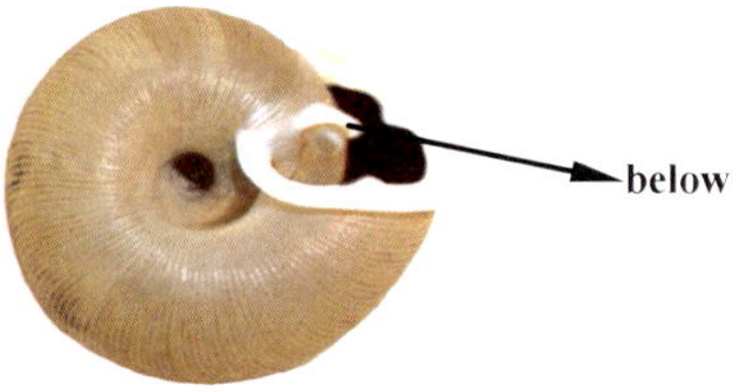

Shell 19-25 mm, 3 teeth, <u>imperforate</u> (*Xolotrema*)

<u>Xolotrema</u> are very similar to *Triodopsis* in terms of size, build, and number of teeth but are imperforate, having a completely sealed umbilicus **pg. 317-319**

**Xolotrema denotatum*
**Xolotrema fosteri*
**Xolotrema obstrictum*

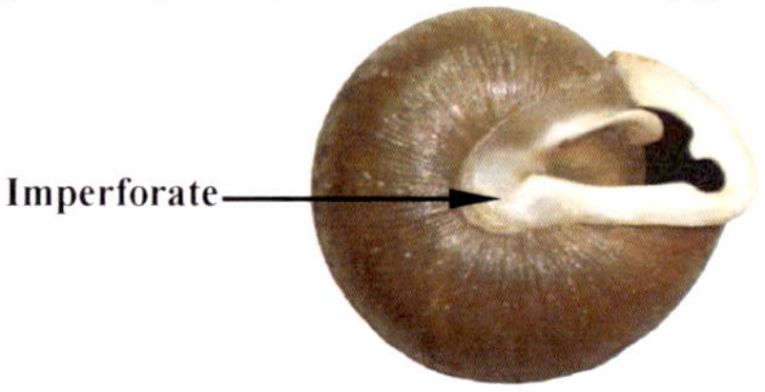

Dished Threetooth

Polygyridae

Triodopsis vulgata Pilsbry, 1940

Diameter: 13.5-19.5 mm, height 7.3-10.4 mm

Description: Depressed heliciform; lip reflected and abruptly turning downward (a); shell with 5-6 whorls; umbilicate; parietal tooth large and curving (b), points well above the palatal tooth; basal tooth small; palatal tooth larger and with a wider base, sitting deeper in the aperture; the three teeth crowd the aperture in frontal view; cream-buff to cinnamon buff; transverse striae are well developed on the top, sides and bottom of the shell; scattered papillae are present but generally a weak feature, being strongest around the umbilicus; adults without hairs; periphery rounded.

Similar Species: The lip of most other *Triodopsis* species lacks the steep downward curve seen in side view (a).

Habitat: Usually found on soil and rubble among limestone in mixed hardwood forests.

Status: G5/S5; Common; a widespread species in Kentucky.

Specimen: Kentucky, Powell County, Furnace Mountain (author's collection).

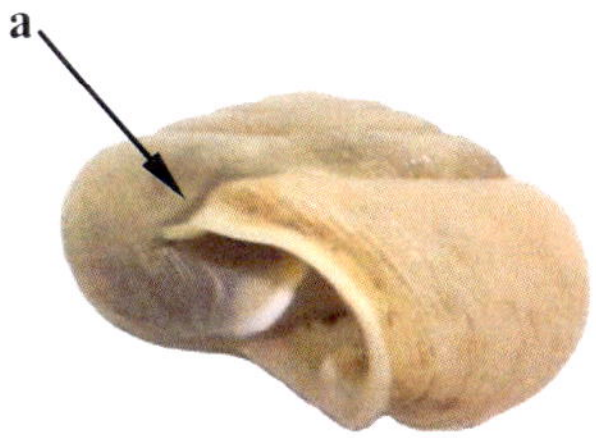

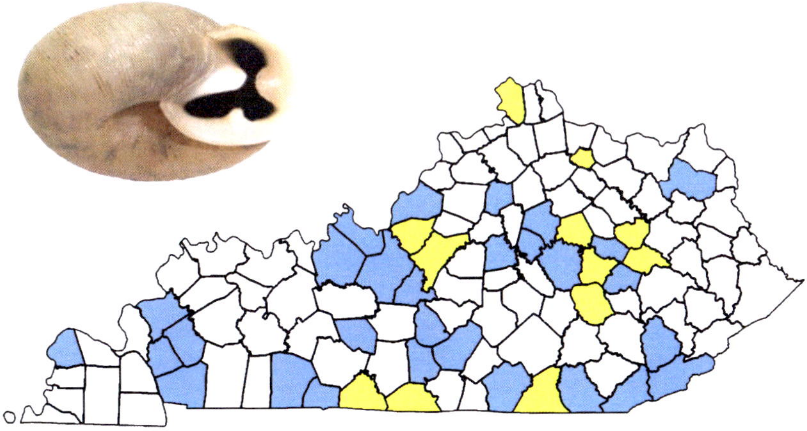

Magnolia Threetooth

Triodopsis hopetonensis (Shuttleworth, 1852)

Diameter: 9.2-13 mm, height 5.5 mm

Description: Depressed heliciform, shell rather compact for a *Triodopsis*; lip reflected; shell with 4.5-5.5 whorls, the last whorl only slightly expanding; umbilicate; the curving parietal tooth (a) is thin and points well above the palatal tooth; palatal tooth not situated on a buttress and is recessed in aperture; olive-horn; transverse striae are well developed on the top and sides of the shell continuing well into the umbilical region; adult shells without hairs; periphery rounded.

Similar Species: *Triodopsis tridentata* is larger, its parietal tooth points at or below the palatal tooth; *T. vulgata* is larger with a thicker shell.

Habitat: A lowland species of waste places, roadsides, also found in and around scrap piles of construction lumbers (one mode of transportation of the species to new locations, pers. obs.); natural habitats include low wet woodlands.

Status: G4/S5; Common; in Kentucky this mostly southern species of the coastal plains is being reported from more and more counties.

Specimen: Kentucky, Martin County, a motel parking lot in the town of Inez (author's collection).

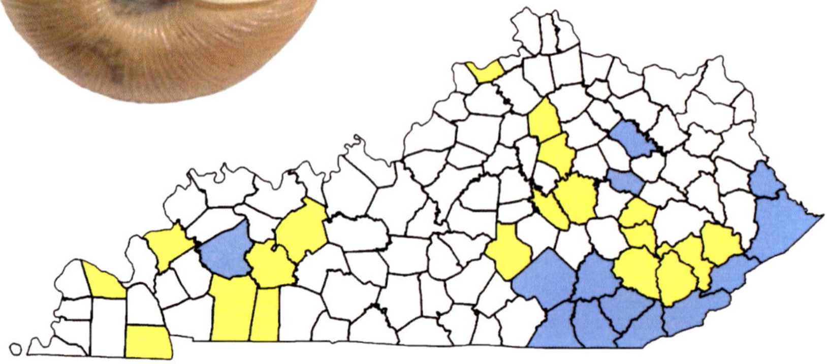

Claiborne Threetooth

Polygyridae

Triodopsis claibornensis Lutz, 1950

Diameter: 10-12 mm, height 7 mm

Description: Depressed heliciform, shell more compact than other *Triodopsis* species; lip reflected and abruptly turning downward; shell with 5 whorls; umbilicate; parietal tooth large, points well above the palatal tooth; basal tooth small; palatal tooth larger and wide, set deep into the aperture with a small notch (a) this occurring on around 90% of specimens examined (pers. comm. Ron Caldwell, 2020), this feature is apparently absent from *T. vulgata* and *T. hopetonensis*; cream-buff; transverse striae are well developed on the top, sides and bottom of the shell; scattered papillae are present.

Similar Species: *Triodopsis vulgata* averages a larger size and has a slightly higher shell profile and in frontal view has a more crowded aperture; *T. hopetonensis* has a thinner lip and teeth.

Habitat: An upland species under leaf litter in mixed rocky hardwood forests; also found along roadsides and under kudzu patches.

Status: G2/S1; State Threatened; A globally rare and range restricted gastropod; in Kentucky it is known only from Whitley County near Cumberland Gap National Park. (Branson 1973).

Specimen: Kentucky, Claiborne County, Harrogate under Kudzu patch (author's collection).

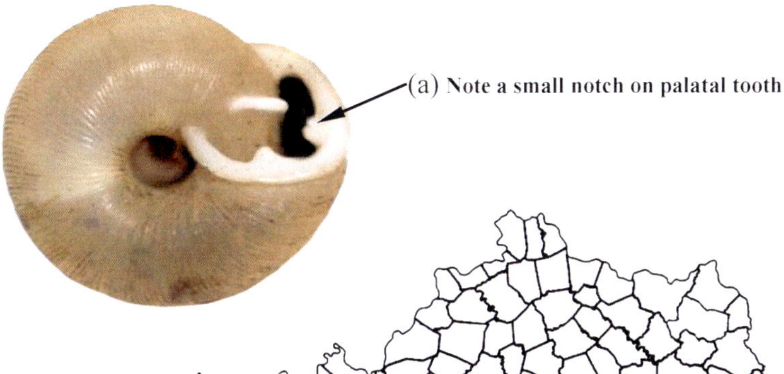

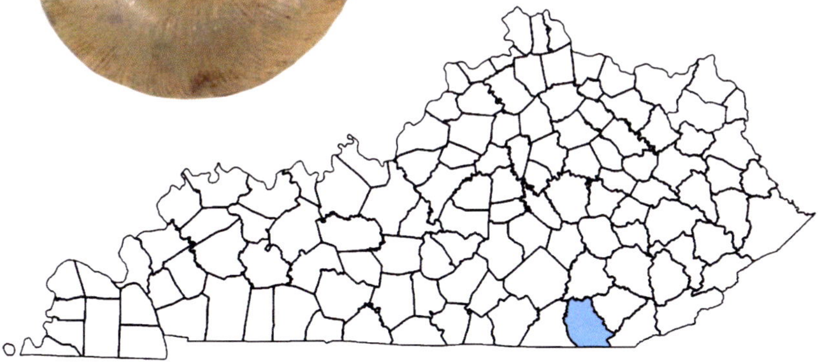

Carter Threetooth

Triodopsis anteridon Pilsbry, 1940

Diameter: 11-14 mm, height 6.6-7.2 mm

Description: Depressed heliciform; lip reflected; shell with 5-5.5 whorls; umbilicate; large parietal tooth, points at or above (mostly above) the palatal tooth; basal tooth and palatal tooth substantially smaller; palatal tooth situated on a wide buttress (a), a widening of the outer lip; shell brown; transverse striae are well developed, closely spaced on the top and sides of the shell continuing well into the umbilicus region; scattered papillae are generally a strong feature especially near sutures and the umbilicus; periphery is rounded.

Similar Species: *Triodopsis tridentata* is larger, with a smaller parietal tooth that points at or below the palatal tooth.

Habitat: Found in mixed hardwood forests under leaf litter and close to logs; occasionally under logs and rocks.

Status: G3/S4; Fairly common; an infrequent gastropod of Kentucky woodlands.

Specimen: Kentucky, Grayson County, around Grayson Lake WMA (author's collection).

***T. anteridon*, Grayson County, Kentucky**

***T. aff. anteridon*, Harlan County Kentucky**

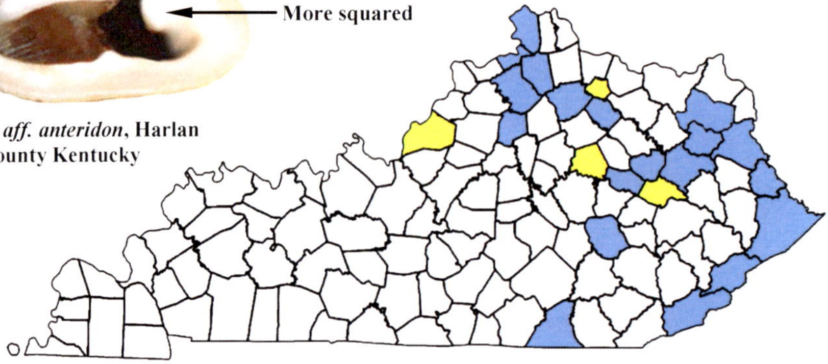

Black Mountain Threetooth

Polygyridae

Triodopsis aff. anteridon Pilsbry, 1940

Diameter: 15-17 mm, height 8-9 mm

Description: Depressed heliciform; lip reflected; shell with 5-5.5 whorls; umbilicate; large parietal tooth, points at or above (mostly above) the palatal tooth; basal tooth and palatal tooth substantially smaller; palatal tooth situated on a wide buttress (a), a widening of the outer lip; shell dark brown; transverse striae are well developed, closely spaced on the top and sides of the shell continuing well into the umbilicus region; scattered papillae are generally a strong feature especially near sutures and the umbilicus; periphery is rounded.

Similar Species: *Triodopsis anteridon* is smaller by around 3-4 mm, lighter in color, has a smaller umbilicus and has a peg-like palatal tooth (the palatal tooth of the Black Mountain Threetooth is larger and more squared, but this is variable).

Habitat: Found in mixed hardwood forests under leaf litter; occasionally under logs and rocks.

Status: Not ranked; Rare; only known from the north side of Black Mountain in Harlan County where it patiently waits for further scientific scrutiny; may be endemic to Kentucky (?).

Specimen: Kentucky, Harlan County, north side of Black Mountain (author's collection).

Shells proportionate

Triodopsis aff. anteridon,
Black Mountain, Harlan Co. KY

Triodopsis anteridon,
Grayson Co KY

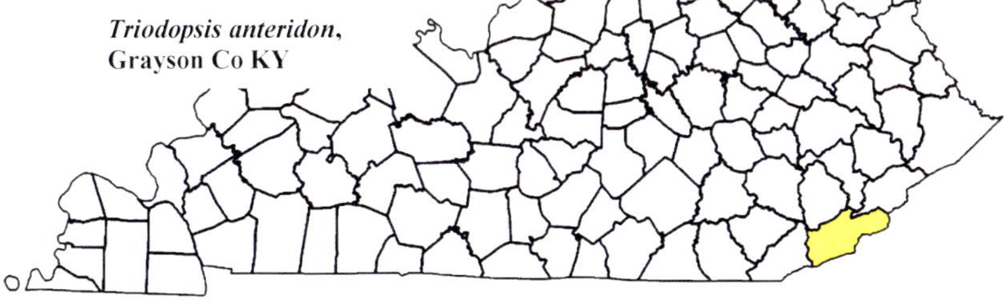

Rivercliff Threetooth

Triodopsis discoidea (Pilsbry, 1904)

Diameter: 14-19.5mm, height 9 mm

Description: Depressed heliciform; lip reflected; shell with 5-5.5 whorls; umbilicate; large parietal tooth present, points at or above (mostly above) the palatal tooth; basal tooth and palatal tooth smaller; palatal tooth not situated on a buttress; dilute cream to very dilute ecru-olive; transverse striae are moderately developed on the top and sides of the shell continuing well into the umbilicus region; some minute papillae can usually be seen around the umbilicus; specimens from Fulton County are smaller, averaging 14-16 mm in diameter.

Similar Species: *Triodopsis tridentata* parietal tooth points below the palatal tooth; *T. complanata* has smaller basal and palatal teeth and less defining transverse striae; *T. tennesseensis* has better developed transverse striae, a duller surface and most importantly, the palatal and basal teeth are a weaker feature.

Habitat: A species found along river bluffs in mixed hardwood, under leaf litter, logs and rocks that provide interstitial space for snails to move.

Status: G3/S3; Uncommon; this species is reported from counties that largely border the Ohio and Mississippi Rivers.

Specimen: Figures (a & b) from Kentucky, Breckenridge County, Yellow Bank WMA, figure (c) from Kentucky, Fulton County, Wilson's Hill (all in author's collection).

a

b

A lighter color morph

c

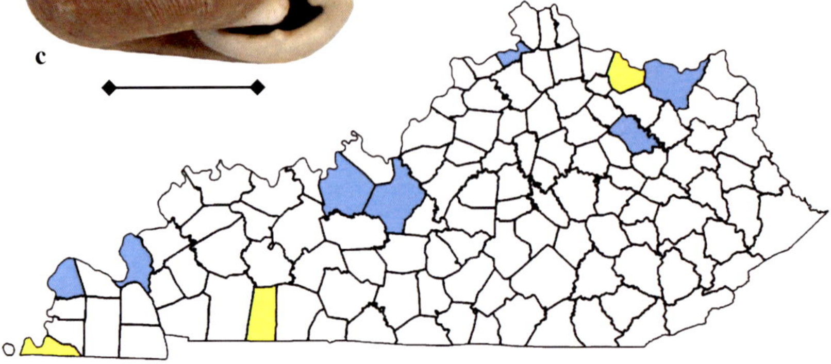

Glossy Threetooth

Triodopsis complanata (Pilsbry,1898)

Diameter: 18-25.3 mm, height 9 mm

Description: Depressed heliciform; lip reflected; shell with 6 whorls; glossy; umbilicate, umbilicus wide; the palatal tooth is positioned higher up on the outer lip, making the distance between the basal and palatal greater than seen in other *Triodopsis* species; parietal tooth small, points at or below (mostly below) the palatal tooth (when present); basal and palatal teeth tiny, hardly perceptible in many shells; cream-buff to light-brown; transverse striae are poorly developed, nearly effaced in places; there are some minute papillae around the umbilicus region and on the reverse side of the lip and near it (Pilsbry 1940); without hairs; the periphery is well rounded.

Similar Species: In Kentucky, *T. tridentata* is smaller and has stronger transverse striae; *T. tennesseensis* has a wider more expanded lip, better developed transverse striae, a duller surface and most importantly, the palatal and basal teeth are a stronger feature.

Habitat: A species found among rocky outcrops of limestone under larger rocks and litter.

Status: G2/S2; Rare; a globally and regionally restricted land snail found in only a few Kentucky counties and in Tennessee one county.

Specimen: Figures (a) from Kentucky, Pulaski County, Limestone cliffs above Buck Creek and figure (b) *T. complanata* from Kentucky, Wayne County (author's collection).

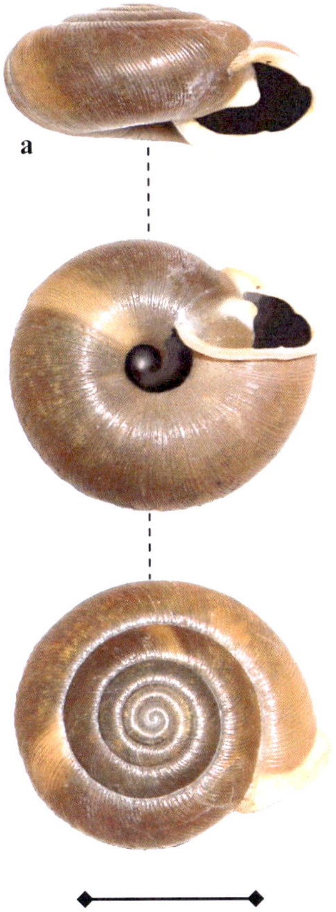

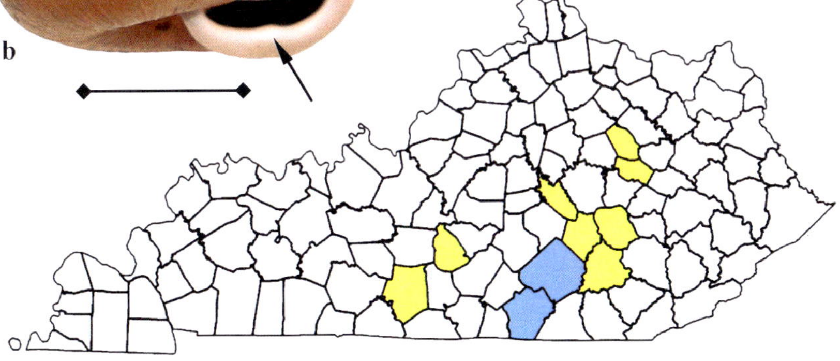

Budded Threetooth

Triodopsis tennesseensis (Walker & Pilsbry, 1902)

Diameter: 19-24 mm, height 10-10.4 mm

Description: Depressed heliciform; lip reflected and wide; shell with 5-5.5 whorls; umbilicate; parietal tooth moderate in build, points directly at or below (mostly below) the palatal tooth; basal and palatal tooth small, roughly the same size and do not crowd the aperture in frontal view; shell snuff brown; transverse striae are well developed on the top, sides and bottom of the shell; scattered papillae mostly along the transverse striae and around the umbilicus; adults without hairs; periphery rounded.

Similar Species: *Triodopsis tridentata* is typically smaller, more compact in build, has a smaller umbilicus and has larger basal and palatal teeth; *Triodopsis aff. tennesseensis* has a smoother shell surface, notably wider umbilicus and usually without basal and palatal teeth.

Habitat: Found under leaf litter in mixed hardwood forests on hillsides and in ravines.

Status: G4/S5; Common; this species becomes larger and more common in southeastern counties of Kentucky.

Specimen: Figures (a) from Kentucky, Pike County, Breaks Interstate Park and figure (b) *Triodopsis tennesseensis* from Kentucky, Powell County (author's collection).

a

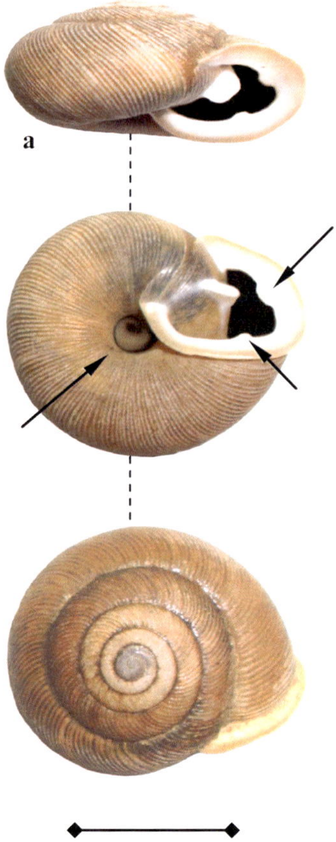

b

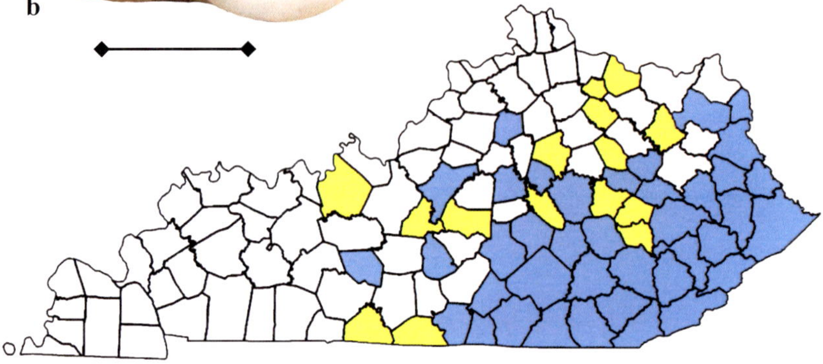

Pike County Threetooth

Triodopsis aff. tennesseensis (Walker & Pilsbry, 1902)

Diameter: 19-21 mm, height 8-9 mm

Description: Depressed heliciform; lip reflected and wide; shell with 5-5.5 whorls; widely umbilicate (the widest of any Kentucky *Triodopsis*); parietal tooth moderate in build, points directly at or below (mostly below) the often absent palatal tooth; in approximately half the specimens examined only a hint of a basal and palatal tooth exists while other shells have completely omitted these two teeth (see below); shell brownish; transverse striae are well developed on the top, sides and bottom of the shell; scattered papillae mostly along the transverse striae and around the umbilicus; adults without hairs; periphery rounded.

Similar Species: Most similar to *T. tennesseensis* but has a notably wider umbilicus, glossier shell surface, absent or nearly absent basal and palatal teeth and has lower more finely developed transverse striae; differs from *T. complanata* by having a courser shell and wider umbilicus.

Habitat: Found under leaf litter in mixed hardwood forests around rocky sandstone outcrops.

Status: Not ranked; Rare; further DNA study is warranted on this interesting ecological form or possibly new species; endemic to Kentucky.

Specimen: all figures from Kentucky, Pike County, (author's collection).

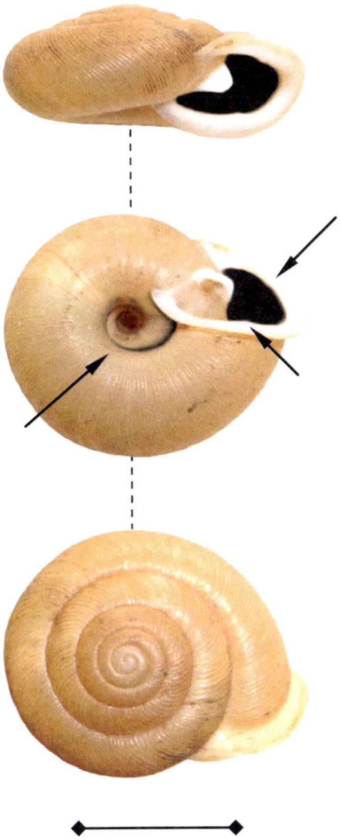

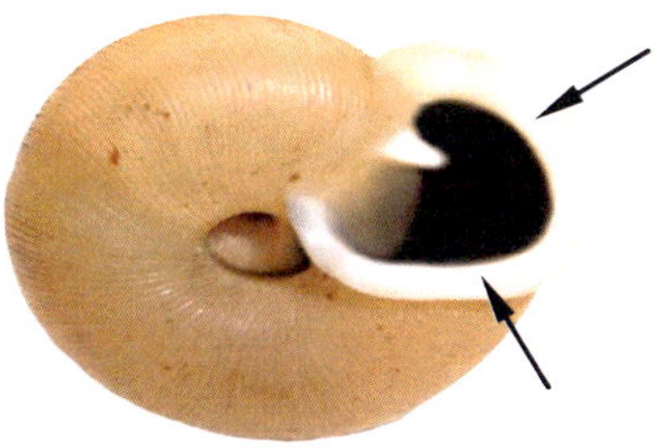

Many missing the basal and palatal teeth

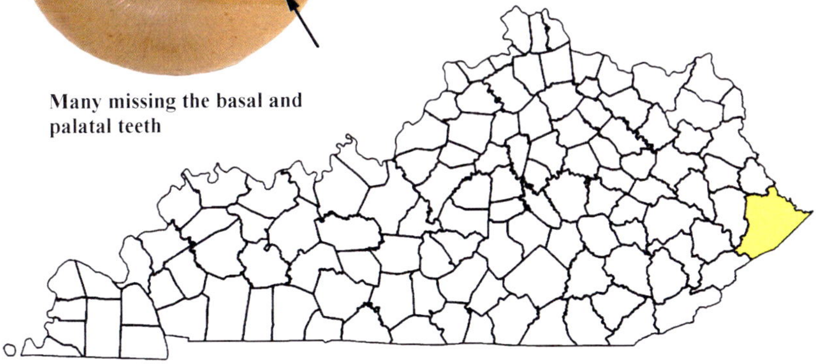

Northern Threetooth

Triodopsis tridentata (Say, 1816)

Diameter: 12-20 mm, height 6.4-7 mm

Description: Depressed heliciform and variable; lip reflected, not abruptly turning downward; shell with 5-6 whorls; umbilicate; parietal tooth points mostly at the palatal tooth; basal and palatal teeth small, not crowding the aperture in frontal view; light cream to pale cinnamon-buff; transverse striae are well developed on the top, sides and bottom of the shell; like most *Triodopsis* species there are scattered papilla on top and on the base; without hairs in adult shells; periphery rounded.

Similar Species: *Triodopsis vulgata* has larger more crowded teeth and an aperture that turns abruptly downward; *T. anteridon* is smaller and its palatal tooth is positioned on a buttress.

Habitat: Found in rich upland woods under forest litter, rocks and logs, but also a snail of roadsides and urban areas.

Status: G5/S5; Common; one of the most widespread and frequent *Triodopsis* in Kentucky central and eastern woodlands.

Specimen: West Virginia, Greenbrier County near Lost Cave (author's collection).

***Triodopsis tridentata*, Pike County, Kentucky**

***Triodopsis tridentata*, Lewis County, Kentucky**

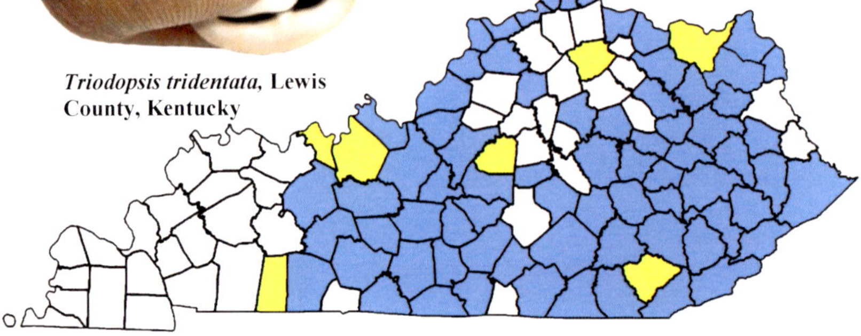

Triodopsis tridentata Shells Compared (proportionate)

These two shells of *Triodopsis tridentata* collected from the same hillside, near Summersville Dam, below the campground, Nicholas County, West Virginia. Note that the specimen on the right is missing its basal tooth, while the one on the left is with the archetypal tooth. Finding uncharacteristic shells is not uncommon and can bewilder the investigator (speaking from my own experience, of course). This is why collecting a series of shells is so central to a correct identification.

Land Shell Freaks!

Mesodon zaletus

Mesodon thyroidus

Land shell "freaks" can be the result of falls from high places or attacks from predators, which can cause everything from simple hair-line fractures to severe structural damage to the shell. If the snail survives, it will repair the breach by using calcium carbonate. Small fractures are sealed with little effort, but a severely damaged shell will require a great deal of time and skill to fully restore. If the damage is around the aperture, the renovated peristome will take on some rather interesting and bizarre shapes, as seen above. Both shells from below a limestone outcrop on Furnace Mountain, Powell County, Kentucky.

Velvet Wedge

Xolotrema denotatum (Fèrussac, 1821)

Diameter: 19-25.6 mm, height 10-13.7 mm

Description: Depressed heliciform; lip reflected; shell with 5.5 whorls; imperforate; parietal tooth large with a wide base, points well above the palatal tooth; basal tooth small, may be absent or so poorly defined it is barely detectable; palatal tooth of medium build; tawny olive to snuff brown; transverse striae weakly developed, shell surface with close set periostracal processes or thickened hairs which pick up forest debris (see next page) typically seen on live and fresh dead shells only; periphery rounded in most specimens, but in some populations the shell periphery may be weakly angular (see below).

Similar Species: Although *Triodopsis* species also contain three teeth, they differ by having an open, not completely sealed umbilicus.

Habitat: Often associated with rotting hardwood trees in advanced stages of decay in a variety of upland mixed hardwood sites; species appears indicative of quality habitat.

Status: G5/S5; Common; this species which is reported from across the state is rarely ever found in large numbers.

Specimen: Figures (a) typical form, from Tennessee, Sevier County, Roaring Fork, GSMNP (GSMNP collection), figure (b) bluntly angular form, from Red River Gorge, Powell County, Kentucky and figure (c) a more angular form, from Livingston County, Kentucky (author's collection).

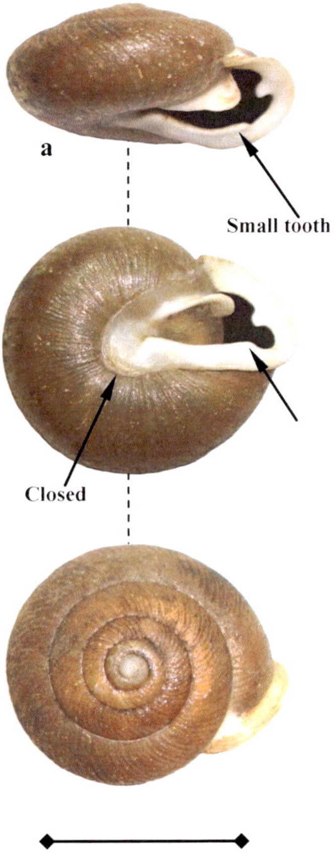

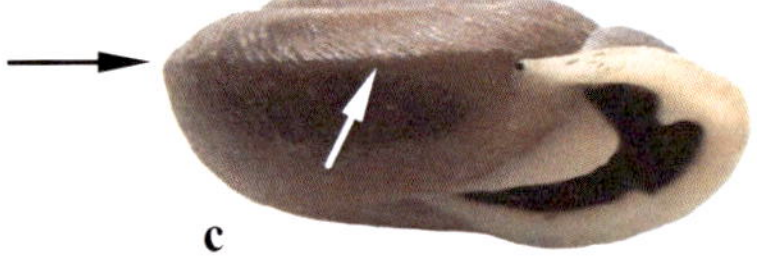

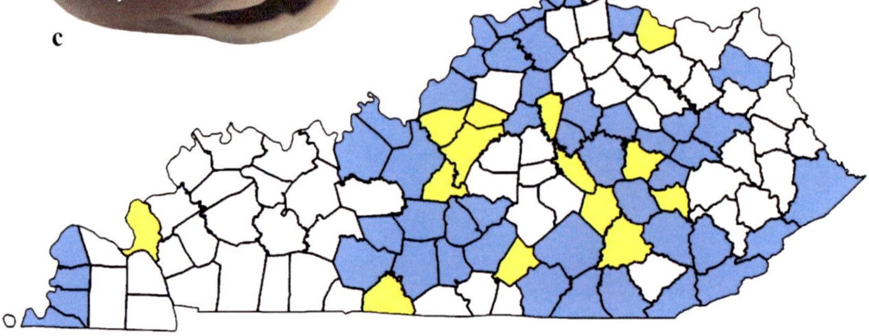

Sharp Wedge

Xolotrema obstrictum (Say, 1821)

Diameter: 17-26.6 mm, height 10-12 mm

Description: Depressed heliciform; lip reflected; shell with 5.5 whorls; imperforate; the large parietal tooth points well above the palatal tooth; basal tooth small and may be absent or so poorly defined that it is not detectable; palatal tooth relatively large; pale brown; transverse striae are well developed on top and on the base; fine wrinkles and close-set periostracal processes or hairs; periphery distinctly carinate to strongly angular the entire periphery (a).

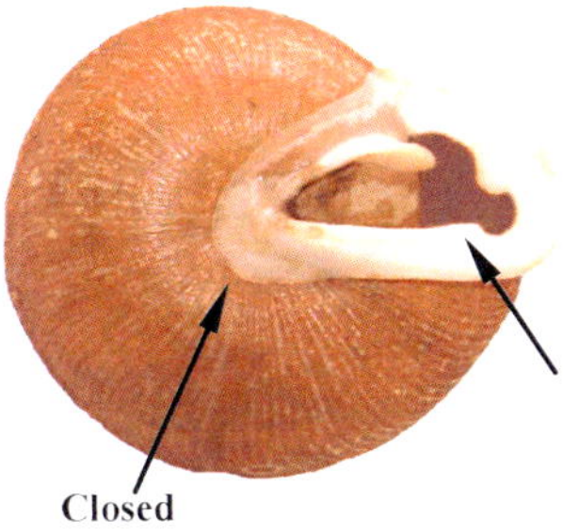

Similar Species: *Xolotrema denotatum* usually has short stiff hairs and has a rounded periphery, although specimens from Red River Gorge occasionally have slightly angular peripheries.

Habitat: Limestone bluffs and outcrops, live snails are pressed into recesses and crawl around the rock at night, sometimes in cave entrances, also found on forested loess bluffs at John James Audubon State Nature Preserve in Henderson County

Status: G4/S3; An uncommon species in Kentucky restricted to central and western potions of the state.

Specimen: Kentucky, Green County near town of Greensburg (author's collection).

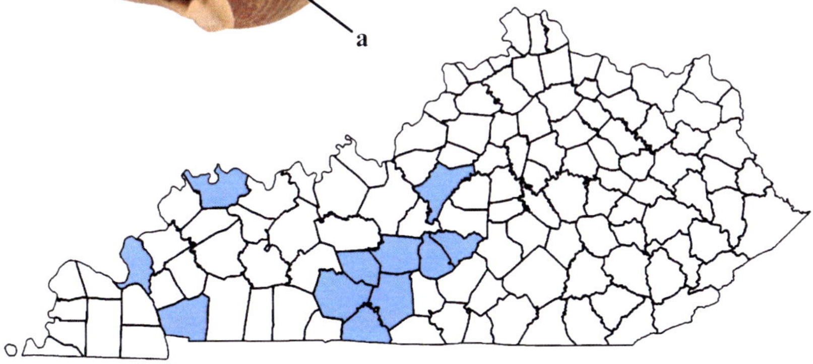

Bladetooth Wedge

Xolotrema fosteri (F. C. Baker, 1921)

Diameter: 17-19 mm, height 8-9 mm

Description: Depressed heliciform; lip reflected; shell with 4.5-5 whorls; imperforate; parietal tooth large (points well above the palatal tooth); basal tooth small, the palatal tooth on many specimens may be absent or so poorly defined that it is not detectable, (for this reason it is important to collect a number of shells); cinnamon-buff to brownish-horn; transverse striae moderately developed; without hairs; periphery rounded to weakly angular.

Similar Species: *Triodopsis* species are umbilicate; *P. appressa* is without a palatal tooth and has spirally arranged papillae. *P. laevior* has same spiral striae but lacks the outer palatal tooth.

Habitat: Often found in large numbers on and around large decayed logs, in rich bottomland hardwood forests and swampy woodlands, degraded habitats such as roadsides and along railroad corridors.

Status: G5/S4; Most common in bottomland forests and swampy woodlands bordering the Ohio and Mississippi Rivers.

Specimen: Figures (a) from Kentucky, Livingston County , West Vaco land on Scotts Chapel Road and figures (b & c) darker color morphs, one missing the palatal tooth, both from Kentucky, Ohio County near Rockport (author's collection).

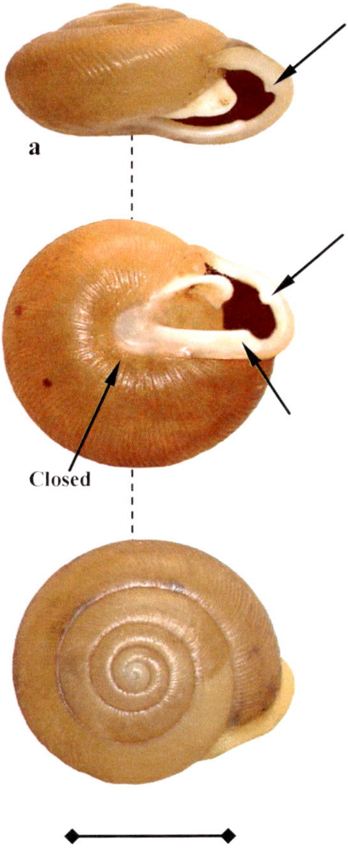

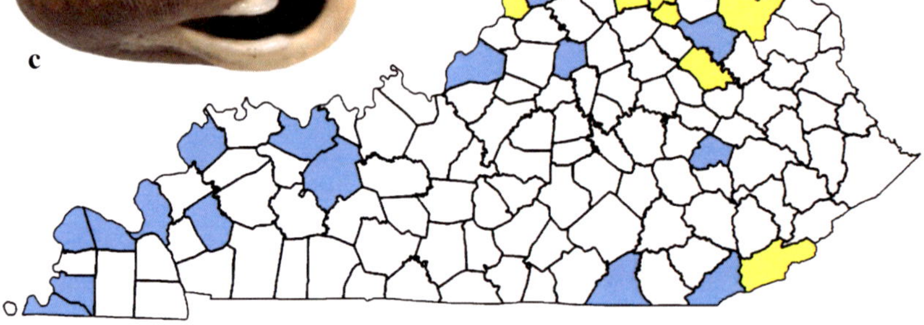

Dead Trees are Exceptional Land Snail Habitat

It's been said that there's more life in dead trees than live ones. Without question, dead trees in advanced stages of decay are outstanding habitats for several species of common land snails, especially slugs. These snails can be found on both standing or downed trees or under exfoliating bark. As rotting trees age, so does land snail cuisine including fungi, slime-molds, and lichens. The above species are commonly associated with dead trees.

Shells mostly large and heliciform, 15-45 mm, most over 20 mm, imperforate, with or without a parietal tooth, no basal or palatal teeth, reflected lip (Fumonelix, *Mesodon, Neohelix,* and *Webbhelix*)

Fumonelix, Mesodon, Neohelix and *Webbhelix* species are among the largest and most conspicuous land snails in Kentucky although several species are medium in size. Mostly heliciform and always having reflected lips in mature shells, several species develop a small parietal tooth but none contain basal or palatal teeth. Transverse and spiral striation are a constant and well-developed features, and the umbilicus is generally imperforate. Top views of these genera are not typically diagnostic for separating species. The following taxa are arranged according to some of these features and are as follows; see page 322 for more detailed key.

Shells medium to large, mostly globose, imperforate, pg. 323-332

Fumonelix wetherbyi
Mesodon mitchellianus
Mesodon elevatus
Mesodon zaletus
Mesodon normalis
Neohelix albolabris
Neohelix dentifera
Webbhelix multilineata (uncommon)

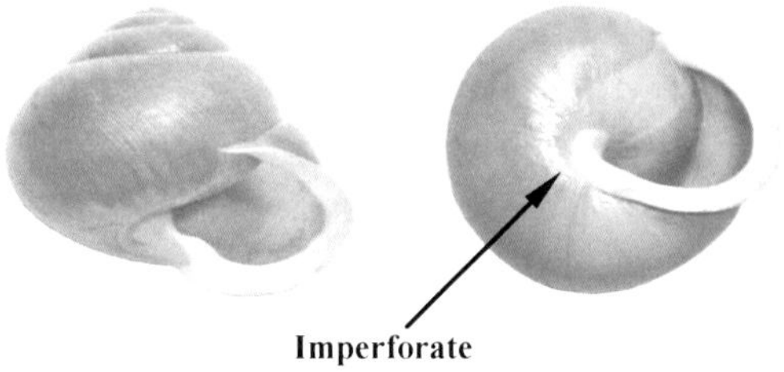

Shells large, depressed heliciform, 20-35 mm, umbilicate, with or without a parietal tooth, basal tooth maybe present, reflected lip (*Allogona* and *Appalachina*)

Allogona and *Appalachina* are also large land snails, but differ from the land snails above by being umbilicate and having more depressed shells. They always have reflected lips at maturity, and both species below have a basal tooth; see page 322 for more detailed key.

Shells large, depressed, umbilicate, pg. 333-336

Allogona profunda
Appalachina chilhoweensis (rare)
Appalachina sayana
Appalachina sayana kentucki (uncommon)

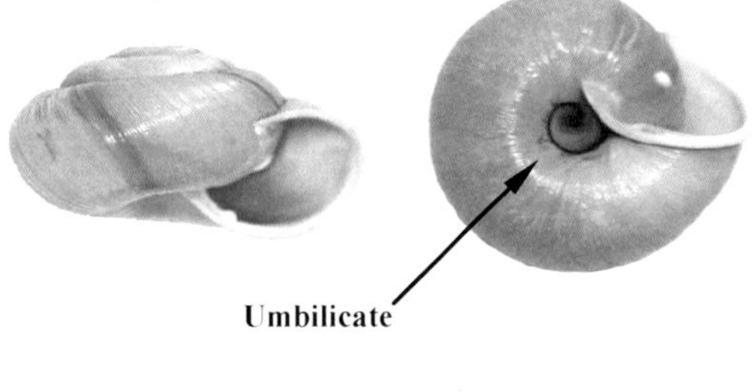

Key to Large Shells with Reflected-Lips (proportionate)

30 mm

1 Shell <u>imperforate</u>, globose, no color bands (*Fumonelix & Mesodon*)

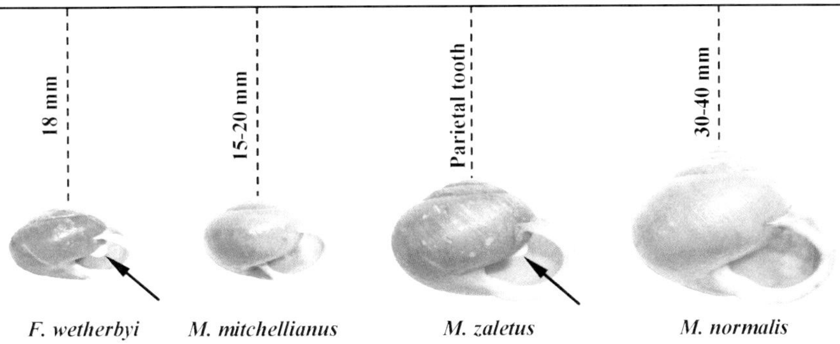

2 Shell <u>imperforate</u>, somewhat depressed, no color bands (*Neohelix*)

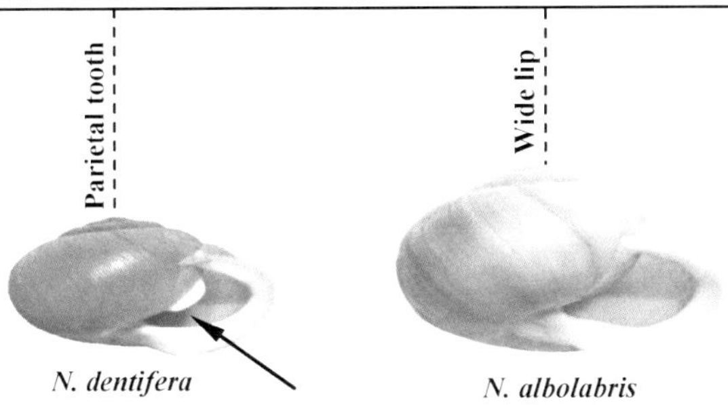

3 Shell <u>imperforate</u>, globose, color bands (*Webbhelix*)

4 Shell <u>umbilicate</u>, depressed (*Allogona* and *Appalachina*)

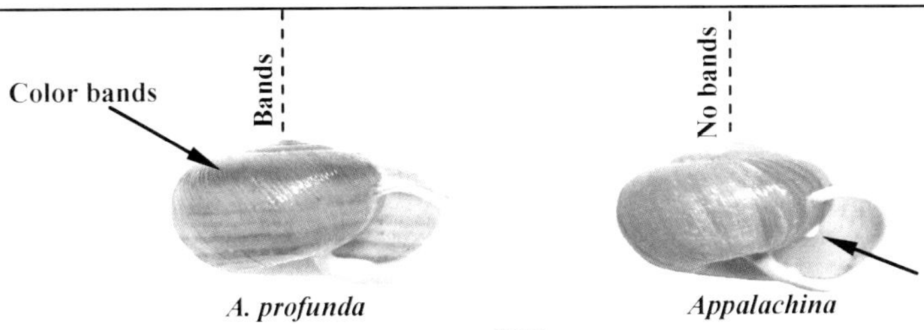

Clifty Covert **Polygyridae**

Fumonelix wetherbyi (Bland, 1873)

Diameter: 17.5-18.3 mm, height 10-11 mm

Description: Heliciform; lip widely reflected; shell with 5-6 whorls; imperforate; parietal tooth size variable; dull cinnamon-buff; sculptured with oblique, prostrate hairs but these are often lost in old shells; the rough shell texture can easily be felt in the fingers and these hairs often trap a thin layer of soil, making the shell more cryptic to predators; periphery well rounded.

Similar Species: Most *Mesodon* species are larger and have smoother shell surfaces with distinct transverse and spiral striae on their shell surface and are without prostrate hairs; *Patera* species, excluding *P. pennsylvanica,* have more compressed shells and larger parietal teeth.

Habitat: Most common on steep slopes below sandstone outcrops or in woods strewn with sandstone talus or blocks of rock;

Status: G2/S2; State Threatened; this uncommon snail is limited to southeastern Kentucky and northern Tennessee; the only limestone site for this snail is along rock outcrops above the Cumberland River.

Specimen: All figures from Kentucky, McCreary County, Yahoo Falls, DBNF (author's collection).

A lighter color morph with larger palatal tooth

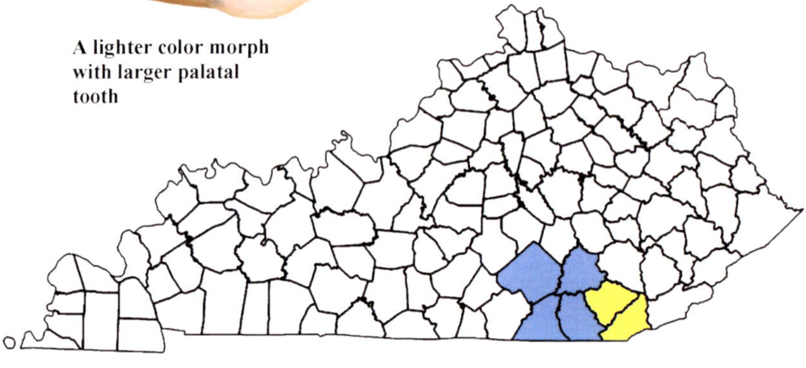

Sealed Globelet

Mesodon mitchellianus (I. Lea, 1839)

Diameter: 15-17.2 mm, height 11.7– 13 mm

Description: Heliciform; lip reflected; shell with 5 whorls; imperforate, umbilicus always completely sealed; no teeth present; pale yellowish-tan; adults without hairs; transverse and spiral striae always present; shell periphery well rounded; river island specimens tend to be thinner and translucent, figures (b and c).

Similar Species: *Mesodon clausus* has a rimate umbilicus that is slightly open whereas the umbilicus in *M. mitchellianus* is always completely closed (a key feature); *M. thyroidus* is larger, usually has a parietal tooth and has a rimate umbilicus.

Habitat: Found in low places such as floodplains, meadows, and roadsides occurring under leaf litter and other forest debris.

Status: G4/S3; Uncommon; reported from only a few widely scattered Kentucky counties.

Specimen: Figures (a) from West Virginia, Raleigh County, Mill Creek, figure (b) from Kentucky, Powell County, Red River Gorge and figure (c) from West Virginia, Wood County, Muskingum Island, note the thin lip and shell, a result of a yet fully adult (figures b and c) (author's collection).

a

Completely sealed

b

Fully adult lip (thick)

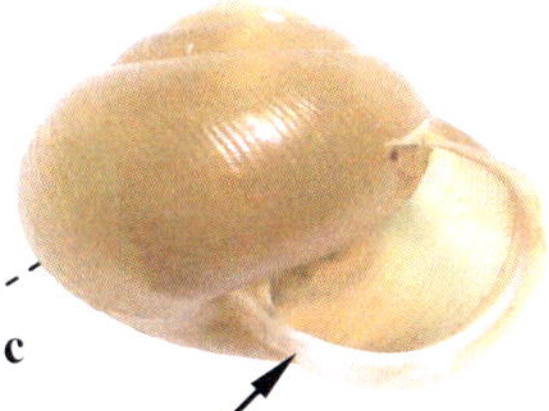

c

Nearly adult lip (thin)

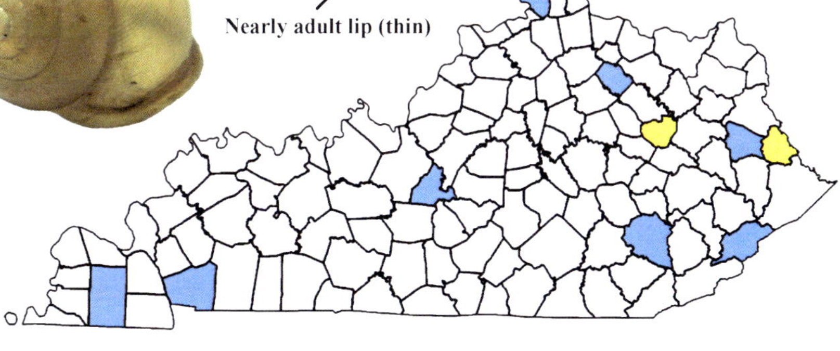

Proud Globe

Polygyridae

Mesodon elevatus (Say, 1821)

Diameter: 19.8-26.3 mm, height 16-20 mm

Description: Heliciform; lip reflected and thickened; shell with 6-7 whorls; shell solid; imperforate; large parietal tooth usually present; pale yellow to light olive; no hairs; transverse and minute spiral striae always present; shell periphery well rounded; young and adult shells with or without a color band (a).

Similar Species: *Mesodon clausus* is smaller, has a thinner shell and is without a parietal tooth; *M. zaletus* has a notably smaller parietal tooth and has a more compressed shell.

Habitat: A species of limestone river buffs and rich wooded slopes in mixed hardwood forests, also found in upland sites under forest litter around caves.

Status: G5/S5;Common; most records are from central Kentucky counties.

Specimen: Figures (b) from Kentucky, Bell County and figure (c) a bleached and more elevated shell from same location (author's collection).

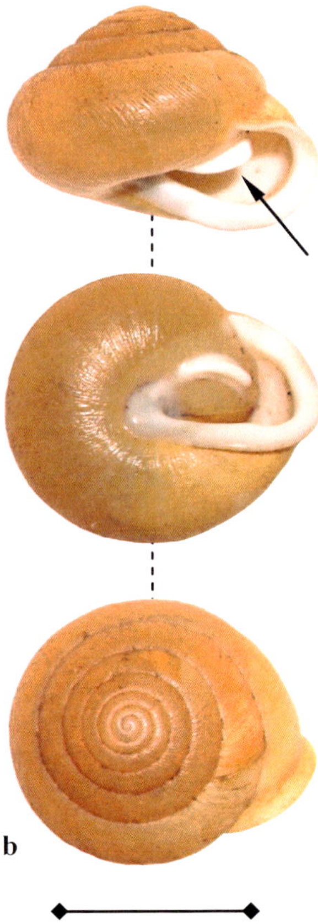

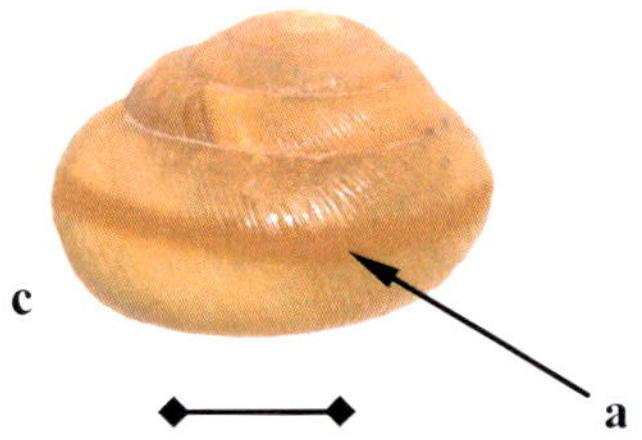

A more elevated form

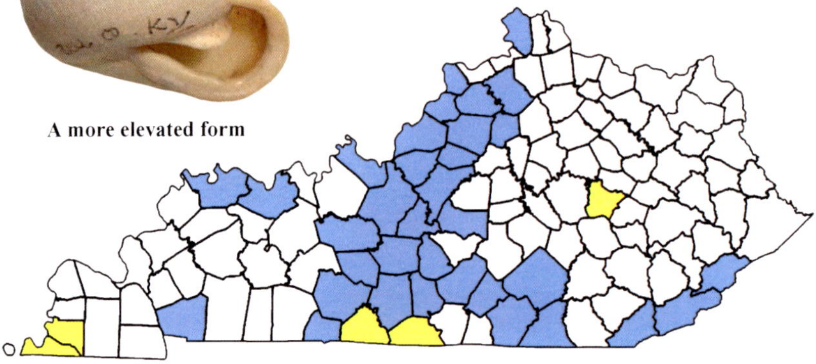

Toothed Globe **Polygyridae**

Mesodon zaletus (A Binney, 1837)

Diameter: 19-31 mm, height 16-18 mm

Description: Heliciform; lip reflected; shell with 5.5-6 whorls; imperforate; small parietal tooth typically present, but may be lacking in some individuals in the same populations; shell solid (thick walled); cream color to cinnamon-buff; no hairs in adults; transverse striae are well developed; spiral striae always present and well defined; shell periphery rounded.

Similar Species: *Mesodon normalis* is larger with a slightly higher profile, has a thinner shell, and never has a parietal tooth; *M. thyroidus* has a thinner shell and is rimate to perforate; *Neohelix albolabris* is generally larger, having a thicker shell and is usually without a parietal tooth.

Habitat: Most common on rich, mesic, wooded slopes with mature to old growth forest cover.

Status: G5/S5; Common; reported from across Kentucky.

Specimen: Tennessee, Sevier County, Roaring Fork, GSMNP (GSMNP collection).

***M. zaletus*, Lost Cave, Greenbrier County, West Virginia**

***M. zaletus*, McCreary County, Kentucky**

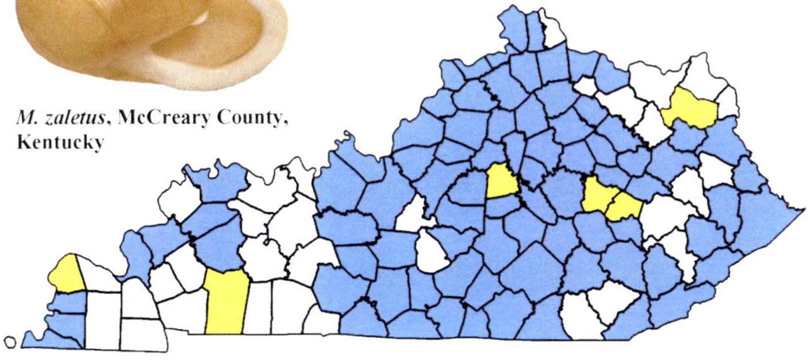

Grand Globe

Mesodon normalis (Pilsbry, 1900)

Diameter: 30-39 mm, height 26-32 mm

Description: Heliciform, shells varying considerably in height; lip reflected; shell with 5.5-6 whorls; imperforate; no teeth present; horn colored to tannish olive; without hairs in adult shells; transverse and spiral striae always present; shell periphery well rounded; shell height will vary between different populations; despite its large size, *M. normalis* is reported to be an annual species (Hubricht, 1985).

Similar Species: *Mesodon thyroidus* has a rimate umbilicus that is slightly open; *M. zaletus* has a thicker shell and has a parietal tooth; *N. albolabris* has a thicker, more compressed shell and wider lip.

Habitat: A species of acidic soils; found in upland habitats (up to 1400 m), usually in mixed hardwoods, but also found in pine woods around logs.

Status: G5/S3; Uncommon, in Kentucky, this more southern species is restricted to eastern and southeastern counties

Specimen: Tennessee, Cocke County, Cosby Campground, GSMNP (GSMNP collection).

2 extremes

***M. normalis*, Macon County, North Carolina**

***M. normalis*, Carter County, Tennessee**

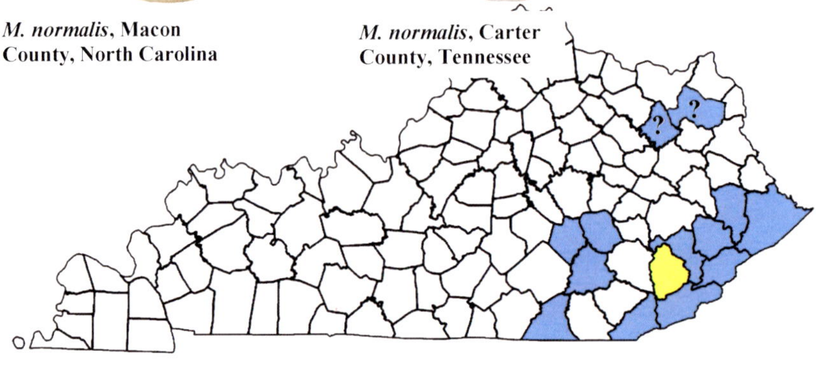

Chrissy Mann

Above image of Grand Globes, *Mesodon normalis*, exchanging sperm during copulation and below the same species feeding on fresh black bear dung; Yummy! Both images from the Great Smoky Mountains Natural Park of Tennessee and North Carolina.

Whitelip

Neohelix albolabris (Say, 1817)

Diameter: 18-45 mm, height 20-32 mm

Description: Depressed heliciform to heliciform, a variable shell in terms of size and form; lip reflected, thickened and wide, varying in shape; shell with 5-6 whorls; shell solid with a dull surface; imperforate; typically without teeth or very occasionally, with a weak parietal tooth (a); cream buff to pale tan; transverse striae well developed, impressed wavy spiral striae also well developed; periphery rounded; specimens from eastern Kentucky are twice as large as specimens from central and western Kentucky and may represent different species (see below).

Similar Species: *Mesodon normalis* has a higher shell profile and thinner shell; *M. zaletus* has a higher shell profile and a parietal tooth.

Habitat: A species found in a wide range of upland mixed hardwood sites; at the base of limestone cliffs, but also found on dry acidic ridge tops, in waste places and urban areas.

Status: G5/S5; Common; one of the most common and widespread large land snails in Kentucky, reported from nearly every county.

Specimen: West Virginia, Monroe County, along Wayside Creamery, Indian Draft Creek (author's collection).

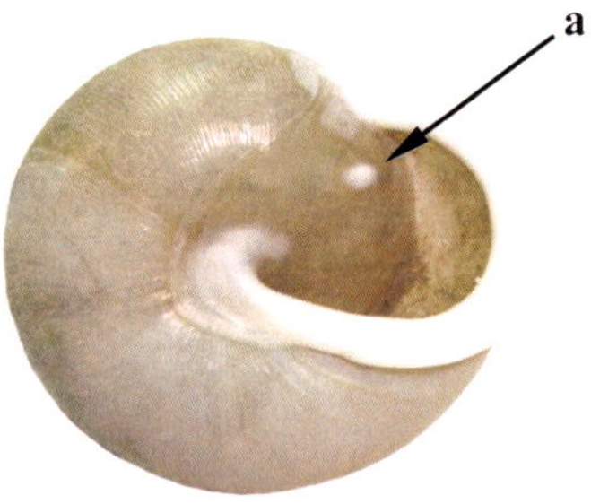

***Neohelix albolabris*, occasionally found with a weak parietal tooth, Hampshire County, West Virginia**

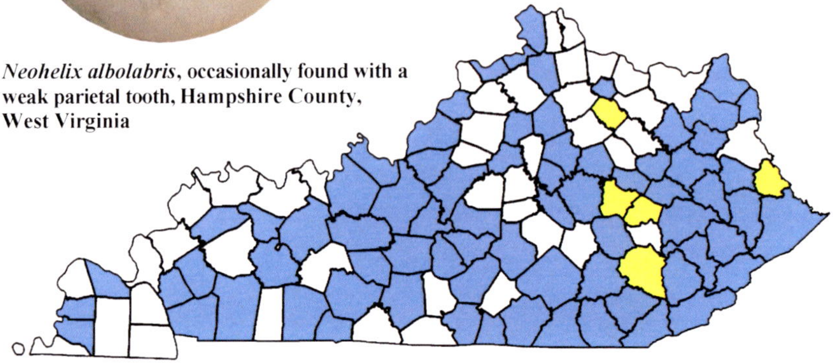

Neohelix albolabris size and shape variability (proportionate)

Neohelix albolabris (44 mm), near town of Campton, Wolfe County, Kentucky

Neohelix albolabris (35 mm), North Carolina, Swain County, around Fontana Lake (GSMNP collection)

Neohelix albolabris (23 mm) Edmonson County, Kentucky

Neohelix albolabris (26 mm) Estill County, Kentucky

Neohelix albolabris (33 mm), Powell County, Kentucky

Big-tooth Whitelip

Neohelix dentifera (A. Binney, 1837)

Diameter: 20-30.5 mm, height 11.4-16.5 mm

Description: Depressed heliciform; lip reflected; shell with 5-5.5 whorls; imperforate; shell surface dull; parietal tooth low and wide; shell color variable from pale olive to chestnut brown; transverse striae are well developed, impressed spiral striae present; no hairs in adult shells; periphery rounded.

Similar Species: *Neohelix albolabris* is larger, with a higher shell profile and no parietal tooth; *Mesodon normalis* has a thinner, higher profile shell and no parietal tooth; *M. zaletus* has a higher shell profile and contains a smaller parietal tooth.

Habitat: Generally restricted to higher elevation mixed northern hardwood forests occurring under leaf litter, around rocks and logs on acidic soils.

Status: G5/S1; State Threatened; this mostly northern land snail is reported from Pine and Black Mountain only where it is scarce.

Specimen: Kentucky, Letcher County, Pine Mountain (author's collection).

***N. dentifera*, Red Spruce Knob Trail, Pocahontas County, West Virginia**

***N. dentifera*, Stuart Recreation Area, Randolph County, West Virginia**

Striped Whitelip

Polygyridae

Webbhelix multilineata (Say, 1821)

Diameter: 14.5-32 mm, height 17-20 mm

Description: Heliciform; lip thin and reflected; shell with 5.5-6 whorls; imperforate; ivory yellow to olive buff, with multiple dark reddish brown bands that vary in width, shell glossy and thin; transverse striae are well developed, impressed spiral striae present; adults shells without hairs; periphery rounded; one of the most colorful native land snails in West Virginia.

Similar Species: *Allogona profunda* is flatter, umbilicate and with a small basal tooth.

Habitat: A species of low moist habitats including floodplains, marshes, and margins of lakes and ponds; also occurs on forested islands in the Ohio River where it can be surprisingly common.

Status: G5/S2; State Threatened; in Kentucky this interesting gastropod is reported from Ohio River counties and islands only.

Specimen: Kentucky, Carlisle County, island in the Mississippi River (author's collection).

Multiple color bands

It's believed that color streaks in shells act as disruptive patterns, making snails more difficult to spot by predators.

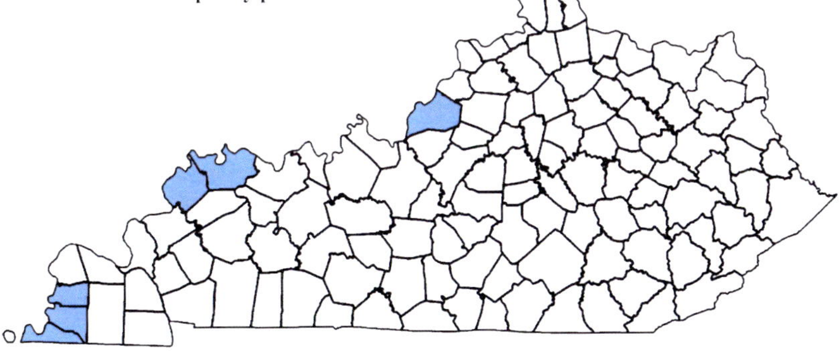

Broad-banded Forestsnail

Allogona profunda (Say, 1821)

Diameter: 19-34 mm, height 15-17.1 mm

Description: Depressed heliciform; lip reflected; shell with 5.5 whorls; umbilicate; a small basal tooth usually present; shell light tan with a few reddish-brown bands that vary in their width, but shells are also found without color bands; shell glossy and thick; transverse striae well developed, spiral striae are better developed on top, but nearly absent on the base; adults shells without hairs; periphery rounded; live animal tricolor.

Similar Species: *Appalachina sayana* is similar in size and build, but is without color bands and has a parietal tooth, in limestone sites both species often occur together.

Habitat: A species of both limestone and acidic mountainsides in mixed hardwood forests; on Bickle Knob in West Virginia live animals have been found climbing on sunflowers and nettles in July at 1189 m (C. Stihler pers. comm.).

Status: G5/S5; Fairly Common; this beautiful land snail is reported from scattered counties across Kentucky; in general found sparsely where it occurs.

Specimen: All specimens including from Kentucky, Powell County, Furnace Mountain (author's collection).

Unbanded form

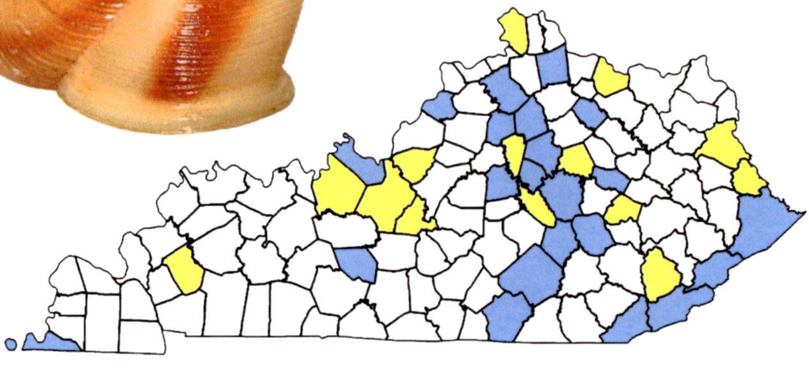

Queen Crater

Polygyridae

Appalachina chilhoweensis (Lewis, 1870)

Diameter: 26-42 mm, height 21.5 mm

Description: Heliciform; lip reflected and wide; shell with 6-6.5 whorls; shell thin for its large size; umbilicate; without teeth; cream-color; no hairs; transverse and minute spiral striae always present; shell periphery well rounded; this is one of Kentucky's largest terrestrial land snails.

Similar Species: *Appalachina sayana* has a thinner wire-like lip and has both a parietal (a) and basal tooth (b), while *A. sayana kentucki* has a parietal tooth only and is notably smaller.

Habitat: A species of upland mixed hardwood forests found under leaf litter, also found on wooded slopes with sandstone boulders or talus; often found with *M. wetherbyi*.

Status: G4/S2; State Threatened; reported only from McCreary and a small portion of Wayne County, found in numbers near Yahoo Falls (John MacGregor pers. comm.); shells are frequently in poor shape and broken.

Specimen: Kentucky, McCreary County, Yahoo Falls, DBNF (author's collection).

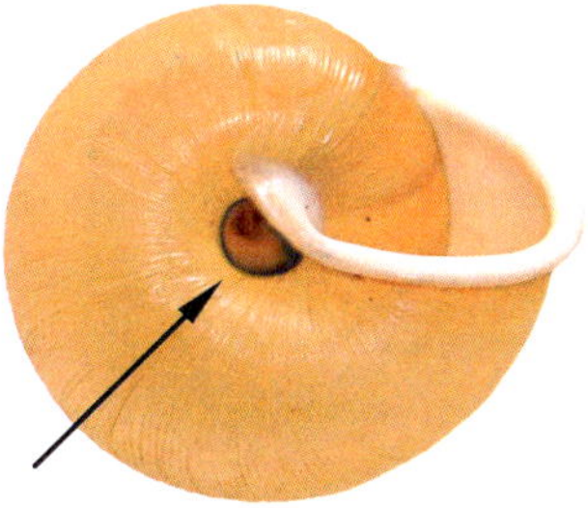

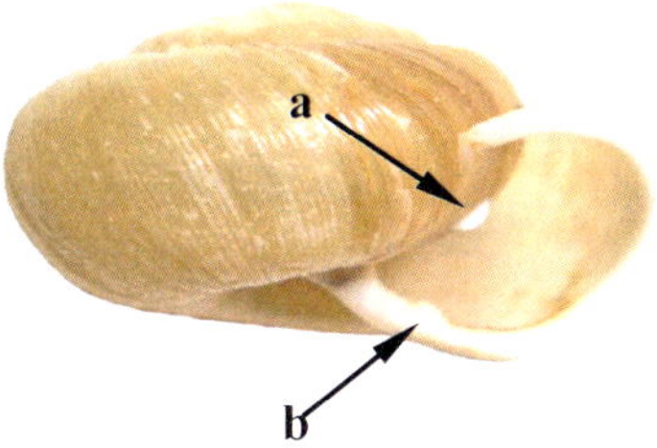

***Appalachina sayana*, Furnace Mountain, Powell County, KY**

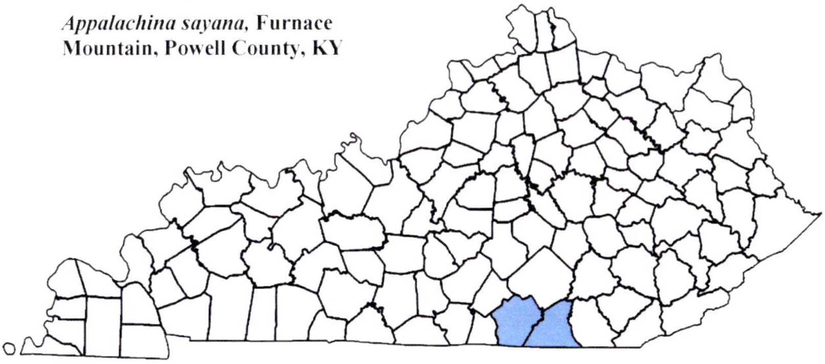

Spike-lip crater

Appalachina sayana (Pilsbry, 1906)

Diameter: 19-27 mm, height 14.5 mm

Description: Depressed heliciform to heliciform; lip reflected, thin and wire-like; shell with 5.5 whorls; umbilicate; shell thin for its large size; parietal tooth variable in size; basal tooth small (a); pale yellow to pale olive-tan, sometimes with darker streaks; no hairs in adult shells; transverse and minute spiral striae always present; periphery well-rounded.

Similar Species: *Mesodon normalis* is larger with a closed umbilicus and thicker shell; *Appalachina sayana kentucki* is smaller and more compact, has a smaller umbilicus and a larger basal tooth and is typically without a parietal tooth (see opposite page comparison).

Habitat: A common species of rich upland mixed hardwood forests under leaf litter and other forest debris on both limestone and acidic sites, the species is generally indicative of intact and quality forests.

Status: G5/S5; Common; found throughout central and eastern Kentucky.

Specimen: Typical form illustrated to the upper right (3 views), from Kentucky, Powell Co, Furnace Mountain (author's collection).

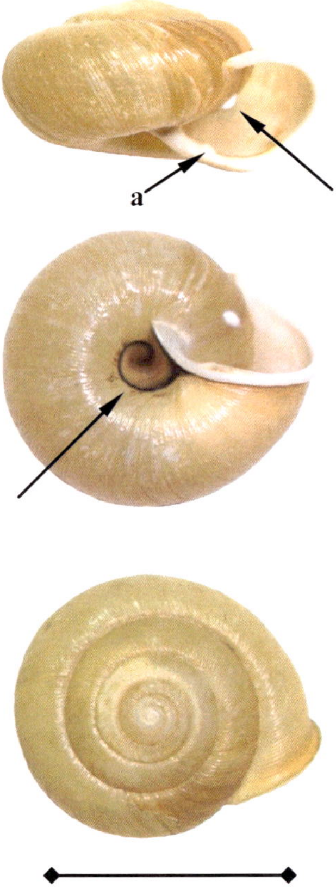

Seneca Rocks, Pendleton County, West Virginia

Yahoo Hollow, Randolph County, West Virginia

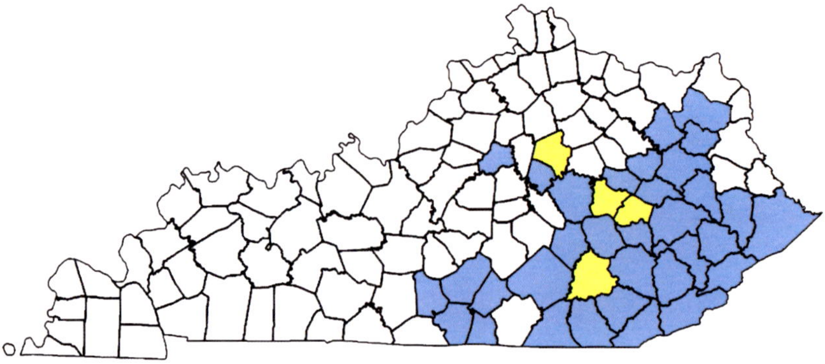

Pine Mountain Crater

Appalachina sayana kentucki Dourson 2011

Diameter: 18-22 mm, height 11.3 mm

Description: Heliciform; lip reflected; shell with 5.5 whorls; shell thin; umbilicate or rimate; usually without a parietal tooth; basal tooth present; pale-yellow to pale olive-tan; no hairs; transverse and minute spiral striae always present; shell periphery well rounded.

Similar Species: *Appalachina sayana* is larger, has a wider umbilicus, has a parietal tooth and smaller basal tooth; *A. chilhoweensis* is much larger and is without teeth.

Habitat: An uncommon species of rich upland and higher elevation mixed hardwood forests; it is generally found under moist leaf litter and other forest debris, becoming less common in drier sites such as ridgetops and Virginia pine forests; the highest numbers of shells are found in mixed-mesophytic sites, suggesting a preference for this habitat.

Status: G1/S1; State Threatened; restricted to Pine and Black Mountains.

Specimen: Pictured to the right is *Appalachina sayana kentucki* from Bad Branch, Pine Mountain, Letcher County, Kentucky (FMNH 328318, Holotype specimen) and below the 3 Kentucky *Appalachina* taxa compared and proportionate.

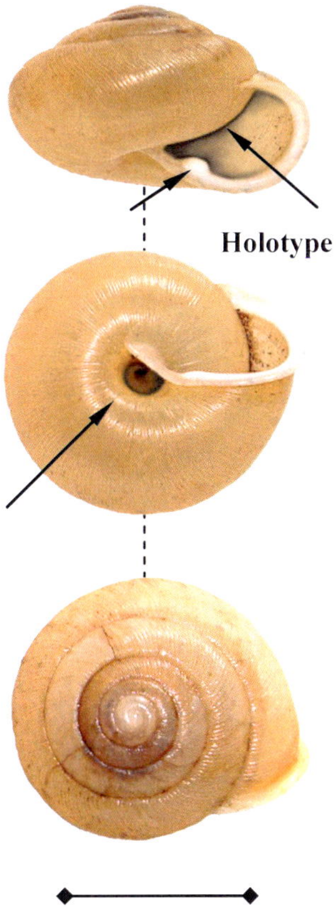

Appalachina sayana

Appalachina sayana kentucki

Kentucky *Appalachina* shells illustrated and are proportionate; note their size differences, all shells from Kentucky

Appalachina chilhoweensis

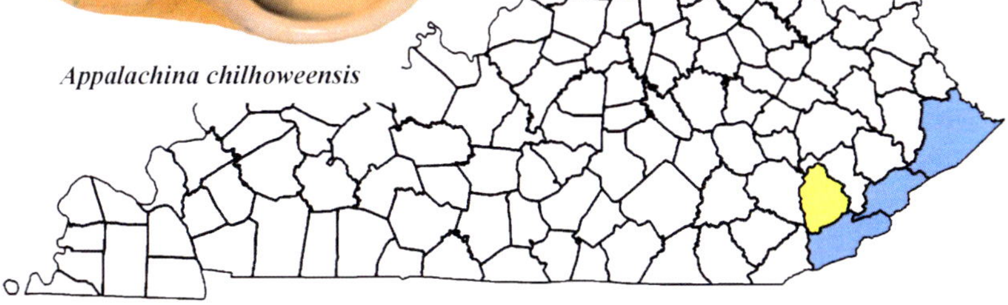

Symbiosis Between Land Snails and Slime Molds?

Above image of a Spike-lip Crater, *Appalachina sayana,* feeding on immature slime mold sporangium in Red River Gorge, Kentucky. There is some speculation that symbiotic relationships exist between land snails and particular species of slime molds (a hypothesis proposed during a discussion between the author, Adam Rollins and Ron Caldwell in Belize, 2010). For example, slime molds in the order Physarales precipitate amorphous calcium carbonate on some portion of their mature fruiting bodies (Townsend et al. 2005). *Badhamia utricularis* (figure a) not only coats the outer surface of its fruiting body with calcium, but has a system of calcareous tubes running throughout its core (figure b). This may be coincidence or a clever strategy for attracting land snails who require the mineral for shell building. What if Physarales slime molds sequester and store calcium for the singular purpose of attracting land snails? To ensure the spores are eaten by the visiting snail, the calcium deposits are intermingled with the spores (b). Pushing the hypotheses a bit further, what if slime molds were one of several conduits that help facilitate land snail migration into more acidic environments? These hypotheses await investigation. Figures a & b images from Eumycetozoan Research Project, University of Arkansas.

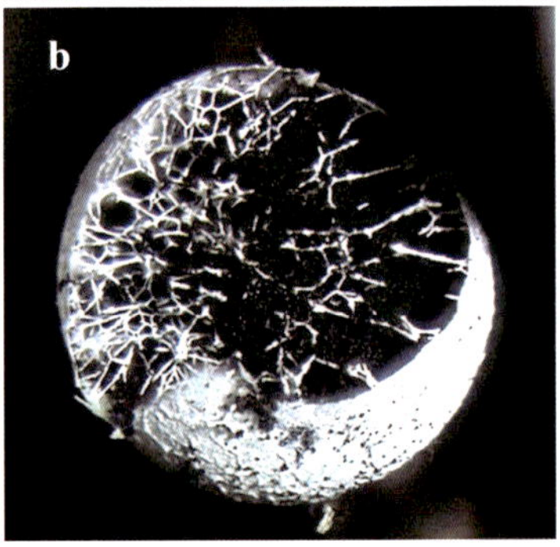

In general, native slugs are the least studied land snails in many regions of the world, and Kentucky is no exception. These under-appreciated gastropods are often all categorized as pests, but nothing could be farther from the truth. Native slugs are rarely a nuisance and typically become scarce or disappear altogether when the natural vegetation has been eliminated. Even in ecosystems that are more or less pristine, native slugs can be hard to find. In contrast, introduced slugs are quick to overpopulate an area, particularly in degraded habitat and without natural controls in place. These inadvertent guests are at home in the most degraded habitats such as small isolated woodlots, yards, roadsides, piles of trash and other waste places.

Considered excellent indicators of a healthy forest, native slugs by and large are only common where there are significant decomposing structure (logs), a habitat also used by other groups of organisms such as small mammals, salamanders, micro-invertebrates, fungi, and slime molds. All slugs, native or otherwise, can vary considerably in color and mantle patterns, even within the same population. Many gastropods in the family *Philomycidae* can only be reliably separated by dissection or DNA sequencing (Fairbanks pers. comm. 2010). *Philomycus* are some of the more interesting slugs a result of the presence of a dart sac and dart (referred to as love darts) they use to stimulate copulation between two individuals.

Genera Included:

(in order of appearance in text)

Philomycus
Megapallifera
Pallifera
Deroceras

Pictorial Key to the Family Philomycidae

Philomycidae are native terrestrial slugs found primarily in the eastern half of the United States, and even today, there remain a number of undescribed species (Fairbanks 1998). Many of the more conspicuous PHILOMYCIDAE are large, reaching crawling lengths of more than 100 mm. There are three North American genera in this family: *Philomycus, Megapallifera,* and *Pallifera* (see below). ***Philomycus*** species are large (75-100 mm), obese slugs. In terms of their reproductive anatomy, the species of *Philomycus* are the most interesting, characterized by the presence of a dart sac and dart (referred to as love darts). Some *Philomycus* species, in isolation, become self-fertilizing (Nicklas and Hoffman 1981). In *Philomycus* species, the mantle covers the entire body including the head (a). ***Megapallifera*** species are smaller in size (approximately 80 mm long or less), are also obese but lack the dart sac and dart (Fairbanks 1998) and the head is not entirely covered by the mantle (b). ***Pallifera*** species are the smallest among the group (50 mm or less), are slender and, like *Megapallifera,* lack the dart sac and dart. The head remains uncovered by the mantle. Great color variation exists in Philomycidae slugs making their ID very difficult. Young slugs of all species will generally have the same mantle patterns as adults. One last detail for separating slugs includes their defensive slime. To view this feature, first gently irritate the animal (in one spot) with a small stick until a small mass of slime accumulates (d). In natural lighting, determine the color of the slime. This is diagnostic mark is useful and appears to vary among different species. The crawling slime of slugs is typically clear.

Defense mucus on a stick

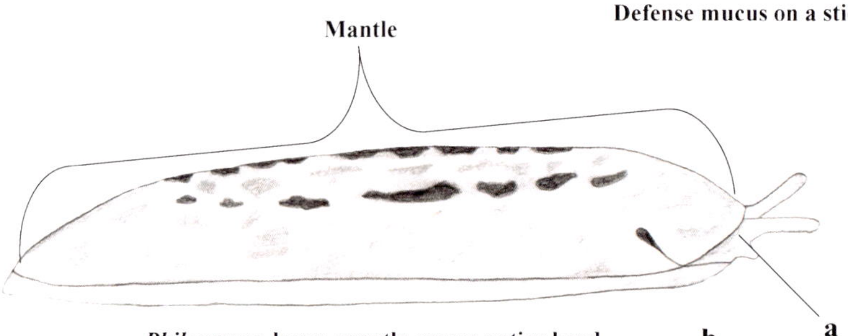

***Philomycus*, large, mantle covers entire head**

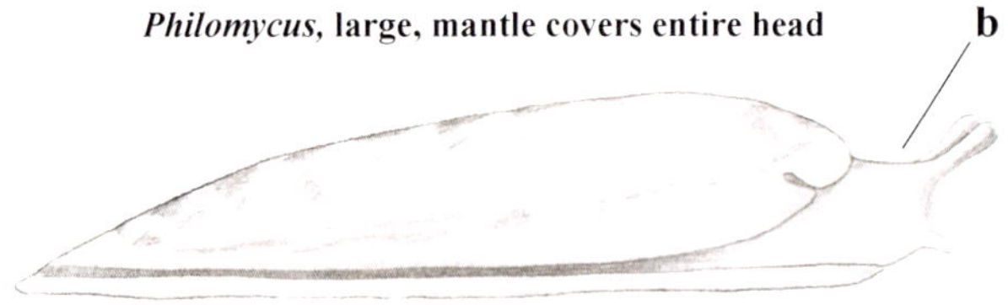

***Megapallifera*, medium, mantle does not cover entire head**

***Pallifera*, small, mantle does not cover head area**

Native Slugs of Kentucky Compared (Proportionate)

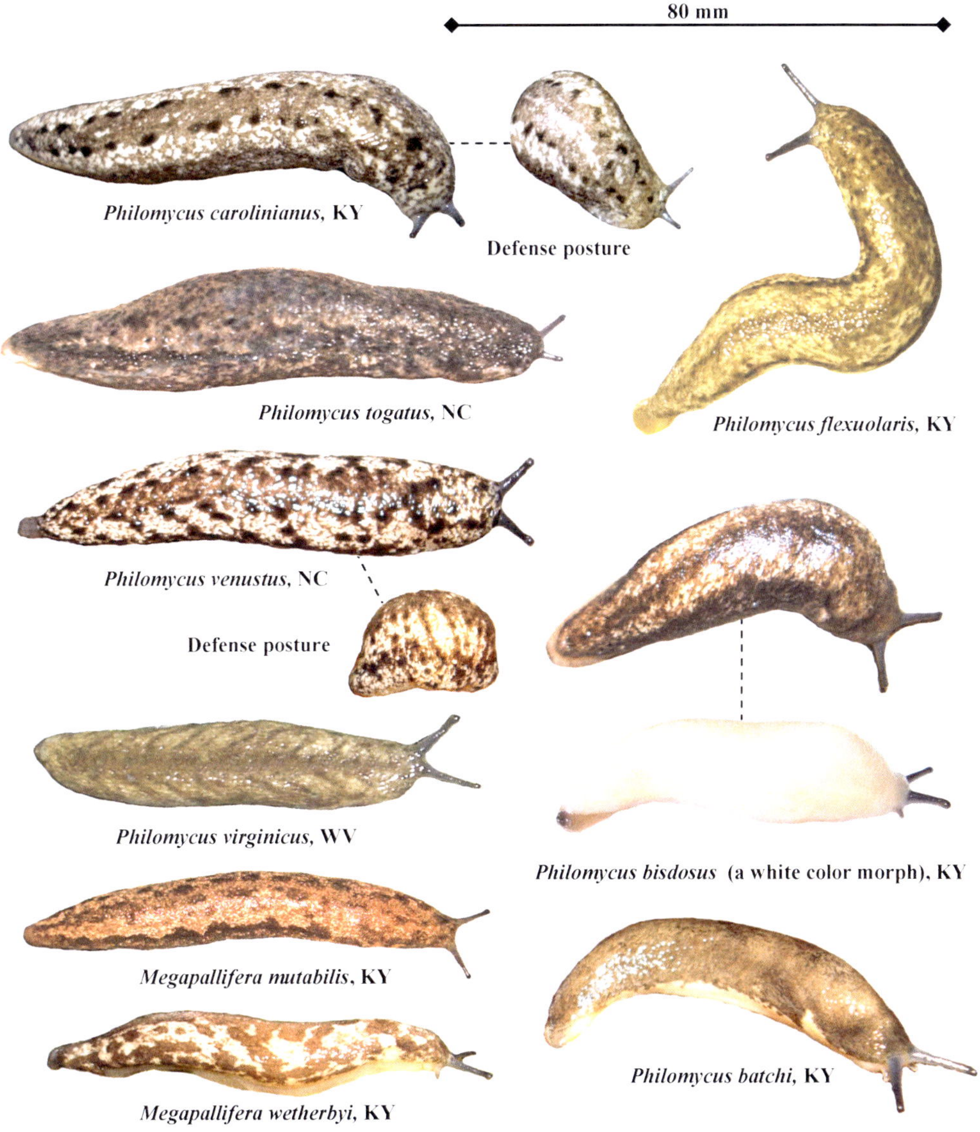

Small Native Slugs of Kentucky

Carolina Mantleslug **Philomycidae**

Philomycus carolinianus (Bosc, 1802)

Length: Adults 50-100 mm while crawling

Description: Mantle uniformly mottled with brown; usually two dorsal longitudinal rows of black spots, the best diagnostic feature for the species; animal carries a dart sac and dart (see opposite page figure b); defense mucus whitish when touched with a dry finger, the mucus immediately seeps out of the entire mantle having a bitter taste much like the skin secretions of a spring peeper (MacGregor pers. comm. 2010); sole of foot white (opposite page figure c).

Similar Species: *P. flexuolaris* is smaller and without the dark rows of black spots; the two species are often found together.

Habitat: Found in floodplain forests over much of its range, it becomes an upland species in the mountains, but does not occur much over 600 meters; found hiding under exfoliating bark of large rotting hardwood logs in advanced stages of decay, especially logs forming log bridges over small ravines, also found inside hollow trees such as large American beech.

Status: G5/S5; Common; reported from scattered locations across the state probably occurring in every Kentucky county.

Specimen: All slug images from Powell County, Red River Gorge, Kentucky (author's photo collection); figure (b) page 340 SEM of *P. carolinianus* love dart by Jodi White-McLean.

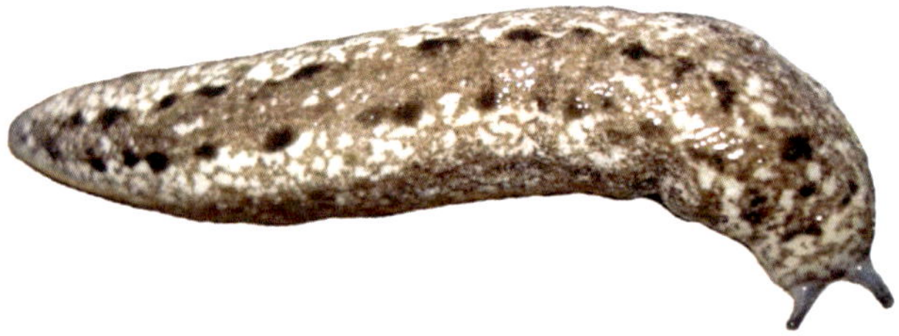

Size of slug crawling

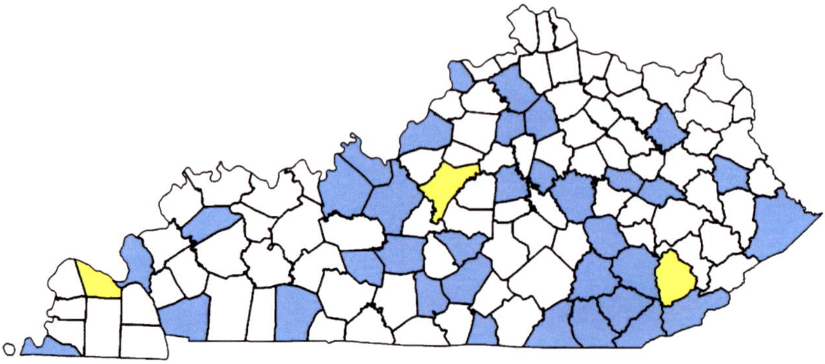

Philomycus carolinianus

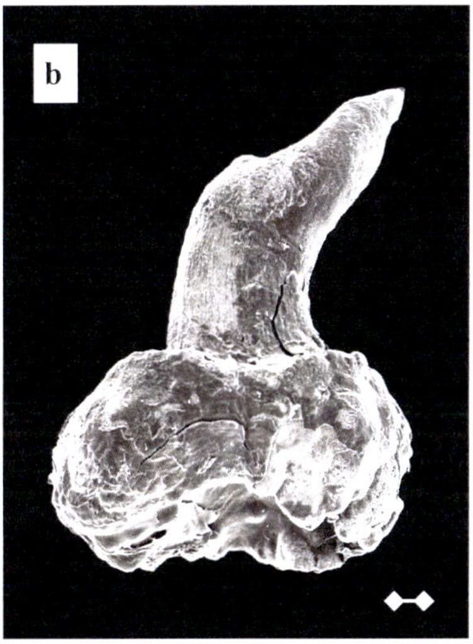

Above image of two Carolina Mantleslugs, *Philomycus carolinianus,* illustrating two color morphs from the same log; below the same slugs in defensive posture and stick showing the color of their whitish defense mucus. Both images from the Nolichucky River Gorge, Unicoi County, Tennessee.

Two Carolina Mantleslugs, *Philomycus carolinianus,* interested in sperm exchange for the purpose of fertilizing the internal eggs each individual slug carries. Red River Gorge, Powell County, Kentucky.

Winding Mantleslug — Philomycidae

Philomycus flexuolaris Rafinesque, 1820

Length: Adults 50-80 mm while crawling

Description: Three longitudinal mottled bands (dorsal and lateral), are usually distinct, but sometimes markings are so pale as to be scarcely discernible, and in some specimens, they may be nearly lost in a general dusky shade (figure b page 346); although the base color of the mantle varies considerably (as can be seen in the images on next three pages) the darker color patterns remain more or less the same; this species possesses a dart sac and dart; defense mucus pale-yellow (figure b, page 346); foot margin pale, sometimes with a hint of yellow; sole of foot cream to flesh color (figure c, page 344); for images of copulating *P. flexuolaris* see pages (347-348).

Similar Species: *Philomycus carolinianus* is generally larger, having two longitudinal rows of black or brown spots running down the center of mantle or back; *P. carolinianus* and *P. flexuolaris* can often be found living on the same large American beech log.

Habitat: A species of upland mixed hardwood forests up to 1500 m; as with most native slugs, during the day it goes into hiding under the exfoliating bark of large rotting logs in advanced stages of decay, especially large American beech trees.

Status: G5/S4; Fairly Common; appears to be restricted to eastern Kentucky; like most native slugs, *P. flexuolaris* has been under reported, due in part to a lack of interest by collectors who often ignore these interesting but obscure creatures.

Specimen: Figures (a, b and c, page 346) from Powell County, Red River Gorge, Kentucky; figure (d) from Pike County, Kentucky (author's photo collection).

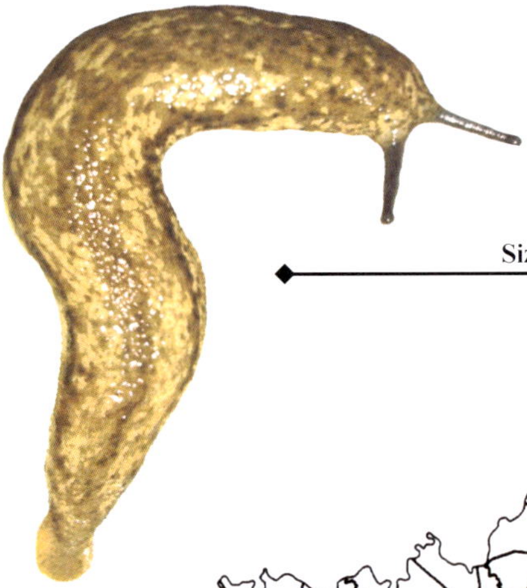

Size of slug crawling

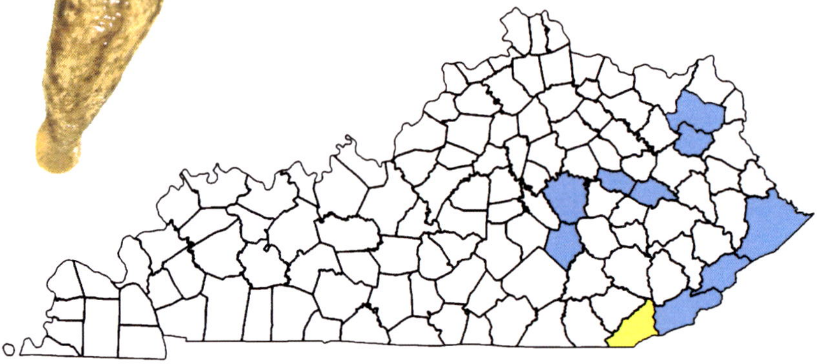

Philomycus flexuolaris

Copulating *Philomycus flexuolaris*

Two Winding Mantleslugs, *P. flexuolaris,* seeking a sperm exchange in Red River Gorge, Powell County, Kentucky

Copulating *Philomycus flexuolaris*

Variable Mantleslug — Philomycidae

Philomycus togatus (Gould, 1841)

Length: Adults 60-100 mm while crawling

Description: Top of animal heavily mottled , consisting of a broad dorsal band (sometimes broken into two bands) and narrower lateral bands (bands in some specimens may be hard to detect at times) on each side and scattered, irregular, small spots between the bands (Hubricht 1951); as can be seen on the next page (figures a-d, page 350) there is considerable variation in the species; defense mucus orange (figure b, page 350); foot margin orange-red to orange (figure c, page 350), probably one of the best external features of the species.

Similar Species: Differs from *P. carolinianus* in being darker brown without the two rows of black spots (Hubricht 1951) and having orange not whitish defense mucus; *P. flexuolaris* is smaller, has a yellowish defense mucus and its foot margin is pale not orange-red.

Habitat: An upland (to 1200 m) species found on wooded hillsides and in ravines, under loose bark of logs; in wet weather found on trunks of smooth-barked trees at night.

Status: G5/S3; Uncommon; most records are reported from scattered locations across eastern Kentucky.

Specimen: Figures (a & b) are from Dickenson County, Breaks Interstate Park, Virginia (author) and figure (d) from Surry County, North Carolina (Wayne Van Devender).

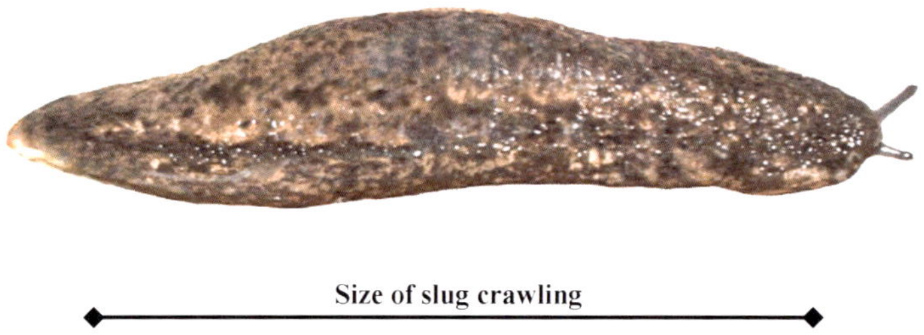

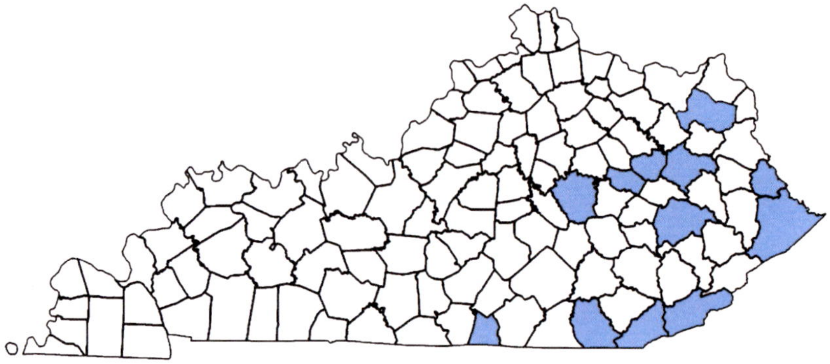

Philomycus togatus

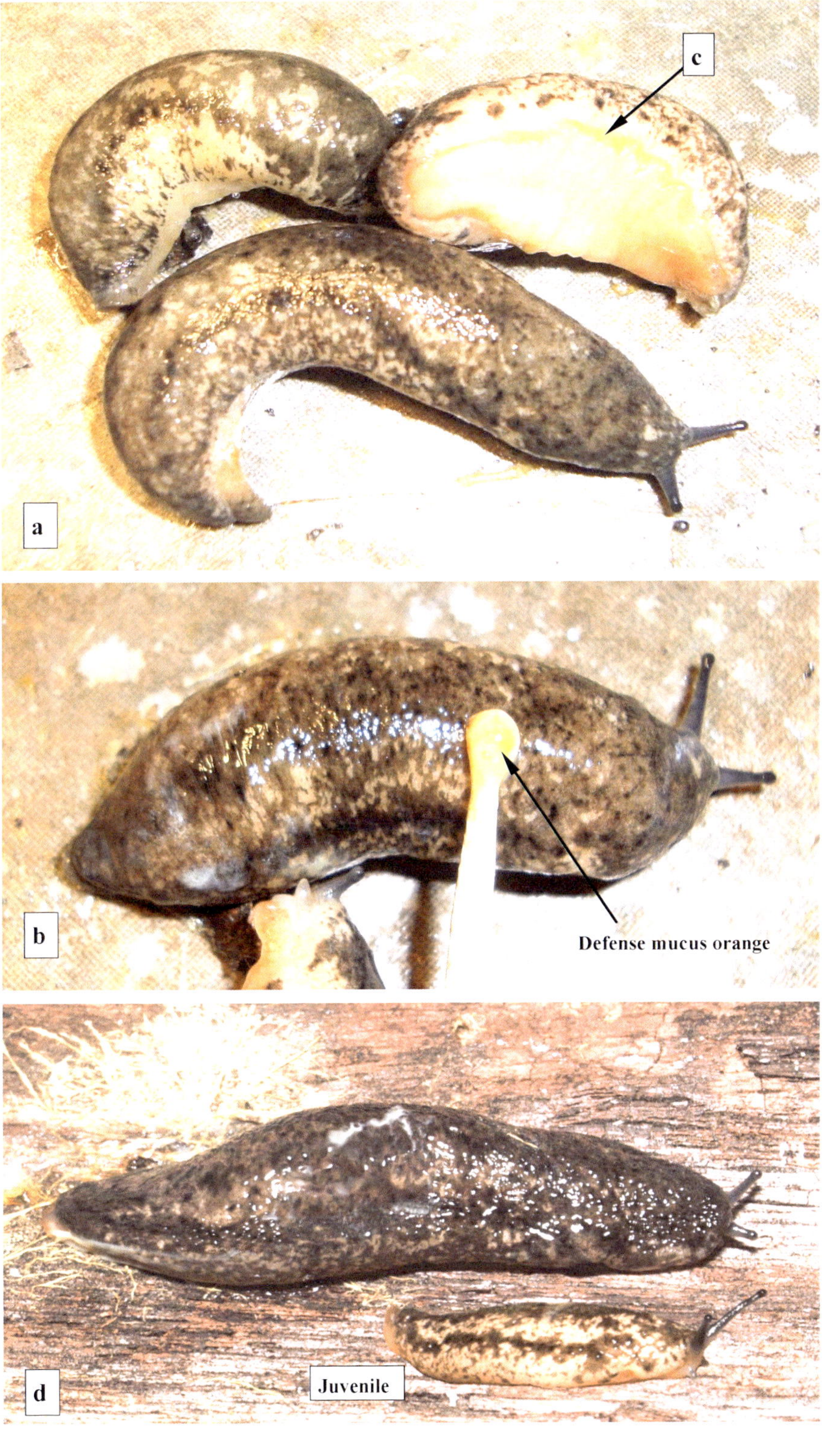

Brown-spotted Mantleslug **Philomycidae**

Philomycus venustus Hubricht, 1953

Length: Adults 75-120 mm while crawling

Description: The color pattern varies from individuals having two contrasting dark chestnut brown to black dorsal bands, a narrow lateral band on each side, connected to the dorsal bands by a series of oblique bands or spots (figures a and c, page 352), forming chevrons; to individuals in which this pattern is broken up into a series of spots (figure d, page 352); defense mucus white; sole of foot creamy white (figure b, page 352).

Similar Species: *Philomycus virginicus* is darker brown in color and has lighter longitudinal and oblique bands; *P. carolinianus* does not have the oblique bands or spots (forming chevrons) seen in *P. venustus*.

Habitat: An upland species (up to 1800 m on Roan Mountain in North Carolina) found on wooded hillsides and mountains; during the day it can be found hiding under the exfoliating bark of large rotting logs in advanced stages of decay; also found in crevices and cavities of beech trees.

Status: G4/S3; Uncommon, in Kentucky known mostly from Pine and Black Mountains.

Specimen: Figure (a) from Forney Creek, GSMNP, North Carolina; figures (b and c) from Pike County, Kentucky and figure (d) from Mitchell County, Roan Mountain, North Carolina.

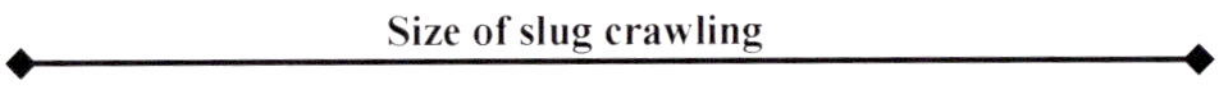

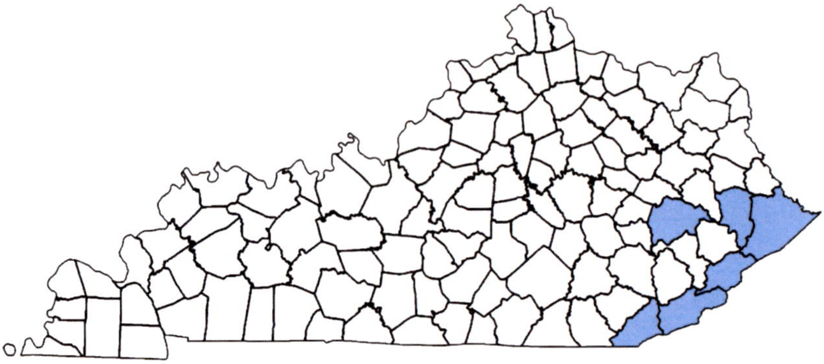

Philomycus venustus

Virginia Mantleslug **Philomycidae**

Philomycus virginicus (Hubricht, 1953)

Length: Adults 75-100 mm while crawling

Description: Base color of mantle is a grayish-white; color pattern consisting of a darker (brownish) broad dorsal band, sometimes bordered by a row of darker spots and a narrow lateral band on each side, connected to the dorsal band by a series of faint oblique (diagonal) stripes (figure d, page 354), the whole pattern obscured by a general fine flecking (figures a-c, page 354); young slugs with the same pattern, but brownish-gray, becoming chestnut-brown with age (Hubricht 1953); defense mucus whitish; foot margin off-white; sole of foot gray to whitish.

Similar Species: *Philomycus venustus* has three darker longitudinal bands often times broken into spots which are also connected by darker oblique bands or spots, all of which are notably darker on a usually lighter colored animal, than is seen in *P. virginicus*.

Habitat: An upland species found in mixed hardwood under the exfoliating bark of large rotting hardwood logs in advanced stages of decay.

Status: G3/S2; Rare; this rare slug is reported from Pine and Black Mountains in extreme southeastern Kentucky.

Specimen: Figure (a & b) from Coopers Rock State Forest, Monongalia County, West Virginia (author's photo collection) and figure (c) is from Lutherock, Avery County, North Carolina (Wayne Van Devender).

Size of slug crawling

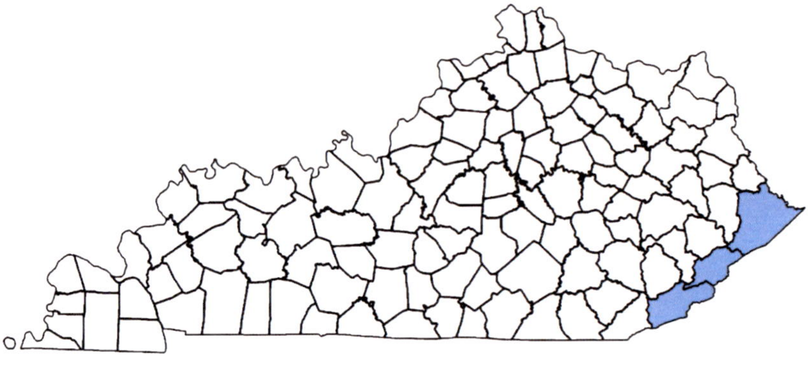

Philomycus virginicus

Breaks Mantleslug — Philomycidae

Philomycus bisdosus Branson, 1968

Length: Adults 50-85 mm while crawling

Description: Ground color of mantle pale-tan; starting mid-dorsally a single or pair of closely approximated, longitudinal, brown wavy bands and on each side by a thinner, wavy band of the same color; posteriorly, the dorsal bands fuse into a single one; anteriorly all 4 bands are broken into small, brown spots; no oblique bands connecting the longitudinal bands; defense mucus whitish; margins of the foot gray (a); foot nearly smooth, white except for a very short black area at extreme posterior end; head is white except for blue-black tentacles (Branson, 1968); picture to the right appears to be a rare white form of *P. bisdosus* (Breaks Interstate Park, Virginia side along Camp Branch).

Similar Species: Based upon external appearance and size, *P. bisdosus* is most like *P. togatus* (Fairbanks 1989) however the foot margin of *P. bisdosus* is gray (a), whereas the foot margin of *P. togatus* is red/orange (b); *P. bisdosus* differ from *P. venustus* by lacking the transverse oblique bands or chevrons and is a smaller slug; *P. flexuolaris* has a pale yellow defense mucus not whitish as in *P. bisdosus*.

Habitat: A species of mixed hardwood forests above the Big Sandy River.

Status: G1/S1; State Threatened; In 1974, Hubricht synonymized *P. bisdosus* with *P. venustus,* considering that the latter was a variable species and that *P. bisdosus* represented one extreme of the variation, but without supporting data. In 1998, Fairbanks re-established *P. bisdosus* as a distinct species, based on sound anatomical work; a rare and local endemic, currently known only from Breaks Interstate Park, at the Kentucky/Virginia state border.

Specimen: Both images from Breaks Interstate Park, Pike County, Kentucky (author's photo collection).

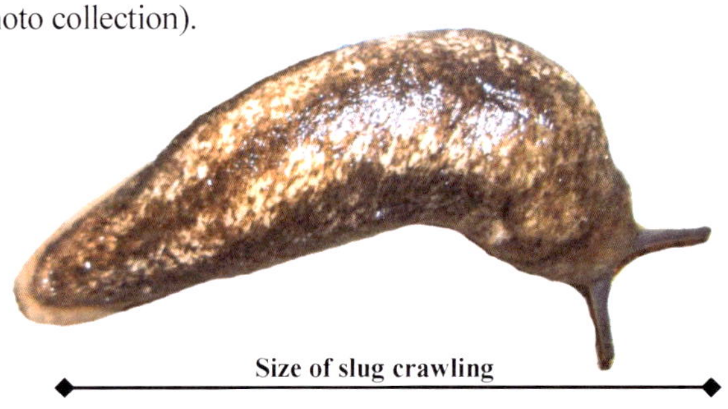

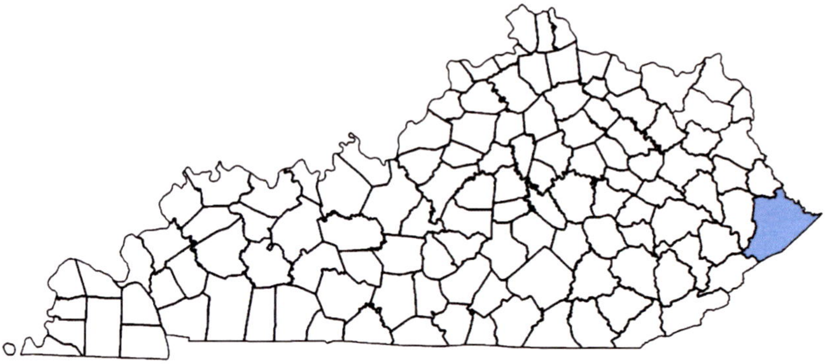

Philomycus bisdosus

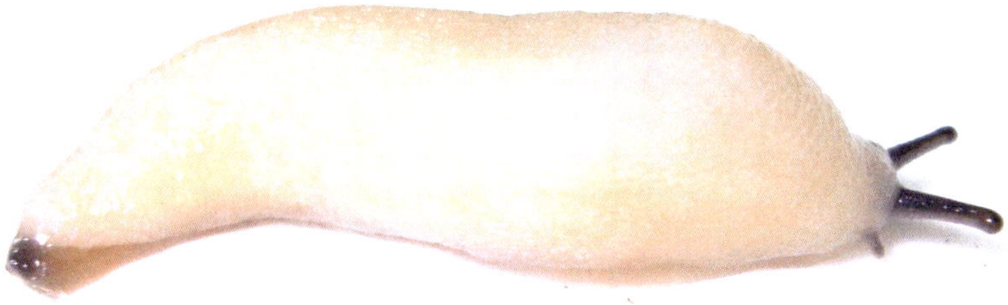

A rare white color morph, *Philomycus bisdosus* (?)

Kentucky Mantleslug **Philomycidae**

Philomycus batchi Branson, 1968

Length: Adults 50-80 mm while crawling

Description: A variable species; the mantle is usually a dark taupe to black but lighter individuals are not uncommon; the lower edges of the mantle are dead-white laterally and anteriorly (c), this feature alone appears to be the best defining external character for the species; occasionally with white blotches dorsally; the light, lateral bands (if present) are more easily seen on pale specimens (d); two *P. batchi* courting showing different color morphs, Red River Gorge (e); tail below mantle yellowish (f), marked on either side by 3 rows of black dashes; defense mucus pale yellow/orange (g); sole of foot white to light -peach.

Similar Species: Differ from other *Philomycus* by largely lacking strong longitudinal stripes, being more uniform in color, having no distinct mantle patterns.

Habitat: Found in floodplains, hillsides and in ravines, under decaying wood and loose bark, especially beech trees.

Status: G3/S3; Uncommon; Although Hubricht (1974) considered *P.* batchi nothing more than a melanistic color form of *P. carolinianus* (without supporting data), Fairbanks (1998) re-established *P. batchi* as a distinct species, based on sound anatomical work; possibly endemic to Kentucky.

Specimen: All images except figure (d) are from Raven Run Nature Sanctuary, Fayette County, Kentucky (author's photo collection).

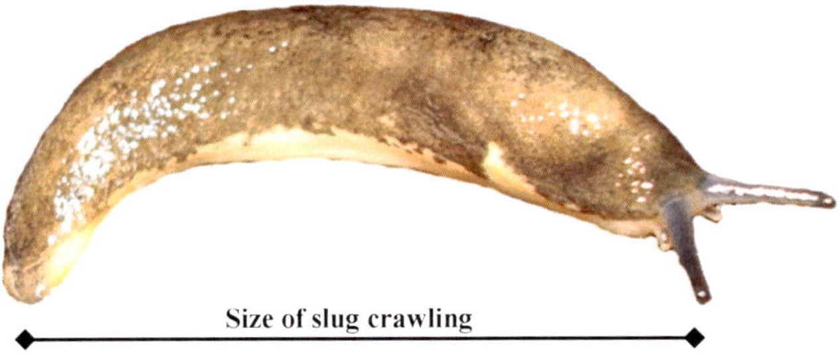

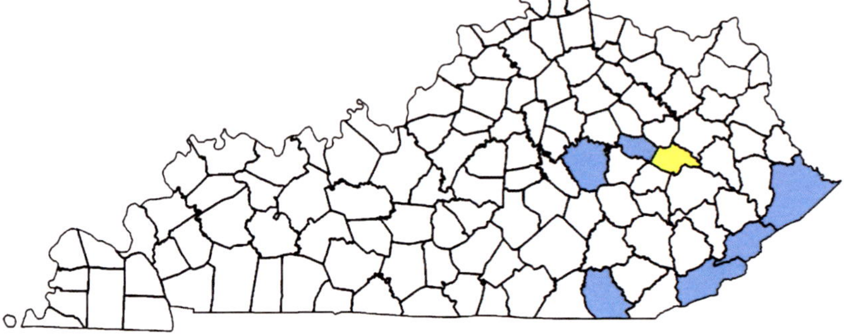

Philomycus batchi

P. batchi courting, Red River Gorge

Changeable Mantleslug — Philomycidae

Megapallifera mutabilis (Hubricht, 1951)

Length: Adults 60 mm while crawling

Description: Mantle color is fawn or tan, not covering entire head; but this feature may be difficult to discern, requiring an observer to rely on the use of a hand lens of at least 10X power; mantle usually has two (on occasions three) interrupted dorsal dark gray-brown longitudinal pigment bands (more often spots (figures a-c page 360) and a continuous wavy lateral band on each side (figure d page 360), small scattered pigment spots between the bands; spots sometimes producing vague chevrons on the mantle; defense mucus tannish-white or white (figure b, page 360); margins of the foot olive, gray or sometimes whitish; sole of foot cream.

Similar Species: *Philomycus* species are slightly larger in length and notably more robust when crawling and the mantle covers the entire body.

Habitat: Found in upland woods; during warm wet weather at night, it can be found crawling on the trunks of smooth bark trees like the American Beech; also a species of urban settings.

Status: G5/S5; this species is wide-ranging across the eastern US; in Kentucky reported from eastern and southern counties of the state; in Whitesburg, Kentucky it has become a common urban species seen crawling on sidewalks and retaining walls on warm rainy nights.

Specimen: Figure (a) crawling on limestone cliff face, above Greenbrier River, Greenbrier County, West Virginia; figure (b) defense mucus on a stick, Rowan County, Kentucky, figure (c) crawling on sandstone cliff, Monongalia County, Coopers Rock State Forest, West Virginia (author's photo collection).

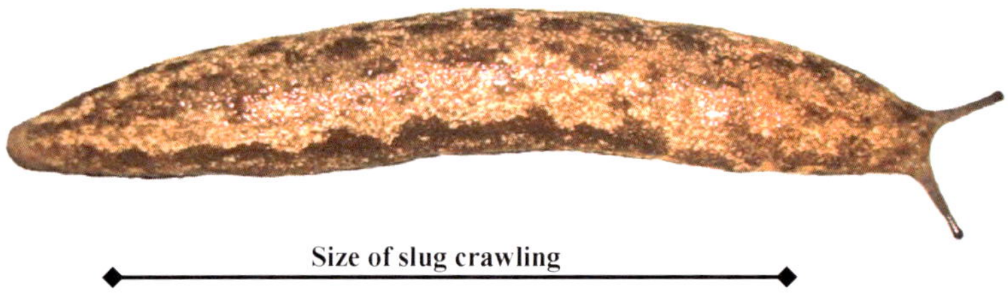

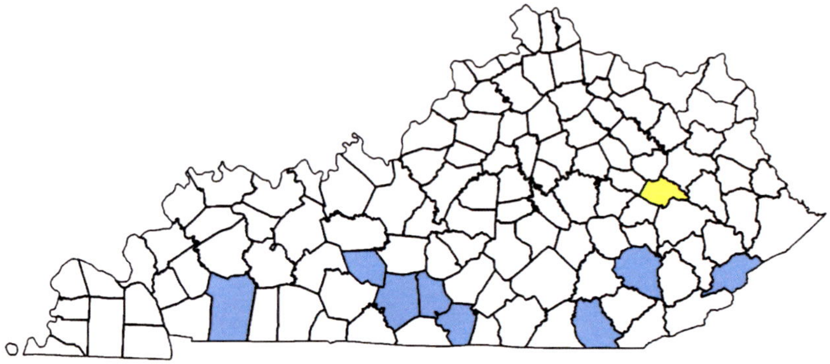

Megapallifera mutabilis

Blotchy Mantleslug **Philomycidae**

Megapallifera wetherbyi (W. G. Binney, 1874)

Length: Adults around 75 mm while crawling

Description; Mantle color is gray, white to cream; does not cover entire head, (as that of *Philomycus* species); mantle with a wide, uninterrupted, dorsal band running down the slug's back and broken, thinner lateral bands or spots, connected by broad, sometimes random chevron shaped bands or spots; defense mucus whitish; foot olive-gray or white.

Similar Species: *Megapallifera mutabilis* is only slightly smaller and has spots between the longitudinal bands instead of oblique connecting bands of color.

Habitat: This is a species of river bluffs and ravines. At night during warm wet weather, it can be found crawling on the trunks of smooth bark trees like the American Beech (Hubricht 1985); it appears to prefer areas in which sandstone outcrops and cliffs are common (Fairbanks 1990).

Status: G2/S2; Rare to uncommon slug is reported from only five southeastern counties in Kentucky and is often found in company with *P. flexuolaris* and *P. bisdosus*.

Specimen: Figures (a and b) from Kentucky, Pike County; Kentucky (author's photo collection), figure (c and d) ©Jason M. Crockwell – .

a

Size of slug crawling

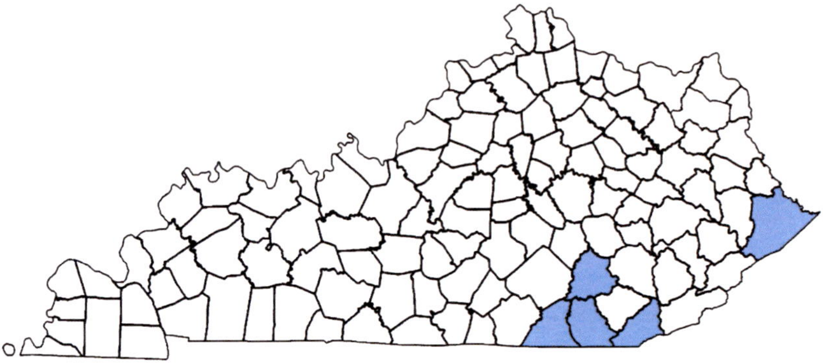

Megapallifera wetherbyi

Pale Mantleslug **Philomycidae**

Pallifera dorsalis (A. Binney, 1842)

Length: Adults 20-30 mm while crawling

Description: The live animal is ashy blue, gray or brownish with or without an interrupted black line down the center of the mantle (in my experience, more often without this feature); defense mucus translucent amber to whitish; margins and sole of the foot are white (figure c, page 364); adult slug shown in a typical defensive posture (figure b, page 364).

Similar Species: *Pallifera fosteri* has distinct black spots, which are more prominent at the mantle edges, its mantle slightly humped in front making the neck and head longer than seen in *P. dorsalis*; *P. ohioensis* has distinct red edges running along the sole of its foot (figure d, page 364).

Habitat: A slug of humid forests, generally found in deep, moist leaf litter in and around rock talus or large piles of rotting wood; like other *Pallifera,* very secretive, rarely seen crawling about on the surface.

Status: G5/S3; Uncommon, only reported from scattered locations in eastern Kentucky but likely more common and widespread than current records indicate.

Specimen: Figure (a, b & c) from Buncombe County, North Carolina); figure (d) comparing *P. dorsalis,* from Powell Co, Kentucky and *P. ohioensis* from Blackwater Falls State Park, West Virginia.

Size of slug crawling

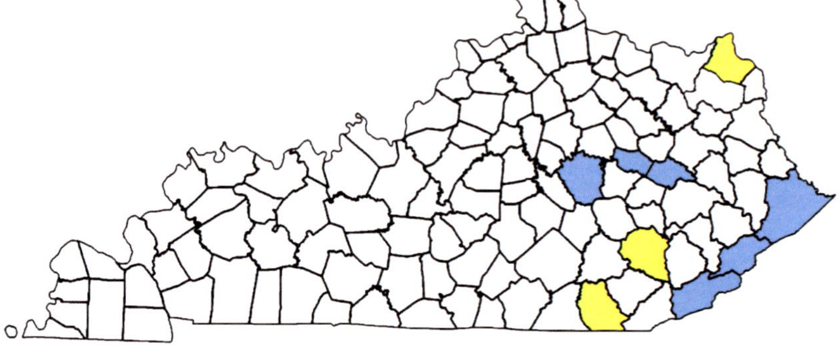

Pallifera dorsalis

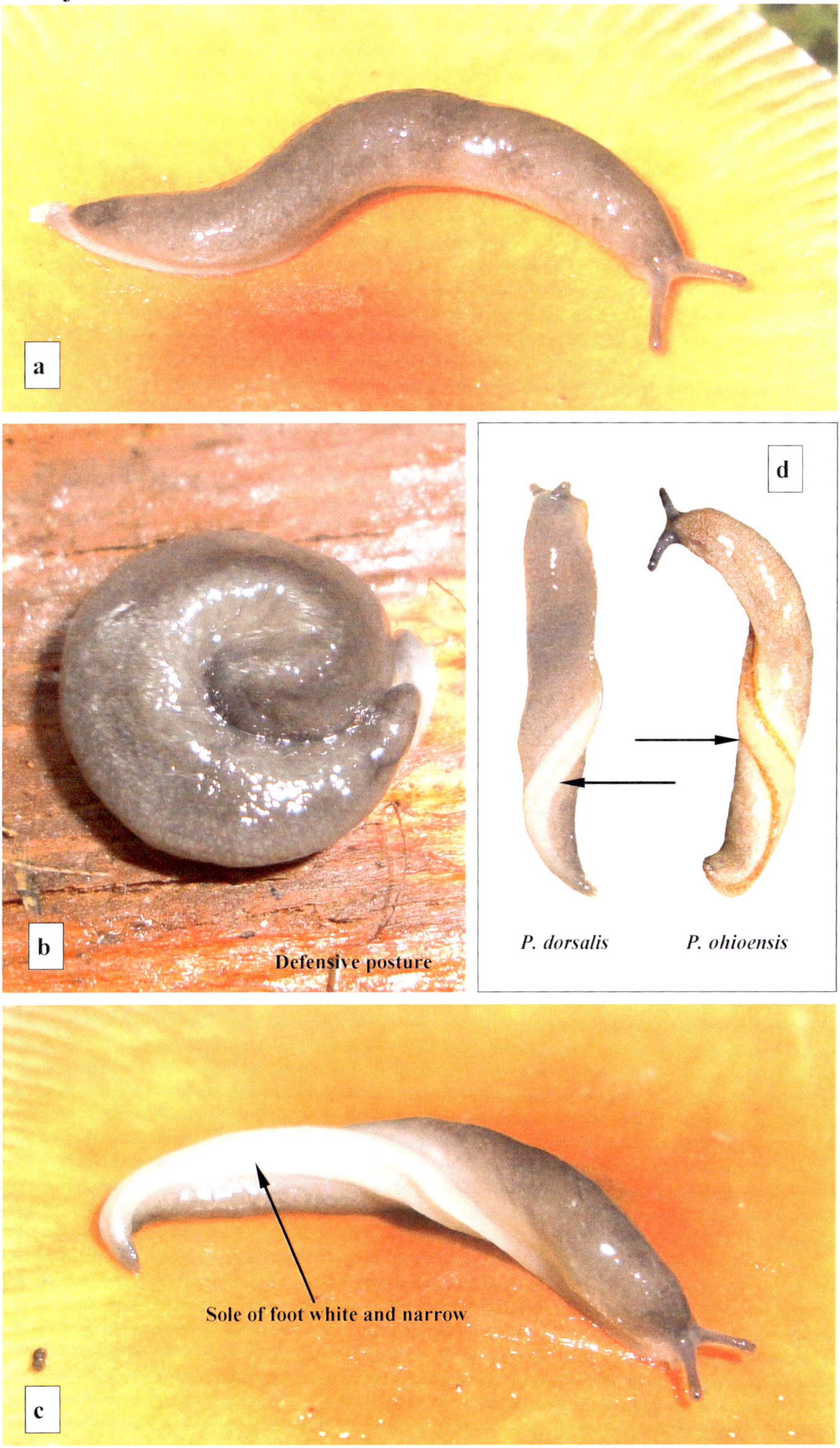

Redfoot Mantleslug — Philomycidae

Pallifera ohioensis (Sterki, 1908)

Length: Adults 15-30 mm while crawling

Description: The live animal has a blue-speckled grayish back with or without an interrupted black line down the center of the mantle; sole of foot with two red to orange bands running lengthwise along the edge (figure c, page 366); defense mucus whitish.

Similar Species: *Pallifera dorsalis* is around the same size, but without the red bands, and on occasion, an interrupted set of black spots running down the center of its back.

Habitat: A slug of humid forests, usually found under bark, deep leaf litter under decaying logs or in rock talus; in Kentucky it is only known in upper elevation red spruce forests around Blackwater Falls State Park.

Status: G3/S1; Rare; this species is not recognized by some authors and it may only represent an interesting race of *P. dorsalis*; *P. ohioensis* is also reported from Ohio, Indiana, and North Carolina (Dourson 2013).

Specimen: Figures a and b from Kentucky, Russell County, Route 1058 just past Creelsboro, above creek flowing into cave; GPS location given for follow up investigations on this dubious species (36.89184, -85.203810).

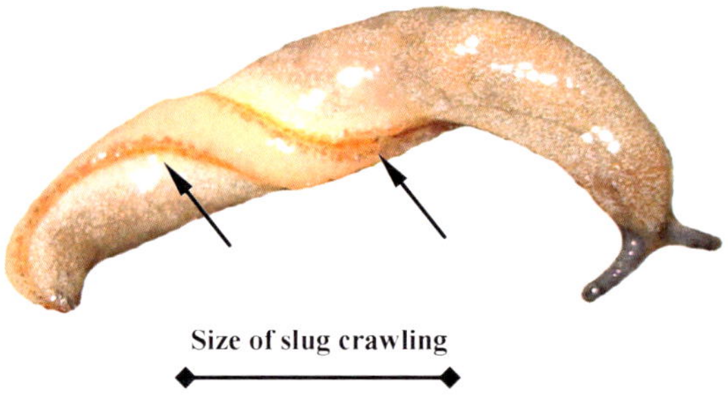

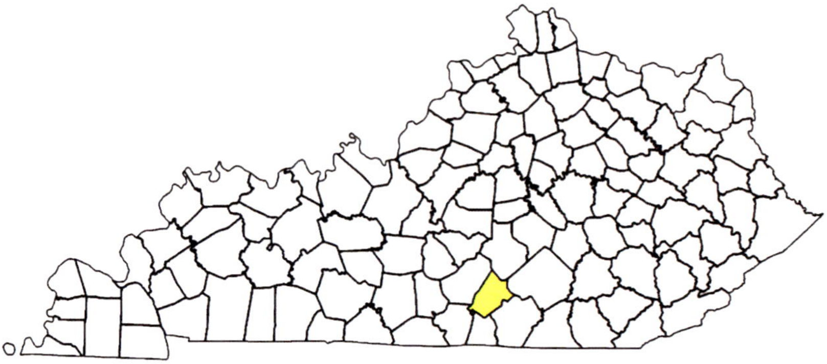

Pallifera ohioensis

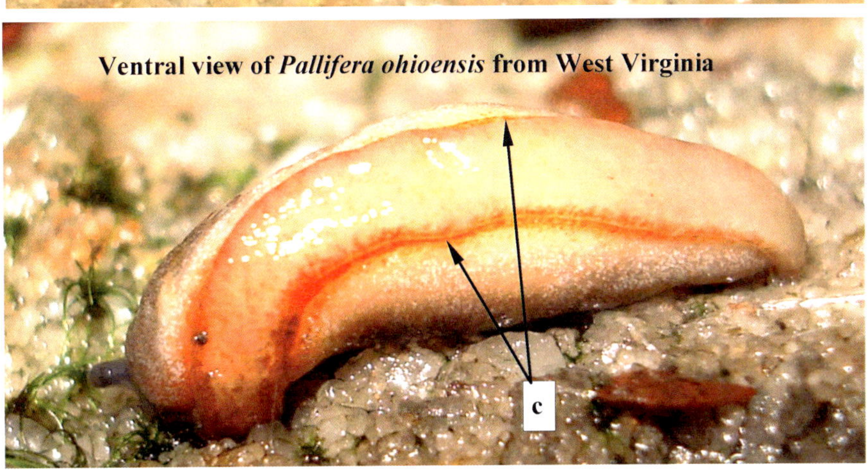

Foster Mantleslug **Philomycidae**

Pallifera fosteri F. C. Baker, 1939

Length: Adults15-25 mm while crawling

Description: The species may be recognized by the blackish spots on a whitish or flesh colored mantle, these spots or blotches often form interrupted, irregular longitudinal lines, especially near the base of the mantle, or irregularly spaced clusters of small dot like spots scattered over the dorsal surface; in some specimens the black spots form coalescing blotches elongated in form (Pilsbry 1948); Grimm (1961) reported that the mantle is light tan, spotted and reticulate with dark brownish gray, reticulations heaviest in the middle of the back (figure d, page 368), but no solid, dorsal line is formed; at the sides, the reticulations often form two broken lines; tentacles are slate gray and the anterior margin of the foot is brownish red (figure b, page 368 three white arrows); other characters include a slightly humped mantle (e); defense mucus transparent amber; in juvenile slugs (below image) the tentacles are proportionally large for their body size, but mantle patterns are like that of adults (this is true for most slug species).

Similar Species: *Pallifera dorsalis* is smaller, is a different color, and most importantly, is without the spots on the mantle edges.

Habitat: A species found in a variety of habitats from floodplains to mountain tops under forest litter and around rotting logs in advanced stages of decay.

Status: G5/S4; Uncommon; this secretive slug is reported from central and eastern Kentucky counties.

Specimen: Figures (a, b and c) from North Carolina, Swain County, Nantahala Gorge (all images © Wayne Van Devender)

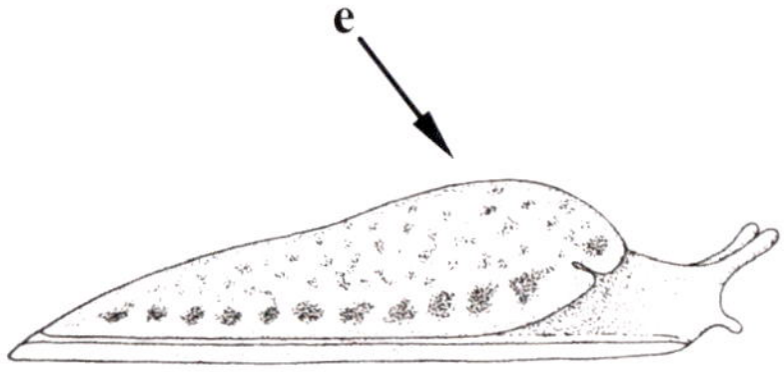

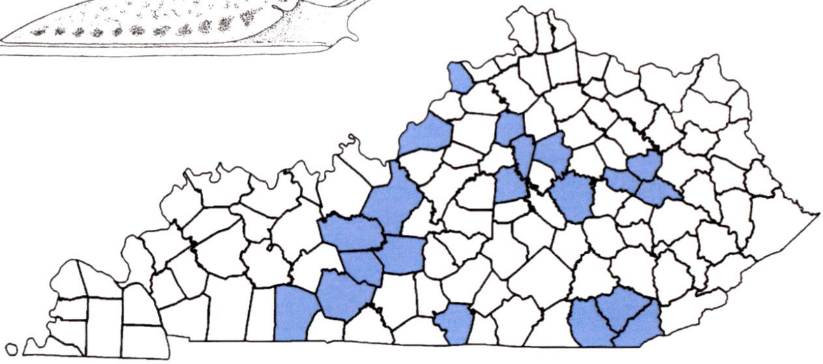

Pallifera fosteri

Severed Mantleslug **Philomycidae**

Pallifera secreta (Cockerell, 1900)

Length: Adults 20-30 mm while crawling

Description: Mantle very dark gray or gray blue with numerous and well scattered small, round, black spots, often more profuse at the mantle edges near the foot margins and on the anterior end of the mantle (figures a-c, page 370); sole whitish with an ochreous (yellowish) tint (Cockerell 1900); defense mucus whitish.

Similar Species: *Pallifera dorsalis* is smaller, with or without an interrupted black line down the center of the mantle; *P. fosteri* is smaller, flesh colored, with interrupted clusters of spots down the center (dorsal) of the mantle and the anterior margins of its foot is brownish red.

Habitat: This slug lives deep down in drifts of damp leaves next to logs, leaf-filled depressions, and rock talus in upland mixed hardwood forests (up to 1500 m in the southern Appalachians); the species reported to rarely come to the surface (Pilsbry 1948).

Status: G4/S4; Uncommon, like other species of *Pallifera,* these slugs are secretive and remain well-hidden, as a result, few records exist for this small slug in Kentucky.

Specimen: Figures (a and b) from North Carolina, Transylvania County, Rosman (©John Slapcinsky) and figure (c) from North Carolina, Mitchell County around Carvers Gap, Roan Mountain (author photo collection).

Size of slug crawling

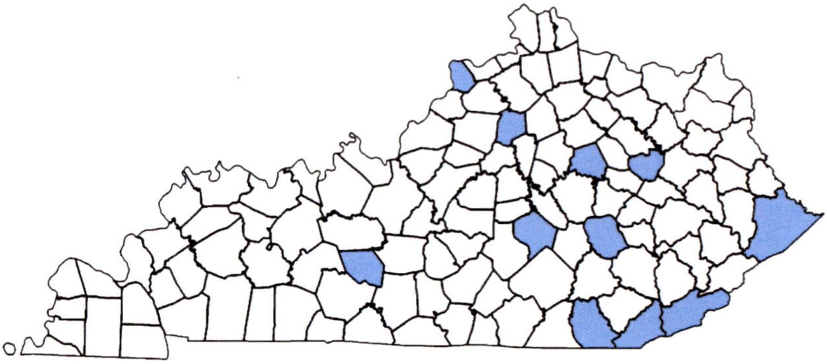

Pallifera secreta

Meadow Slug **Agriolimacidae**

Deroceras laeve (Müller, 1774)

Length: Adults 25 mm while crawling

Description: Color of animal varies from yellowish to pale gray, dark gray to nearly black, sometimes flecked with gray; the mantle (figure b, page 371) is situated on the anterior (front) end of the dorsal surface (back); pneumostome (breathing hole) behind middle of mantle (as opposed to those of *Arion* species, which are in front of the middle of mantle); defense mucus colorless and watery; margins of the foot tan; this slug is a favorite food of the brown snake, *Storeria dekayi* (MacGregor pers. comm. 2010).

Similar Species: Exotic *Arion* species are similar (see page 371); native slugs of the eastern US have mantles that cover or nearly cover the entire animal back.

Habitat: A Holarctic species often found under cardboard and rubbish along roadsides and in vacant lots, also found in meadows and natural glades; even though *D. laeve* is considered native to eastern North America, it has become somewhat of a garden pest.

Status: G5/S5; Common; reported from scattered locations across Kentucky but like most native slugs it has been under reported.

Specimen: Figures (a, c and d) from Madison County, Kentucky (all images © Wayne Van Devender)

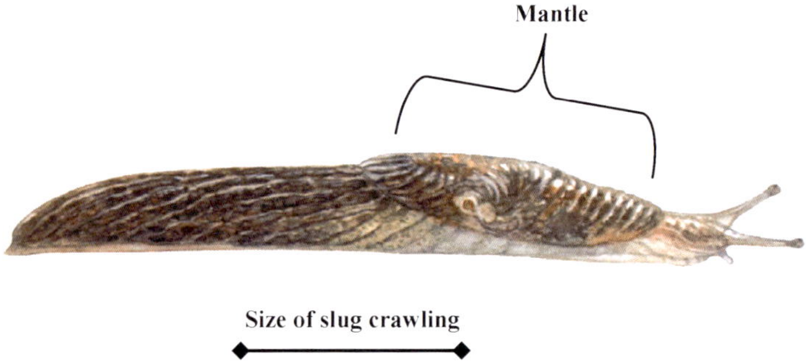

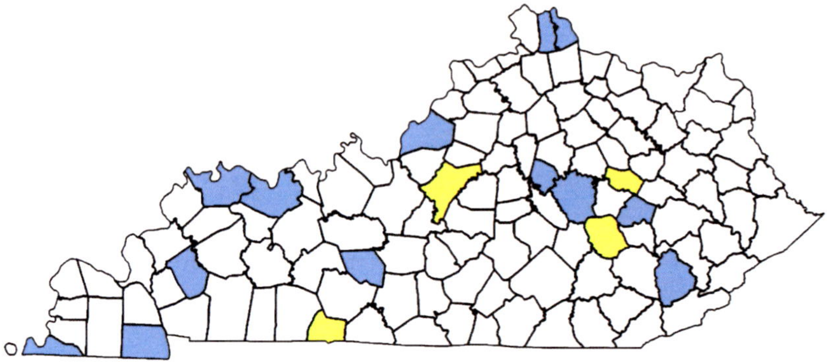

Deroceras laeve

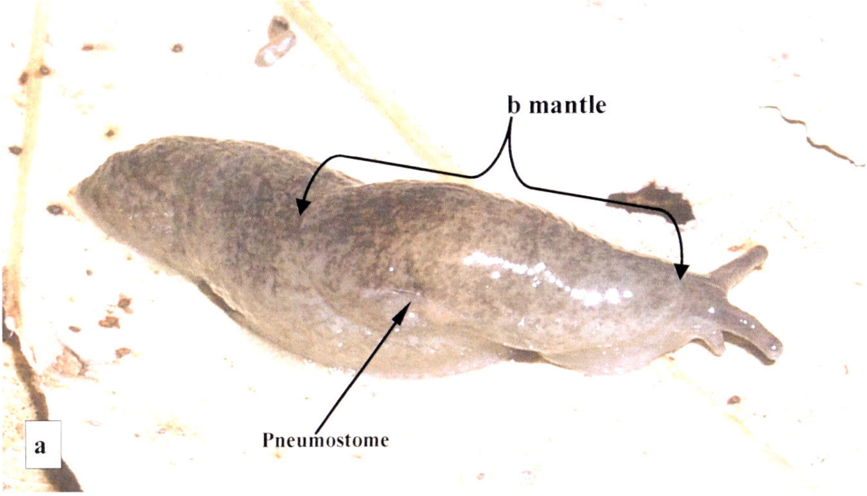

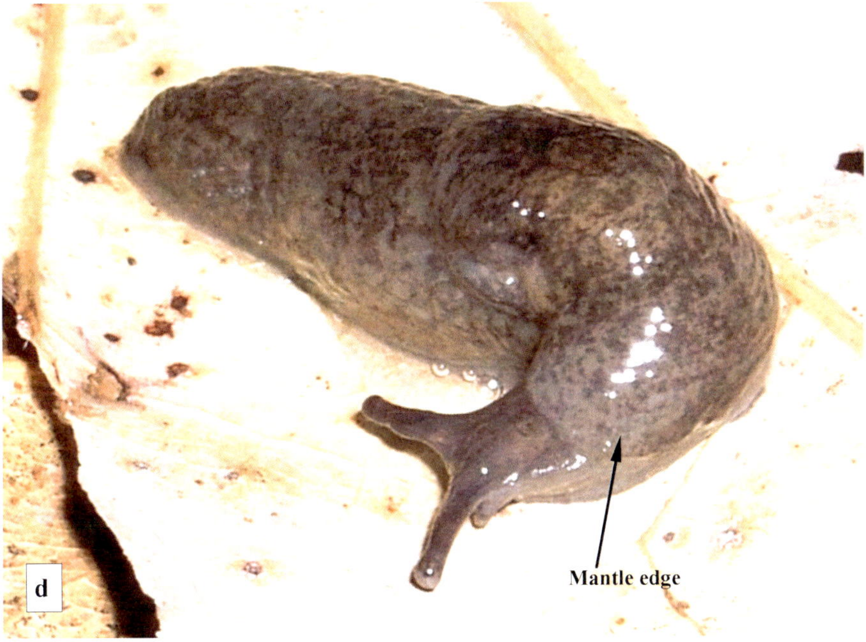

Unknown species of ***Philomycus***, Pike County, Kentucky. This single specimen was discovered on a rotting hardwood log in advanced states of decay on a wooded hillside above Russell Fork. It is not known if the species is just an anomaly of a common and known species or perhaps a new species of slug. Additional specimens will be needed to make a final determination.

Part II
Pleistocene (Fossil) Land Snails of Kentucky

Pleistocene glaciatation (also known as quaternary glaciation) is an alternating series of glacial and interglacial periods. The Quatenary periods began 2.5 million years ago. Since the Earth still has ice sheets, geologists consider the Quaternary glaciation to be ongoing with the Earth now experiencing an interglacial period. During glacial periods, the ice expanded, and during interglacial periods the ice contracted. The only surviving ice sheets from the Pleistocene epoch are the Antarctic and Greenland Ice sheets. Other ice sheets completely melted during the interglacials. The major effects of this expansion then retraction is the erosion of land and deposition of material (silt, sand) over large swaths of land, the modification of river systems, the creation of millions of lakes, changes in sea level, flooding, and abnormal winds. The soils and minerals left behind during the melting (interglacial) periods produced a sediment called loess. The deposition of these sediments along with climatic changes brought about by the cooling then warming of the planet produced effects that have shaped entire environments on land and in the oceans and their associated biological communities.

Road construction and development for cities and towns in the corridor along the Ohio River from near Cincinnati all the way to the Mississippi River have exposed loess deposits that contained both extinct and extant land snail species. Some of the species documented are still extant today while other species have only been documented by relics left behind from the last glacial period. Land snails now considered extinct in Kentucky are still extant in cooler northern regions. As we continue to attempt to understand the climate changes occurring today, scientists are now using land snail shells found in loess deposits to reconstruct previous glacial and interglacial periods.

Summary of Pleistocene Epoch (2.5 million to 1200 years) Land Snail Records from Loess Deposits in Jefferson County, Kentucky (1960) and Hubricht's Records (1985)

Xeric Ambersnail

Succineidae

Succinea indiana Pilsbry, 1905

Height: 10-11 mm, Diameter: 6-7 mm

Description: Shell obesely ovate, thin, brownish amber or raw sienna colored; apex is reddish. Sculpture of fine growth lines and wrinkles becoming rather coarse on last half whorl; 3.5 whorls; very convex with the last whorl very inflated. Aperture large. Lip simple. Columella thin,

Similar Species: Closely related to *S. campestris* of Florida and Georgia and *S. unicolor* of New Orleans. It differs in both color and texture.

Habitat: Generally found in rather dry, sunny situations on bare ground.

Status: G5/SX. All records in Kentucky are from Pleistocene fossil deposits along the Ohio River in Jefferson County. Extant populations are known from Posey County, Indiana (across the river from Henderson and Union Counties in Kentucky).

Specimen: Indiana, New Harmony, Type specimen (ANSP 12378).

Type Specimen

Apex of shell reddish

Type Specimen

Loess Ambersnail

Succineidae

Succinea bakeri Hubricht, 1963

Height: 13.1-15.3 **Diameter:** 6.7-7.0 mm

Description: Shell with a little over 3 whorls, thin but firm; elongate, ovate, sculpture of unevenly spaced growth lines and wrinkles; spire acute, moderately long; sutures well marked, periphery well-rounded.

Similar Species: The only other species that *S. bakeri* might be confused with is *S. ovalis pleistocenica,* but *S. bakeri* is smaller in size, more slender in form with less impressed sutures.

Habitat: Found in loess deposits of the upper Mississippi Valley, actual habitat unknown.

Status: GX/SX. All records in Kentucky are from Pleistocene fossil deposits along the Ohio River in Jefferson County. Extant populations are unknown, considered globally extinct.

Specimen: Illinois, St. Clair, Type Specimen (FMNH 116914).

Type Specimen

Type Specimen

Frigid Ambersnail

Succineidae

Mediappendix gelida (F.C. Baker, 1927)

Diameter: 3.3-4.5 mm, **Height**: 6.6-8.0 mm

Description: Shell small, elongated, rather narrow; 3.5 whorls, convex, separated by deep sutures, last whorl comparatively small, flat-sided, spire long, aperture rounded about half as long as the shell; columella straight curving into the parietal wall in a gentle curve; there is a slight callus which is spread over the parietal wall, sculpture rather fine, vertical striae.

Similar Species: This species is smaller, rounder, less elongated than *Mediappendix vermeta.* This species was once considered a form of *Succinea grosvenori.*

Habitat: Occurs in loess formations.

Status: GX/SX. All records in Kentucky are from Pleistocene fossil deposits along the Ohio River in Jefferson County. Extant populations are unknown. The species is presumed globally extinct.

Specimen: Illinois, Boone County, Irene, Paratype Specimen (ANSP 142712).

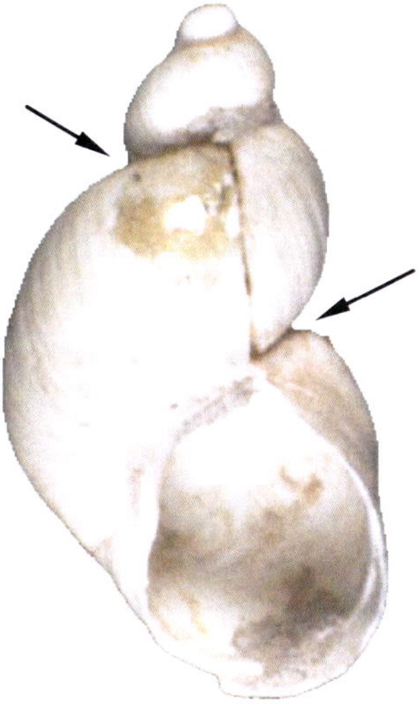

Paratype Specimen

Paratype Specimen

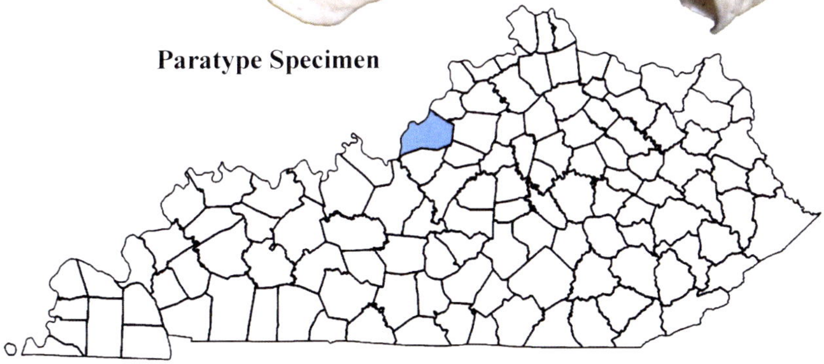

Blue Glass Snail

Oxychilidae

Perpolita binneyana (E. S. Morse, 1864)

Diameter: 3.5-4.3 mm, height 1.75 mm

Description: Heliciform; lip simple; shell with 3.5-4 whorls; umbilicate; no teeth present; transverse striae are closely spaced but poorly developed, missing on the base; no spiral striae present; shell nearly colorless or with a greenish tinge, translucent; live animal almost white with darker tentacles; periphery is rounded; deceased and dried animal clearly seen through bottom shell view.

Similar Species: Formerly known as *Nesovitrea binneyana*, this species resembles *P. electrina,* but is one third smaller and has a different color being nearly white with a greenish tinge; *G. wheatleyi* is less compact, larger, and has fine spiral striae.

Habitat: G5/SX. A species of upland mixed hardwood forests living among the leaf litter.

Status: All records in Kentucky are from Pleistocene fossil deposits along the Ohio and Mississippi Rivers. The species is extant in northern states including Minnesota, Michigan, Maine and Vermont.

Specimen: Michigan, Keweenaw County, Isle Royale NP (FMNH 58444).

Perpolita binneyana

Perpolita electrina

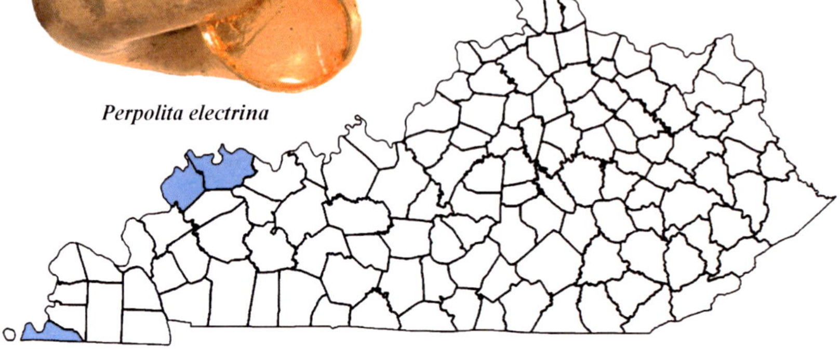

Amber Glass Snail

Oxychilidae

Perpolita electrina (Gould, 1841)

Diameter: 4.6-5.2 mm, height 2.5-2.8 mm

Description: Heliciform; lip simple; shell with 3.5-4.5 whorls; umbilicate; no teeth present; transverse striae are crowded but poorly developed, not reaching the base, which is generally smooth; without spiral striae or very rarely very weak and seen only as mere traces; shell faintly yellow or pale greenish and glossy, translucent; periphery is rounded, specimen shown here with an epiphragm (figures a and b); this is a snail's way of conserving moisture during dry periods; deceased and dried animal seen through the translucent shell in all three views.

Similar Species: Separation between *Glyphyalinia* and *Perpolita* can be an arduous task; the primary difference between the two genera is that *Perpolita* species are usually more compact in form; *P. electrina* most resembles *G. wheatleyi,* but is without the fine spiral striae.

Habitat: A species of low wet ground, margins of ponds and marshes and floodplains of rivers.

Status: G5/SX. All records in Kentucky are from Pleistocene fossil deposits along the Ohio and Mississippi Rivers. Extant populations however occur along Kentucky's border with West Virginia and southern Ohio.

Specimen: New York, Schoharie County, 0.7 miles E of Sharon, roadside (FMNH 240400).

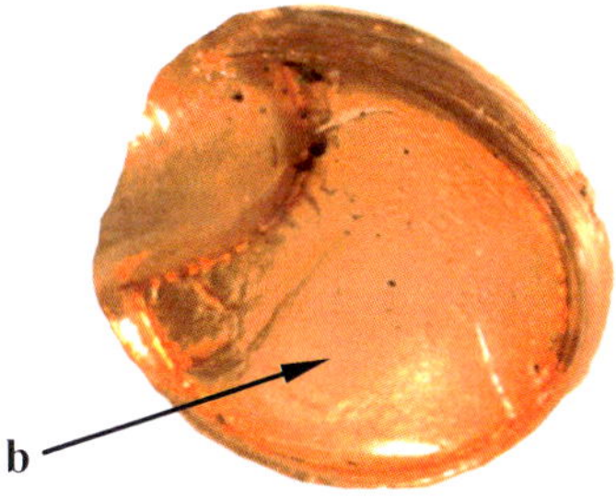

Aperture with an epiphragm

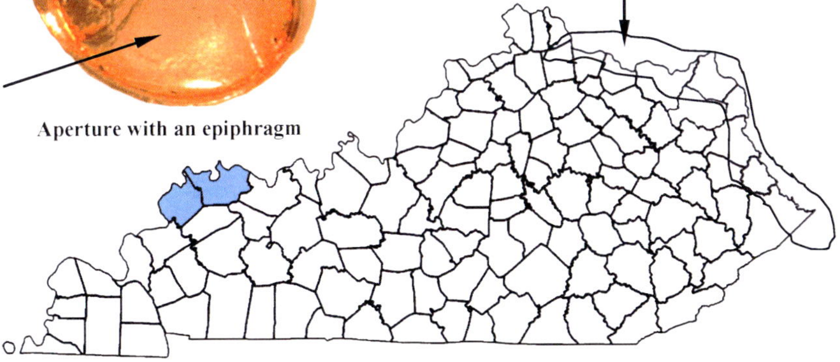

Trumpet Vallonia

Vallonia parvula Sterki, 1893

Diameter: 1.6-2 mm, height 1.0 mm

Description: Depressed heliciform; lip reflected, does not cross the body whorl (a & c) and strongly thickened within; shell with 3 loosely coiled whorls, transverse striae are more rib-like and paper thin (can be seen with a hand lens of 10X), the last whorl has between 30-38 ribs and is greatly expanded; widely umbilicate; horn or pale gray to nearly white; no teeth present; periphery roundish.

Similar Species: *Vallonia costata* averages fewer ribs (20-35 ribs on the last whorl) and is larger in diameter; *V. perspectiva* has a wider umbilicus, a notably thinner lip, and the aperture lip is continuous across the body whorl (b).

Habitat: A species of dry weed and grass openings.

Status: G4/SX. In Kentucky, this species is known only from fossil records found in river deposits along the Ohio River; it is presumed extinct in the state, however extant populations are reported from northern Ohio and Illinois west to Texas.

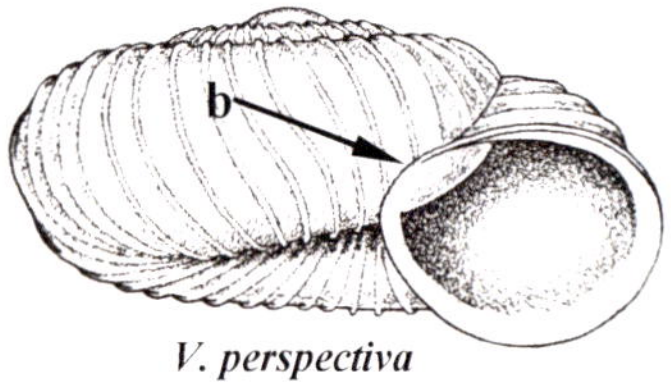

V. perspectiva

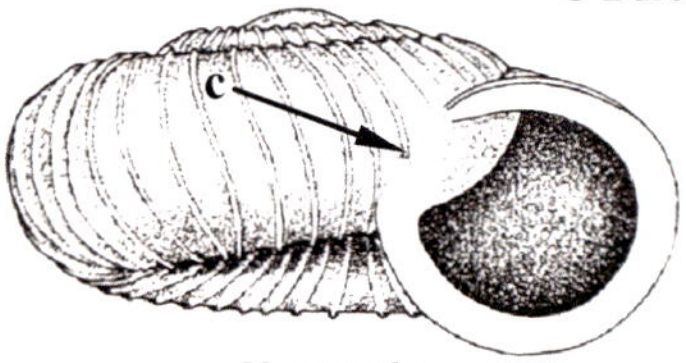

V. parvula

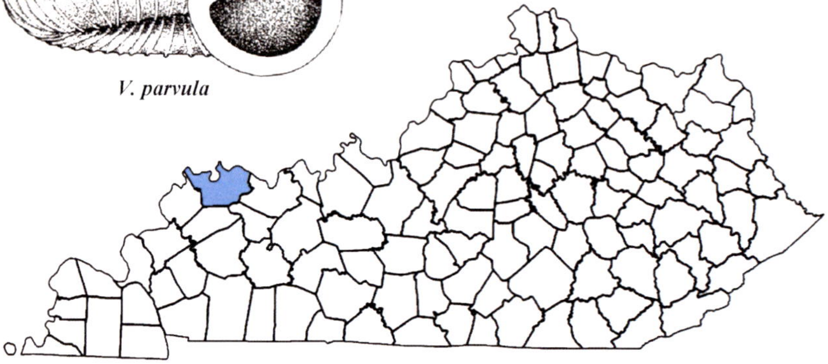

Multirib Vallonia

Vallonia gracilicosta Reinhardt,1883

Diameter: 2.5-3 mm, height 1 mm

Description: Depressed heliciform; lip reflected and thickened; shell with 3.5 loosely coiled whorls, the last whorl greatly expanded; widely umbilicate; pale tan, corneous, or whitish-gray; no teeth present; shell surface with 45 -60 paper-thin ribs in the last whorl (see page 218 for counting ribs); periphery is well rounded.

Similar Species: *Vallonia costata* is slightly smaller and has only 20-35 ribs on the last whorl.

Habitat: Reported from river drift (Hubricht 1985).

Status: G5/SX. Presumed extinct in the state of Kentucky, only known from fossil records found in river deposits along the Ohio River; there are, however, extant populations reported from North and South Dakota, New York and Maine.

Specimen: Robert Forsyth photo collection.

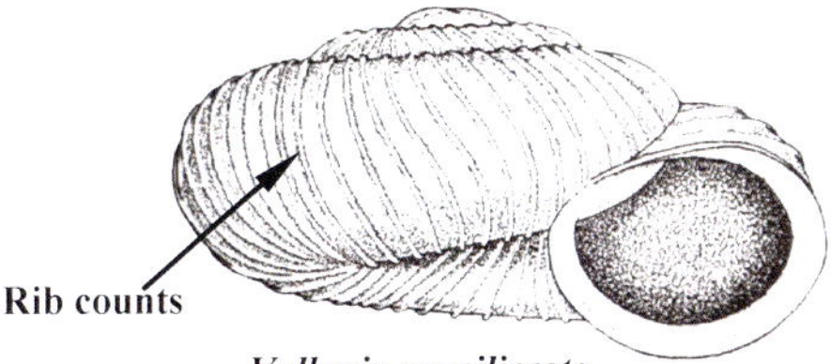

Vallonia gracilicosta

© Burch, 1962

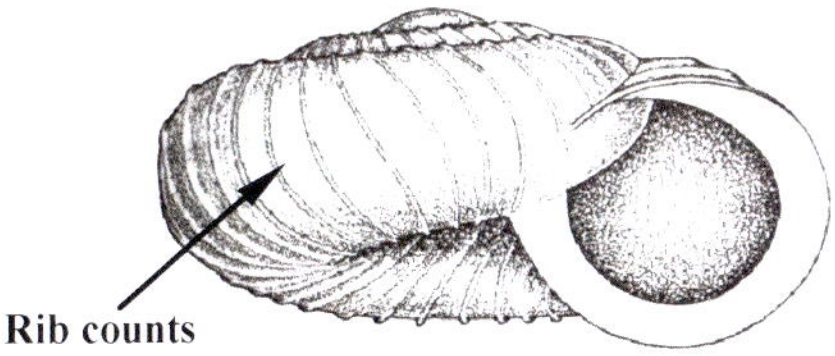

Vallonia costata

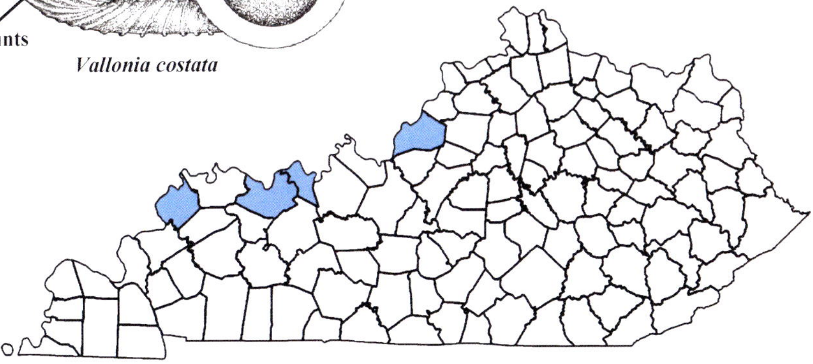

Temperate Coil

Helicodiscus shimeki Hubricht, 1962

Diameter: 4.2 mm, height 1 mm

Description: Depressed heliciform-discoidal; lip simple; shell with 5-6 well rounded whorls; fresh shells somewhat shiny and translucent; widely umbilicate; embryonic whorl smooth; in the last whorl there are usually three pairs of small conical teeth on the outer and basal walls (seen in bottom view of juvenile shells below, figures a and b); spiral striae clearly present as in most *Helicodiscus* species on all whorls (dissecting scope required).

Similar Species: Both *Helicodiscus parallelus* and *H. notius* have a narrower and deeper umbilicus.

Habitat: A land snail found in upland mixed hardwood forests living among the leaf litter on acid soils, also in low places and in rock talus.

Status: G4/SX. All records in Kentucky are from Pleistocene fossil deposits along the Ohio and Mississippi Rivers. Extant populations are known in West Virginia north to Maine.

Specimen: West Virginia, Monongalia County, Snake Hill WMA (author's collection).

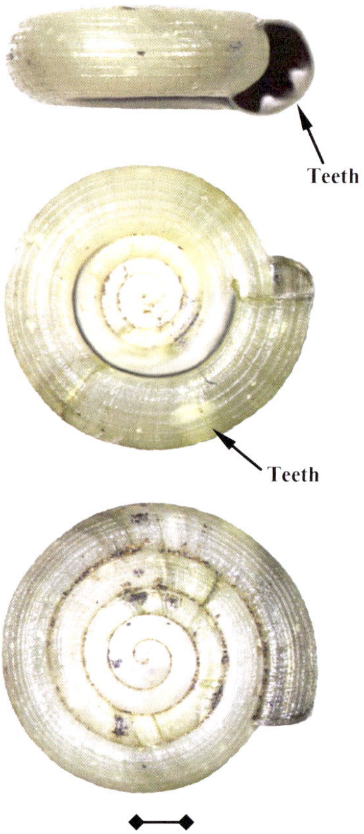

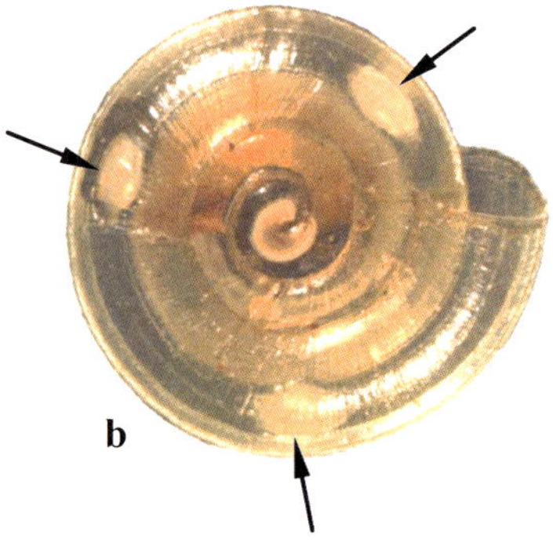

b

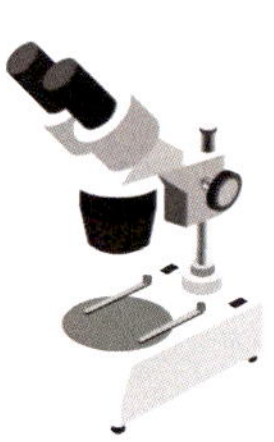

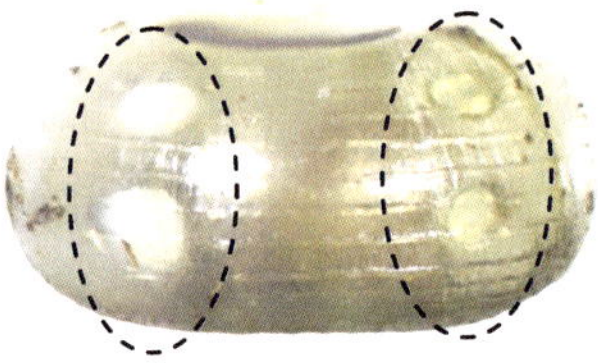

a

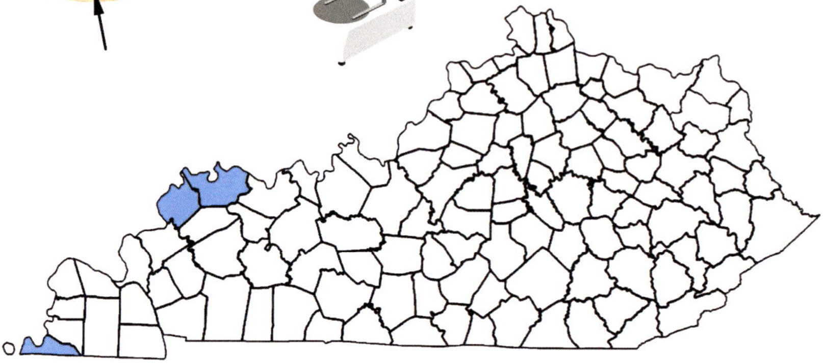

Angular Disc

Discus catskillensis (Pilsbry, 1896)

Diameter: 5 mm, height 2.5 mm

Description: Depressed heliciform; lip simple; shell thin with 4 whorls; umbilicate; no teeth present; transverse striae are well-developed and are rib-like (a) and are distinct on the entire shell including the umbilicus region; periphery bluntly angular.

Similar Species: *Discus patulus* is larger, has a small rounded tubercle (figure a, page 173) and generally a wider umbilicus; *D. whitneyi* is larger and has a more rounded aperture; *D. nigrimontanus* (page 174) has a smaller aperture, is flatter in profile and has an angular periphery (not bluntly angular as in *D. catskillensis*).

Habitat: A species of upper slopes in mixed hardwood under leaf litter and rotting hardwood logs in advanced stages of decay. In New York, it often occurs in low ground in company with *D. whitneyi.*

Status: G5/SX. All records in Kentucky are from Pleistocene fossil deposits along the Ohio and Mississippi Rivers.

Specimen: West Virginia, Tucker County, Canaan Valley (author's collection).

Characteristic rib sculpture of *Discus* shells

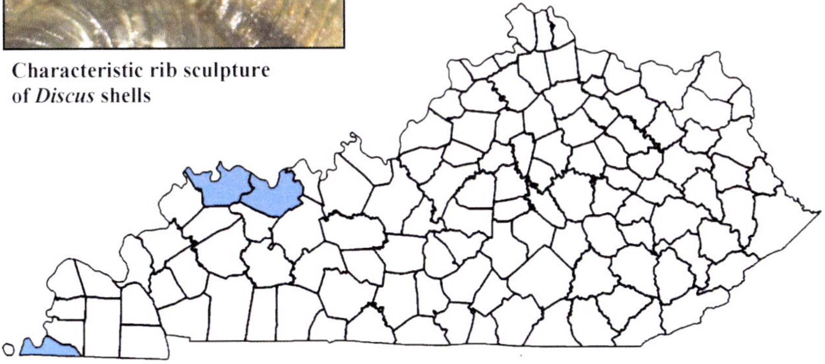

Lowland Pillsnail

Euchemotrema leaii leaii (A. Binney, 1840)

Diameter: 6.7-9.4 mm, height 3.9-5.7 mm

Description: Depressed Heliciform; lip reflected; shell with 5.5-6.5 closely coiled whorls; umbilicate (two variations illustrated); large straight parietal tooth that does not curve toward umbilicus (b); shell somewhat glossy; rust color minute hairs present in last whorl, but hairs are often lost in older shells; transverse striae poorly developed; shell periphery rounded; the internal Fulcrum is short and best observed in fresh shells (c).

Similar Species: Like *Stenotrema* species but without the basal notch and interdenticular sinus; *Euchemotrema fraternum* is larger and has a closed umbilicus, *E. leai aliciae* is around the same size but is more elevated in form and has a rimate to nearly closed umbilicus.

Habitat: A species of low wet areas, marshes; swamps and floodplains, but also found in meadows and along roadsides.

Status: G5/SX. All records in Kentucky are from Pleistocene fossil deposits along the Ohio and Mississippi Rivers.

Specimen: Figure (a) *Euchemotrema leai* from Michigan, Genesee County, Fentonville and figure (b) *Euchemotrema leai aliciae* from Louisiana, Calcasieu Parish (all images from Pilsbry, 1940).

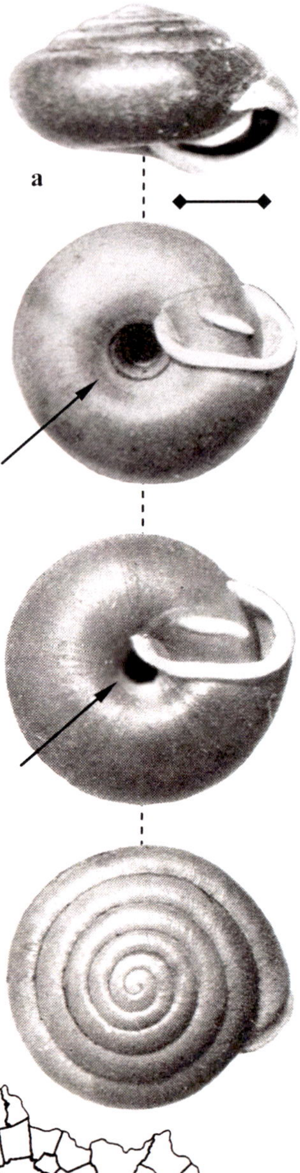

a

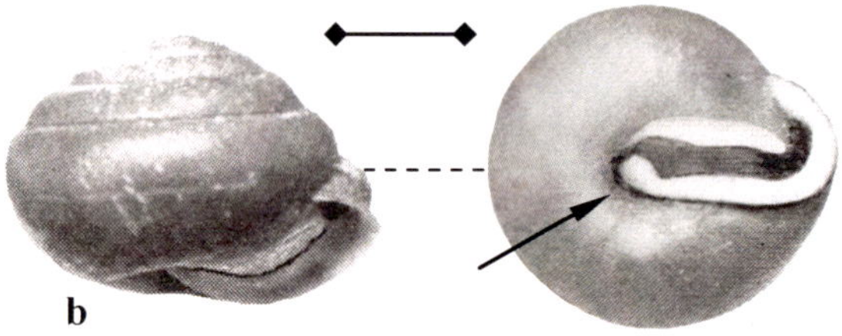

b

Euchemotrema leai aliciae

Callused Vertigo

Vertigo arthuri E.von Martens, 1882

Height: 1.5 mm, width 0.9 mm

Description: Pupa shaped; left side of lip reflected while the right side of the lip is simple containing a small dent in the aperture wall; shell with 5-5.5 whorls; 5 teeth, the upper palatal not long and entering; lower palatal is long and deeply set into aperture. Formerly *V. hubrichti* but synonymized with *V. arthuri* in 2018.

Similar Species: It is most like *V. gouldii* in terms of shell sculpture, form and size but has two deeply set curving outer palatal teeth (figures a & b) whereas the same teeth in *V. gouldii* are less curving.

Habitat: Found in cool leaf litter pockets, in particular northern white cedar groves on carbonate bedrock ledges (Nekola and Coles 2010).

Status: G5/SX. In Kentucky, known only as a Pleistocene fossil found in river deposits along the Ohio and Mississippi Rivers; extant populations are known however from Wisconsin, Minnesota and Iowa.

Specimen: Iowa, Winneshiek County, Blue Springs East (Nekola and Coles image 2010).

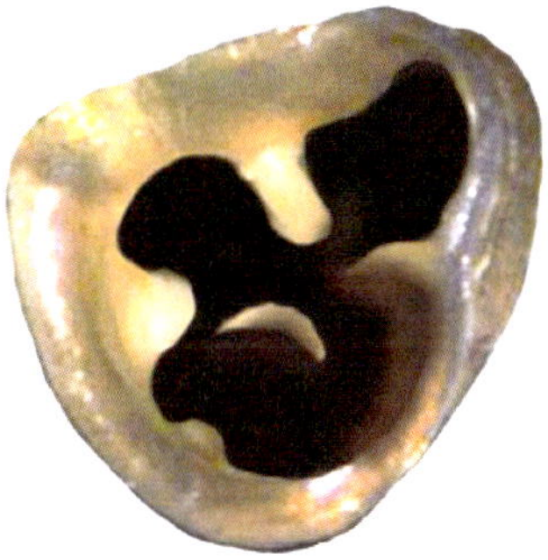

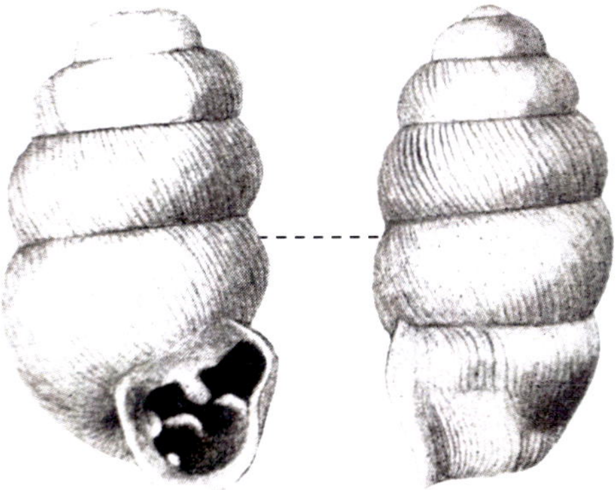

Vertigo arthuri
© Pilsbry, 1948

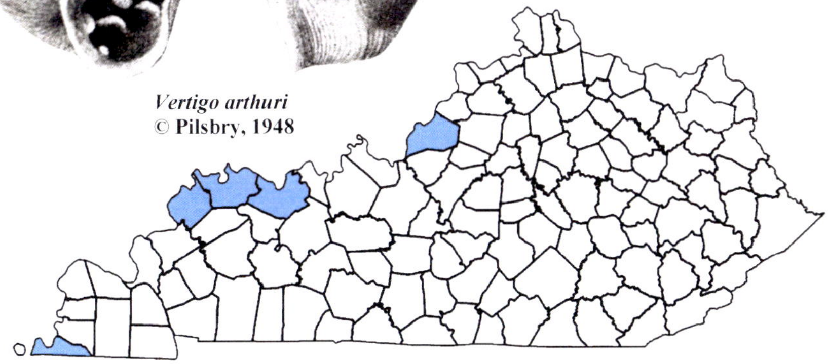

Cross Vertigo

Vertiginidae

Vertigo modesta (Say, 1842)

Height: 2.2-2.7 mm, width 1.35 mm

Description: Pupa shaped; left side of lip reflected while the right side of the lip is simple containing a small dent in the aperture wall; shell with 4.5-5.5 whorls; the shell has reduced striae and a presence of a crest; 4 teeth arranged in the form of a cross, occasionally with 3 or 5 teeth.

Similar Species: All members of the *V. modesta* clade (i.e. *V. cristata, V. coloradensis*) also have teeth arranged in a cross configuration as does *V. meramecensis* of the *V. gouldii* clade but *V. modesta's* large size (>2.25 mm height) separates it from other *Vertigo* species in Kentucky (Nekola and Coles 2010).

Habitat: Individuals occur in accumulations of humid leaf litter in mesic to wet tundra notably shrub carr dominated by willow, alder, or birch (Nekola and Coles 2010).

Status: G5/SX. All records in Kentucky are from Pleistocene fossil deposits along the Ohio and Mississippi Rivers, a primarily Canadian species (Hubricht 1985).

Specimen: Alaska, South Fork Koyukuk River (Nekola and Coles 2010).

Vertigo modesta
© Pilsbry, 1948

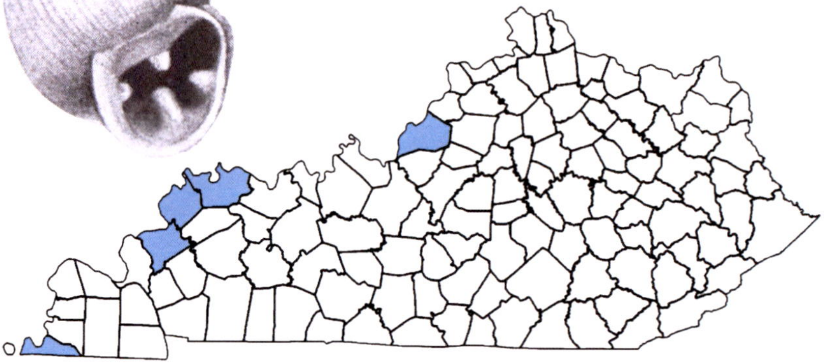

Deep-throat Vertigo

Vertigo nylanderi Sterki, 1909

Height: 1.6-1.8 mm, width 0.8-0.9 mm

Description: Pupa shaped; left side of lip reflected while the right side of the lip is simple containing a small dent in the aperture wall; cinnamon color; shell with 4.5-5 whorls; the outer margin has an indentation barely above its middle, forming a well-marked sinulus; behind it the trace of a crest and behind that a long deep furrow-like impression over the palatal folds; palatal folds long with the lower-palatal deep-seated, emerging only about to the inner end of the upper (Pilsbry 1948); the shell with notable striae and a presence of a crest; 4 teeth arranged in the form of a cross, occasionally with 3 or 5 teeth.

Similar Species: This species and *V. arthuri* possess the only shells in the subgenus Vertigo for which the lower palatal lamella is more deeply inserted into the aperture than the upper.

Habitat: Currently limited to wooded wetlands dominated by northern white cedar, tamarack or alder (Nekola and Coles, 2010).

Status: G3/SX. All records in Kentucky are from Pleistocene fossil deposits along the Ohio River, a primarily Canadian species (Hubricht 1985).

Specimen: Canada, Ontario, Leeds & Granville Counties; Wolford Bog (image from Nekola).

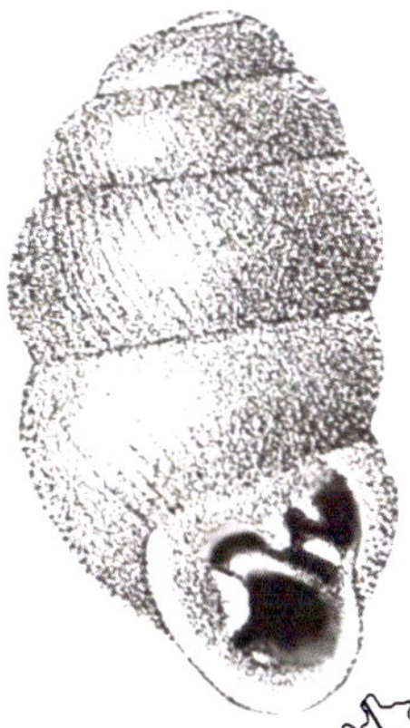

Vertigo nylanderi © **Pilsbry, 1948**

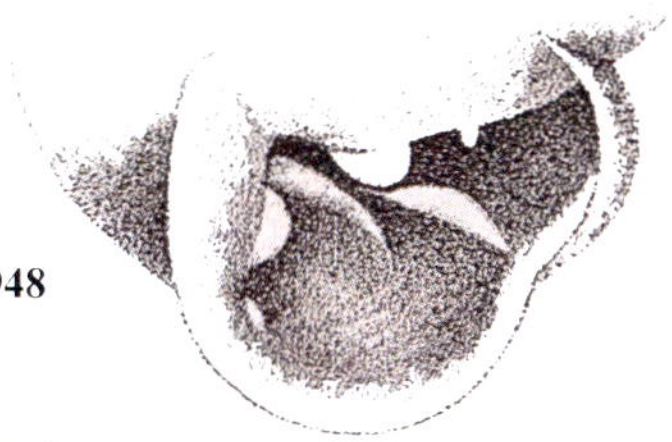

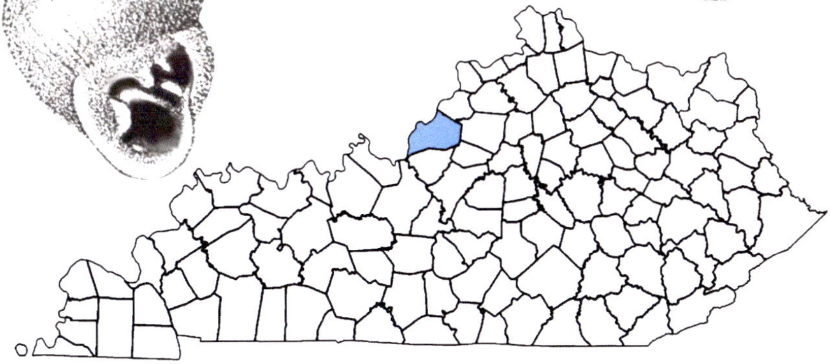

Great Lakes Snaggletooth

Gastrocopta similis (Sterki, 1909)

Height: 3-4 mm, width 1.8 mm

Description: Pupa shaped; lip reflected; shell with 6.5-7.5 whorls; angulo-parietal lamella notably forked; aperture crowded with large teeth; spiral of shell tapering; aperture more or less roundish in shape.

Similar Species: *Gastrocopta armifera* is slightly larger, generally more obese in form, but most importantly *G. similis* has a sheet-like columellar lamella (not pyramidal) and a lower palatal (b) that is as wide as it is tall (not taller than wide); *G. contracta* is smaller and its aperture has larger more crowded teeth in the aperture.

Habitat: A calciphile most frequently encountered in dry, gravelly prairie and bedrock glades (Nekola and Coles 2010). In Kentucky, this snail is known only as a Pleistocene fossil from river deposits along the Ohio River.

Status: G5/SX. All records in Kentucky are from Pleistocene fossil deposits along the Ohio and Mississippi; extant populations however are currently known from New York, Michigan, Indiana, Illinois and Minnesota.

Specimen: Iowa, Jones County, Beams Cabin (Nekola and Coles image collection 2010).

G. armifera *G. clappi* *G. similis*

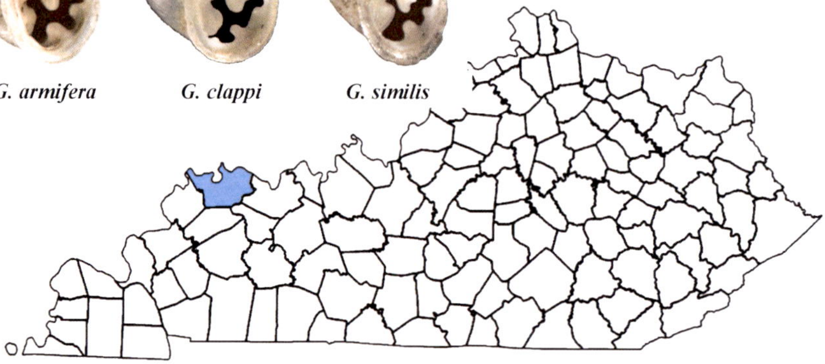

Part III
Exotic Land Snails in Kentucky

Exotic land snails including slugs that have been reported from Kentucky are included in this section. Most have been accidently introduced by way of nursery plants, construction lumber, potting soils, and fruit trees. In addition, a full color plate of exotic slugs that may eventually find their way to degraded lands of Kentucky can be found on page 402. Map locations of invasive species by counties are also included here, representing the extent of which non-native snails have taken hold in the state. It is by no means an accurate accounting of the true number of counties that have been invaded. It is expected that any long established town or city in Kentucky will harbor at least a few exotic species, some benign, while others may be quite problematic in home gardens and/or agricultural areas.

Grovesnail or Woodsnail **Helicidae**

Cepaea nemoralis Linnaeus, 1758

Diameter: 22-25 mm, height: 12-22mm

Description: Heliciform; lip slightly reflected and dark; shell with 5 whorls; imperforate; without teeth in all stages of growth; shell yellow, olive or red with 1-5 solid reddish-brown or black bands, although some specimens are without color bands; transverse striae moderately developed.

Similar Species: *Cepaea hortensis* is slightly smaller and has a white lip.

Habitat: Found in mostly degraded habitats of cities and urban areas, usually living among or under stones, old wood piles, window wells, but seems especially fond of old limestone fences; the species is less problematic in gardens than exotic slugs due to their preference for dead material (Whitson 2005), *C. nemoralis* can reach large numbers.

Status: G5/EX. A European species, reported from scattered anthropogenic locations across the state.

Specimen: Ohio, Hamilton County, near Mt. Airy Forest in Cincinnati (author's collection).

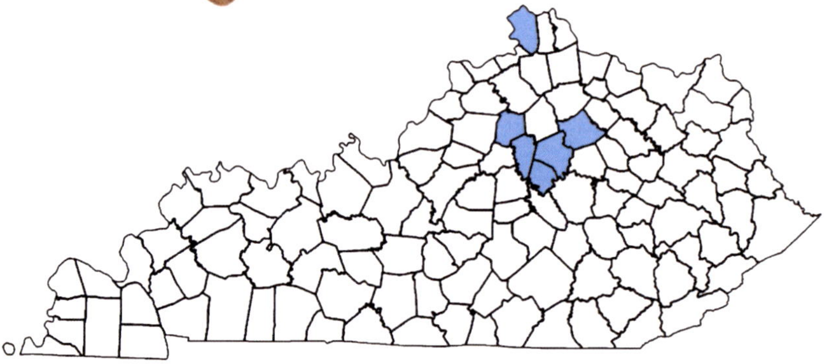

White-lip Gardensnail **Helicidae**

Cepaea hortensis Linnaeus, 1774

Diameter: 16-21 mm, height: 10-18 mm

Description: Heliciform; lip slightly reflected and white; shell with 5 whorls; imperforate; without teeth in all stages of growth; shell yellow, with 1-5 solid reddish-brown or black bands, although some specimens are without color bands.

Similar Species: *Cepaea nemoralis* is slightly larger and has a dark lip.

Habitat: Found in mostly degraded habitats of cities and urban areas, usually living among or under stones, old wood piles, window wells,

Status: G5/EX. There is some disagreement on whether this species is native or introduced to the US; Both Burch (1962) and Hubricht (1985) considered it native to the offshore islands of the New England Coast, Maine, New Hampshire and Massachusetts; it was in North America before the Vikings (Pearce 2010); although not yet reported in Kentucky, the species has been introduced to cities across the US and will likely find its way to urban areas of the state.

Specimen: (FLMNH collection).

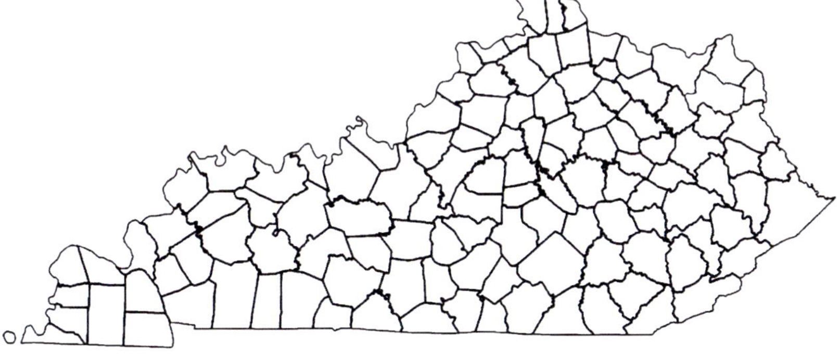

Milk Snail

Helicidae

Otala lactea Müller, 1774

Diameter: 27-36 mm, height: 16-25 mm

Description: Heliciform; lip slightly reflected, elongate and dark brown, abruptly turning downward (a); the interior of shell also dark; shell with 5 whorls; imperforate; without teeth in all stages of growth; shell white, with broken reddish-brown spiral color bands, transverse striae are well developed. This is the species that is most often consumed as escargot in fancy French restaurants.

Similar Species: *Cepaea nemoralis* is smaller and has a dark lip that is less reflected. *C. hortensis* is smaller and has a white lip; C. aspersum is smaller but has a higher shell profile.

Habitat: Found in mostly degraded habitats of cities and urban areas, usually living among or under stones, old wood piles and window wells; the species is less problematic in gardens than the exotic slugs, but can sometimes reach large numbers.

Status: G5/EX. In 1973, Branson found this species living in in old parking lots and other waste places within the city limits of Louisville, Kentucky.

Specimen: (FLMNH 158833).

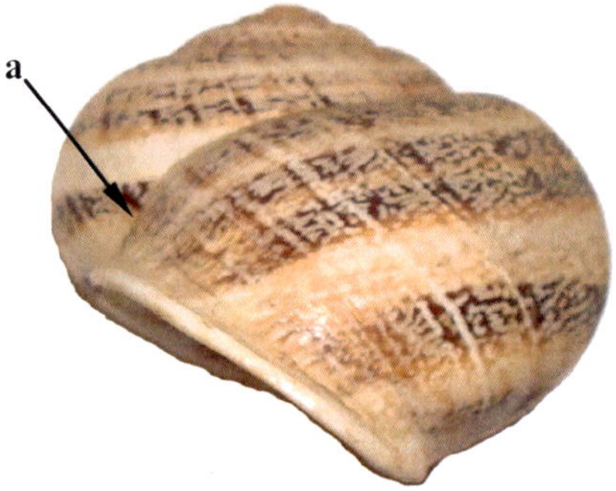

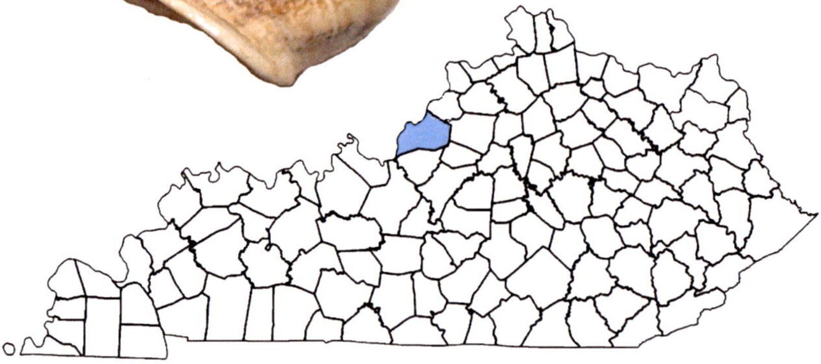

Brown Gardensnail

Helicidae

Cornu aspersum (Müller, 1774)

Diameter: 25-35 mm, height: 25-35 mm

Description: Heliciform; lip exceptionally large and slightly reflected; shell with 5 whorls; imperforate; without teeth in all stages of growth; shell usually pale brown, occasionally yellow, with dark spiral bands, variable and often flecked with white, although some specimens are without color bands.

Similar Species: *Cepaea hortensis* and *C. nemoralis* are smaller, have bolder color bands and have notably smaller apertures.

Habitat: Found in mostly degraded habitats of cities and urban areas, old wood piles, window wells; the species can be problematic in gardens.

Status: G5/EX. A European species, reported from scattered locations including Maine, Massachusetts, Michigan, South Carolina, Texas, Utah, Indiana, Ohio and Virginia. Reported from Kentucky based on Nature Serve Data, but without location information. There are currents records for the species from Montgomery County, Tennessee, adjacent to Christian County Kentucky.

Specimen: California, in the city of Los Angeles (author's collection).

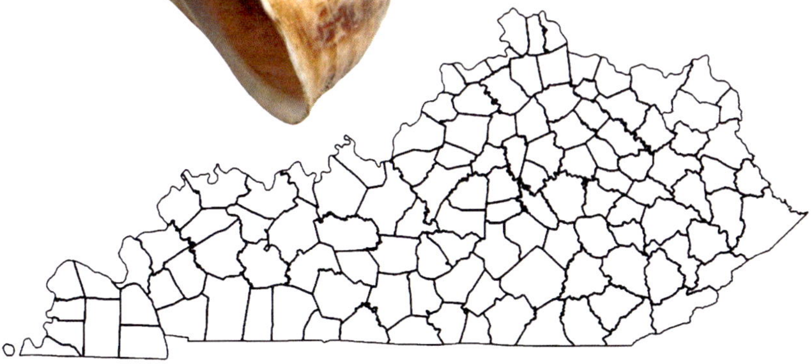

Vineyard Gardensnail

Cernuella virgata (Da Costa, 1778)

Diameter: 8-25 mm, height 6-19 mm

Description: heliciform; lip simple with a slight reflection in mature shells; a low callus bridge just inside lip; shell with 5.5-6 whorls; perforate to umbilicate; shell whitish with light color bands (a); no teeth at any stage of growth; transverse striae poorly developed, no hairs; periphery well rounded.

Similar Species: Similar to *Anguispira kochi* but in miniature.

Habitat: Found in dry grasslands and agriculture fields; in West Virginia this exotic species from Europe was recently discovered (2014) by Jeff Hajenga along a roadway cut where it was especially abundant.

Status: G5/EX. Not yet reported from Kentucky but known from nearby in West Virginia.

Specimen: West Virginia, Fayette County, Exit 66 off I-77 (author's collection).

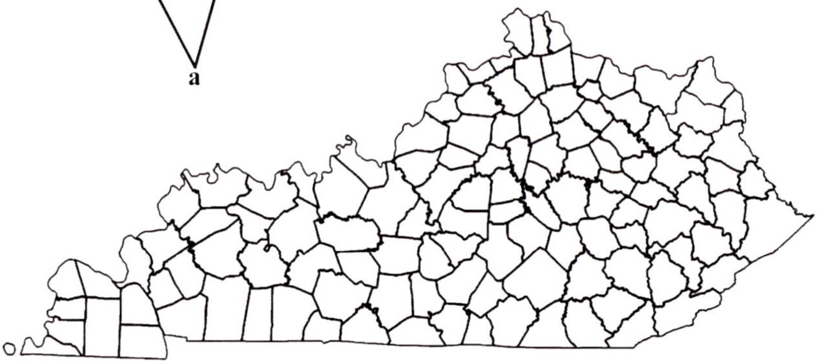

Heath Snail **Geomitridae**

Xerolenta obvia **(Menke, 1828)**

Diameter: 14-20 mm, height 7-10 mm

Description: depressed heliciform; lip simple with a slight reflection in mature shells; shell with 5.5-6 whorls; umbilicate; shell whitish with dark color bands (a); no teeth at any stage of growth; transverse striae poorly developed, no hairs; periphery well rounded.

Similar Species: Few other land snail In Kentucky is have a similar body shape with such dark color bands; A

Habitat: In its native habitat this land snail is a specie of dry grassy areas; in Kentucky the species was found around an old building foundation and parking lot with low weeds near the Ohio River in Louisville (pers. comm. John MacGregor).

Status: G5/EX. The species was reported to be especially abundant at the location described above.

Specimen: Château de Peyrepertuse, Aude, France © H. Zell, CC BY-SA 3.0 <https://creativecommons.org/licenses/by-sa/3.0>,via Wikimedia Commons

Multiple dark bands

Dark–bodied Glass-snail

Oxychilidae

Oxychilus draparnaudi (Beck, 1837)

Diameter: 12-16.5 mm, height: 6.5-8.0

Description: Depressed heliciform; lip simple; shell with 5-5.5 whorls; umbilicate; shell glossy, thin and semi-transparent; no teeth at any stage of growth; transverse striae poorly developed, no spiral striae present; no hairs; live animal cobalt blue, with gray mantle; periphery well rounded; all *Oxychilus* species are carnivorous.

Similar Species: Young shells of *Mesomphix inornatus* are similar, but are with minute spirally arranged papillae and a less open umbilicus; *O. alliarius* is much smaller and has a higher shell profile.

Habitat: Found in mostly degraded habitats and waste places of cities and urban areas; this species can reach considerable numbers.

Status: G5/EX. Reported from a few counties in Kentucky where it is found in and around flower gardens, transported to new location via potting soils and potted plants.

Specimen: Virginia, Norfolk County, vacant lot in Norfolk (FMNH 251405).

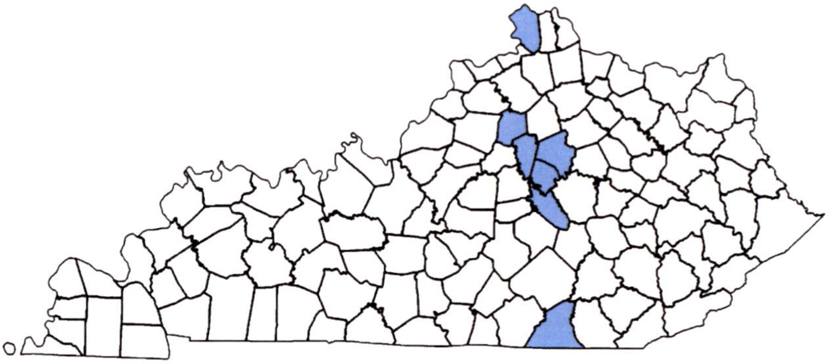

Cellar Glass-snail

Oxychilus cellarius (Müller, 1774)

Diameter: 9-14 mm, height: $.5-6 mm

Description: Depressed heliciform; lip simple; shell with 5.5-6 whorls; umbilicate; shell glossy, thin and semi-transparent; no teeth at any stage of growth; transverse striae poorly developed, no spiral striae present; no hairs; live animal pale bluish-gray, but dark forms also occur; periphery well rounded.

Similar Species: Young shells of *Mesomphix inornatus* are similar, but have minute spirally arranged papillae and a less open umbilicus; other *Mesomphix* species are more globose in overall form; in *O. draparnaudi,* the last whorl is notably wider.

Habitat: Found in mostly degraded habitats of urban areas; under rubbish; frequently found in caves in its native continent of Europe.

Status: G5/EX. Although only reported from one Kentucky county this species is no doubt hiding in many more locations anxiously awaiting their unpretentious discovery! Like other *Oxychilus* species, it is transported to new location via potting soils and nursery plants.

Specimen: England (FMNH 90855).

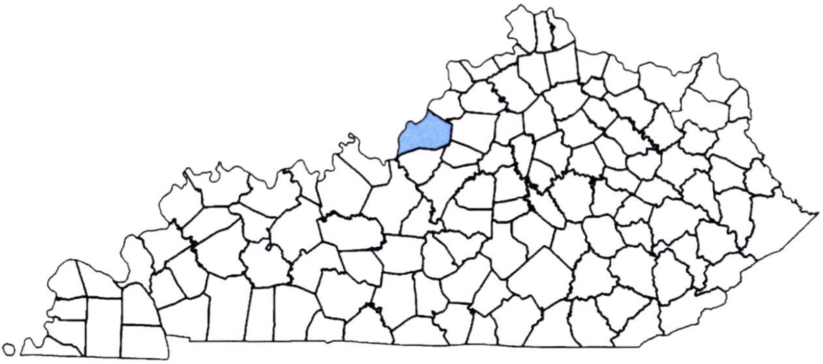

Garlic Glass-snail **Oxychilidae**

Oxychilus alliarius (Miller, 1822)

Diameter: 5-7 mm, height: 3.5-4 mm

Description: Depressed heliciform; lip simple; shell with 4-4.5 closely coiled whorls; umbilicate; shell glossy, thin, pale yellowish-brown or greenish, often white below; semi-transparent; no teeth at any stage of growth; transverse striae poorly developed; no spiral striae present; no hairs; live animal dark bluish-gray, smelling strongly of garlic if disturbed; periphery well rounded.

Similar Species: Young shells of *Mesomphix inornatus* are similar, but have minute spirally arranged papillae and a less open umbilicus; other *Mesomphix* species are more globose in overall form; *O. cellarius* has 1-2 more loosely coiled whorls.

Habitat: As with most *Oxychilus* species, it is found in degraded habitats and waste places of cities and urban areas; tolerant of poor acidic places (Kerney and Cameron 1979).

Status: G5/EX. Although the species has not yet been reported in Kentucky, it is no doubt hiding in some over manicured rose garden.

Specimen: England, High Beach (FMNH 45218).

Oxychilus draparnaudi

Oxychilus alliarius

© Gordon Riley, 1979

5 mm

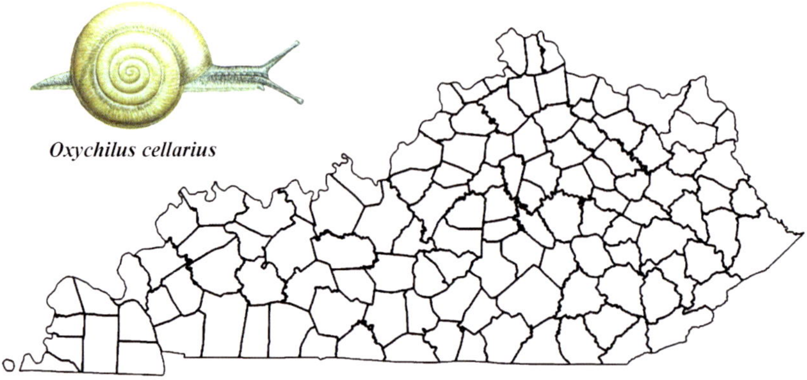

Oxychilus cellarius

Southern Flatcoil — Polygyridae

Polygyra cereolus Muhlfeldt, 1818

Diameter: 11.5-18.2 mm, height 3.5-4.6 mm

Description: Discoidal (flat) or with a slightly raised spire; lip only slightly reflected; shell with 7-9 whorls; widely umbilicate; small parietal tooth; uniformly wood brown to white with radial streaks of pale brown; transverse striae well developed; periphery is angular to subcarinata.

Similar Species: Very similar to the native *Daedalochila* species in Kentucky but differs mainly in its notably smaller parietal tooth but it does have the wide umbilicus seen in *Daedalochila.*

Habitat: *Polygyra cereolus* is native to the Coastal Plains of the US but in Kentucky it has been found living among topical vegetation in the Fayette Mall (Branson and Batch, 1988). It might also be expected to thrive in greenhouses and other similar indoor situations throughout the state.

Status: G5/EX. In Kentucky currently known only from planting boxes within Fayette Mall (Branson and Batch, 1988), but may occur in other similar environments (i. e. greenhouses).

Specimen: Puerto Rico, Utuado Campus, Images ©Abraham Breure.

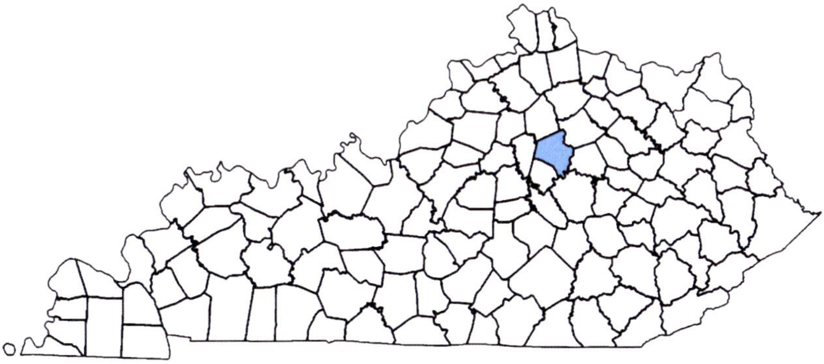

Sharp Awlsnail

Achatinidae

Opeas pyrgula Schmacker & O.Boettger, 1891
Height: 5-8 mm, width 2 mm
Description: Shell cylinder-shaped; lip simple; shell with 6-7 whorls; nearly umbilicate; shell surface glossy and translucent in live animals and fresh shells; color pale corneous to almost white; with live animal inside, shell greenish-yellow (figure a); densely sharply sculptured with irregular and rather strong transverse striae which are very deeply curved; the retraction of the outer lip at the suture is deep (figure a); outer lip thin; last whorl rounded.
Similar Species: Other similar looking native species of West Virginia are much smaller.
Habitat: A tropical and sub-tropical species; in temperate climates like Kentucky, a snail of green houses and indoor gardens, not capable of surviving outside in Kentucky's harsh winters.
Status: G5/EX. Likely already occurs in green houses and inside flower gardens of shopping malls in Kentucky.
Specimen: *Opeas pyrgula* shell and penny by Larry Watrous ©.

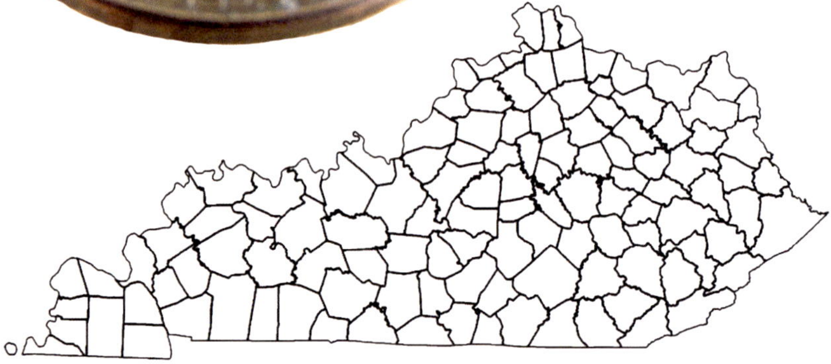

Introduced &Potential Species in Kentucky (proportionate)

30 mm

Cornu aspersum

Otala lactea

Cernuella virgata

Cepaea nemoralis

Cepaea hortensis

Polygyra cereolus

Xerolenta obvia

Oxychilus cellarius

Gray Fieldslug — Agriolimacidae

Deroceras reticulatum (Müller, 1774)

Length: Adults 35-50 mm while crawling,

Description: Color ranges from nearly white or pale tan to mottled brown or nearly black; mantle or saddle farther back on the slug's body than seen in *Arion* species; posterior end strongly keeled (a); tubercles large and distinct; a whitish border surrounds the pneumostome (b); defense mucus milky and adhesive-like; margins and sole of foot are white, pale and or gray.

Similar Species: Distinguished from *D. laeve* and *D. panormitanum* by its larger size, thicker skin, pale grayish color, slower movements, and its ability to produce sticky, milky-colored mucus when bothered (Grimm et al. 2009); it is most reliably separated by differences in their reproductive organs.

Habitat: *Deroceras reticulatum* is likely the most serious molluscan plant pest in North America, a result of its insatiable appetite for live plant material. The species is ubiquitous and generally abundant in fields and gardens, but also in practically every other anthropogenic habitat, including roadsides, forest campsites, and wasteland. It is rarely found in completely undisturbed habitats. Toads, frogs, shrews, brown snakes, and ground beetles are known predators.

Status: G5/EX. Reported from scattered counties in Kentucky, but likely much more widespread than current records suggest.

Specimen: Illustrations © Aleta Karstad (Grimm et al 2009)

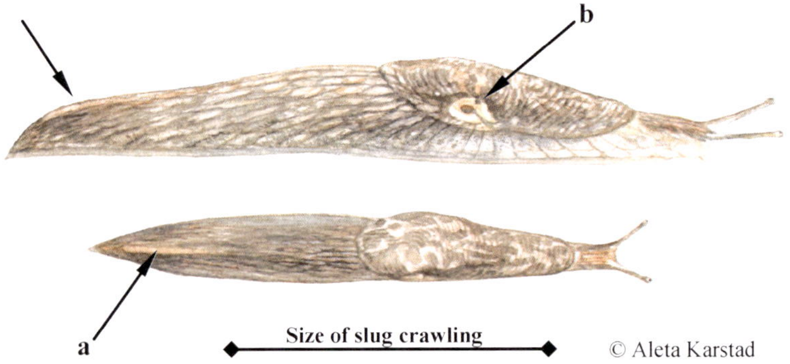

Two color morphs

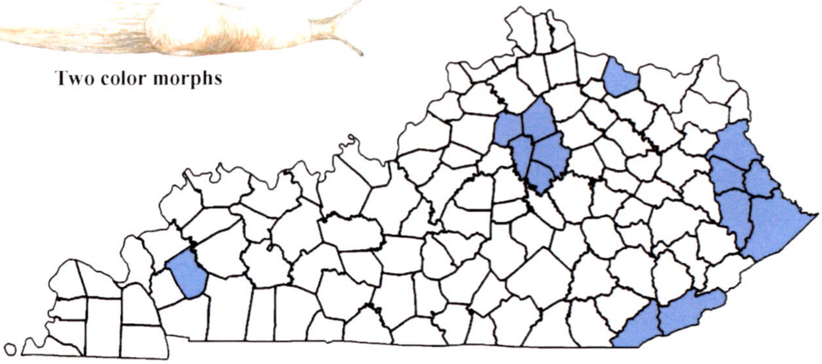

Orange-banded Arion — Arionidae

Arion fasciatus (Nilsson, 1823)

Length: Adults 30-40 mm while crawling

Description: Gray overall, gray above, fading to gray-white on the sides below dark lateral bands; gray to yellowish saddle-like mantle covers only the anterior (front) part of the slugs dorsal region (back); breathing pore is in the front half of the mantle in all *Arion* (as opposed to *Deroceras* and *Limax*); reproductive pore in front of breathing pore; defense mucus clear; margins of the foot grayish-white.

Similar Species: Native slugs in West Virginia, except for *Deroceras laeve,* have mantles that nearly cover the entire body; the mantle of *D. laeve* is longer and has no lateral bands; *A. subfuscus* is larger, the reproductive pore is located behind the breathing pore, and the defense slime is yellow or orange.

Habitat: Found in mostly degraded habitats of cities and urban areas, usually living among or under stones, old wood piles and window wells; this pest species is especially problematic in gardens and can reach large numbers.

Status: G5/EX. Although not yet reported from Kentucky, this species is likely already in the state or will be soon enough.

Specimen: Illustrations © Aleta Karstad (Grimm et al 2009)

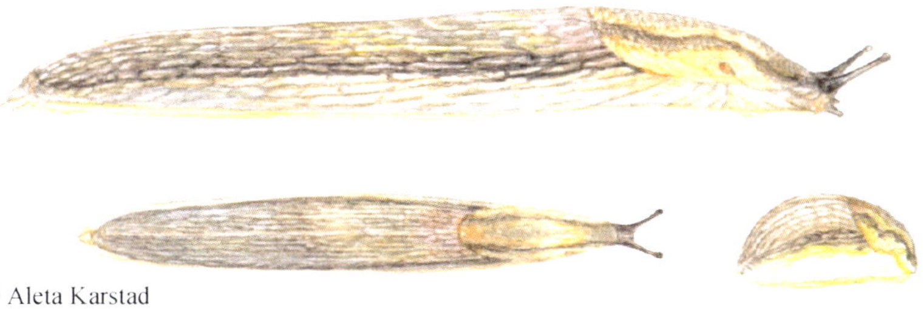

© Aleta Karstad

Size of slug crawling

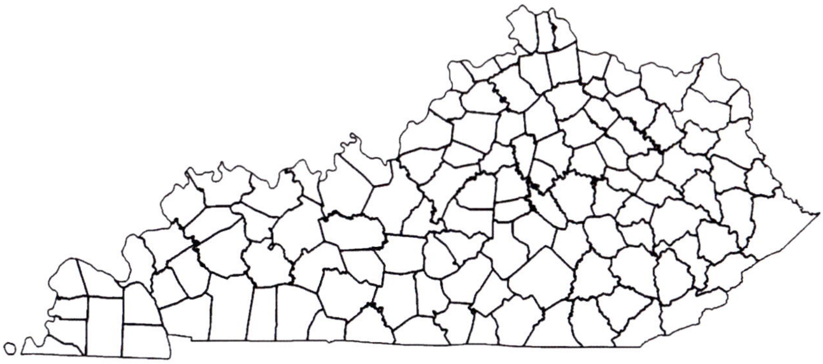

Dusky Arion Arionidae

Arion subfuscus (Draparnaud, 1805)

Length: Adults 50-60 mm while crawling

Description: Orange and brown with lateral light brownish bands; orange saddle-like mantle which covers only the anterior (front) part of the dorsal region; breathing pore is in the front half of the mantle in all *Arion* (but not in *Deroceras* and *Limax*); reproductive pore located behind the breathing pore; defense posture of slug (a); defense mucus bright yellow or orange; foot pale yellow.

Similar Species: *A. fasciatus* is smaller and has a colorless defense mucus; *L. maximus* is much larger and has a different body color and clear defense slime.

Habitat: Found in mostly degraded habitats of cities and urban areas, usually living among or under stones, old wood piles, plastic, discarded carpet, and in window wells; this pest species is especially problematic in gardens and can reach substantial numbers.

Status: G5/EX. *Arion subfuscus* is one of the more common and widespread invasive slugs in Kentucky and is likely in most cities and towns that have been long established. The species is easily and efficiently transported to new location by way of construction materials (i.e. lumber), potting soils and nursery plants.

Specimen: Illustrations © Aleta Karstad (Grimm et al 2009)

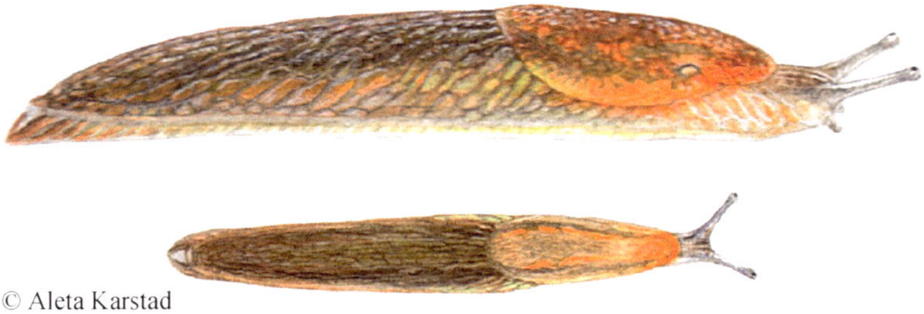

Size of slug crawling

a

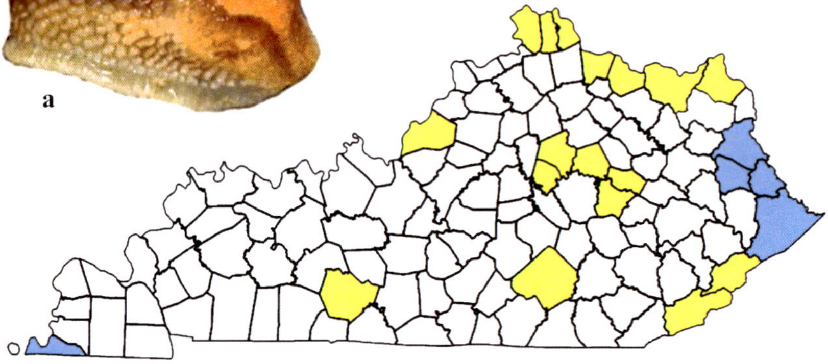

Giant Gardenslug **Limacidae**

Limax maximus Linnaeus 1758

Length: Adults 100-200 mm while crawling

Description: Mantle and body black spotted or mottled rather than with continuous bands; body color yellowish-gray or light brownish-gray; tentacles reddish-brown; posterior end of slug keeled, but not up to the mantle; defense mucus clear; margins of the foot and sole whitish.

Similar Species: This sizeable slug looks like no other native or exotic slug reported in West Virginia.

Habitat: Very common in degraded habitats of cities and urban areas, under railroad ties used for landscaping; the species feeds mostly on dead vegetation and fungi; other foods include roots, fruit, leafy crops and carrion (Grimm et al. 2009); reported to crawl onto porches to consume pet food (MacGregor pers. comm. 2010); especially problematic in gardens and can reach large numbers.

Status: G5/EX. Although reported from only ten counties in Kentucky this species is no doubt more common in the state, in particular, areas around city parks and playgrounds (the species often transported to new locations via construction lumber).

Specimen: Illustrations © Aleta Karstad (Grimm et al 2009)

© Aleta Karstad

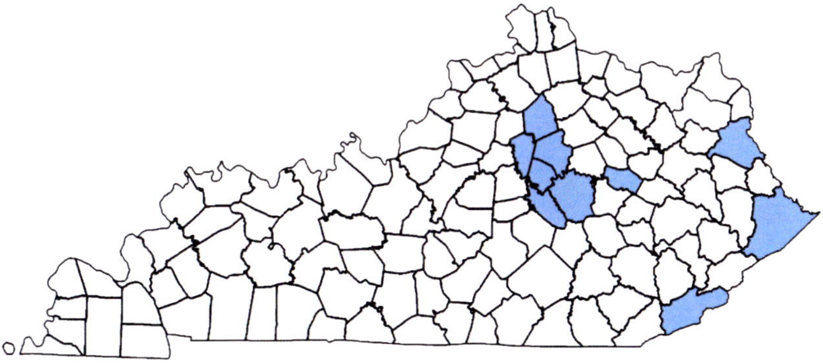

Size of slug crawling

Threeband Gardenslug **Limacidae**

Ambigolimax valentianus (A. Férussac, 1821)

Length: Adults up to 60 mm while crawling

Description: Yellow-gray body; short keel. dark stripes or reticulations on sides of mantle and tail; body bands near to mid-dorsal line; lyre pattern on mantle; a light stripe down mid-dorsum; dark head; pale sole; clear mucus (Wiktor *et al.* 2000)

Similar Species: Very similar to *L. nyctelia* externally; distinguished from this species by internal anatomy; like *L. marginata* but lighter yellowish body, body bands nearer middorsal band. (Quick 1960).

Habitat: Found in mostly degraded habitats of cities and urban areas, in vacant lots and waste places, in greenhouses, around meadows, and in hedge rows; in their natural habitat these slugs prefer moist, shady locations with lots of vegetation.

Status: G5/EX. This species of invasive slug is reported from a few Kentucky counties but is likely far more common than current records indicate.

Specimen: Illustrations © Aleta Karstad (Grimm et al 2009)

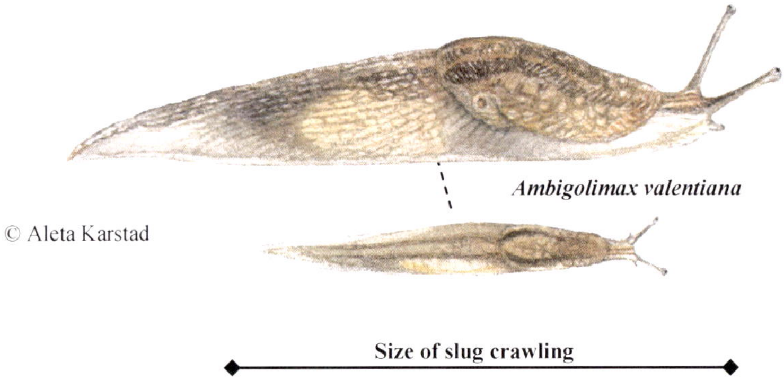

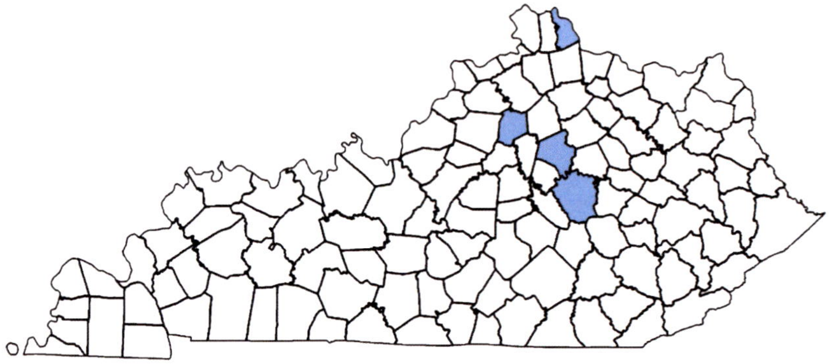

Introduced & Potential Slugs in Kentucky (proportionate)

All images © Aleta Karstad

Arion ater (100-200 mm)

Limax maximus (100-200 mm)

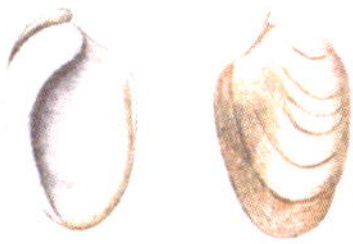

Shell of slug

Testacella haliotidea (100 mm)

Limacus flavus (75-100 mm)

Arion circumscriptus (30-40 mm)

Arion subfuscus (50-70 mm)

Arion fasciatus (30-40 mm)

Ambigolimax valentianus (70 mm)

Deroceras panormitanum (30-40 mm)

Milax gagates (50-60 mm)

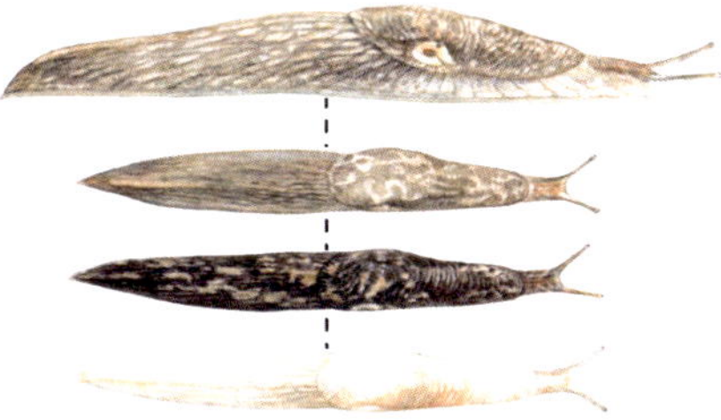

Deroceras reticulatum (35-50 mm)

Arion intermedius (15-20 mm)

Boettgerilla pallens (30-40 mm)

Part IV

Land Snails of Bordering States: West Virginia, Ohio, Illinois, Indiana Tennessee, Missouri & Virginia

This section provides detailed information on land snails that have a reasonable chance of occurring in Kentucky due to their close proximity or similar ecological requirements. If you find species that do not match gastropods reported from Kentucky, this section should cover possible candidates, unless the species is new to science or is a peculiar and isolated ecological form. For each species illustrated, a Kentucky map is included with a dashed circle indicating where the species would most likely occur in the state. These potential land snail additions were extracted from several sources including Hubricht's (1985) range maps, other researchers' distribution records, and my own knowledge of these species in the adjacent states.

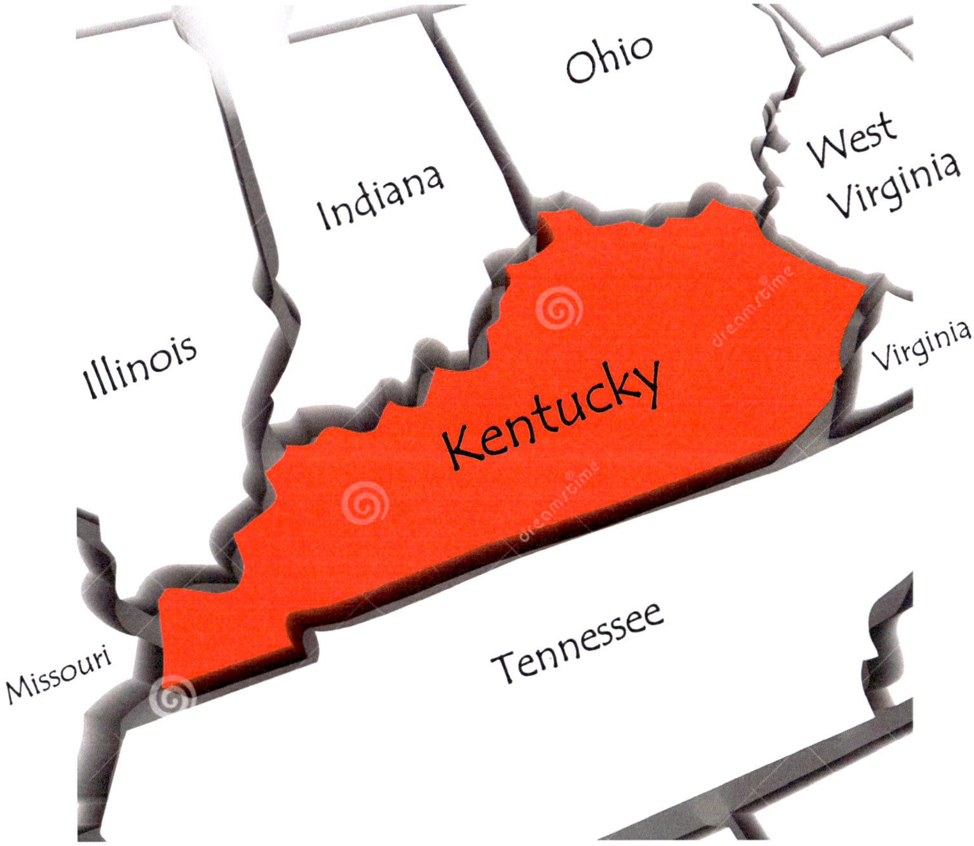

Louisiana Ambersnail

Succineidae

Oxyloma salleanum (Pfeiffer, 1849)

Height: 14.4-19.8 mm, width 7.4-10.7 mm

Description: Succiniform; lip simple; shell with 3 whorls; perforate; no teeth or lamella present in the aperture at any stage of growth; shell surface weakly striate and marked with irregular spiral lines (Pilsbry 1948); the spiral sculpture may be hard to detect; live animal straw-yellow.

Similar Species: *Oxyloma retusum* is around the same size but has a more elongate form.

Habitat: Found in the same habitat as *Oxyloma retusum.*

Status: Not yet recorded from Kentucky but known near its border with Tennessee to the south and Indiana and Illinois to the north. Area of possible occurrence in shaded dotted line below.

Specimen: Unknown

Shells proportionate to each other

Novisuccinea ovalis

Oxyloma retusum

Oxyloma salleanum

Succinea concordalis

Catinella oklahomarum

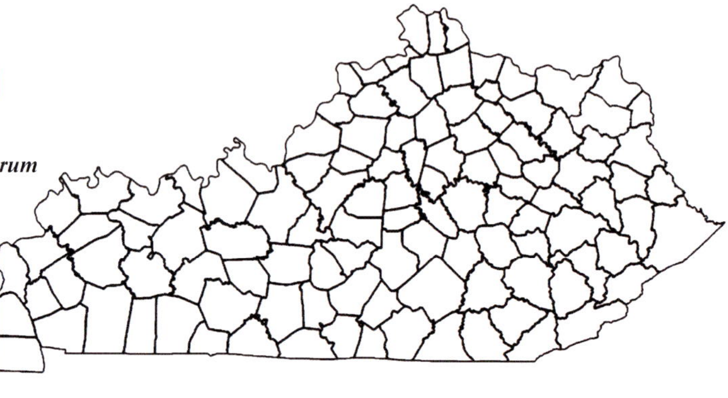

Plains Snaggletooth

Gastrocopta abbreviata (Sterki, 1909)

Height: 3.3-3.8 mm, width 1.0-2.2 mm

Description: Pupa shaped; lip reflected; shell with 5-6 whorls; penultimate whorl comparatively broader, the last rounded at the base; slightly flattened over the palate; milky-whitish, dullish gloss, striae fine and slight.

Similar Species: *Gastrocopta armifera* has columellar lamella with both forward and basally pointing components, appearing more or less pyramidal in aperture view (a) while *G. abbreviata* has a columellar lamella that lacks a basal lobe (b and c), but a larger basal fold (d).

Habitat: Found under stones within loess bank fissures and in thin litter accumulations in riparian forests and grasslands (Nekola and Coles 2010).

Status: This largely mid-western species is reported from along the Mississippi River in counties close to Kentucky. Area of possible occurrence in shaded dotted line below.

Specimen: Iowa, Jones County, Beams Cabin (Nekola and Coles image 2010).

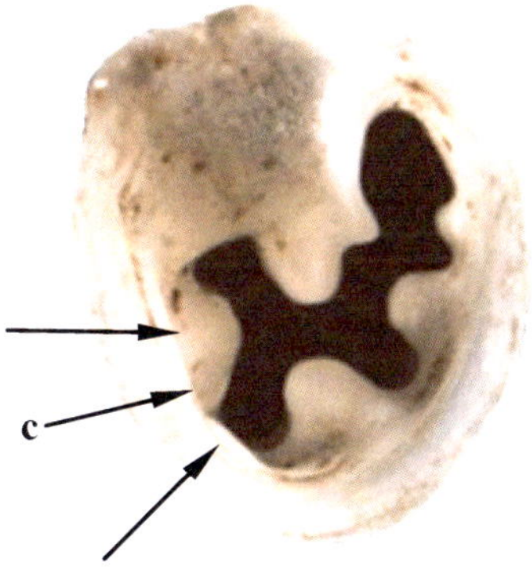

G. armifera

© Pilsbry, 1948

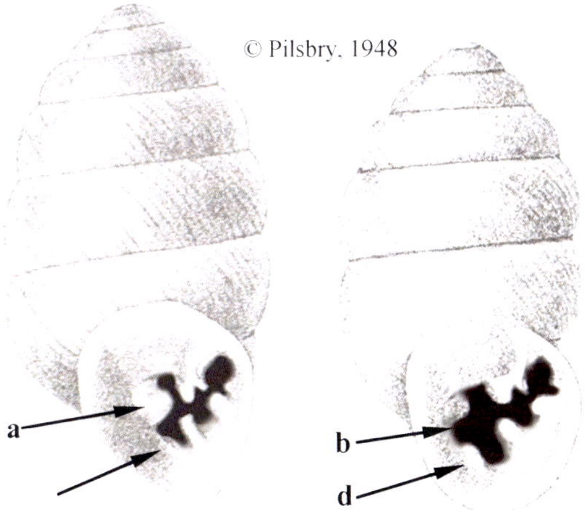

G. armifera *G. abbreviata*

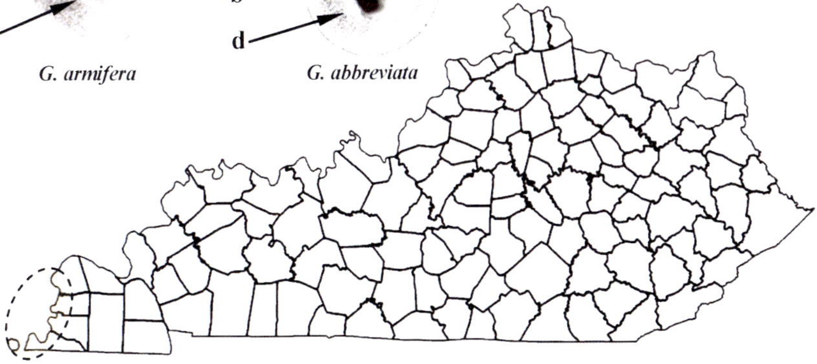

Rogers Snaggletooth

Gastrocoptidae

Gastrocopta rogersensis Nekola and Coles, 2001

Height: 1.7-2.5 mm, width 0.81-1.05 mm

Description: Pupa shaped; lip reflected; shell with 6 whorls; angulo-parietal lamella notably forked; aperture crowded with large teeth; shell has a notable taper; aperture more or less rectangular in shape.

Similar Species: Similar in form and size to *G. procera* but distinguished primarily by the shape of the angulo-parietal lamella seen in pictures below (a and b).

Habitat: A calciphile found in open areas such as cedar glades, under stones and in leaf litter and grass thatch on dry bedrock cliffs (Nekola and Coles 2010).

Status: A largely mid-western species is also reported from Adams County, Ohio near Edge of Appalachia (Nekola and Coles 2010), area of possible occurrence in shaded dotted line below.

Specimen: Iowa, Jones County, Beams Cabin (Nekola and Coles image 2010).

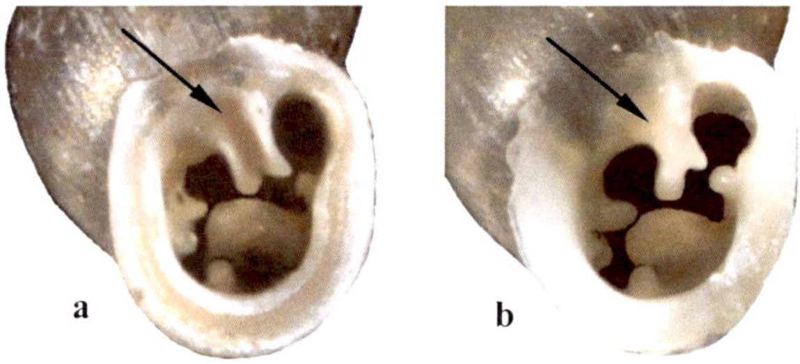

G. rogersensis *G. procera*

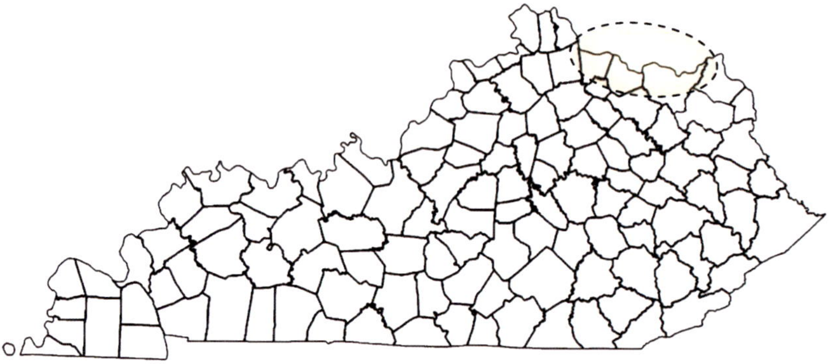

Club Supercoil

Pristilomatidae

Paravitrea bellona Hubricht, 1978

Diameter: 5.8 mm, height 2.1 mm

Description: Depressed heliciform; lip simple; shell with 6-7 tightly coiled whorls; shell thin, glossy and translucent in fresh dead and live shells; pale brownish above; umbilicate; umbilicus diameter 1.3 mm; transverse striae closely but irregularly spaced, distinct above but becoming obsolete below; two or three pairs of teeth are present in immature shells of less than 5 whorls (a), but are absent in adults, these teeth seen through the bottom of fresh shells; live animal with a bluish-gray foot.

Similar Species: *Paravitrea bellona* is most like *P. capsella* differing in having at least one pair of teeth in juvenile shells and having a smaller umbilicus, it differs from *P. seradens* by the absence of teeth in adult shells (or shells under 5 whorls).

Habitat: Found in pockets of moist leaf litter on wooded river bluffs and in ravines.

Status: Endemic to West Virginia but known in West Virginia counties close to eastern Kentucky.

Specimen: West Virginia, Logan County, Horsepen Mountain 4.5 miles south of Sarah Ann (FMNH 249562).

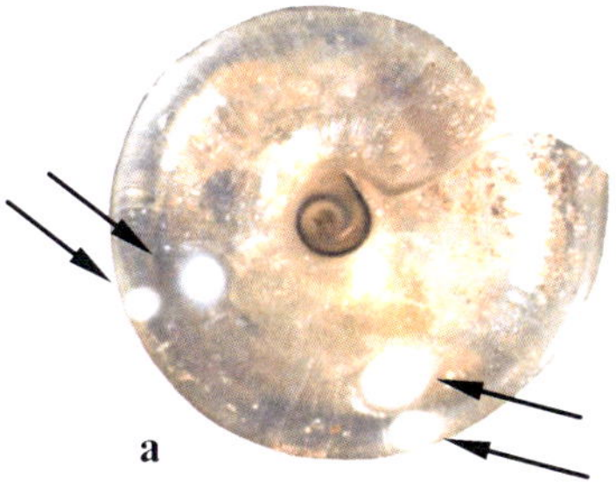

a

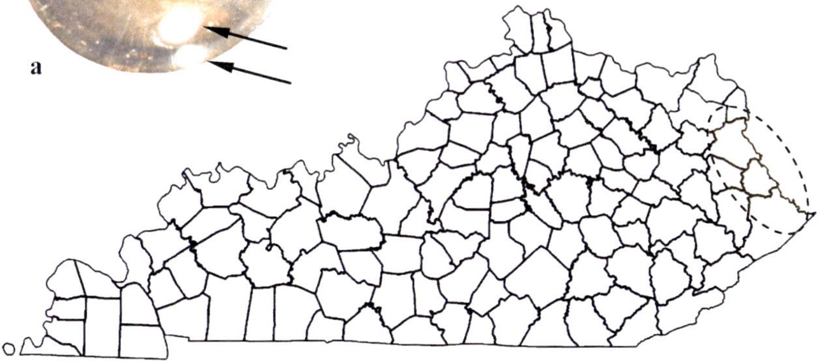

Domed Supercoil

Pristilomatidae

Paravitrea significans (Bland, 1866)

Diameter: 4.5 mm, height 2 mm

Description: Shell umbilicate, depressed heliciform, thin; lip simple; shell with 6 tightly coiled whorls; shell thin, glossy and translucent in fresh dead and live shells; shinning pale horned colored; umbilicate; umbilicus one fifth diameter of shell; transverse striae irregularly spaced and strong on top becoming obsolete on the base; in juvenile shells with less than 4 whorls, there are sometimes 1 to 2 pairs of rather small teeth, which are always absent in adults; periphery of shell angled and widest at the bottom, not rounded as seen in most *Paravitrea* species.

Similar Species: *Paravitrea significans* is most like *P. seradens* differing in having an angled, not rounded periphery, smaller umbilicus, and lacking teeth in adult shells; *P. bellona* of West Virginia is larger and has a more rounded periphery.

Habitat: Found in tall grass prairies and brushy thickets in both upland and bottomland areas with cottonwood and sycamores.

Status: This largely mid-western species is reported from along the Mississippi River in counties in Tennessee, Missouri and Illinois that border Kentucky. Area of possible occurrence in shaded dotted line below.

Specimen: Oklahoma, Fort Gibson, Indian Territory (ANSP 11984 Lectotype).

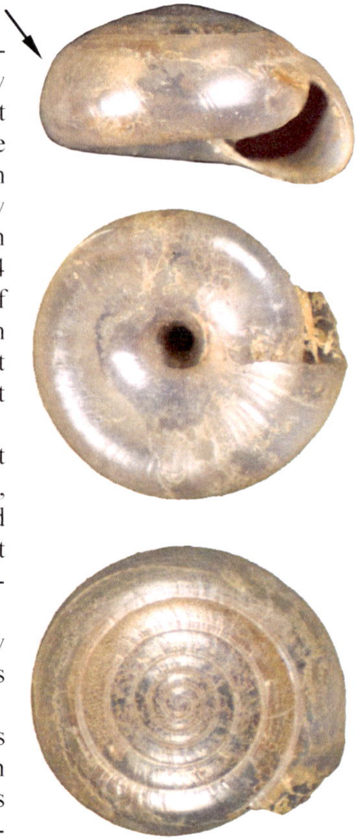

Lectotype

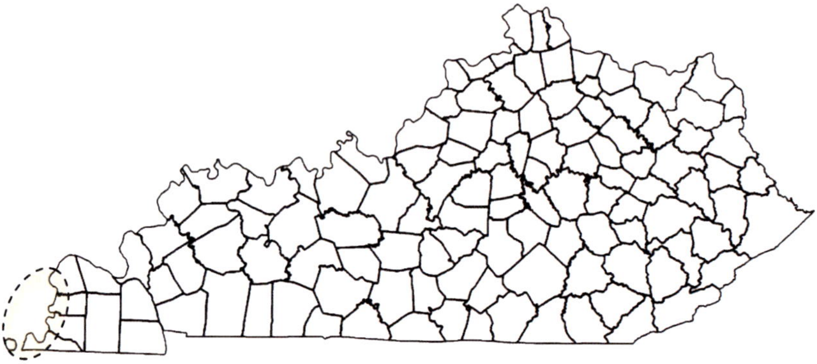

Barred Supercoil

Pristilomatidae

Paravitrea seradens Hubricht, 1972

Diameter: 5.5 mm, height 2.8 mm

Description: Depressed heliciform; lip simple; shell with 6-7 tightly coiled whorls; shell thin, glossy and translucent, sometimes with a hint of yellow in fresh and live shells; umbilicate, deep and well-like, exhibiting all the whorls; transverse striae numerous and irregularly spaced; whorls slowly expanding, last whorl expanding more rapidly and deflecting downward to the aperture; internal armature of one to three pairs of rather large teeth on the outer and basal walls (see below) at all stages of growth, however these teeth occasionally missing in very old shells; teeth can be seen through the bottom of fresh shells of both juveniles and adults (Hubricht 1972).

Similar Species: *Paravitrea seradens* is most like *P. capsella* differing in having at least one pair of teeth at all stages of growth and having a larger umbilicus in immature shells.

Habitat: Pockets of moist leaf litter located in mixed hardwood forests in ravines.

Status: Uncommon; most records of this infrequent land snail are from eastern West Virginia counties that border Kentucky; area of possible occurrence in shaded dotted line.

Specimen: Figures (a) from West Virginia, Mingo County, Guyandotte River, opposite Justice (FM 249218) and figure (b) from West Virginia, Fayette County (FMNH 173020 Holotype).

a

b Holotype

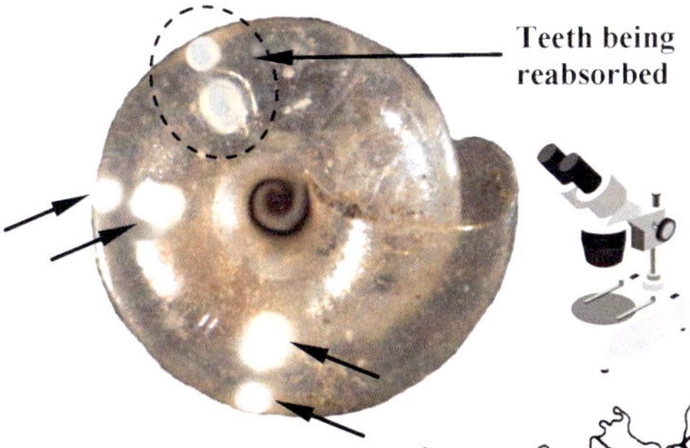

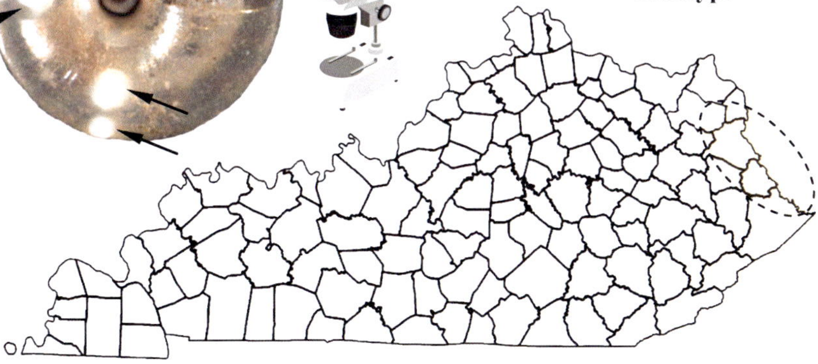

Caneyfork Supercoil

Paravitrea metallacta Hubricht, 1963

Diameter: 3.9 mm, height 2.1 mm

Description: Depressed heliciform; lip simple; shell with 6.5 whorls; the last quarter whorl conspicuously expanded in mature shells (b); shell thin and translucent in fresh and live shells; pale yellowish, subhyaline, shining; umbilicate, umbilicus deep and well-like; live animal white; transverse striae are irregularly spaced, distinct above but becoming obsolete on the base; in young shells there are one or two pairs of small teeth within the last whorl (c); adults are without teeth..

Similar Species: *Paravitrea lapilla* is larger and has teeth in adult shells, similar to *P. tantilla* but without teeth in adult shells; *P. capsella* has a higher shell profile and typically without teeth at any stage of growth; *P. placentula* is larger (fully 3 mm) with larger paired teeth in young shells (d).

Habitat: Found under moist leaf litter on wooded hillsides and in ravines.

Status: Not yet reported from Kentucky but known from counties in north central Tennessee; area of possible occurrence in dotted line below.

Specimen: Tennessee, White County, 3.2 miles south of Doyle (FMNH 249459).

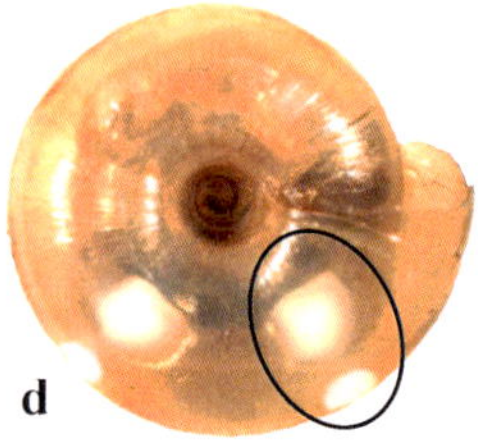

P. placentula

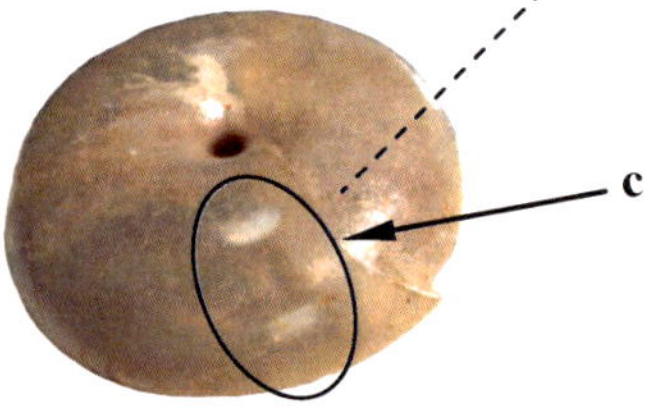

Paravitrea metallacta

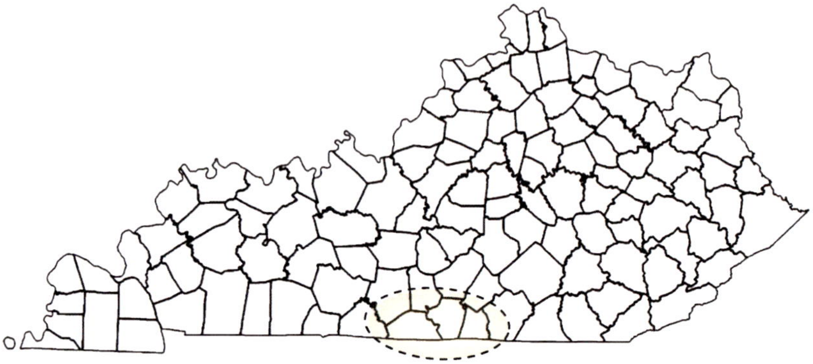

Natural Bridge Supercoil

Paravitrea pontis Baker,1931

Diameter: 5.12-6.5 mm, height 2.8 mm

Description: Depressed heliciform; lip simple; shell with 6-7 tightly coiled whorls, except for the last which is the widest and quite suddenly expanded; shell thin, hyaline; umbilicate; umbilicus wide, deep and well-like exhibiting all the whorls; transverse striae numerous and closely but irregularly spaced, distinct above but becoming weaker below; the internal armature (only present in shells of 2-3 whorls and diameter of 1.5 mm) consisting of smooth, radial barriers extending through the middle half of the periphery in final whorl (a & b); deceased and dried animal visible through shell.

Similar Species: *Paravitrea ceres* has a similar build, but is smaller by 1.5-2 mm, has a narrower umbilicus, and is believed to be without internal armature in all stages of growth.

Habitat: A calciphile found in pockets of moist leaf litter in wooded ravines and on hillsides; the type locality was in limestone talus in a canyon just above the base of Natural Bridge, Rockbridge County, Virginia.

Status: Known from West Virginia near Pike County Kentucky.

Specimen: West Virginia, Pendleton County, 3.4 miles NW Onego (FMNH 249591).

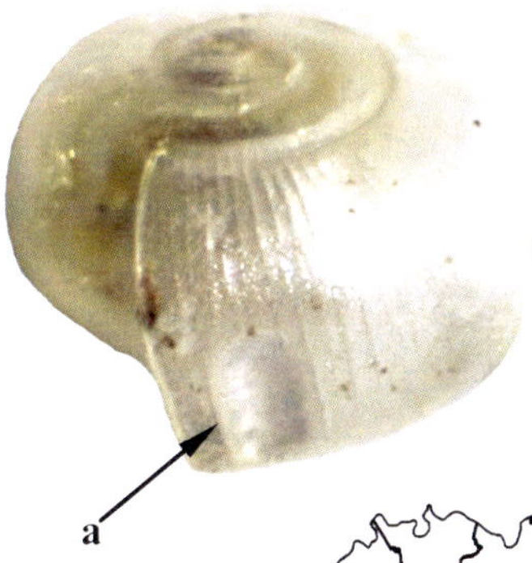

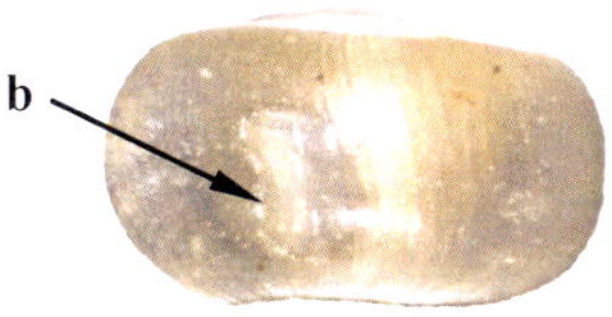

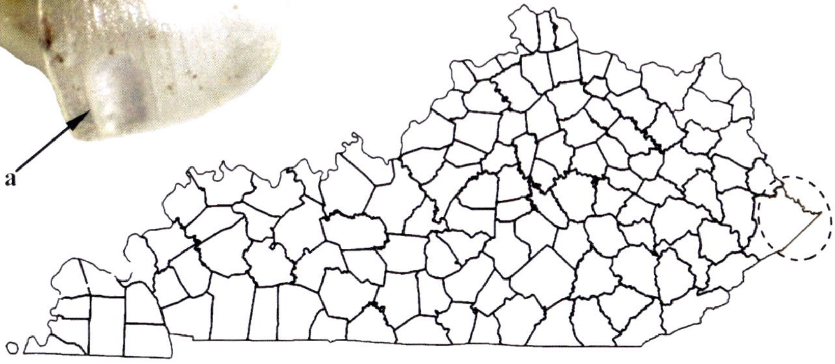

Black Gloss Gastrodontidae

Zonitoides nitidus (Müller, 1774)

Diameter: 6-8 mm, height 3.6-4 mm

Description: Heliciform; lip simple not reflected; shell with 4.5–5 whorls; umbilicate; olivaceous-yellow, semi-transparent and very glossy; transverse striae are poorly developed; without spiral striae; without teeth in any stages of growth; shell periphery roundish.

Similar Species: *Z. arboreus* is smaller, typically more depressed (although this is somewhat variable among populations), glossier with spiral striae, and the aperture is more oval in shape, but this feature also varies considerably between populations.

Habitat: A species of low ground; found around logs and leaf litter near floodplains, marshes, and wet roadsides; a Holarctic species (Hubricht 1985); may be transported on construction lumber to new locations.

Status: This species is reported from Ohio and West Virginia. Could show up just about anywhere in Kentucky on construction materials.

Specimen: Figures (a) from Massachusetts, Suffolk County, Great Brewster Island (author's collection).

a

Zonitoides arboreus

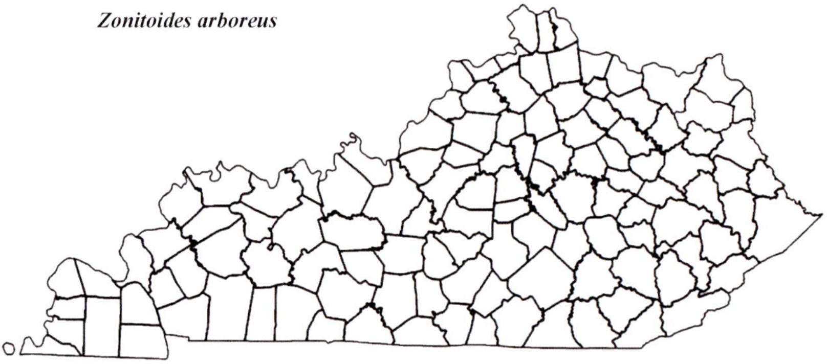

Eightfold Pinecone

Strobilops affinis Pilsbry, 1893

Diameter: 2.7 mm, height 2.5 mm

Description: Dome shape; lip reflected; shell with 6 tightly coiled whorls; perforate to umbilicate; light to dark brown, glossy; elongated teeth or lamellae present in the aperture; elongated lamellae present in the aperture; within the interior are a series of rather short, nearly equal lamellae, that are in line (this page, figures a & b; these lamellae are easily seen through the bottom of live and fresh dead shells or through the opened top; transverse striae modified into minute ribs.

Similar Species: *Strobilops labyrinthicus* has a smaller umbilicus, notably lower shell profile and a more roundish periphery; *S. texasianus* has stronger developed riblets on the base.

Habitat: Found in mixed hardwood forests and in both open and shaded glades.

Status: This mostly northern species is known from Meigs County, Ohio along the Ohio River, a few counties east of Kentucky.

Specimen: New York, Duchess County, upper Red Hook (FM 234948).

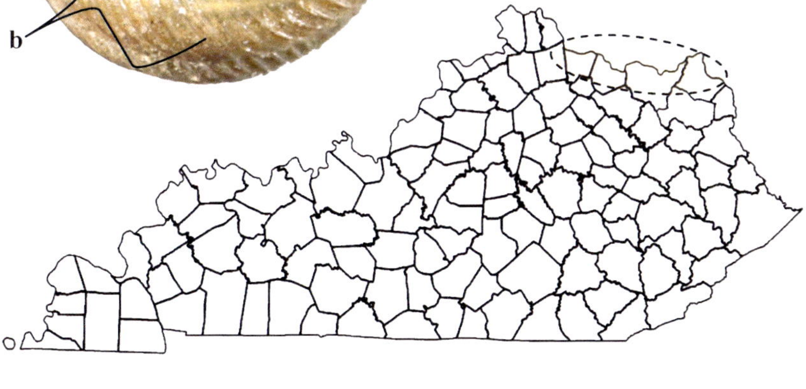

Twilight Coil

Helicodiscus multidens Hubricht, 1962

Diameter: 4.75 mm, height 1.55 mm

Description: Depressed heliciform; lip simple, aperture roundish; shell with 4.5-5 whorls; shell pale greenish-yellow, dull, opaque; widely umbilicate; in the last whorl there are usually three pairs of teeth on the outer and basal wall (a), figure shows one pair of teeth (other sets are deeper)—these teeth are radially elongate and are about twice as broad as high (unlike the teeth of other *Helicodiscus* species which are more knob-like); alternating with these paired teeth are usually three teeth on the parietal wall (b) figure shows only one (others are deeper in shell); under a strong lens the spiral striae (about 28-35 on the body whorl) are fringed (figured on opposite page) and present on all whorls, but this feature is typically lost in aging shells.

Similar Species: *Helicodiscus parallelus* and *H. notius* have knob-like teeth and are without a wide parietal tooth.

Habitat: A calciphile found under rocks and leaf litter on river bluffs and in caves.

Status: Not yet recorded from Kentucky, but known from Tennessee and Virginia counties that boarder the state; area of most likely occurrence in dashed circle below.

Specimen: Tennessee, Blount County, Gregory Cave area, GSMNP (author's collection).

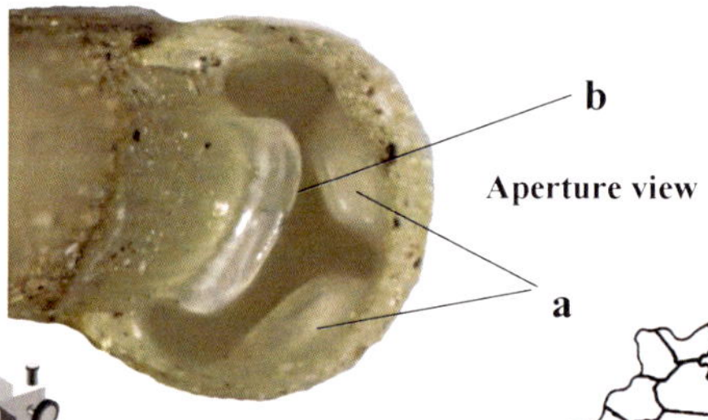

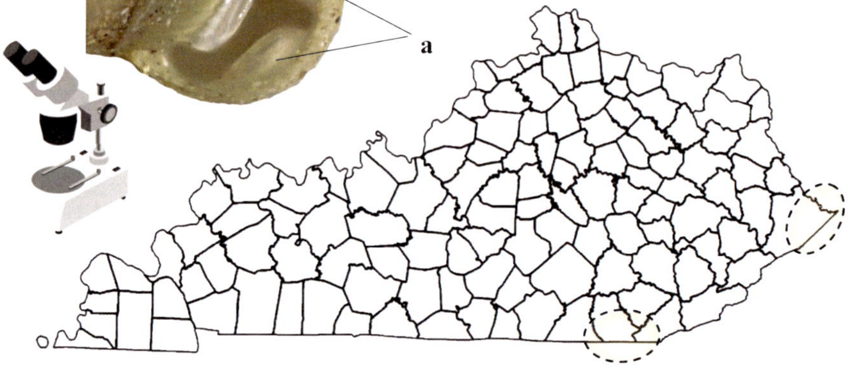

Glossy Dome

Ventridens acerra (J. Lewis, 1870)

Diameter: 12.6-19 mm, height 11-12 mm

Description: Heliciform; lip simple; shell with 7-8 moderately coiled whorls, the last whorl only slightly expanded; perforate; shell light yellowish-olive and very glossy; transverse striae are poorly developed, strongest on the top but fade away on the sides and base; without notable spiral striae; no teeth at any stage of growth; periphery rounded.

Similar Species: *Ventridens arcellus* is smaller, has a larger umbilicus, is more tightly wound (the last whorl not expanding greatly), and is usually found above 1200 m; *V. ligera* is less glossy and found in low places.

Habitat: A habitat generalist found in a variety of mixed hardwood forests on hillsides under leaf litter usually under 1200 m in elevation.

Status: Not yet reported from Kentucky but reported from a Tennessee county that boarders Bell County, Kentucky; area of most likely occurrence in dashed circle below.

Specimen: North Carolina, Swain County, Nantahala National Forest (author's collection).

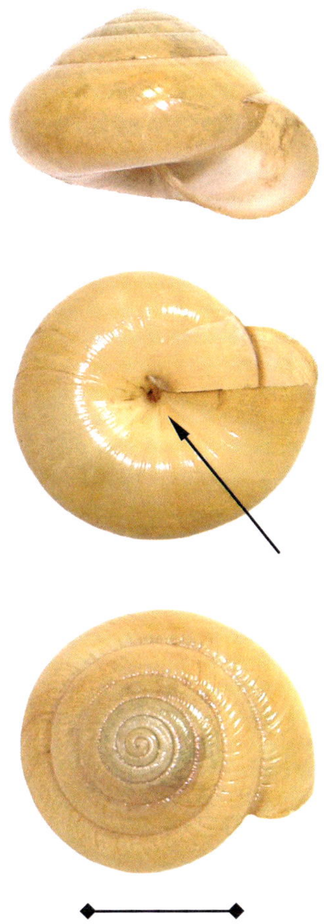

***V. acerra*, Graham County, North Carolina**

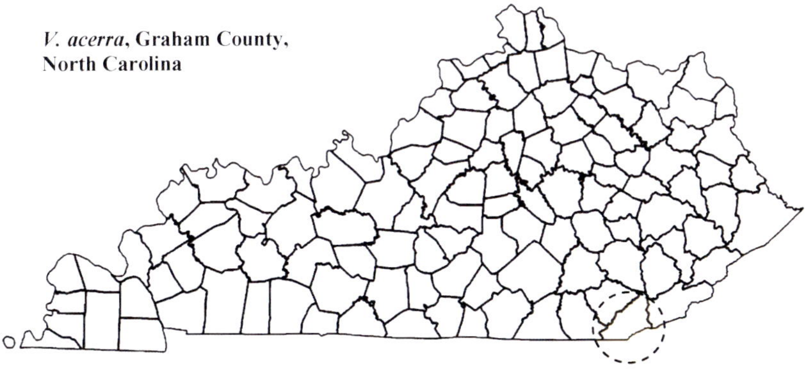

Carinate Dome

Gastrodontidae

Ventridens eutropis Pilsbry, 1946

Diameter: 14-15 mm, height 7.7 mm

Description: Depressed heliciform; lip simple; shell with 5.5-6 whorls; perforate; shell yellowish-horn to olive-buff and dull; broken transverse striae are well developed; spiral striae are also well developed; without teeth or basal lamellae in any stage of growth; shell periphery sharply carinate.

Similar Species: *V. intertextus* periphery is slightly angular or rounded but has the same unique broken transverse striae and spiral sculpture; *Euchemotrema hubrichti* is smaller and more neatly compressed, has a long blade-like parietal tooth and notably reflected lip.

Habitat: Found in mixed oak forests on hillsides under leaf litter, also under flat limestone slabs around brushy vegetation.

Status: Not yet recorded in Kentucky; area of possible occurrence in the state in dotted line.

Specimen: Tennessee, Cheatham County, 1.5 miles southeast of Pleasant View (FMNH 246996).

Ventridens intertextus

Euchemotrema hubrichti

Alabama Tigersnail

Anguispira alabama (G. H. Clapp, 1920)

Diameter: 18-21.5 mm, height 9.5-10 mm

Description: Depressed heliciform; lip simple; shell with 6 whorls; widely umbilicate; transverse striae are modified into distinct ribs that are well spaced; periphery sharply carinate; color light with radiating streaks on top and a single row of squarish flames just below the carina on the base that are usually distinct, however is not well illustrated in the bottom view of the specimen displayed here.

Similar Species: *Anguispira alternata* is larger in size and more finely sculptured; the transverse striae are less rib-like; *A. mordax* is about the same diameter but is less compressed and does not have a sharply carinate periphery.

Habitat: A calciphile; found climbing on rocky limestone ledges and cliffs at night and in wet weather (Hubricht 1985).

Status: Common only on the Cumberland Plateau of Tennessee including Tennessee counties that border Kentucky; area of possible occurrences in dotted line below.

Specimen: Tennessee, White County, Cumberland Mountain (FMNH 238002 image by Jochen Gerber).

Anguispira alternata

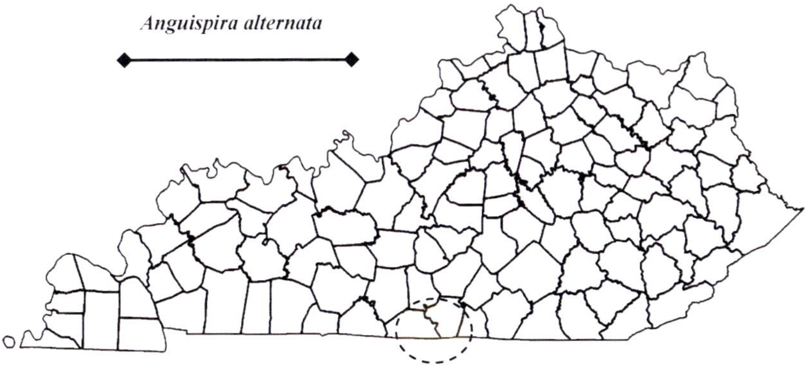

Sequatchie Slitmouth **Polygyridae**

Stenotrema edgarianum (I. Lea, 1841)

Diameter: 9-9.4 mm, height 5 mm

Description: Depressed heliciform; shell lens-shaped; lip reflected and closed; shell with 5 whorls; imperforate; cinnamon-buff; top of shell sculptured with scattered, periostracal ridges, the base with stiff hairs; basal notch and the interdenticular sinus are barely perceptible; fulcrum well developed with a slightly convex edge; shell periphery sharply shouldered and angular to carinate.

Similar Species: *Stenotrema spinosum* is larger, more compressed and has a wider, more open aperture; *S. edvardsi* is slightly smaller with a less compressed shell and is not as abruptly shouldered as in *S. edgarianum*.

Habitat: Found on logs and under leaf litter on wooded hillsides; never found in numbers (Hubricht 1985).

Status: Reported from Tennessee counties that border Kentucky, the main distribution of this interesting species occurring along the Cumberland Plateau in Tennessee; area of possible occurrence in the state, especially along Kentucky's Cumberland Plateau region.

Specimen: Figures (a) from Tennessee, Bledsoe County (CM 94378) and figure from Bledsoe County (image by www.jaxshells.org)

Stenotrema edvardsi

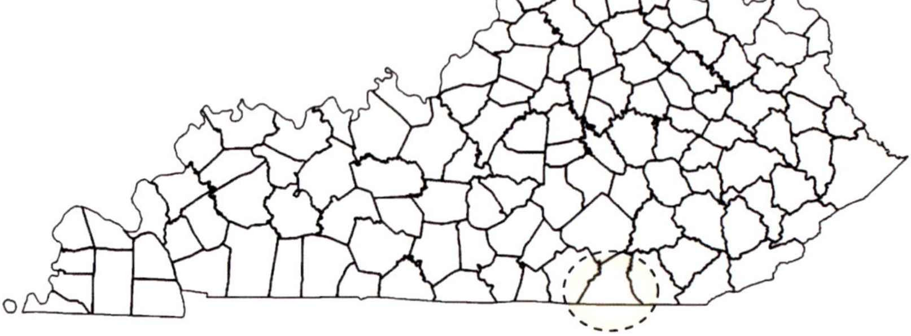

Carinate Slitmouth

Stenotrema spinosum (I. Lea, 1830)

Diameter: 11.8-15 mm, 5-6.5 mm

Description: Depressed heliciform; shell lens-shaped; lip reflected; shell with 5.5-6 whorls; imperforate; cinnamon-brown; without stiff hairs; spiral striae poorly developed on base; basal notch and the interdenticular sinus are barely perceptible; fulcrum long (a); shell periphery sharply and notably angular to carinate.

Similar Species: *Stenotrema edvardsi* is not nearly as compressed and is with stiff hairs.

Habitat: Usually found on limestone rock faces where it hides in narrow crevices by day.

Status: A more southern snail, as far south as southern Alabama; the species is found in counties from Tennessee and Virginia that border Kentucky around Cumberland Gap National Park; area of possible occurrence in Kentucky.

Specimen: Tennessee, Franklin County (author's collection).

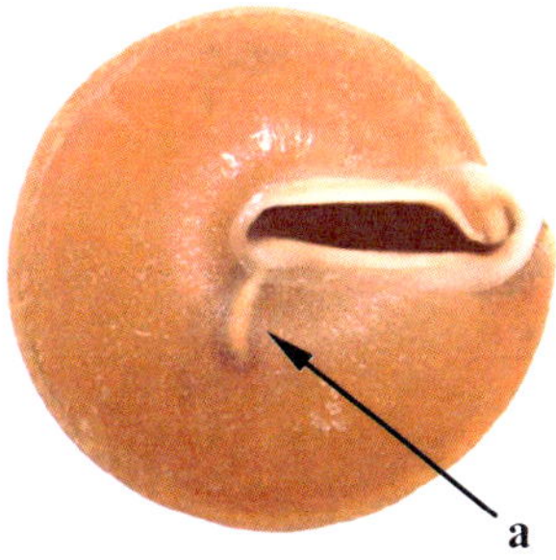

Stenotrema edgarianum

Euchemotrema hubrichti

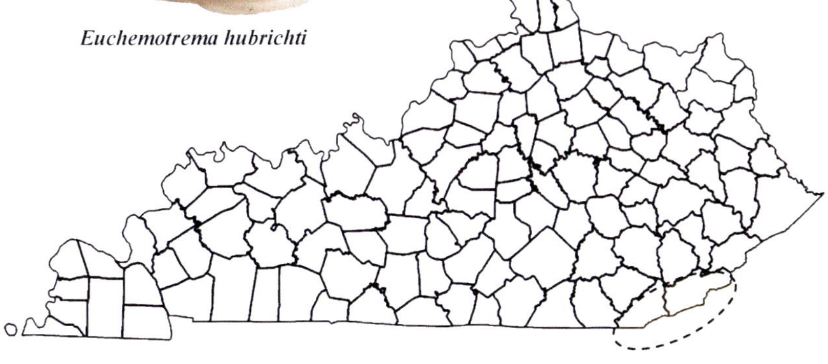

Carinate Pillsnail

Polygyridae

Euchemotrema hubrichti (Pilsbry, 1940)

Diameter: 8.9-9.7 mm, height 3.8-4.7 mm

Description: Depressed heliciform; shell lens-shaped and acutely carinate; lip reflected and closed; shell with 5 whorls; imperforate; cream to buff; spire low conic, embryonic whorls mostly granulose; later whorls lightly striate; base with fainter striae and densely minutely granulose; aperture resembles *E. fraternum* but is narrower, the parietal tooth nearer to the basal lip.

Similar Species: *Stenotrema spinosum* is larger, more compressed and has a wider, more open aperture.

Habitat: Known only from patchy limestone bluffs in the LaRue-Pine Hills region above the globally famous Snake Road. Sites have one or more relatively flat shelves, often covered with thin, loose limestone slabs of various sizes, area often shaded by eastern red cedars.

Status: Endemic to Illinois, although not likely living in Kentucky, the species might be discovered in debris along the Mississippi River during or after a major flood events.

Specimen: Illinois, Union County, top of bluff off S Larue Road (author's collection)

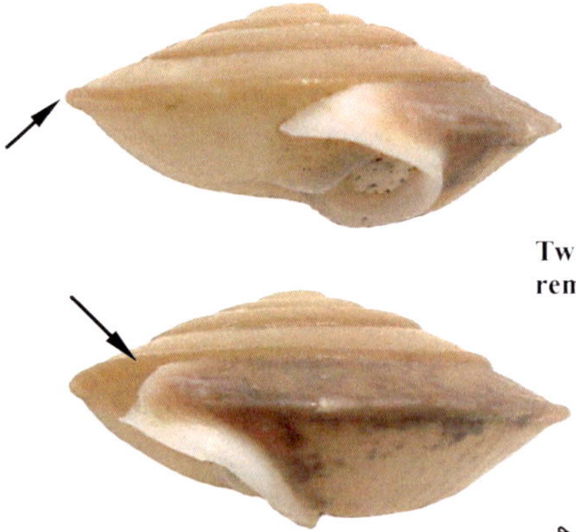

Two additional views of this most remarkable and handsome species

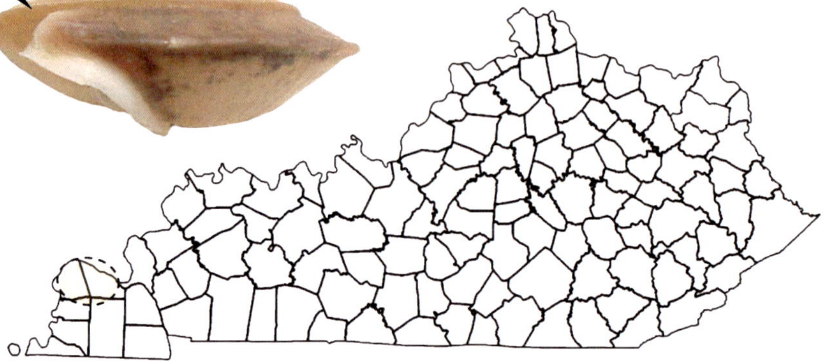

Atlantic Threetooth

Polygyridae

Triodopsis juxtidens (Pilsbry, 1894)

Diameter: 11.9-19.1 mm, height 6.6-10.2 mm

Description: Depressed heliciform; lip reflected; shell with 5-5.5 whorls; umbilicate; the large parietal tooth points directly at or in most cases above the palatal tooth; basal tooth and palatal tooth smaller; palatal tooth not situated on a buttress; reddish-brown; transverse striae are well developed on the top and sides of the shell continuing into the umbilicus region; scattered papillae are on top and base of shell, strongest around the umbilicus region; adults without hairs; periphery rounded.

Similar Species: *T. tennesseensis* has a notably smaller parietal tooth, is less compact, and the basal and palatal teeth are typically smaller; *T. juxtidens robinae* has a wider lip, larger umbilicus, and is restricted to the upper Bluestone River watershed in Mercer and Summer Counties, West Virginia.

Habitat: A species found under leaf litter, logs and rocks of upper elevation in a variety of mixed hardwood forests; also a snail of waste places, pastures, roadsides and urban areas.

Status: Not yet recorded from Kentucky but found in eastern West Virginia in counties that border Kentucky; area of most likely occurrence in dashed circle below.

Specimen: West Virginia, Pendleton County, near Franklin (author's collection).

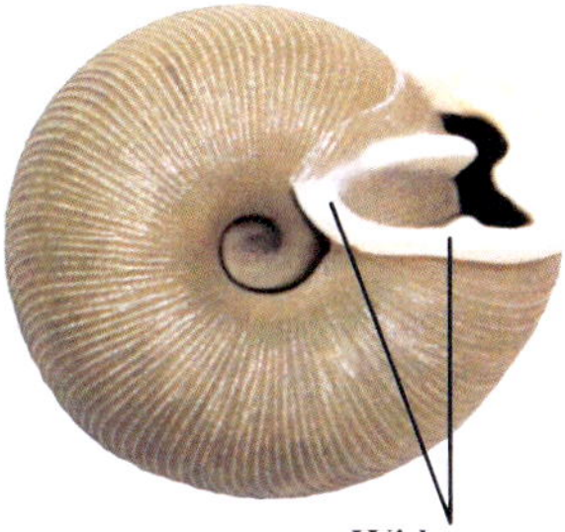

Triodopsis tennesseensis

Buttressed Threetooth **Polygyridae**

Triodopsis rugosa Brooks and MacMillan, 1940

Diameter: 9.2-10.9 mm, height 5.6 mm

Description: Depressed heliciform; lip reflected and thickened within; shell with 5-5.5 whorls; umbilicate; large parietal tooth, points above the palatal tooth; basal tooth and palatal tooth substantially smaller, of nearly equal size; palatal tooth situated on a wide buttress; shell brown; the best diagnostic feature of the shell is the well developed and widely-spaced transverse striae observed on the top and sides, continuing as strongly to the base (see page 429); scattered papillae; adults without hairs; periphery is rounded or shouldered.

Similar Species: *Triodopsis tridentata* is larger, the parietal tooth is smaller and points at or below the palatal tooth; *T. anteridon* has a larger shell with a slightly wider umbilicus and finer, more closely set transverse striae.

Habitat: Found in damp ravines and on hillsides of mixed hardwood forests under leaf litter and close to logs.

Status: G1; Not yet reported from Kentucky but found close to eastern Kentucky borders; area of most likely occurrence in dashed circle below.

Specimen: West Virginia, Logan County, Blair Mountain (FMNH 264690).

Triodopsis tridentata

Triodopsis anteridon

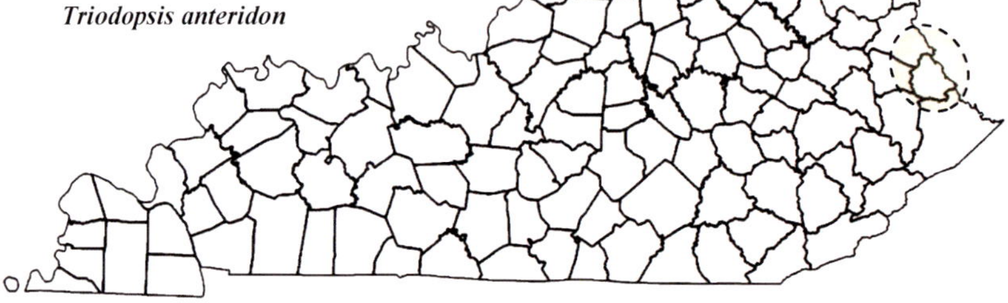

Triodopsis rugosa and *Triodopsis anteridon* Compared

Western Whitelip **Polygyridae**

Neohelix alleni (Wetherby in Sampson, 1883)

Diameter: 20.7-31.5 mm, height 13.2-19.2 mm

Description: Heliciform; lip reflected; shell with 5-5.5 whorls; umbilicate; very rarely with a small parietal tooth; color varies from brown to honey yellow; glossy; transverse striae are well developed on the top and sides of the shell continuing into the umbilicus region; spiral striae strongest around the umbilicus region; adults without hairs; periphery rounded.

Similar Species: Most similar to *Neohelix albilabris* in size and body form but is more notably glossy with a thinner more delicate lip; periphery well rounded.

Habitat: Usually found on wooded hillsides and in ravines, under rocks and logs, in leaf litter, along roadsides and waste ground in urban areas. Found in same habitat as *N. albolabris* but not with it (Hubricht, 1985).

Status: G5; Not yet reported from Kentucky but found in Scott County, Missouri that borders Ballard County Kentucky; area of most likely occurrence in Kentucky in shaded dotted line below.

Specimen: Missouri, Adair County, Kirksville, ©Claude and Amandine Evanno on Flickr 11-VI-2007.

Neohelix albolabris

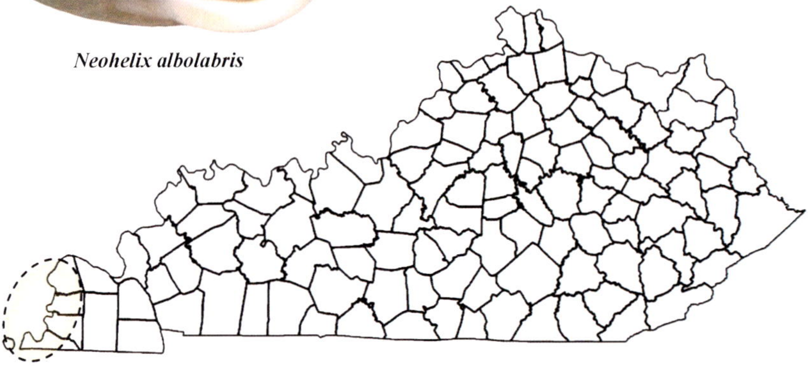

Engraved Bladetooth **Polygyridae**

Patera perigrapta (Pilsbry, 1894)

Diameter: 15.8-23.4 mm, height 8.5-12.6 mm
Description: Depressed heliciform; lip reflected; shell with 5.5-6 whorls; imperforate; parietal tooth small with a narrow base; basal tooth small and sometimes poorly defined; tan colored; shell surface glossy; no hairs; transverse striae well developed; spiral striae (not papilla) are engraved into the shell and are a strong and constant feature.
Similar Species: *Patera appressa* is generally smaller, more compressed, has spirally arranged papilla and a long blade-like partial tooth; *P. laevior* is smaller and has spirally arranged striae that often times is barely perceptible; the parietal tooth of *P. panselenus* has a wider base and its shell is more compressed.
Habitat: Commonly found on downed logs on their undersides, sometimes in hollow trees in mixed hardwood forests.
Status: The status of this species in Kentucky remains unclear; Branson (1973) reported it in Pike and Bell counties; Hubricht (1985) records do not show it in the state, but shows it in Tennessee Counties that border Monroe and Cumberland County, Kentucky; area of most likely occurrence in Kentucky below.
Specimen: North Carolina, Swain County, Nantahala National Forest (author's collection).

Patera appressa

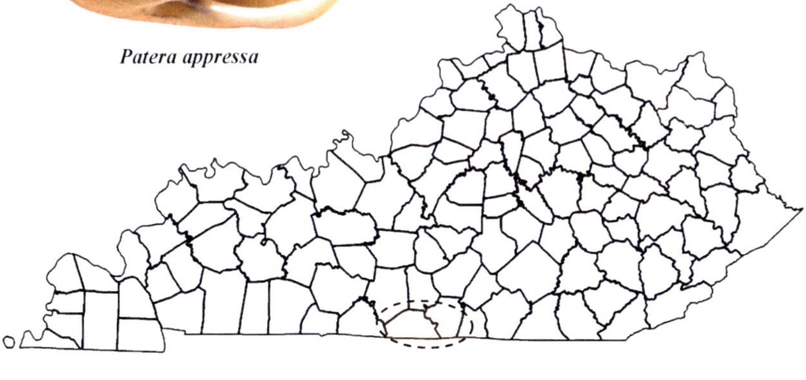

Part V
IV Ecoregions of Kentucky

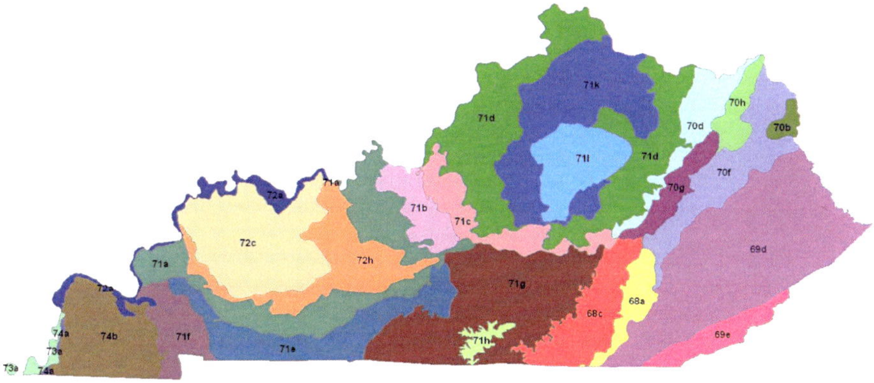

Land snail distributions as they relate to specific Ecoregions in Kentucky are poorly understood. Consequently, I felt that including maps and detailed descriptions of **Level IV Eco-Regions** might be of use. As snail inventories continue, perhaps land snail patterns will emerge.

The Level IV Ecoregions section below was provided by Woods, A.J., Omernik, J.M., Martin, W.H., Pond, G.J., Andrews, W.M., Call, S.M, Comstock, J.A., and Taylor, D.D., (2002). Maps occurring with each Level IV Ecoregion created by Piper Roby.

Ecoregions denote areas of general similarity in ecosystems and in the type, quality, and quantity of environmental resources. They are designed to serve as a spatial framework for the research, assessment, management, and monitoring of ecosystems and ecosystem components. By recognizing the spatial differences in the capacities and potentials of ecosystems, ecoregions stratify the environment by its probable response to disturbance (Bryce *et al.* 1999). Ecoregions are general purpose regions that are critical for structuring and implementing ecosystem management strategies across federal agencies, state agencies, and nongovernment organizations that are responsible for different types of resources in the same geographical areas (Omernik *et al.* 2000).

The approach used to compile these maps is based on the premise that ecological regions can be identified through the analysis of the spatial patterns and the composition of biotic and abiotic phenomena that affect or reflect differences in ecosystem quality and integrity (Omernik 2000). These phenomena include geology, physiography, vegetation, climate, soils, land use, wildlife, and hydrology. The relative importance of each phenomenon varies from one ecological region to another regardless of ecoregion hierarchical level. A Roman numeral hierarchical scheme has been adopted for different levels of ecological regions. Level I is the coarsest level, dividing North America into 15 ecological regions. Level II divides the continent into 52 regions (Commission for Environmental Cooperation 1997). At level III, the continental United States contains 104 ecoregions and the conterminous United States has 84 ecoregions. Level IV is a further subdivision of level III ecoregions. Explanations of the methods used to define the USEPA's ecoregions are given in Omernik *et al.* (2000).

In Kentucky, there are 7 level III ecoregions and 25 level IV ecoregions; all but four level IV ecoregions continue into ecologically similar parts of adjacent states (Griffith *et al.* 1998; Woods *et al.* 1998). Ecological and biological diversity in Kentucky is very strongly related to regional physiographic, geologic, land use, and soil characteristics. Deciduous forests widely covered Kentucky at the time of European settlement. About half is still forested. Extensive "barrens" (i.e. bluestem prairies) were once maintained by fires set by Native Americans on rolling to flat parts of the Interior Plateau (71) and Mississippi Valley Loess Plains (74). Today, these areas are dominated by cropland and pastureland and the historic "barrens" are nearly extinct. Major coal fields occur in the Southwestern Appalachians (68), Central Appalachians (69), Western Allegheny Plateau (70), and Interior River Valleys and Hills (72).

This work is part of a collaborative project primarily between USEPA Region 4, USEPA National Health and Environmental Effects Research Laboratory (Corvallis, Oregon), and the Kentucky Natural Resources and Environmental Protection Cabinet–Department for Environmental Protection (KDEP). Collaboration and consultation also occurred with the U.S. Department of Agriculture–Forest Service (USFS), U.S. Department of Interior–Geological Survey (USGS), The Nature Conservancy, Eastern Kentucky Earth Resources Observation Systems Data Center, Kentucky Geological Survey, and other Commonwealth of Kentucky agencies.

The project is associated with an interagency effort to develop a common framework of ecological regions (McMahon *et al.* 2001). Reaching that objective requires recognition of the differences in the conceptual approaches and mapping methodologies applied to develop the most common ecoregion-type frameworks, including those developed by the USFS (Bailey *et al.* 1994), the USEPA (Omernik 1987, 1995), and the U.S. Department of Agriculture–Soil Conservation Service. As each of these frameworks is further refined, their differences are becoming less discernible. Each collaborative ecoregion project, such as this one in Kentucky, comprises a step toward attaining consensus and consistency in ecoregion frameworks for the entire nation.

Description of Level IV Ecoregions of Kentucky

(excerpted from EPA's Level IV Ecoregions of Kentucky Poster with permission)

68a. Cumberland Plateau

The Cumberland Plateau (68a) is composed of low hills, ridges, rolling uplands, and valleys. It is underlain by flat-lying, Pennsylvanian sandstone, shale, siltstone, conglomerate, and coal; Mississippian carbonates are absent in contrast to the nearby Plateau Escarpment (68c), Northern Forested Plateau Escarpment (70g), and Eastern Highland Rim (71g). Ecoregion 68a is mostly forested. Forest composition is highly variable. Logging, coal mining, and livestock grazing are common and limited cropland occurs on broader ridge tops and in valleys. Cleared land, as a percentage of the total, is greater than in more rugged Ecoregions 68c and 69d but less than in Ecoregion 70f. Streams have moderate to low gradients and have broader floodplains and more riparian wetlands than in the Plateau Escarpment (68c). Acidic drainage and sedimentation from coal mining have decreased the biological productivity of many streams. As a result of sedimentation, streams are generally more turbid than in Ecoregion 68c.

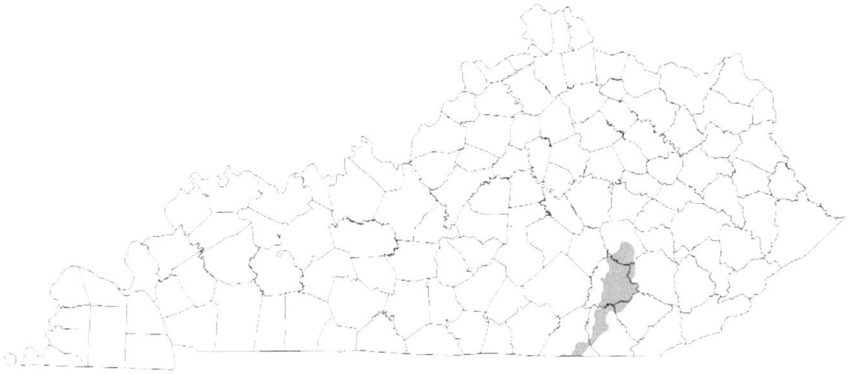

68c. Plateau Escarpment

The Plateau Escarpment (68c) contains narrow ridges, cliffs, and gorges. It is more rugged, dissected, and forested and has higher average stream gradients than Ecoregions 68a and 71g. Uplands are underlain by Pennsylvanian strata including cliff-forming sandstone and coal. Lower slopes and western valleys are usually underlain by Mississippian carbonates that are absent from Ecoregions 68a and 70f. Some of the highest quality streams in Kentucky occur here and have high gradients, riffles, pools, and boulder or bedrock substrates. They are home to many rare or endangered fishes and mussels and certain segments have been designated as Kentucky Wild Rivers. Other streams have cut down to Mississippian limestone and have wider valleys, lower gradients, warmer thermal regimes, higher nutrient and alkalinity levels, and a higher number of fish and mussel species. Logging and coal mining occur. Acidic drainage and sedimentation from coal mining have lowered the biological productivity of many stream reaches.

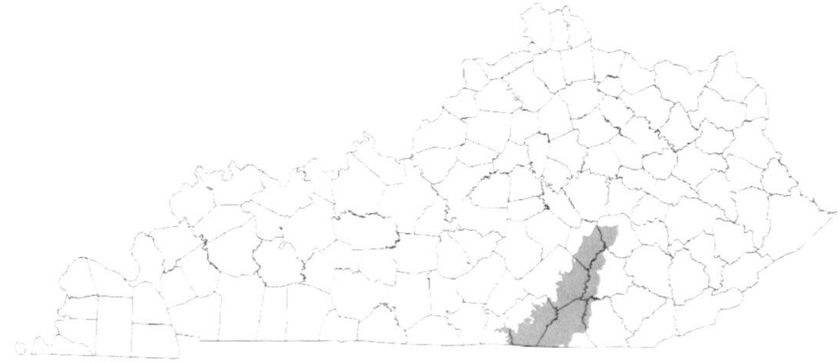

69d. Dissected Appalachian Plateau

The Dissected Appalachian Plateau (69d) is composed of narrow ridges, deep coves, and narrow valleys and is mostly forested. Cool, high gradient streams with cobble and boulder substrates and extensive riffles are common. Ecoregion 69d is more rugged, more extensively forested, and has higher stream gradients than the Cumberland Plateau (68a) and the Ohio/Kentucky Carboniferous Plateau (70f). Forest composition is controlled by aspect, slope position, degree of topographic shading, and past usage and, thus, is highly variable. Ecoregion 69d is underlain by flat-lying Pennsylvanian shale, siltstone, sandstone, and coal. Surface and underground coal mining, logging, and both gas and oil production are common and have degraded surface waters. Acidic drainage and sedimentation from coal mines have decreased the biological productivity of many streams and, in some reaches, all but the most tolerant aquatic biota have been eliminated. However, gradual improvement in the control of acidic mine drainage is occurring. Nutrient levels in streams are very low and are a reflection of the ecoregion's low population density, limited agriculture, and non-carbonate rocks.

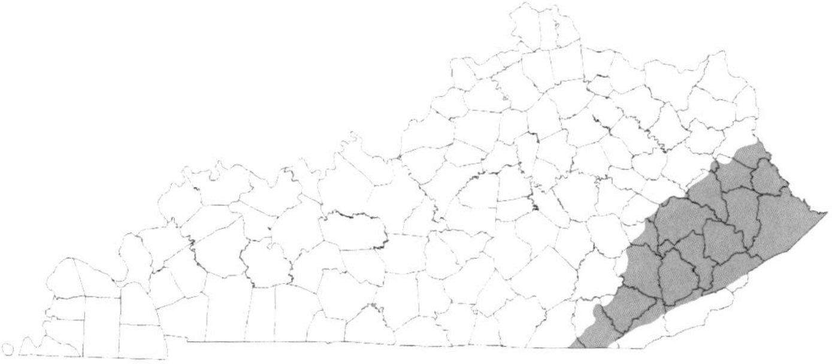

69e. Cumberland Mountain Thrust Block

The mostly forested Cumberland Mountain Thrust Block (69e) contains high, steep ridges, hills, coves, narrow valleys, and the Pine Mountain Overthrust Fault. Maximum elevation is greater than elsewhere in Kentucky, Black Mountain reaching 4,150 feet. Forests are usually more mesophytic than in the Dissected Appalachian Plateau (69d) but forest composition is highly variable and controlled by aspect, slope position, past usage, and degree of topographic shading. Components of the bird, amphibian, small mammal, land snail and plant assemblages are also distinct from Ecoregion 69d. This region harbors the most distinctive and rarest land snail fauna in the state containing many northern and Southern Appalachian species. The Cumberland Mountain Thrust

Block (69e) is mostly underlain by Pennsylvanian shale, siltstone, sandstone, conglomerate, and coal. Sedimentation from coal mines, coal washing, and logging as well as acidic mine drainage have decreased the biological integrity and productivity of surface waters. Small streams are common and have high gradients, waterfalls, many riffles, few pools, and cobble or boulder substrates. Nutrient and alkalinity levels are lower, thermal regimes are cooler, and fish populations are less diverse than in Ecoregion 69d.

70b. Monongahela Transition Zone

The Monongahela Transition Zone (70b) is differentiated from other ecoregions by its clayey regolith, erosion-prone soils, land slips, and rough, broken terrain. Steep, clayey slopes can be unstable when saturated and, as a result, streams can be more turbid than elsewhere in Ecoregion 70. Vandalia and Upshur soils are common and are easily erodible after disturbance. They have developed from Pennsylvanian clay shale, siltstone, and sandstone of the Monongahela and Conemaugh formations which are not widely exposed in other parts of Kentucky. Mixed deciduous– evergreen forests dominate steeper hills and ridges, pastureland is found on gentler slopes, and some cropland occurs in valleys. Streams have cobble, gravel, and sand substrates and moderate gradients.

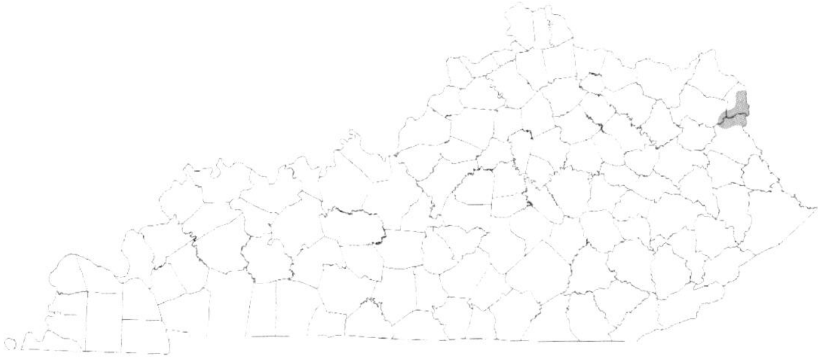

70d. Knobs–Lower Scioto Dissected Plateau

The Knobs–Lower Scioto Dissected Plateau (70d) contains rounded hills and ridges, narrow valleys with high gradient streams, and a few wide, locally swampy, bottoms underlain by weak shales. Cliffs occur especially in the south. High amounts of topographic and geologic variation are typical and create substantial ecological diversity. On a per site basis, these Knobs in the vicinity of Powell County and neighboring ecoregions contain the highest reported land snail faunas in North America (Dourson 2006). Ecoregion 70d is underlain by a mixture of Pennsylvanian-age through Silurian-age sedimentary rocks that is absent from the rest of Ecoregion 70. Ecoregion 70d is geographically adjacent and ecologically connected to the Western Allegheny Plateau (70) and, as such, is not a part of the Interior Plateau (71). Uplands knobs are forested and oak and oak–pine forests predominate. Broad valleys are mostly covered by bottomland forests but some are used for livestock or general farming. Elevation, local relief, and forest density are much greater than in Ecoregions 71d, 71k, and 71l. Nutrient and ionic concentrations in streams are lower than in Ecoregions 71d and 71l. No coal mining or related stream acidity problems occur.

70d. Knobs–Lower Scioto Dissected Plateau

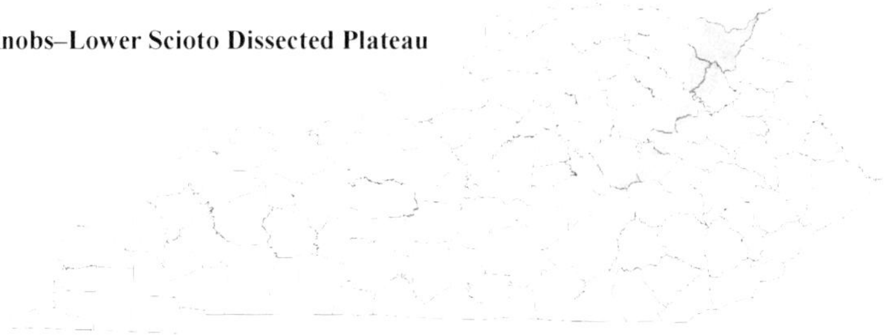

70f. Ohio/Kentucky Carboniferous Plateau

The hilly Ohio/Kentucky Carboniferous Plateau (70f) is a mosaic of woodland, pastureland, and cropland. Mixed deciduous–evergreen forests characterized by oaks and pines occur. Ecoregion 70f is lower and less rugged than adjacent portions of the Dissected Appalachian Plateau (69d) and Northern Forested Plateau Escarpment (70g). Characteristically, cleared land, as a percentage of the total, is greater than in Ecoregions 69d, 70g, and 70h. Valleys are less meandering, contain higher gradient streams, and have fewer wetlands than Ecoregion 68a. They lack the karst of Ecoregions 68c, 70g, and 70h. Streams are cool and have sand or boulder substrates. Fish and mussel assemblages are unlike the Cumberlandian fauna of Ecoregion 68a. Fish, macroinvertebrate, and diatom diversity is high in good quality streams. However, water quality in many stream reaches has been degraded by underground and surface coal mining, logging, agriculture, and oil production.

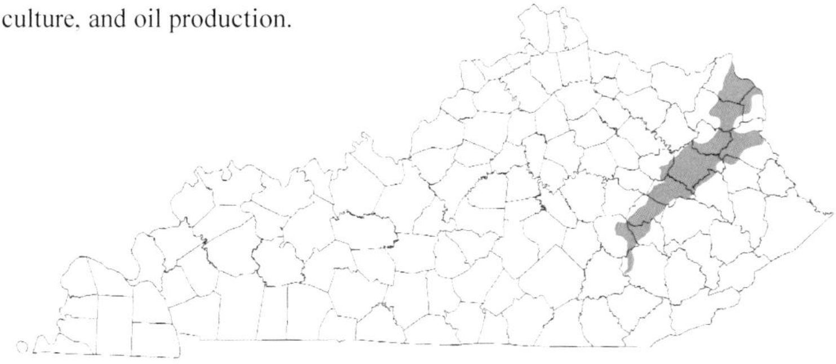

70g. Northern Forested Plateau Escarpment

The hills and ridges of the Northern Forested Plateau Escarpment (70g) are very rugged and highly dissected. Cliffs, narrow valleys, and ravines are common and characteristic.

Thick, resistant Pennsylvanian quartzose sandstone caps the ridges and Mississippian limestone, shale, and siltstone are exposed on lower slopes and in valleys. Karst is not extensive but does occur containing some of the most important bat caves in Kentucky. Ecoregion 70g is physiographically and lithologically unlike Ecoregion 70d and is higher and more rugged than adjacent portions of Ecoregions 70f and 71d. In addition, cliffs are more numerous and higher than in Ecoregion 70h. Streams are cool and typically have moderate to high gradients, rock bottoms, and low amounts of turbidity. Fish assemblages are diverse but lack the endemic Cumberlandian species of Ecoregion 68c but does contain the endemic white-haired goldenrod (a cliffline species). Mixed oak and oak–pine forests are common and bottomland hardwood forests grow along streams. Logging and recreation are important land uses and dwellings are found on valley sides and on bottomlands. There is less agriculture and cleared land than in Ecoregion 70f.

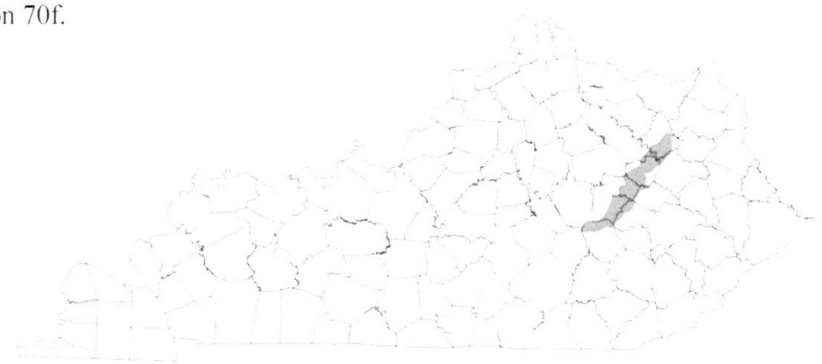

70h. Carter Hills

The Carter Hills (70h) is composed of hills, ridges, and narrow valleys. It is covered with mixed deciduous forests containing successional cedar and pine. Resistant Pennsylvanian sandstone and conglomerate cap the ridges and Mississippian limestone is exposed in some valleys. Limestone is thicker than in Ecoregion 70g and is associated with karst topography; karst is more extensive than in Ecoregion 70g. Ecoregion 70h is physiographically and lithologically unlike Ecoregion 70d and is higher and more rugged than adjacent portions of Ecoregions 70f and 71d. In addition, cliffs are lower and fewer in number than in Ecoregion 70g. Most stream segments have upland characteristics including moderate to high gradients, rock substrate, extensive riffles, and low waterfalls. Important land uses include logging, clay mining, recreation, and livestock farming.

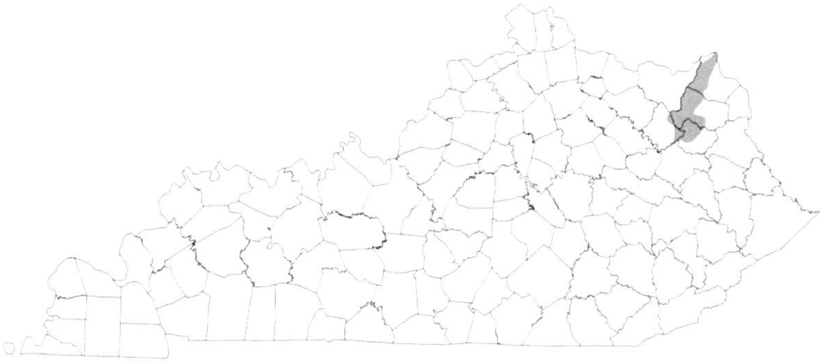

71a. Crawford–Mammoth Cave Uplands

The hilly Crawford–Mammoth Cave Uplands (71a) is higher and more rugged than neighboring Ecoregions 71b and 71e. Sandstone cliffs, dissected shale valleys, and less dissected limestone valleys with well developed karst occur. Resistant Mississippian sandstone forms the uplands and cavernous Mississippian limestones are often basal; Pennsylvanian carboniferous sedimentary rocks are absent in contrast to Ecoregions 72c and 72h. In valleys underlain by thick cavernous limestone, stream density is low and sinkholes, caverns, subterranean drainage, and springs are common. Elsewhere, surface drainage is significant and streams have higher gradients than in Ecoregions 71b and 71e. Upland streams are rocky, cool and clear. Rivers are all through- flowing, meandering, and deeply incised into bedrock. Fish assemblages vary among river systems. A mixture of forests, pastureland, and cropland occur; farming is more widespread than in the Caseyville Hills (72h) but not nearly as extensive as in Ecoregions 71b or 71e.

71b. Mitchell Plain

The rolling Mitchell Plain (71b) is underlain by Mississippian limestones and is characterized by well developed karst, low relief, and extensive agriculture. Sinkholes, ponds, springs, sinkhole wetlands, subterranean drainage, and dry valleys occur. Stream incision is typically limited except along master streams. Drainage density is lower than in Ecoregions 71a and 71c but higher than in Ecoregion 71e. Mean elevation, relief, and stream gradient are lower than in the lithologically distinct Ecoregions 71a, 71c, and 71g. Potential natural vegetation is a mosaic of bluestem prairie and oak–hickory forest. Today, cropland and pastureland is extensive, mixed oak forests are found on steep slopes, and pin oak, swamp white oak, and sweet gum grow in poorly-drained areas. Sinkhole wetlands are common. Water quality has been degraded by municipal effluent, agricultural discharge, and bank erosion following riparian forest removal. However, nutrient concentrations are not typically as high as in Ecoregion 71e.

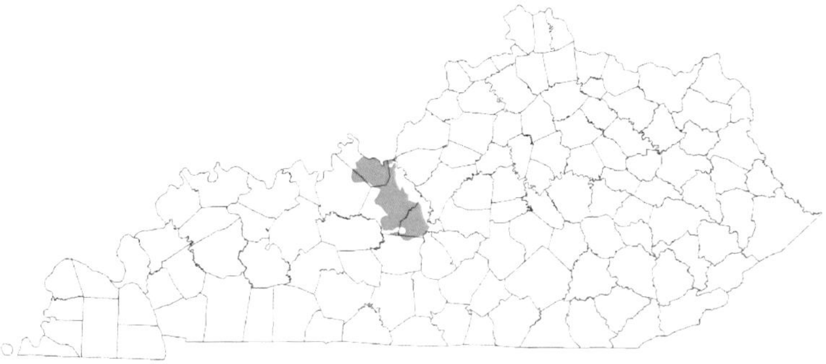

71c. Knobs–Norman Upland

The Knobs–Norman Upland is underlain by Pennsylvanian-age through Silurian-age sedimentary rocks. Its rounded hills and ridges are mostly forested and divide the Bluegrass (Ecoregions 71d, 71k, and 71l) from the rest of the Interior Plateau (71). Inceptisols and Ultisols occur on slopes and support mixed deciduous forests. Narrow, high gradient valleys are also common. In addition, a few wide, locally swampy valley floors occur and are used for livestock farming, general farming, and woodland. Ecoregion 71c is characterized by large amounts of geological, topographical, and ecological diversity. Overall, however, physiography, soils, lithology, and land use are distinct from the limestone-and Alfisol-dominated agricultural plains of Ecoregions 71b, 71d, 71e, and 71l. The density of perennial upland streams is far greater than on nearby limestone plains. Nutrient and ionic concentrations are much lower in streams that originate in Ecoregion 71d than outside it in heavily populated, agricultural ecoregions underlain by limestone. Fish and macroinvertebrate diversity is in between that of the Bluegrass (Ecoregions 71d, 71k, and 71l) and that of Ecoregions 71b, 71g, and 70g.

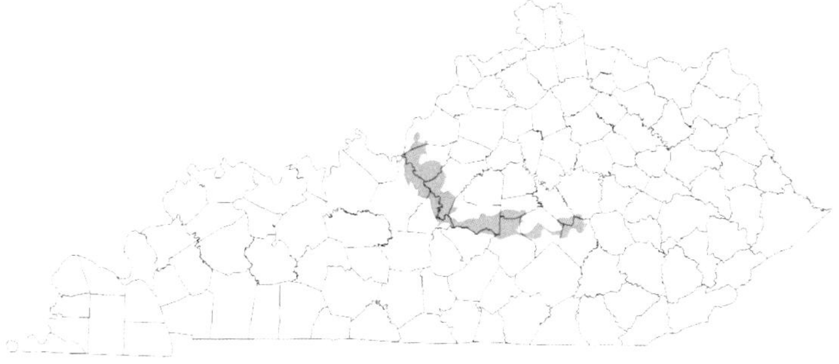

71d. Outer Bluegrass

The rolling to hilly Outer Bluegrass (71d) contains sinkholes, springs, entrenched rivers, and intermittent and perennial streams. Local relief is variable but is usually less than in the geomorphically distinct Knobs–Norman Upland (71c). Discontinuous glacial outwash and leached, pre-Wisconsinan till deposits occur in the north from Louisville to Covington. Glacial deposits do not occur elsewhere in Kentucky. Ecoregion 71d is mostly underlain by Upper Ordovician limestone and shale. Natural soil fertility is higher than in the shale-dominated Hills of the Bluegrass (71k). Today, pastureland and cropland are widespread and dissected areas are wooded. At the time of settlement, open savanna woodlands were found on most uplands. On less fertile, more acidic soils derived from Silurian dolomite, white oak stands occurred and had barren openings. Cane grew along streams and was especially common in the east. Distinct vegetation grew in areas underlain by glacial drift (see summary table). Upland streams have moderate to high gradients and cobble, boulder, or bedrock substrates. Mean stream density is greater than in Ecoregion 71l but less than in Ecoregion 71k. Mean summer stream temperatures are much warmer than in Ecoregions 71b, 71c, and 71e. Concentrations of suspended sediment and nutrients can be high.

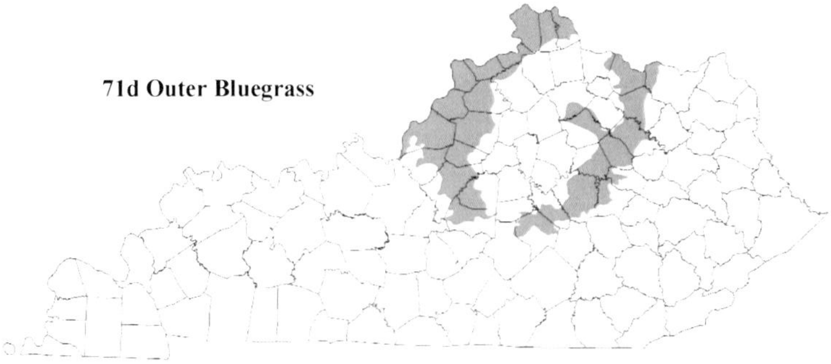

71e. Western Pennyroyal Karst Plain

The weakly dissected Western Pennyroyal Karst Plain (71e) is underlain by Middle Mississippian limestones and is extensively farmed; it is both physiographically and lithologically distinct from surrounding ecoregions. Sinkholes, ponds, springs, sinking streams, and dry valleys occur. Underground drainage is well developed, stream density is low, and soils are quick to dry. Most upland streams have limited discharge or are intermittent or ephemeral; they become laden with suspended sediment after heavy rains. Deeper, more intense dissection occurs near incised master streams which are fed by cool, nitrate-rich groundwater. Fish assemblages vary among river systems. Potential natural vegetation is mapped as a mosaic of bluestem prairie and oak–hickory forest. Barrens (i.e. bluestem prairies) were once more widespread than elsewhere in Kentucky. Today, extensive tobacco, livestock, corn, soybean, and small grain farming occurs.

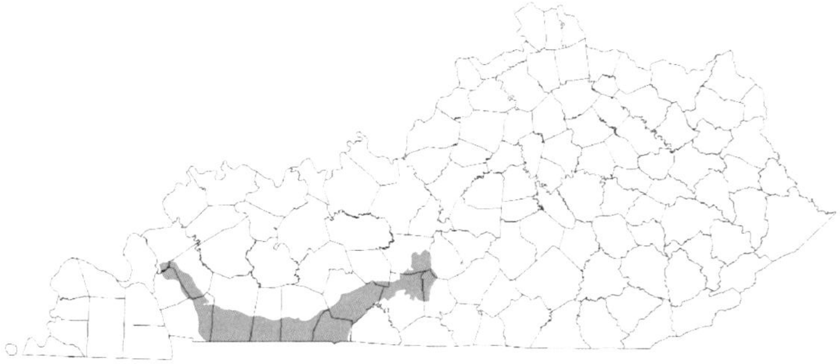

71f. Western Highland Rim

The Western Highland Rim (71f) is a hilly area that is bisected by the Tennessee River and the Cumberland River valleys. Ecoregion 71f is much more wooded and rugged than the nearby agricultural plains of Ecoregions 71e and 74b. It is underlain by Mississippian limestone and shales and, in the west, by Cretaceous-Paleocene shale, siltstone, and sandstone. Ridges and hills are often capped by cherty gravels and veneered by thin loess. Karst valleys underlain by limestone also occur. Ecoregion 71f is lithologically distinct from Ecoregions 71a, 71e, 72h, and 74b. Potential natural vegetation is oak–hickory forest; it lacks the barrens (i.e. bluestem prairies) of Ecoregions 71e and 74b.

Upland soils tend to be cherty, droughty, low in fertility, and are mostly covered by mixed oak forests. Some agriculture occurs on flatter interfluves and in valleys. Recreation is an important land use in the Land-Between-the-Lakes area. Streams are cool and clear. They have moderate gradients and gravel and sand substrates. Stream alkalinity and hardness vary from east to west but are usually greater than in Ecoregion 74. Fish assemblages vary among river systems.

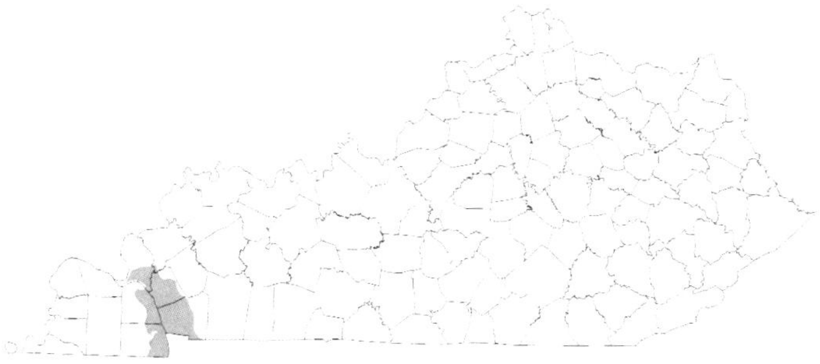

71g. Eastern Highland Rim

The Eastern Highland Rim (71g) is a diverse ecoregion with undulating plains, hills, and karst. Near the Cumberland River, steep bluffs, springs, cascades, and wide bottomlands occur. The degree of dissection is variable. Overall, nearly level terrain is more extensive than in Ecoregion 71f and stream density is greater than in the lower, less rugged Ecoregion 71e. Ecoregion 71g is mostly underlain by Mississippian limestone, chert, shale, siltstone, and sandstone; it is lithologically distinct from the Pennsylvanian carboniferous sedimentary rocks of Ecoregions 68, 69, and 70. Potential natural vegetation is mapped as oak–hickory forest but, in ravines near the Cumberland Plateau (68a), forests are mixed mesophytic in character. Today, white oak dominates upland forests and bottomland trees grow along streams. Streams are nutrient-rich and moderate in gradient. Riffle substrates are composed of cobble, gravel, or bedrock. Fish, macroinvertebrate, and mussel biodiversity is greater than in the Bluegrass (Ecoregions 71d, 7lk, and 71l).

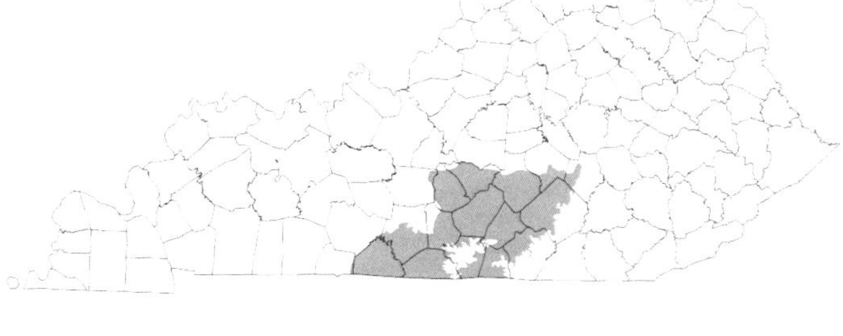

71h. Outer Nashville Basin

The Outer Nashville Basin (71h) in Kentucky contains untillable steep ridges and bluffs and cultivated terraces and floodplains along the meandering, downcutting Cumberland River. Springs occur in bluffs and feed waterfalls. Soluble limestone, dolomite, and weak shales of Mississippian-age through Ordovician-age are typically exposed. Potential natural vegetation is oak–hickory forest and, in moister areas, mixed mesophytic forest. Today, forests, pastures, and hay, tobacco, corn, soybean, and small grain farming occur. Cumberland River tributaries are productive, nutrient-rich, and mostly moderate in gradient. Macroinvertebrate communities more closely resemble those of the Bluegrass (Ecoregions 71d, 71k, and 71l) than the adjacent Eastern Highland Rim (71g). However, fish assemblages are Cumberlandian and are similar to those of Ecoregion 71g. Flow on the Cumberland River in Ecoregion 71h is dam-controlled; cold reservoir discharge has eliminated nearly all native fish and mussel fauna downstream to the Tennessee border.

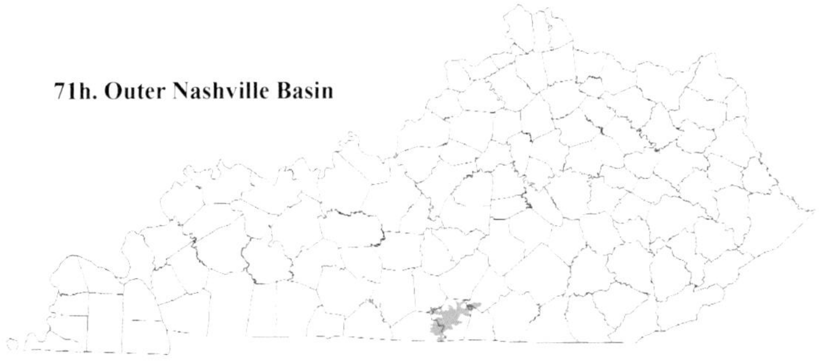

71k. Hills of the Bluegrass

The mostly forested Hills of the Bluegrass (71k) is underlain by Upper Ordovician calcareous shale, siltstone, and limestone. It is lithologically unlike the Knobs–Norman Upland (71c), Outer Bluegrass (71d), and Inner Bluegrass (71l). Upland soils are fairly high in phosphorus, potassium, and lime but are not as naturally fertile as Ecoregions 71d and 71l; they support young, mixed forests rich in white oak, hickory, and cedar. The Hills of the Bluegrass (71k) has steeper terrain, droughtier soils, lower soil fertility, higher drainage density, and is more erosion-prone than Ecoregions 71d and 71l. As a result, less than ten percent of Ecoregion 71k is suited to row crop agriculture and the rest is wooded, pastureland, or hayland. Stream nutrient levels are generally lower than in Ecoregions 71d and 71l. Upland streams are often intermittent and have cobble, boulder, or bedrock substrates. Gradients are steeper than in Ecoregion 71l. Fish and macroinvertebrate assemblages are similar to Ecoregions 71d and 71l but have elements that are distinct from Ecoregion 71c.

71l. Inner Bluegrass

The nearly level to rolling Inner Bluegrass (71l.) is a weakly dissected agricultural plain containing extensive karst, intermittent streams, and expanding urban-suburban areas that originally developed near major springs. Deep, forested gorges also occur along the Kentucky and Dix rivers. The Inner Bluegrass (71l) is characteristically underlain by Middle Ordovician Lexington Limestone and is lithologically distinct from the rest of Ecoregion 71. Very fertile Alfisols and Mollisols have developed from the residuum of underlying phosphatic limestone; natural soil fertility is greater than in Ecoregion 71k. The original open woodlands, savannas, and swamp forests have been largely replaced by agriculture and urban-suburban-industrial areas. However, deciduous forests containing eastern redcedar still occur in ravines, along the Kentucky River, and near streams. Thoroughbred horse farms, cattle grazing, tobacco, alfalfa, and hay farming are common land uses. Some upland streams are very warm and have seasonally variable flows but others, fed by major springs, are colder and have plentiful perennial flow. In either case, they have moderate to low gradients, cobble or bedrock substrates, and fish assemblages that are similar to the Outer Bluegrass (71d) and the Hills of the Bluegrass (71k). Higher gradient streams draining into the Kentucky River gorge have macroinvertebrate and fish assemblages that are more typical of the Knobs–Norman Upland (71c) than the rest of Ecoregion 71l. Agriculture contributes sediment, nutrients, pesticides, and pathogens to surface water; algal blooms and low concentrations of dissolved oxygen occur especially where the riparian tree canopy has been removed. Wastewater discharge and runoff downstream of urban areas release trace metals into some streams. Package waste treatment plants for small residential subdivisions often discharge into dry valleys, produce effluent-dominated streams, and have a high failure rate. The Kentucky River has some of the highest nitrite plus nitrate and phosphate concentrations in Kentucky. It has been impounded by a series of locks and dams, causing the number of pool-inhabiting fish to increase at the expense of upland habitat species.

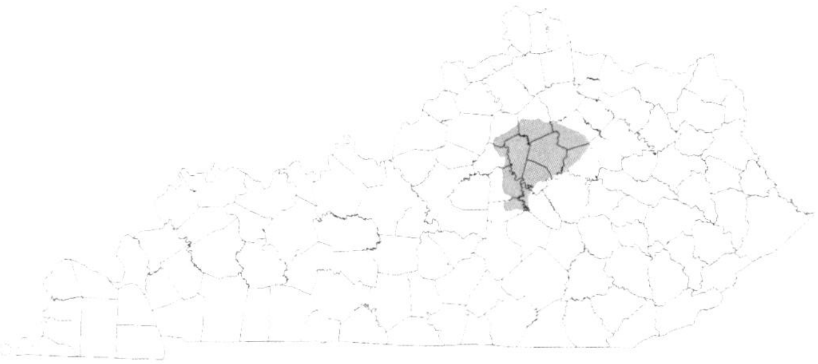

72a. Wabash–Ohio Bottomlands

The Wabash–Ohio Bottomlands ecoregion is composed of nearly level, poorly-drained floodplains and undulating terraces. Wetlands, ponds, abandoned channels, oxbow lakes, and low ridges occur. Potential natural vegetation is mapped as southern floodplain forest. Ecoregion 72a is lower, more poorly-drained, and has a different natural vegetation than other parts of Ecoregion 72. Today, some woodlands remain but livestock, alfalfa, corn, soybean, and wheat farming is extensive. Land use is affected by seasonally high water tables and localized flooding. Low gradient streams with silt or sand bottoms occur and are inhabited by Ohio River-type fish fauna. Channelization and drainage ditches are common.

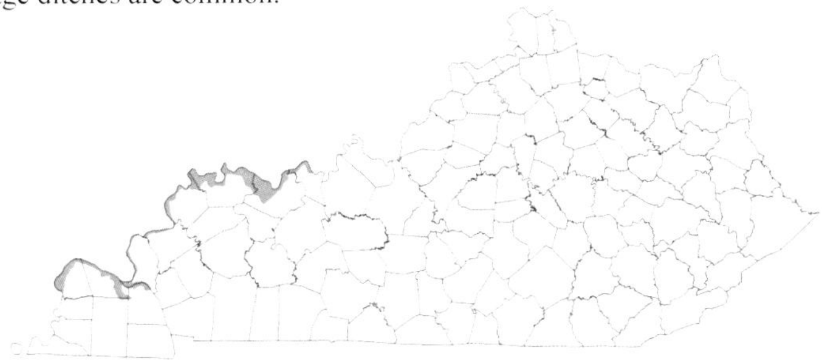

72c. Green River–Southern Wabash Lowlands

The Green River–Southern Wabash Lowlands (72c) is dominated by agriculture and coal mining. Wide, poorly-drained, low gradient valleys filled with alluvial and lacustrine deposits are extensive and low hills mantled with loess occur. Ecoregion 72c is largely underlain by Pennsylvanian carboniferous sedimentary rocks of the Sturgis and Carbondale formations that are not exposed in the higher, more rugged, and more wooded Ecoregion 72h. Bottomland forests were once common and oak–hickory forests grew on the better-drained upland sites. Today, some forests and wetlands remain but cropland, pastureland, and both underground and surface coal mining are now extensive. Siltation from mining and agriculture has increased flooding and prompted remedial channelization projects. Channelized streams lack riparian forests and have very warm water, high turbidity, and limited concentrations of dissolved oxygen. Acid coal mine runoff has decreased biological productivity in streams; many tributaries have low numbers of fish and fish species while others are entirely devoid of fish. Macroinvertebrate and fish communities are similar to those in Ecoregion 72a but are less diverse than in the upland streams of Ecoregion 72h.

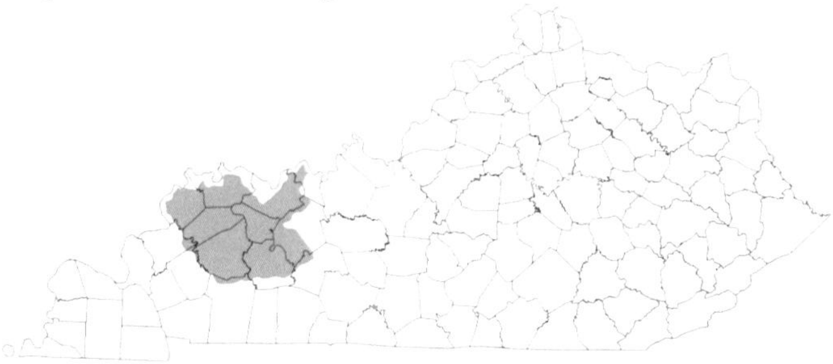

72h. Caseyville Hills

The dissected Caseyville Hills (72h) is dominated by forests and pastureland and is mantled by thin loess. Valleys are narrower and are much less extensive than in Ecoregion 72c but are wider, deeper, and more numerous than in Ecoregion 71a. Ecoregion 72h is underlain by Pennsylvanian sandstones, siltstones, shales, and coal of the Tradewater and Caseyville formations and Mississippian Chesterian limestones, sandstones, siltstones, and shales. These formations are absent from the lower, less rugged, and less wooded Ecoregions 72a and 72c. Limestone is much less common than in the Interior Plateau (71). Potential natural vegetation is oak–hickory forest and forests remain common. Today, livestock and hay farming, logging, oil production, and coal mining are the dominant land uses. However, coal mining is less extensive than in Ecoregion 72c. Upland perennial streams are cooler and have higher gradients, rockier substrates, better water quality, more diverse habitats, and more productive fish and macroinvertebrate communities than Ecoregion 72c.

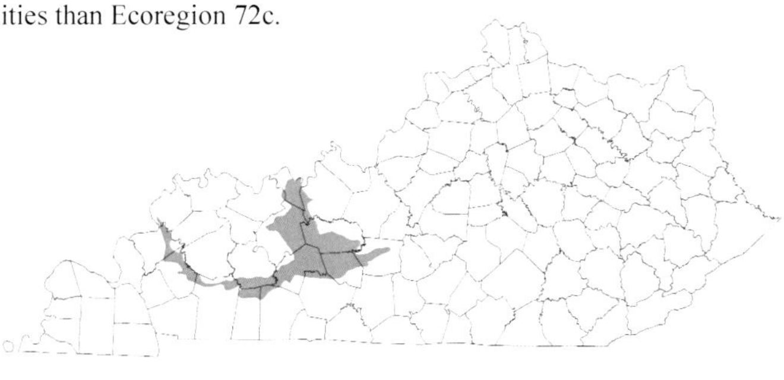

73a. Northern Holocene Meander Belts

The Holocene Meander Belts(73a) contains floodplains, low terraces, levees, abandoned channels, oxbow lakes, wetlands, and bayous. Soils are derived from deep Quaternary alluvium and are naturally fertile. They support a potential natural vegetation of southern floodplain forest that is distinct from the oak–hickory forest of nearby upland ecoregions. Aquatic communities are largely lentic in nature and Mississippian-type biotic assemblages occur. Undrained areas are subject to seasonal flooding. Elsewhere, drainage canals and field drains are common. Extensive wetland drainage and bottomland forest clearance has occurred in Ecoregion 73a. Today, the ecoregion is dominated by soybean, corn, small grain, and livestock farming. Only limited areas of bottomland forests and swamps still occur. The remaining bottomland forests are dominated by water-tolerant oaks, maples, sweetgum, sugarberry, sycamore, American elm, and pecan. The remaining swamps contain many southern species that are near their northern distributional limits including bald cypress and tupelo. Water and marsh birds are abundant.

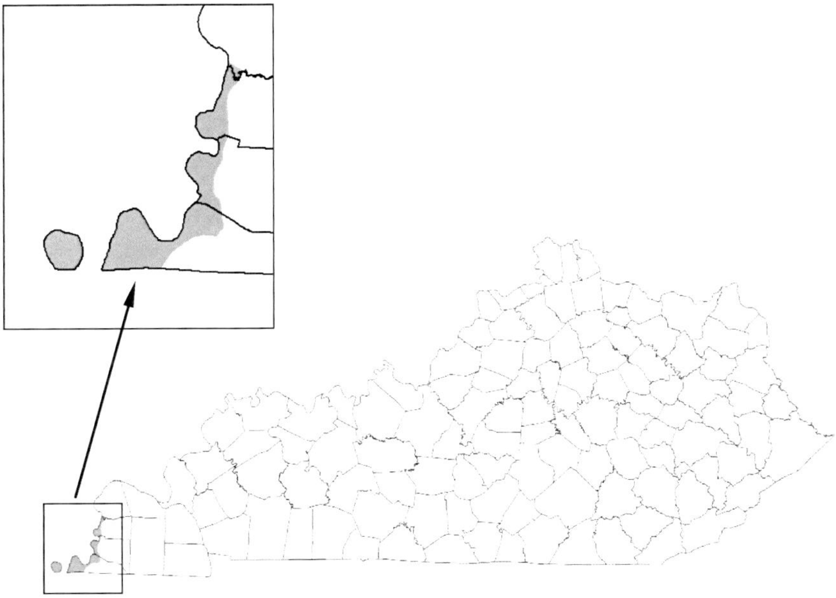

74a. Bluff Hills

The highly dissected Bluff Hills (74a) is a narrow belt of rugged terrain that is mantled by thick loess and underlain by unconsolidated coastal plain sediments. The Bluff Hills (74a) ecoregion is highly susceptible to landslides and erosion if disturbed. Ecoregion 74a is higher and more rugged than adjacent ecoregions. Steep hills, bluffs, winding ridges, and narrow valleys occur and have a mosaic of macroenvironments. Drier uplands are covered in oak–hickory forest and mixed oak forests. Mesic slopes have closed forests that are dominated by sugar maple, bitternut hickory, yellow-poplar, and beech. Aquatic habitats and fish assemblages are more typical of uplands than of nearby Ecoregions 73a and 74b. Intermittent and ephemeral streams are common and have silty or sandy bottoms. Higher gradient stream reaches with gravel substrates also occur.

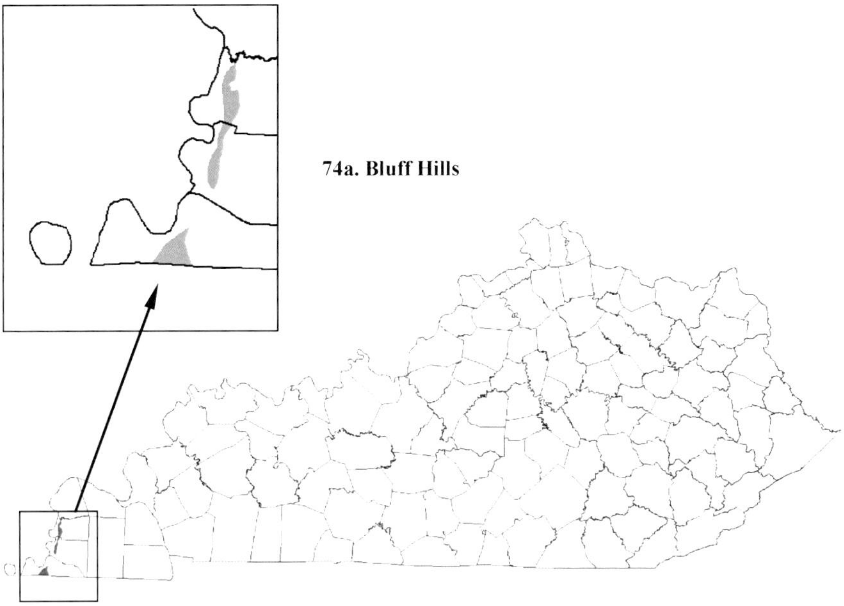

74a. Bluff Hills

74b. Loess Plains

The Loess Plains (74b) is a productive agricultural area that is composed of gently rolling uplands, broad bottomlands, and terraces. It is mantled by thick loess and alluvium and is underlain by weak, unconsolidated coastal plain sediments. Ecoregion 74b is lithologically distinct from higher, more easterly ecoregions. Potential natural vegetation is a mosaic of oak– hickory forest and bluestem prairie and is unlike the southern floodplain forest of Ecoregions 72a and 73a. Grasslands and forested wetlands were once widespread here and in the Western Pennyroyal Karst Plain (71e). Most of the original vegetation has now been replaced by cropland. Extensive corn, soybean, wheat, hay, tobacco, livestock, and poultry farming occurs. Agricultural runoff has degraded surface water quality. High turbidity and siltation are common in the streams and rivers of Ecoregion 74b. Many channelized streams occur.

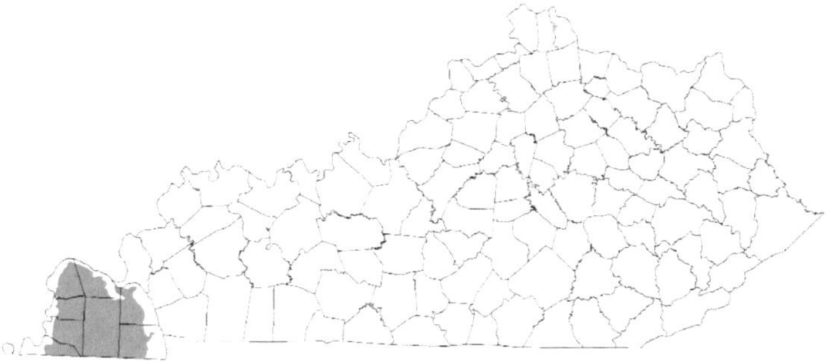

Map of Level IV Ecoregions of Kentucky

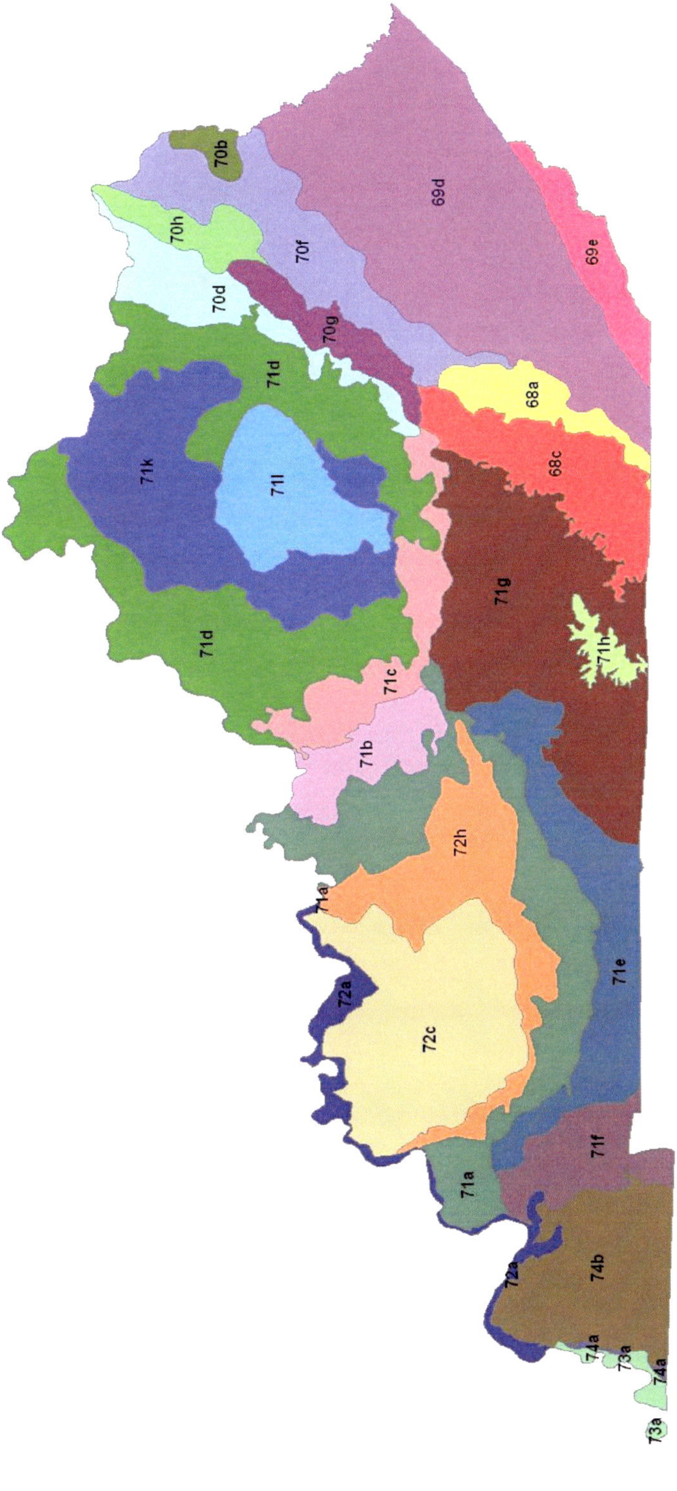

Legend of Level IV Ecoregions in Kentucky

Legend

74a	Bluff Hills
70h	Carter Hills
72h	Caseyville Hills
71a	Crawford-Mammoth Cave Uplands
69e	Cumberland Mountain Thrust
68a	Cumberland Plateau
69d	Dissected Appalachian Plateau
71g	Eastern Highland Rim
72c	Green River-South. Wabash Lowlands
71k	Hills of the Bluegrass
71l	Inner Bluegrass
70d	Knobs-Lower Scioto Dissected Plateau
71c	Knobs-Norman Upland
74b	Loess Plains
71b	Mitchell Plains
70b	Monongahela Transition Zone
70g	Northern Forested Plateau Escarpment
70g	Northern Holocene Meander Belts
70f	Ohio/Kentucky Carboniferous Plateau
71d	Outer Bluegrass
71h	Outer Nashville Basin
68c	Plateau Escarpment
72a	Wabash-Ohio Bottomlands
71f	Western Highland Rim
71e	Western Pennyroyal Karst Plain

Non-Toxic Invasive Land Snail Control

Commercial slug baits and repellents contain a molluscicide that kills gastropods. Although they are widely used, they may be toxic to pets, fish, wildlife, and humans. Less toxic methods have been used by organic gardeners for years and are effective in controlling persistent slugs and shelled snails. Included here are several suggestions.

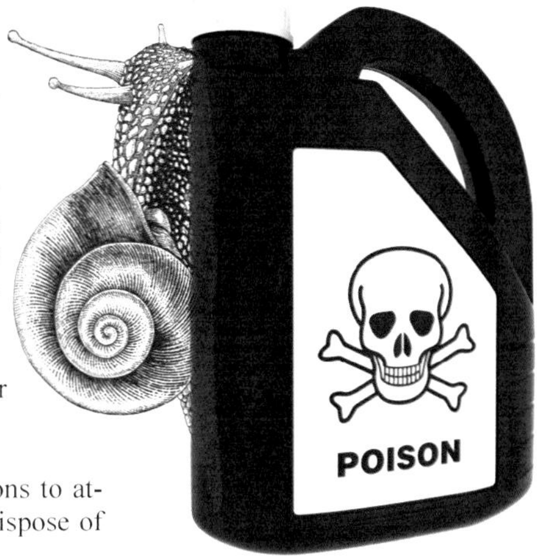

Gather individuals and place in boiling water or drench them in salt water.

Maintain a compost heap near the garden (which attracts the slugs).

Place boards, moist newspaper, cardboard, or old scraps of carpet in the garden to provide cover for the slugs during daytime hours; the slugs can then be easily harvested and euthanized.

Place cups in the garden filled with beer or grape juice to attract and drown the slugs.

Place wet dog food in piles close to infestations to attract slugs, check in the evening and simply dispose of the food and slugs.

Scatter inverted orange or grapefruit rinds among the vegetables to collect slugs which can then be harvested, mashed and composted.

Release ducks (ducks may also eat young plants) and chickens (Rhode Island Reds are best) in the garden to keep the plot clear of slugs and snails. Other birds reported to eat pesky gastropods include blackbirds, crows, jays, owls, robins, seagulls, starlings, and thrushes.

Lava rock, the spiny fruits of sweet gum trees, wood ash, oak leaves, pine needles, coffee grounds, cedar chips, Epsom salts, builders sand, nut shells, oat bran, lime, and talcum power are reported to act as barriers or repellents to slugs or are toxic to them.

Watering the soil with liquid seaweed extract has just enough of an alkaline effect that slugs detest it, however, not enough to significantly change soil pH.

Human hair, pet fur, and horsehair will entangle slugs. Ground beetles (particularly *Carabidae* beetles), box turtles, toads, frogs, lizards, salamanders, lighting bug larvae, and garter snakes are reported to eat slugs and shelled snails, so providing cover for these organisms in and around your garden will help in controlling unwanted gastropods.

Springtime is the best time of year to deal with these uninvited guests. Spring cultivation of the soil will help kill hibernating slugs and their eggs.

Removing slug slime from your hands can be accomplished with white vinegar and warm water or consider using chopsticks for picking up those slimy gastropods.

Deer/Elk Brainworm Disease and Land Snails

Land snails and white-tailed deer play a significant role in the transmission of a parasitic nematode known as brainworm, (*Parelaphostrongylus tenuis*). The initial host of the nematode is the white-tailed deer and the intermediate hosts are several species of snails and slugs. Interestingly, brainworm nematodes do not appear to significantly affect white-tailed deer populations yet may cause debilitation or even fatality in elk, moose, goats, and sheep.

The life cycle of the nematode begins with adult worms normally located between the membranes (meninges) that cover both the brain and in the spinal cord of the white-tailed deer, with most frequent occurrences in the meninges. Eggs are deposited either on these membranes or directly into blood vessels. Those deposited on the membranes hatch and the larvae enter small blood vessels to be carried to the lungs where they enter the alveoli. Eggs deposited into blood vessels are carried to the lungs and eventually hatch with larval penetration of the alveoli. Activity in the lung tissue produces an interstitial pneumonia. The larvae pass up the respiratory tract from the alveoli, are swallowed and then eliminated in the feces of the deer. Larvae appear in the feces about three months after the host becomes infected.

The larvae then enters into a snail or slug (the intermediate host) through the gastropods foot while crawling over infected deer feces or direct ingestion of deer feces by the gastropod. Development of the larvae in the gastropod to a stage when they are infective to the vertebrate host takes about three weeks. Cervids become parasitized by ingesting infected gastropods that are clinging to grass or other browse foods.

In the final host, other cervids, development of the larva to the adult worm takes place in tissues of the central nervous system, particularly the spinal cord. Parasites leave the tissues of the spinal cord after about 20-40 days and locate between the spinal membranes where they mature. Subsequently, they tend to accumulate in the cranial region. The adult worms are about 50 mm in length and may be seen fairly readily when free in the cranial cavity. From one to 20 worms have been found in the crania of infected deer, yet as previously stated, *P. tenuis* seldom causes damage in white-tailed deer.

In other cervids, there is often extensive damage to tissues of the brain and spinal cord. The resulting neurologic disease is characterized by weakness, fearlessness, lack of coordination of movement, circling, deafness, impaired vision, paralysis and subsequent death. When in moose this disease is often called "moose sickness" or, "moose disease".

A correlation to prevalence of the disease in elk, moose, and other Cervids has been linked to an increase in population of the white-tailed deer.

In Canada, studies by Anderson and Prestwood (1981) have substantiated the importance of this problem in management of big game and have given some indication of the dynamics of the host/parasite relationships among wild populations.

Land snail species that have been implicated in the spread of brainworm disease infestation include: *Anguispira alternata, Deroceras leave* (a slug), *Deroceras reticulatum* (a slug), *Discus catskillensis, Euchemotrema fraternum, Philomycus carolinianus* (a slug), *Neohelix albolabris, Pallifera dorsalis* (a slug), *Pseudohyalina exigua, Triodopsis tridentata, Ventridens intertextus,* and *Zonitoides arboreus.* Any species of land snail that occurs where there is a high prevalence of white-tailed deer could potentially be a host and contribute to the spread of this disease.

GLOSSARY

Abcission: when a plant drops one or more of its parts like leaves, fruit or seeds.

Aff.: abbreviation for the word affinity, refers to a species whose identity is in question; means similar to a specific species, but is likely not that species.

Amorphous: shapeless.

Algific: cold-producing as in talus, a loose-rock slope affected by the movement of cold air.

Angular periphery: shell having an angular rather than a round contour (see carinate).

Angular lamella: the tooth on the parietal wall of the aperture to the right of the parietal lamella in dextral shells.

Aperture: opening or mouth of the snail shell.

Apex: the tip of a gastropod shell farthest away from its aperture.

Axis: The imaginary line around which the whorls of a coiled shell are formed.

Basal: pertaining to, situated at or forming, the base; the part of the shell furthest from the apex; the ventral portion of the aperture lip.

Basal lamella: tooth on the bottom left side of the aperture below the columellar lamella.

Basal tooth: calcareous deposit on the basal apertural lip.

Bifid: divided by a deep cleft or notch in two parts.

Body Whorl: the last whorl of a shell, measured from the outer lip back to a point immediately above the outer lip.

Breathing pore: opening in mantle or mantle edge for passage of air into the air sac or lung cavity.

Calcareous: composed of or containing calcium carbonate.

Callus: a deposit of lime or shell material, often as a thickening near the umbilicus.

Calciphile: A calciphile plant is a plant that grows exclusively in soils rich in calcium carbonates, in other words in calcareous soil.

Carinate: the periphery of the shell is compressed, creating a sharper angular appearance with a ridge-like rim.

Cations: ions with a positive charge such as many minerals contained in the soil such as magnesium, aluminum and calcium.

Caudal: situated in or near the tail or posterior end.

Cf.: (from the Latin word meaning "refers to") a species whose identity is in question; probably that species, but further examination is needed to make a definitive identification.

Columella: the internal column around which the whorls revolve; the axis of a spiral shell.

Columellar lamella: tooth on the columellar wall of the aperture.

Columellar wall: the surface of the columella.

Corneous: horn-like in color.

Crypsis: the ability of one organism to avoid detection by other organism.

Depressed heliciform: shell with flattened spiral.

Dextral: spiraled to the right when the shell is held so that the apex is up and the aperture is facing the observer.

Dioecious: having both male and female organs in same individuals.

Dessication: drying out.

Discoidal: having a flat spiral shell.

Embryonic whorl: the earliest whorls that are formed in the egg.

Epiphragm: a hardened mucous covering sealing the aperture that prevents desiccation during dry weather.

Endemic: native or restricted to a relatively small area such as an area or region.

Foot: the locomotory organ of mollusks; it is often modified for digging or grasping prey; the long broad ventral surface of the animal.

Fulcrum: support or point on which a body moves or turns.

Globose: globular, formed like a globe, spherical.

Growth line: a line on the surface of a shell indicating a rest period during growth.

Heliciform: shell that has an elevated spire or globose form.

Hermaphrodite: possessing both male and female reproductive organs in a single individual.

Hirsute: covered with hairs.

Holarctic: a region which includes the northern parts of North America, Greenland, Europe and Asia.

Imperforate: lacking an opening on the ventral or anterior end of the shell.

Impressed: marked by a furrow.

Infra-parietal lamella: the tooth on the parietal wall to the left of the parietal lamella in dextral shells.

Lamella: a fold, "an elongated tooth", OR raised callus in the aperture of a shell.

Lirae: are fine lines or ridges (much finer than ribs) that are a sculptural feature of the outside of the shells of various animals. The term is commonly applied to the shells of molluscs like gastropods, bivalves and nautiloids.

Loess: periglacial or aeolian (windborn e) sediment, defined as an accumulation of 20% or less of clay and a balance of roughly equal parts sand and silt (with a typical grain size from 20 to 50 micrometers), often loosely cemented by calcium carbonate.

Lip: the edge of the aperture; also called the peristome.

Loosely coiled: having few, widely expanding whorls.

Lower palatal lamella: lowermost of the two major teeth often found on the palatal wall.

Lunate: shaped like a half-moon.

Mantle: a membranous flap or outer covering of the softer parts of the mollusk; it secretes the shell.

Mouth: the opening or aperture of a gastropod shell.

Mucus: a viscid, slippery secretion, slime.

Operculate: bearing an operculum or cover to close the aperture.

Operculum: a horny or calcareous plate that serves the purpose of closing the aperture when the snail withdraws into its shell.

Outer lip: the outer edge of the aperture.

Palatal depression: indentation of the shell surface at the location of the palatal lamellae.

Palatal tooth: refers to the tooth located on the outer lip.

Palatal wall: the right side of the aperture in dextral shells.

Papillae: small calcium deposits that appear as minute bumps on the surface of the shell.

Parietal lamella: major tooth in the middle of the parietal wall of the aperture.

Parietal tooth: refers to the tooth located on inner wall of the aperture or body of the snail shell.

Pellucid: translucent or clear

Penultimate whorl: the next to the last

whorl in an adult shell.

Perforate: Having a minute opening in the umbilicus.

Periostracum: the chitinous external layer covering most mollusk shells.

Periphery: the part of the whorl farthest from the central axis.

Protoconch: the embryonic or larval shell

Radula: is an minutely toothed anatomical structure called a chitinous ribbon used by mollusks for feeding; sometimes compared to a tongue. It scrapes or cuts the food before it enters the esophagus.

Reflected: turned outward, e.g., a portion of the apertural lip of some snail shells.

Ribs: prominent protrusions on the shell surface perpendicular to the spiral of the shell.

Rimate: the lip of the aperture slightly covers the umbilicus.

Rounded periphery: an evenly curved periphery; not angular or carinate.

Slug: a common designation for a snail without an external shell. The shell is either rudimentary and enclosed or is wanting entirely.

Senescence: biological aging.

Spiral: winding, coiling, or circling around a central axis; the form of the shell of most snails.

Spiral striae: surface features that are indented or raised on the shell surface running parallel with the whorls.

Spire: all the whorls above the aperture.

Striae: surface features that are either indented or raised in the shell surface.

Succiniform: shell that is higher than wide with a very large aperture (mouth). The spire is generally brief and the body whorl very expanded. (see *Succinea*)

Suture: indentation of the shell surface where two whorls meet.

Taiga: also known as boreal forest, a northern biome characterized by coniferous forests consisting mostly of pines, spruces, and larches

Tentacles: elongated, flexible organs on the head of snails used for feeling, tasting or sensing light.

Tightly coiled: having many, narrowly expanding whorls.

Tooth: A hard, calcareous nodule or projection in or around the aperture of the shell of many snail species; usually of diagnostic value in snail taxonomy.

Transverse striae: surface features indented or raised in the shell surface running perpendicular with the whorls.

Truncate: cut off at the end; terminating abruptly; ending in a transverse line.

Umbilicate: Having an opening or cavity at the base of a gastropod shell, with the opening more than a narrow perforation.

Umbilicus: an opening in the center of the columella or axis of the shell.

Upper palatal lamella: uppermost of the two major teeth often found on the palatal wall.

Whorl: one complete turn of a gastropod shell.

Xeric: containing very little moisture; dry.

Xerophile: tolerant of dry conditions.

BIBLIOGRAPHY

The following sources were used to develop the species accounts. Specific references are noted in the text as (Abbott 1989). All others may be implied and embedded.

ABBOTT, R.T. 1989. Compendium of Landshells. American Malacologists, Inc. Melbourne, Florida. 240 pp.

ANDERSON, R.C. & A.K. PRESTWOOD. 1981. Lungworms. Pp. 266-317 in W. R. Davidson, F. A. Hayes, V. F. Nettles, and F.E. Kellogg, eds., Diseases and parasites of white-tailed deer. Miscellaneous Publication. No. 7. Tall Timbers Research Station. Tallahassee, FL. 458 pp.

ARTHUR , M. A., L. M. TRITTON, & T. J. FAHEY. 1993. Dead bole mass and nutrients remaining 23 years after clear-cutting of a northern hardwood forest. Canadian Journal of Forest Reserves 23:1298-1305.

ATKINSON, J. W. 1998. Food manipulation and transport by a carnivorous land snail, *Haplotrema concavum.* Invertebrate Biology 117(2):109-113.

BAILEY, R. G.; P. E. AVERS, T. KING; W. H. MCNAB; eds. 1994. Ecoregions and subregions of the United States. Washington, DC: USDA Forest Service.

BONATO, V., K. G. FACURE, & W. UIEDA. 2004. Food habits of bats of subfamily Vampyrinae in Brazil. Journal of Mammalogy 708-713.

BOYCOTT, A. E. 1934. The habitats of land mollusca in Britain. Journal of Ecology 22:1-38.

BRANSON, B. A. 1973. Kentucky land mollusca: checklist, distribution, and keys for identification. Kentucky Department of Fish and Wildlife Resources. Frankfort, KY. 67 pp.

BRANSON, B. A. & D. L. BATCH. 1988. Distribution of Kentucky land snails (Mollusca: Gastropoda). Transactions of the Kentucky Academy of Science 49:101-116.

BRAUN, E. L. 1940. An ecological transect of Black Mountain, Kentucky. Ecological Monographs 10:193-241.

BRYCE, S.A., J.M. OMERNIK, AND D.P. LARSEN. 1999. Ecoregions: a geographic framework to guide risk characterization and ecosystem management. Environmental Practice 1(3):141-155. BURCH, J. B. 1955. Some ecological factors of the soil affecting the distribution and abundance of land snails in eastern Virginia. Nautilus 69:26-29.

BURCH, J. B. 1962. How to know the eastern land snails. Dubuque: William C. Brown Company.

BURCH, J. B. & T. A. PEARCE. 1990. Terrestrial Gastropoda. Pp. 201-309 in D. L. Dindall (ed.), Soil Biology Guide. Wiley and Sons, Inc., New Jersey

CAMERON, R. A. D. 1970. Differences in the distributions of three species Helicid snails in the limestone district of Derbyshire. Proceedings of the Royal Society of London 176:130-159.

CHASE, R. & K. C. BLANCHARD. (2006) The snail's love-dart delivers mucus to increase paternity. Proceedings of the Royal Society Biology. 273:1471-1475.

COMMISSION FOR ENVIRONMENTAL COOPERATION. 1997. Ecological regions of North America: toward a common perspective. Commission for Environmental Cooperation, Montreal, Quebec, Canada. 71pp

CONEY, C. C., W. A. TARPLEY, J. C. WARDEN, & J. W. NAGEL. 1982. Ecological studies of land snails in the Hiwassee River basin of Tennessee, U.S.A. Malacological Review 15:69-106.

COUNTS, C. L. III. 1982. Occurrence distribution of land snails of the family Polygyridae (Mollusca:Gastropoda:Pulmonata) in West Virginia. Brimleyana 8:145-157.

DALLINGER, R. 1993. Strategies of metal detoxification in terrestrial invertebrates. *in*: Dallinger, R. and Rainbow, P. S. (eds.). Ecotoxicology of Metals in Invertebrates. Lewis Publishers. Boca Raton, Florida. Pp. 245-289.

DALLINGER, R., & A. W. WIESER. 1984. Patterns of accumulation, distribution and liberation of Zn, Cu, Cd and Pb in different organs of the land snail Helix pomatia L. Comparative Biochemistry and Physiology-Part C: Toxicity and Pharmacology. 79 (1): 117-124.

DALLINGER, R. & A. W. WEISER. 1984a. Molecular fractionation of Zn, Cu, Cd, and Pb in the midgut gland of *Helix pomatia* L. Comparative Biochemistry and Physiology. 79:125-129.

DALLINGER, R., B. BERGER, C. GRUBER, P. HUNZIKER & S. STUZENBAUM. 2000. Metallothioneins in terrestrial invertebrates: structural aspects, biological significance, and implications for their use as biomarkers. Cellular and Molecular Biology 46:331-346.

DOUGLAS, D. A. 2011. Land snail species diversity and composition among different disturbance regimes in central and eastern Kentucky forests. (MS thesis). Eastern Kentucky University. Richmond, KY.

DOUGLAS, D. A, D. DOURSON, & R. C. CALDWELL. 2013. The land snails of White Oak Sinks, Great Smoky Mountains National Park. Southeastern Naturalist. 13(1): 166-175.

DOURSON, D. 2007. A selected land snail compilation of the Central Knobstone Escarpment on Furnace Mountain in Powell County Kentucky, USA. Journal of the Kentucky Academy of Science 68(2):119-131.

DOURSON, D. 2007. Survey Protocol for Cheat Threetooth (*Triodopsis platysayoides*) (Revised). West Virginia Department of Natural Resources, United States Fish and Wildlife Service.

DOURSON, D. 2008. The feeding behavior and diet of an endemic West Virginia land snail, *Triodopsis platysayoides*. American Malacological Bulletin. 26:153-159

DOURSON, D. 2010. Kentucky's land Snails and their ecological communities. 298 pp. Goatslug Publications, Bakersville, NC.

DOURSON, D. 2013. Land Snails of the Great Smoky Mountains National Park and Southern Appalachians. 336 pp. Goatslug Publications, Bakersville, NC.

DOURSON, D. 2012. Four new land snail species from the southern Appalachian Mountains. Journal of the North Carolina Academy of Science 128 (1):1-10.

DOURSON, D. 2011. Descriptions of three new land snails from Kentucky. Journal of the Kentucky Academy of Science . Vol. 72 (1):39-45.

DOURSON, D. & J. BEVERLY. 2009. Diversity, substrata divisions and biogeographical affinities of land snails at Bad Branch State Nature Preserve, Letcher County Kentucky, USA. Journal of the Kentucky Academy

of Science. 68(2):119-131.
DOURSON, D. & M. GUMBERT. 2004. A survey of terrestrial mollusca in selected areas of the Great Smoky Mountains National Park, North Shore Road Project. Report submitted to Arcadis G & M of North Carolina, Inc. 76 pp.
DOURSON, D. & K. LANGDON. 2012. Land snails of selected rare high elevation forests and heath balds of the Great Smoky Mountains National Park. Journal of North Carolina Academy of Science: Summer 2012, Vol. 128, No. 2, pp. 27-32.
EMBERTON, K. C. 1991. The genitalic, allozymic and conchological evolution of the tribe Mesodontini (Pulmonata: Stylommatophora: Polygyridae). Malacologia 33:71-178.
EMBERTON, K. C. 1994. Polygyrid land snail phylogeny: external sperm exchange, early North American biogeography, iterative shell evolution. Biological Journal of the Linnean Society 52:241-271.
FAIRBANKS, H. L. 1998. Clarification of the taxonomic status and reproductive anatomy of *Philomycus batchi* Branson, 1968. The Nautilus 112 (1):1-5.
FOOTE, B. A. 1959. Biology and life history of the snail-killing flies belonging to the genus Sciomyza. Annals of the Entomological Society of America. 52:31-32.
FOURNIE, J. & M. CHETAIL. 1984. Calcium dynamics in land gastropods. American Zoologist 24:857-870.
GEIGER, R. 1965. The climate near the ground. Harvard University Press. Cambridge.
GETZ, L. L. 1974. Species diversity of terrestrial snails in the Great Smoky Mountains. The Nautilus 88:6-9.
GOSZ, J. R., G.E. LIKENS, & F. H. BORMANN. 1973. Nutrient release from decomposing leaf and branch litter in the Hubbard Brook Forest, New Hampshire. Ecological Monographs 43:173-191.
GRAVELAND, J. R. 1996. Avian eggshell formation in calcium-rich and calcium-poor habitats: importance of snail shells and anthropogenic calcium sources. Canadian Journal of Zoology 74:1035-1044.
GRAVELAND, J. R., & R. VAN DER WAL, J. H. 1994. Poor reproduction in forest passerines from decline of snail abundance on acidified soils. Nature 368:446-448.
GRIFFITH, G.E., J.M. OMERNIK, & S.H. AZEVEDO. 1998. Ecoregions of Tennessee. (2 sided color poster with map, descriptive text, summary tables, and photographs). U.S. Geological Survey, Reston, VA. Scale 1:940,000.
GRIMM, F. W., R. G. FORSYTH, F. W. SCHUELER, & A. KARSTAD. 2009. Identifying land snails and slugs in Canada: introduced species and native genera. Canadian Food Inspection Agency. Ontario, Canada. 168 pp.
GUHA, M. M. & R. L. MITCHELL. 1966. The trace and major element composition of some deciduous trees: II. Seasonal changes. Plant and Soil 24:90-112.
HAMES, R. S., K. V. ROSENBERG, J. D. LOWE, & S. E. DHONDT. 2002. Adverse affects of acid rain on the (Wareborn, 1992) distribution of the

woodthrush, *Hylocichla mustelina*, in North America. Proceedings of the National Academy of Science. 99 (17): 11235-11240

HICKMAN, C. P., L. S. ROBERTS & A. LARSON. 2003. Animal diversity. 3rd ed. McGraw-Hill, NY.

HOTOPP, K. 2002. Land snails and soil calcium in central Appalachian forests. Southeastern Naturalist 1(1): 27-44.

HOTOPP, K. 2015.A new *Triodopsis juxtidens* subspecies (Gastropoda: Pulmonata) from West Virginia, U. S. A. Zootaxa 3914(4):490-494. http://dx.doi.org/10.11646/zootaxa.3914.4.8

HOTOPP, K. & T. A. PEARCE. 2008. Land snail distributions in West Virginia. Report for contract WVFIMS 3203-2008-6305-099-025. Submitted to Michael Welch, Wildlife Resources Section, West Virginia Division of Natural Resources.

HOTOPP, K. , T. A. PEARCE & D.C.DOURSON. 2008. Land Snails of the Cheat River Canyon, West Virginia (Gastropoda: Pulmonata), Banisteria, Number 31, pp. 40-46.

HUBRICHT, L. 1958. New species of land snails from the eastern United States. Transactions of the Kentucky Academy of Science 19:70-76.

HUBRICHT, L. 1964. Land snails from the caves of Kentucky, Tennessee, and Alabama. Bulletin of the National Speleological Society. 26(1):33-36.

HUBRICHT, L. 1985. The distributions of the Native Land Mollusks of the Eastern United States. Fieldiana Zoology New Series, No. 24, Publication 1359: Field Museum of Natural History, Chicago, Illinois. 191 pp.

HUBRICHT L., R. S. CALDWELL, & J. G. PETRANKA. 1983. *Vitrinizonites latissimus* (Pulmonata: Zonitidae) and *Vertigo clappi* (Pupillidae) from eastern Kentucky. The Nautilus 97:20-22.

JACOT, A. P. 1935. Molluscan populations of old growth forests and rewooded fields in the Asheville Basin of North Carolina. Ecology 16:603-605.

JENKINS, M. A., S. JOSE, & P. S. WHITE. 2007. Impacts of an exotic disease and vegetation change on foliar calcium cycling in Appalachian forests. Ecological Applications 17(3):869-881.

KALISZ, P. J. & J. E. POWELL, 2003. Effect of calcareous road dust on land snails and millipedes in acid forest soils of the Daniel Boone National Forest. Forest Ecology and Management 186:177-183.

KARLIN, E. J. 1961. Ecological relationships between vegetation in the distribution of land snails in Montana, Colorado, and New Mexico. American Midland Naturalist 65:60-66.

KELLER, H. & K. SNELL 2002. Feeding activities of slugs on Myxomycetes and macrofungi. Mycologia, 94:757-760.

KERNEY, M. P. AND R. A. D. CAMERON. 1979. Field guide to the land snails of Britain and north-west Europe.

KOENE J. M, T. S. LIEW, K. MONTAGNE-WAJER, M. SCHILTHUIZEN 2013. A syringe-like love dart injects male accessory gland products in a tropical hermaphrodite. PLOS ONE 8(7): e69968. doi:10.1371/journal.pone.0069968

LEE, J. C. 1994. The amphibians and reptiles of the Yucatán Peninsula. Cornell University Press, Ithaca /.

MACHENSTED, U. & K. MARKEL. 2001. Chapter 4: Radular structure and function. Pp. 213-236. In: The biology of terrestrial mollusks. G. M Barker, ed. CABI. New York, New York.
MACMILLAN, G. K. 1949. The land snails of West Virginia. Annals of the Carnegie Museum 31:89-238.
MARTIN, A. C., H. S. ZIM, & A. L. NELSON. 1951. American wildlife and plants: A guide to wildlife food habits. McGraw-Hill, Inc., New York.
MCHARGUE, J. S. & W. R. ROY. 1932. Mineral and nitrogen content of the leaves of some forest trees at different times in the growing season. Botanical Gazette 94:381-393.
MCMAHON, G., S.GREGONIS, S. WALTMAN, J. M. OMERIK, T. D. THORSON, J. A. FREEOUF, A. H. RORICK, J. E. KEYS. 2001. Developing a Spatial Framework of Common Ecological Regions for the Conterminous United States. *Environmental Management* **28**, 293–316 (2001). https://doi.org/10.1007/s0026702429
MOLLUSCABASE EDS. 2022. Molluscabase. Accessed at https://www.molluscabase.org. on 2022-11-27. doi: 10.14284/448
MORITZ, C., K. S. RICHARDSON, S. FERRIER, J. STANISIC, S. E. WILLIAMS, & T. WHIFFIN. 2001. Biogeographic concordance and efficiency of taxon indicators for establishing conservation priority in a tropical rainforest biota. Proceedings of the Royal Society 268:1875-1881.
NATION, R. 2005. The influence of soil calcium on land snail diversity in the Blue Ridge Escarpment of South Carolina. Dissertation presented to Clemson University. Clemson, South Carolina.
NEKOLA, J. C. 1999. Terrestrial gastropod richness of carbonate cliff and associated habitats in the Great Lakes region of North America. Malacologia 41(1):231-252.
NEKOLA, J. C. 2003. Large-scale terrestrial gastropod community composition patterns in the Great Lakes region of North America. Diversity and Distribution. 9:55-71.
NEKOLA, J. C. and M. Barthel. 2002. Morphometric analysis of *Carychium exile* and *Carychium exiguum* in the Great Lakes region of North America. Journal of Conchology 37(5): 515.
NEKOLA, J. C. & B. F. COLES. 2010. Pupillid land snails of eastern North America. American Malacological Bulletin. 28:29-57.
NICKLAS N.L. & R. J. HOFFMAN. 1981. Apomictic parthenogenesis in a hermaphroditic terrestrial slug, *Deroceras laeve* (Müller). Biological Bulletin. 160:123-135.
OMERNIK, J. M. 1987. Ecoregions of the conterminous United States. Map (scale 1:7,500,000). Annals of the Association of American Geographers 77(1):118-125.
OMERNIK, J.M. 1995. Ecoregions: A spatial framework for environmental management. In: Biological Assessment and Criteria: Tools for Water Resource Planning and Decision Making. Davis, W.S. and T.P. Simon (eds.), Lewis Publishers, Boca Raton, FL. p. 49-62.
OMERNIK, J.M. 2004. Perspectives on the nature and definition of ecological regions. Environmental Management 34(Supplement 1):S27-S38.
OMERNIK, J.M. and G.E. Griffith 2014. Ecoregions of the conterminous United States: evolution of a hierarchical spatial framework. Environ-

mental Management 54(6):1249-1266.
OMERNIK, J.M., S.S. Chapman, R.A. Lillie, and R.T. Dumke. 2000. Ecoregions of Wisconsin. Transactions of the Wisconsin Acadamy of Sciences, Arts, and Letters 88:77-103.
PATTERSON, C. M. & J. B. BURCH. 1966. The chromosome cycle in the land snail *Catinella vermeta* (Stylommatophora: Succineidae) Malacologia. 3:309.
PEARCE, T. A. 2008. When a snail dies in the forest, how long will the shell persist? Effect of dissolution and micro-bioerosion. American Malacological Bulletin 26:111-117.
PEARCE, T. A. & A. GAERTNER. 1996. Optimal foraging and mucus-trail following in the snail-eating snail *Haplotrema concavum* (Gastropoda: Pulmonata). *Malacological Review* 29:85-99.
PETRANKA, J. G. 1982. The distribution and diversity of land snails on Big Black Mountain, Kentucky. (Master's thesis). University of Kentucky. Lexington, KY.
PETRANKA, J. W. 1998. Salamanders of the United States and Canada. Washington, D. C., USA: Smithsonian Institution Press. Pp. 587.
PILSBRY, H. A. 1940. Land mollusca of North America (north of Mexico), Vol. I. Part 2. The Academy of Natural Sciences of Philadelphia Monographs.
PILSBRY, H. A. 1946. Land mollusca of North America (north of Mexico), Vol. II. Part 1. The Academy of Natural Sciences of Philadelphia Monographs.
PILSBRY, H. A. 1948. Land mollusca of North America (north of Mexico), Vol. I. Part 2. The Academy of Natural Sciences of Philadelphia Monographs.
POLLARD, E. 1975. Aspects of the ecology of *Helix pomatia* L. Journal of Animal Ecology 44:305-329.
POTTER, C. S., H. L. RAGSDALE, & C. W. BERISH. 1987. Reabsorption of foliar nutrients in a regenerating southern Appalachian forest. Oecologia 73:268-271.
QUICK, H. E., 1960. British slugs (Pulmonata; Testacellidae, Arionidae, Limacidae). Bulletin of the British Museum of Natural History. 6:103-226.
RAWLS, H. & R. YATES. 1971. Fluorescence in Endodontid Snails. The Nautilus 85:17-20.
REID, F. A. 2006. Mammals of North America. Peterson Field Guide. New York: Houghton-Mifflin.
RICHTER, K. O. 1980. Evolutionary aspects of mycophagy in *Ariolimax columbianus* and other slugs. In: D. L. Dindal, ed., Soil Ecology as Related to Land Use Practices. Proceedings of the VII International Colloquium of Soil Biology, Washington D. C.
RICKLEFS, R. E. & K. K. MATTHEWS. 1982. Chemical characteristics of the foliage of some deciduous trees in southeastern Ontario. Canadian Journal of Botany 60:2037-2045.
RIMMER, C. C., K. P. MCFARLAND, D. C. EVERS, E. K. MILLER, Y. AUBRY, D. BUSBY & R. J. TAYLOR. 2005. Mercury concentrations in Bicknell's Thrush and other insectivorous passerines in montane forests of

northeastern North America. Ecotoxicology 14:223-240.
ROODY, W. C. 2003. Mushrooms of West Virginia and the Central Appalachians. The University of Kentucky Press. Lexington, Kentucky 519 p.
ROTH, B. 1988. Identities of two Californian land mollusks described by Wesley Newcomb. Malacological Review 20:129-132.
SALITH, A., A. A. HAMA, A. M. H. KARZAN, A. DITTA. 2021. The land snail, Eobania vermiculata, as a bioindicator of the heavy metal pollution in the urban areas of Sulaimani, Iraq. Sustainability Basel 13(24);137.
SATHEESHKUMAR, P., A. B. KHAN, & D. SENTHILKUMAR. 2010. Marine organisms as potential supply for drug finding-a review study. Middle-East Journal of Scientific Research 5 (6): 514-519.
SCHEIFLER, R., C. SWARTZ, G. ECHEVARRIA, A. DE VAUFLEURY, P. BADOT, & J. MOREL. 2003. "Nonavailable" soil cadmium is bioavailable to snails: evidence from isotopic dilution experiments. Environmental Science Technology 37(1):81-86.
SHEARER, A. & J. W. ATKINSON. 2001. Comparative analysis of food-finding behavior of an herbivorous and a carnivorous land snail. Invertebrate Biology 120:199-205.
SLAPCINSKY, J. & B. COLES. 2004. Revision of the genus *Pilsbryna* (Gastropoda: Pulmonata: Gastrodonatidae) and comments on the taxonomic status of *Paravitrea tridens* Morrison, 1935. The Nautilus 118 (2):55-70.
SYMONDSON, W. O. C. 2004. Chapter 2: Coleoptera (Carabidae, Staphylinidae, Lampyridae, Drilidae, and Silphidae) as predators of terrestrial gastropods. pp. 37-84 in *Natural Enemies of Terrestrial Mollusks*. CABI Publishing. Oxford, UK.
TAYLOR, R. W. AND C. L. COUNTS, III. 1976. Note on some land snails from Blennerhassett Island, West Virginia. Sterkiana 63-64:11.
THABAH, A. & G. LI, Y. WANG, B. LIANG, L. HU, S. ZHANG, & G. JONES. 2007. Diet, echolocation calls, and phylogenetic affinities of the great evening bat (IA IO; Vespertilionidae): another carnivorous bat. Journal of Mammology. 88(3):728-735.
TOWNSEND, J. S., H. C. ALDRICH, L. D. WILSON, & J. R. MCCRANIE. 2007. First report of sporangia of a myxomycete (*Physarum pusillum*) on the body of a living animal, the lizard *Corytophanes cristatus'*. Mycologia. 97(2):246-248.
VANDEVENDER, A. S. & R. W. VANDEVENDER. 2003. Surveying the land snails of the southern Appalachians. Program and Abstracts of American Malacological Society: 61.
VESTERDAL. I., & K. RAULUND-RASMUSSEN 1998. Forest floor under seven tree species along a soil fertility gradient. Canadian Journal of Forest Reserves 28:1636-1647.
WALLACE, M.S., R. RAUCK, R. S. FISHER, G. CHARAPATA, D. ELLIS, & S. DISSANAYAKE. 2008. Ziconotide 98-022 Study Group. Intrathecal ziconotide for severe chronic pain: safety and tolerability results of an open-label, long-term trial. Anesthesia and Analgesia 106: 628-637.
WAREBORN, I. 1970. Environmental factors influencing the distribution of land mollusks of an oligotrophic area in southern Sweden. Oikos 2:285-291.

WIKTOR A., C. DE-NIU. & M. WU. 2000. Stylommatophoran slugs of China (Gastropoda: Pulmonata) – prodromus. Folia Malacologica 8(1): 3-35.
WOODS, A.J., J.M. OMERNIK, W.H. MARTIN, G.J. POND, W.M. ANDREWS, S.M. CALL, J.A. COMSTOCK, AND D.D. TAYLOR. 2002. Ecoregions of Kentucky. (2 sided color poster with map, descriptive text, summary tables, and photographs). U.S. Geological Survey, Reston, VA. Scale 1:1,000,000.

Index of Scientific Names

A

B

C

D

E

Index of Common Names

Common names follow the nomenclature from NatureServe © 2024.

Native Land Snail Species in Kentucky

PUPILLIDAE
Toothless Pupa, *Columella simplex*

COCHLICOPIDAE
Glossy Pillar, *Cochlicopa lubrica*
Thin Pillar, *Cochlicopa lubricella*
Appalachian Pillar, *Cochlicopa morseana*

PUPILLIDAE
White-lip Dagger, *Pupoides albolabris*

POMATIOPSIDAE
Brown Walker, *Pomatiopsis cincinnatiensis*
Slender Walker, *Pomatiopsis lapidaria*

SUCCINEIDAE
Oval Ambersnail, *Novisuccinea ovalis*
Blunt Ambersnail, *Oxyloma retusum*
Detritus Ambersnail, *Mediappendix oklahomarum*
Suboval Ambersnail, *Mediappendix vermeta*
Spotted Ambersnail, *Succinea concordialis*

ORTHALICIDAE
Whitewashed Rabdotus, *Rabdotus dealbatus*

ELLOBIIDAE
Ice Thorn, *Carychium exile*
Appalachian Thorn *Carychium clappi*
Obese Thorn, *Carychium exiguum*
Cave Thorn, *Carychium stygium*
Floodplain Thorn, *Carychium riparium*
File Thorn, *Carychium nannodes*

VERTIGINIDAE
Armed Snaggletooth, *Gastrocopta armifera*
Bluegrass Snaggletooth, *Gastrocopta clappi*
White Snaggletooth, *Gastrocopta tappaniana*
Comb Snaggletooth, *Gastrocopta pentodon*
Bottleneck Snaggletooth, *Gastrocopta contracta*
Bark Snaggletooth, *Gastrocopta corticaria*
Lambda Snaggletooth, *Gastrocopta holzingeri*
Wing Snaggletooth, *Gastrocopta procera*
Gulf Coast Snaggletooth, *Gastrocopta riparia*
Cupped Vertigo, *Vertigo clappi*
Blade Vertigo, *Vertigo milium*
Crested Vertigo, *Vertigo pygmaea*
Ovate Vertigo, *Vertigo ovata*

Capital Vertigo, *Vertigo oscariana*
Smallmouth Vertigo, *Vertigo parvula*
Honey Vertigo, *Vertigo tridentata*
Bluff Vertigo, *Vertigo meramecensis*
Variable Vertigo, *Vertigo gouldii*
Delicate Vertigo, *Vertigo bollesiana*
Widespread Column, *Pupilla muscorum*

PUNCTIDAE
Small Spot, *Punctum minutissimum*
Glass Spot, *Punctum vitreum*
Brown Spot, *Punctum blandianum*
Lamellate Spot, *Punctum smithi*

GASTRODONTIDAE
Ribbed Striate, *Striatura exigua*
Median Striate, *Striatura meridionalis*
Fine-ribbed Striate, *Striatura milium*
Black Striate, *Striatura ferrea*
Green Gloss, *Zonitoides elliotti*
Quick Gloss, *Zonitoides arboreus*
Striate Gloss, *Zonitoides lateumbilicatus*
Dull Gloss, *Zonitoides limatulus*
Brown Bellytooth, *Gastrodonta interna*
Appalachia Bellytooth, *Gastrodonta fonticula*
Imperforate Glyph, *Glyphyalinia solida*
Thin Glyph, *Glyphyalinia cryptomphala*
Furrowed Glyph, *Glyphyalinia luticola*
Carved Glyph, *Glyphyalinia indentata*
Spiral Mountain Glyph, *Glyphyalinia carolinensis*
Suborb Glyph, *Glyphyalinia sculptilis*
Brilliant Glyph, *Glyphyalinia praecox*
Tongued Glyph, *Glyphyalinia rimula*
Sculpted Glyph, *Glyphyalinia rhoadsi*
Bright Glyph, *Glyphyalinia wheatleyi*
Hill Glyph, *Glyphyalinia cumberlandiana*
Hollow Glyph, *Glyphyalinia specus*
Pale Glyph, *Glyphyalinia lewisiana*
Maryland Glyph, *Glyphyalinia raderi*
Flat Dome, *Ventridens suppressus*
Split-tooth Dome, *Ventridens virginicus*
Flat Dome, *Ventridens suppressus*
Throaty Dome, *Ventridens gularis*
Yellow Dome, *Ventridens pilsbryi*
Hollow Dome, *Ventridens lasmodon*
Rounded Dome, *Ventridens lawae*
Copper Dome, *Ventridens suppressus*
Sculptured Dome, *Ventridens collisella*
Flat Dome, *Ventridens demissus*

Pyramid Dome, *Ventridens intertextus*
Tennessee Dome, *Ventridens percallosus*
Glossy Dome, *Ventridens cf. acerra*
Globose Dome, *Ventridens ligera*
Dusky Button, *Mesomphix capnodes*
Smooth Button, *Mesomphix perlaevis*
Common Button, *Mesomphix vulgatus*
Frog Button, *Mesomphix anurus*
Plain Button, *Mesomphix inornatus*
Imitator Button, *Mesomphix rugeli oxycoccus*
Wrinkled Button, *Mesomphix rugeli*
Copper Button, *Mesomphix cupreus*
Globose Button, *Mesomphix globosus*
Brittle Button, *Mesomphix friabilis*
Glassy Grapeskin, *Vitrinizonites latissimus*

PRISTILOMATIDAE
Minute Gem, *Hawaiia minuscula*
Southeastern Gem, *Hawaiia alachuana*
Dimple Supercoil, *Paravitrea capsella*
Shrew Supercoil, *Paravitrea blarina*
Cherokee Supercoil, *Paravitrea petrophila*
Gem Supercoil, *Paravitrea lapilla*
Teasing Supercoil, *Paravitrea tantilla*
Slender Supercoil, *Paravitrea subtilis*
Funnel Supercoil, *Paravitrea mira*
Round Supercoil, *Paravitrea reesei*
Brown Supercoil, *Paravitrea septadens*
Lamellate Supercoil, *Paravitrea lamellidens*
Dentate Supercoil, *Paravitrea multidentata*
Comb Supercoil, *Paravitrea dentilla*
Glossy Supercoil, *Paravitrea placentula*

HELICODISCIDAE
Smooth Coil, *Lucilla singleyana*
Oldfield Coil, *Lucilla inermis*

EUCONULIDAE
Brilliant Granule, *Guppya sterkii*
Brown Hive, *Euconulus fulvus egenus*
Fat Hive, *Euconulus polygyratus*
Wild Hive, *Euconulus chersinus*
Silk Hive, *Euconulus trochulus*
Toothed Hive, *Euconulus dentatus*

OXYCHILIDAE
Black Mountain Bud, *Pilsbryna aff. vanattai*

DISCIDAE
Domed Disc, *Discus patulus*
Black Mountain Disc, *Discus patulus*
Forest Disc, *Discus patulus*
Banded Globe, *Anguispira kochi*
Flamed Tigersnail, *Anguispira alternata*
Pine Mountain Tigersnail, *Anguispira rugoderma*
Mountain Tigersnail, *Anguispira jessica*
Southeastern Tigersnail, *Anguispira strongylodes*
Kentucky Tigersnail, *Anguispira aff. strongylodes*
Craggy Tigersnail, *Anguispira aff. mordax*

HELICODISCIDAE
Compound Coil, *Helicodiscus parallelus*
Tight Coil, *Helicodiscus notius*
Fringed Coil, *Helicodiscus fimbriatus*
Cricket Coil, *Helicodiscus hadenoecus*
Punctuate Coil, *Helicodiscus punctatellus*
Burrowing Coil, *Helicodiscus fimbriatus*

HAPLOTREMATIDAE
Gray-foot Lancetooth, *Haplotrema concavum*

HELICINIDAE
Globular Drop, *Helicina orbiculata*
Cherrystone Drop, *Hendersonia occulta*

STROBILOPSIDAE
Maze Pinecone, *Strobilops labyrinthicus*
Southern Pinecone, *Strobilops cf. texasianus*
Bronze Pinecone, *Strobilops aeneus*

POLYGRYRIDAE
Inland Slitmouth, *Stenotrema stenotrema*
Fraudulent Slitmouth, *Stenotrema macgregori*
Kentucky Slitmouth, *Stenotrema angellum*
Bristled Slitmouth, *Stenotrema barbatum*
Ridge-and-Valley Slitmouth, *Stenotrema edvardsi*
Hairy Slitmouth, *Stenotrema hirsutum*
Bluegrass Liptooth, *Daedalochila fatigiata*
Cumberland Liptooth, *Daedalochila plicata*
Nashville Liptooth, *Daedalochila troostiana*
Gulf Coast Liptooth, *Daedalochila leporina*
Tiny Liptooth, *Lobosculum pustuloides*
Upland Pillsnail, *Euchemotrema fraternum*
Floodplain Pillsnail, *Euchemotrema leaii aliciae*
Yellow Globelet, *Mesodon clausus*
White-lip Globe, *Mesodon thyroidus*
Shagreen, *Inflectarius inflectus*

Deep-tooth Shagreen, *Inflectarius rugeli*
Brown Globelet, *Inflectarius kalmianus*
Proud Globelet, *Patera pennsylvanica*
Flat Bladetooth, *Patera appressa*
Pine Mountain Bladetooth, *Patera aff. appressa*
Carinate Bladetooth, *Patera estillensis*
Smooth Bladetooth, *Patera laevior*
West Virginia Bladetooth, *Patera panselenus*
Dished Threetooth, *Triodopsis vulgata*
Magnolia Threetooth, *Triodopsis hopetonensis*
Claiborne Threetooth, *Triodopsis clairbornensis*
Carter Threetooth, *Triodopsis anteridon*
Black Mountain Threetooth, *Triodopsis aff. anteridon*
Rivercliff Threetooth, *Triodopsis discoidea*
Glossy Threetooth, *Triodopsis complanata*
Budded Threetooth, *Triodopsis tennesseensis*
Pike County Threetooth, *Triodopsis aff. tennesseensis*
Northern Threetooth, *Triodopsis tridentata*
Velvet Wedge, *Xolotrema denotatum*
Sharp Wedge, *Xolotrema obstrictum*
Bladetooth Wedge, *Xolotrema fosteri*
Clifty Covert, *Fumonelix wetherbyi*
Sealed Globelet, *Mesodon mitchellianus*
Proud Globe, *Mesodon elevatus*
Toothed Globe, *Mesodon zaletus*
Grand Globe, *Mesodon normalis*
Whitelip, *Neohelix albolabris*
Big-tooth Whitelip, *Neohelix dentifera*
Striped Whitelip, *Webbhelix multilineata*
Broad-banded Forestsnail, *Allogona profunda*
Queen Crater, *Appalachina chilhoweensis*
Spike-lip Crater, *Appalachina sayana*
Pine Mountain Crater, *Appalachina sayana kentucki*

VALLONIIDAE
Lovely Vallonia, *Vallonia pulchella*
Iroquois Vallonia, *Vallonia excentrica*
Costate Vallonia, *Vallonia costata*
Thin-lip Vallonia, *Vallonia perspectiva*

PHILOMYCIDAE
Carolina Mantleslug, *Philomycus carolinianus*
Winding Mantleslug, *Philomycus flexuolaris*
Variable Mantleslug, *Philomycus togatus*
Brown-spotted Mantleslug, *Philomycus venustus*
Virginia Mantleslug, *Philomycus virginicus*
Breaks Mantleslug, *Philomycus bisdosus*
Kentucky Mantleslug, *Philomycus batchi*
Changeable Mantleslug, *Megapallifera mutabilis*

Blotchy Mantleslug, *Megapallifera wetherbyi*
Severed Mantleslug, *Pallifera secreta*
Pale Mantleslug, *Pallifera dorsalis*
Redfoot Mantleslug, *Pallifera ohioensis*
Foster Mantleslug, *Pallifera fosteri*

AGRIOLIMACIDAE
Meadow Slug, *Deroceras laeve*

Pleistocene (Fossil) Land Snails of Kentucky

SUCCINEIDAE
Xeric Ambersnail, *Succinea indiana*
Loess Ambersnail, *Succinea bakeri*
Frigid Ambersnail, *Mediappendix gelida*

OXYCHILIDAE
Blue Glass Snail, *Perpolita binneyana*
Amber Glass Snail, *Perpolita electrina*

VALLONIIDAE
Trumpet Vallonia, *Vallonia parvula*
Multirib Vallonia, *Vallonia gracilicosta*

HELICODISCIDAE
Temperate Coil, *Helicodiscus shimeki*
Angular Disc, *Discus catskillensis*

POLYGYRIDAE
Lowland Pillsnail, *Euchemotrema leaii leaii*
Callused Vertigo, *Vertigo arthuri*
Cross Vertigo, *Vertigo modesta*
Deep-throat Vertigo, *Vertigo nylanderi*
Great Lakes Snaggletooth, *Gastrocopta similis*

Exotic Land Snails of Kentucky

HELICIDAE
Grovesnail, *Cepaea nemoralis*
White-lip Gardensnail, *Cepaea hortensis*
Milk Snail, *Otala lactea*
Brown Gardensnail, *Cornu aspersum*

GEOMITRIDAE
Vineyard Gardensnail, *Cernuella virgata*
Heath Snail, *Xerolenta obvia*

OXYCHILIDAE
Dark-bodied Glass-snail, *Oxychilus draparnaudi*

Garlic Glass-snail, *Oxychilus alliarius*
Southern Flatcoil, *Polygyra cereolus*
Sharp Awlsnail, *Opeas pyrgula*

AGRI0LIMACIDAE
Gray Fieldslug, *Deroceras reticulatum*
Orange-banded Arion, *Arion fasciatu*
Dusky Arion, *Arion subfuscus*

LIMACIDAE
Giant Garden Slug, *Limax maximus*
Threeband Gardenslug, *Ambigolimax valentianus,*

LAND SNAILS OF BORDERING STATES; WEST VIRGINIA, OHIO, ILLINOIS, INDIANA, TENNESSEE, MISSOURI, & VIRGINIA

SUCCINIIDAE
Louisiana Ambersnail, *Oxyloma salleanum*

GASTROCOPTIDAE
Plains Snaggletooth, *Gastrocopta abbreviate*
Rogers Snaggletooth, *Gastrocopta rogersensis*

PRISTILOMATIDAE
Club Superccoil, *Paravitrea bellona*
Domed Supercoil, *Paravitrea significans*
Barred Supercoil, *Paravitrea seradens*
Caneyfork Supercoil, *Paravitrea metallacta*
Natural Bridge Supercoil, *Paravitrea pontis*

GASTRODONTIDAE
Black Gloss, *Zonitoides nitidus*
Eightfold Pinecone, *Strobilops affinis*
Carinate Dome, *Ventridens eutropis*

HELICODISCIDAE
Twilight Coil, *Helicodiscus multidens*
Glossy Dome, *Ventridens acerra*

DISCIIDAE
Alabama Tigersnail, *Anguispira Alabama*

POLYGRIDAE
Sequatchie Slitmouth, *Stenotrema edgarianum*
Carinate Slitmouth, *Stenotrema spinosum*
Carinate Pillsnail, *Euchemotrema hubrichti*
Atlantic Threetooth, *Triodopsis juxtidens*

Buttressed Threetooth, *Triodopsis rugosa*
Western Whitelip, *Neohelix alleni*
Engraved Bladetooth, *Patera perigrapta*

ABOUT THE AUTHOR

Dan Dourson is a biologist/naturalist/illustrator who has spent most of his adult life dedicated to the preservation, conservation, and understanding of the planet's more obscure animals. As an employee of the United States Forest Service for nearly 20 years, he worked to manage and conserve rare plants, bats, birds, reptiles, amphibians, land snails and freshwater mussels. In addition, Dan has conducted biological inventories for multiple agencies (i.e., Nature Conservancy, State Agencies and National Park Service) for numerous taxa groups and has worked for more than a decade studying the land snails of the Great Smoky Mountains National Park as part of the "All Taxa Biodiversity Inventory" through the Discover Life in America initiative. Dan has also studied land snails and other animals abroad, including Belize, Guatemala, Costa Rica and Panama and in South America, conducted land snail studies in the Tamshiyacu-Tahuayo Communal Reserve in the upper Amazon Basin of northwestern Peru.

Dan has published numerous peer review papers in prestigious journals such as the American Malacological Bulletin, The Nautilus, Southeastern Naturalist, Mesoamericana, Journal of the North Carolina Academy of Science, and others. He has described 10 new land snail species from the states of Kentucky, West Virginia, Tennessee, and North Carolina and in Central America from Belize. Dan has also authored 7 books including "*Wild Yet Tasty*" (his first), "*Kentucky's Land Snails and their Ecological Communities*", *Biodiversity of the Maya Mountains, Belize, Central America*", "*Land Snails of the Great Smoky Mountains National Park*", and is currently working on "*The Land Snails of Belize Central America*". Dan has illustrated nature books such as *"Wildflowers of Kentucky", "Ginseng Dreams", "Reptiles and Amphibians of Kentucky", "How Snakes Work"* and multiple environmental posters used in schools around the country.

Dan remains committed to conservation work protecting the earth's most amazing and underappreciated organisms. His allegiance with the natural world is clearly reflected through his writing and simple lifestyle. He divides his time between the USA and Belize with his wife, Judy.

www.ingramcontent.com/pod-product-compliance
Ingram Content Group UK Ltd.
Pitfield, Milton Keynes, MK11 3LW, UK
UKRC032333051026
15625UKWH00015B/632

9798218680688

9 798218 680688